UNITS AND CONVERS

Energies

$1\ \text{cm}^{-1}$	$k_B \times 1.439$ K
	1.9855×10^{-16} ergs
	1.2395×10^{-4} eV
	258.4 cm/sec (Mössbauer velocity shift)
1 eV	$8068.3\ \text{cm}^{-1}$
	1.602×10^{-12} ergs
	23.06 kcals/mole
	$k_B \times 11{,}610$ K
1 au	27.2116 eV
$e^2/1$ Å	14.4 eV
$(\hbar^2/2m_e)(1/1\ \text{Å})^2$	3.806 eV
$\beta^2/(1\ \text{Å})^3$	1.298×10^4 MHz
1 calorie	4.184 J

The wavelengths of excitations in the infrared are measured in microns (μm)(1 $\mu\text{m} = 10^{-3}$ mm) and in the visible and ultraviolet in angstrom units (Å)(1 Å $= 10^{-7}$ mm $= 10^{-8}$ cm) or in nanometers (nm) or millimicrons (mμm) (1 nm $=$ 1 mμm $=$ 10 Å).

The frequency ν' and wavenumber ν of excitations are connected by Planck's relation $E = h\nu' = h\nu c$, and ν is taken as a measure of energy in spectroscopy. The electron volt (eV) is also used as a measure of energy in spectroscopy and is equivalent to $8068.3\ \text{cm}^{-1}$ or 1.602×10^{-12} erg or 23.06 kcals/mole. A quantum of light of wavelength $L \times 10^2$ mm has energy $(12.395/L)$ eV.

Lengths

1 au	0.5291 Å
1 Å	1.889 au

Pressures and Stresses

$1\ \text{N/m}^2$ (Pa)	10^{-5} bar
	0.98692×10^{-5} atm
	$10\ \text{dyne/cm}^2$
	7.5006×10^{-3} torr

Defects and Defect Processes in Nonmetallic Solids

W. HAYES
Department of Physics
University of Oxford

A. M. STONEHAM
Department of Physics and Astronomy
University College London

Dover Publications, Inc.
Mineola, New York

Copyright

Bibliographical Note

This Dover edition, first published in 2004, is an unabridged republication of the edition published by John Wiley & Sons, New York, 1985.

Library of Congress Cataloging-in-Publication Data

Hayes, William.
Defects and defect processes in nonmetallic solids / W. Hayes, A.M. Stoneham.
p. cm.
Originally published: New York : Wiley, c1985.
Includes bibliographical references and index.
ISBN 0-486-43483-4 (pbk.)
1. Nonmetallic materials—Defects. 2. Point defects. 3. Crystals—Defects. I. Stoneham, A. M. II. Title.

QC176.8.N65H39 2004
530.4'1—dc22

2004043928

Manufactured in the United States of America
Dover Publications, Inc., 31 East 2nd Street, Mineola, N.Y. 11501

Preface

All solids in equilibrium at finite temperatures contain intrinsic point defects. In addition, all solids inevitably contain chemical impurities, which are present inadvertently. The concentration of intrinsic defects may be increased by thermal treatment and by irradiation, and the concentration of chemical impurities may be increased by doping. The presence of defects can affect the usefulness of materials and the performance of solid-state devices in both desirable and undesirable ways. This accounts, to a considerable extent, for the intensive and continuing research activity in the defect solid state.

The study of defects in solids divides naturally into investigations of metals and nonmetals, requiring different experimental and theoretical skills. One could not hope to encompass the mass of information available in a single volume of reasonable size. The present book is confined to nonmetals and, even with this restriction, has to be selective. Thus, the emphasis is on point defects and point-defect processes; the more extended defects (such as dislocations) are discussed, though not in detail.

Chapter 1 deals with electronic properties of solids, including a general outline of some of the theoretical techniques used to calculate electronic band structure and defect electronic states. This, together with discussions of polarons and excitons, provides background for later chapters. Chapter 2 lays out the general philosophy of lattice dynamics and discusses the vibrational modes of crystals containing point defects and also the vibrational modes of mixed crystals and of amorphous solids. Chapter 3 describes the thermodynamics of defect formation and such phenomena as diffusion of defects, lattice polarization and lattice relaxation near defects, and dimensional changes in crystals caused by defects. In addition, this chapter contains accounts of nonstoichiometric materials and fast-ion conductors (also known as superionics). Chapter 4 is concerned in a general way with the interaction of optical excitations of defects with lattice vibrations, including the Jahn–Teller effect. Nonradiative transitions are also discussed, together with some of the diverse processes that they govern. In view of its importance as an experimental technique in the study of point defects in solids, Chapter 4 also contains an account of electron paramagnetic resonance (epr). Chapter 5 deals with the electronic properties of specific types of point defects such as *F* centers and hole centers, vacancies and interstitials, chemical impurities such as oxygen and transition-metal ions (deep-level impurities), shallow donors and acceptors, and impurity-bound excitons. Chapter 6 explains how radiation damage is caused by fast particles, and also by ionizing radiation, and includes a discussion of the photographic effect. This chapter also discusses the phenomenon of recombination-enhanced diffusion. Chapter 7 outlines some properties of surfaces, including surface reconstruction and surface electronic states, and associated properties such as catalysis and corrosion. Chapter 8 ranges over amorphous materials, metal–insulator transitions, intercalates, and polymers, with emphasis on defect properties. It should be said that although much of the material covered in this book is now quite well established, there are areas, including some discussed in Chapters 7 and 8, that are still unsettled.

We emphasize that we have not attempted to assign credit systematically to authors of original work. Our choices of examples were determined to a considerable degree by requirements of illustration and by the extent to which they complement our own treatment.

At the present time the study of the defect solid state is becoming increasingly fragmented into more restricted specializations. This book is intended to provide a background that will help researchers place specialized interests in a more general context. It is aimed primarily at first-year graduate students in solid-state physics and solid-state chemistry, but it should also be useful to undergraduates taking special solid-state options in their final year. A knowledge of the elementary theory of point groups would occasionally be useful, but is not essential.

Because of the diverse nature of the topics discussed and the corresponding variety in usage of units, we have not attempted to impose a single system of units on the book. A table showing some physical constants and some useful relationships between units appears at the end of the book. Also, because of the diversity of topics discussed, a supplementary reading list for each chapter has been included at the end of the book.

We would like to thank P. A. Cox, E. A. Davies, R. H. Friend, A. E. Hughes, A. B. Lidiard, M. J. L. Sangster, L. M. Slifkin, P. W. Tasker and J. M. Vail for helpful comments on the manuscript and Jean Chanter, Pearl Hawtin and Penny Jackson for careful preparation of the typescript.

W. Hayes
A. M. Stoneham

Oxford, England
Dorchester-on-Thames, England
October 1984

Contents

CHAPTER ONE

Electronic Properties

The simplest solids are crystalline, with their atoms in a regularly repeating structure. It is such solids, their defects, and the reactions among the defects that are the main subject of this book. However, many of the important processes and properties we shall encounter are observed more generally. They occur in less-ordered forms of condensed matter, such as polymers and amorphous materials. We shall also be concerned with such systems (Chapter 8), so that where possible we shall avoid restricting our discussion of electronic properties to crystals alone.

Solids can be grouped into a few classes, the main ones being (i) *metals*, (ii) *covalent semiconductors*, such as silicon and germanium; (iii) *ionic crystals*, including most halides and oxides; and (iv) *van der Waals crystals*, including rare gases like argon and molecular crystals like solid methane. These divisions, determined primarily by the nature of interatomic forces, are not completely clear-cut. The nature of interatomic forces is also a factor in determining crystal structure. In covalent semiconductors, the directed bonds favor open structures in which each atom has a small and definite number of neighbors. Van der Waals crystals are held together weakly by short-range forces, which favor close-packed structures. The powerful Coulomb interactions in ionic crystals also encourage close packing, but in such a way that each anion is surrounded by cations, and vice versa. These features carry over into disordered systems so that, in ionic liquids, anions tend to have cation neighbors and in amorphous covalent semiconductors bond angles and bond lengths remain close to their crystal values.

There is a very strong correlation between the physical nature of solids and the positions of their component elements in the periodic table. Interatomic forces tend to be stronger for smaller atoms and ions. Since these are also the lighter atoms or ions, vibrational frequencies and melting points are often higher for compounds which contain elements low in the periodic table. Thus the melting temperature decreases along the sequence MgO, CaO, SrO, and BaO, and also along the sequence MgO, MgS, MgSe, and MgTe.

Figure 1.1 shows the different types of binary compounds which form as a function of position in the periodic table. Any such scheme oversimplifies, for in a compound *AB* the three types of interactions—*AA*, *BB*, *AB*—must all be represented. In some cases, for example, intercalates (Section 8.3), the *AA* interactions form a strong matrix which weakly interacts with the *B* species intercalated between the *A* species. In others, for example, certain glasses (Section 8.1.5), there is sufficient flexibility of bonding to allow a range of compositions and pairings.

The physical ideas of structure are closely related to chemical ideas of bonding. Much of the vast store of systematic structural data now

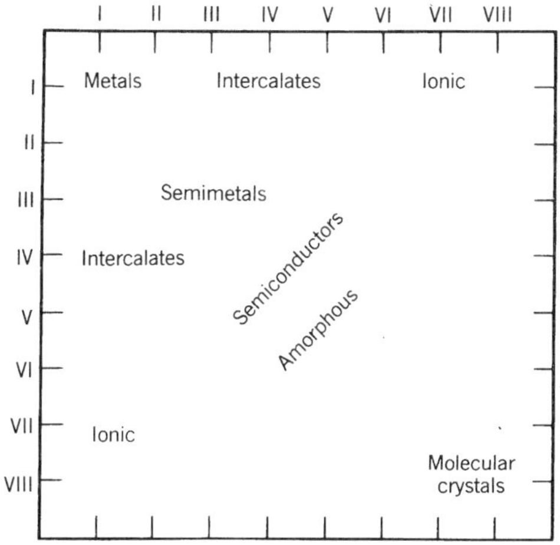

Figure 1.1 Types of solids formed from binary compounds. The two component elements are defined by their column in the periodic table (e.g., I = alkali, VII = halogen).

available can be understood in terms of three properties of each atomic species involved. The first is *electronegativity*, which measures the extent to which an atom in a solid attracts electrons. The second is the *tendency to form covalent bonds*, and the third is *atomic or ionic size*. Qualitatively, the variation within the periodic table is obvious; chlorine is more electronegative than potassium, carbon forms stronger bonds than argon, and sulfur is larger than oxygen. If electronegativity dominates, we have the ionic limit, with electrons transferred from a cation to an anion. If covalent bonding is dominant, there is a large electron density between the two components. Although quantitatively neither covalency nor ionicity is uniquely defined (Catlow and Stoneham 1983), they can be used to correlate broad structural features such as the number of nearest neighbors.

One very important characteristic of a solid is the amount of energy needed to raise it to an excited electronic state. Ignoring for the moment those excitations that merely change electron spin orientations, the minimum excitation energy is the critical factor in deciding whether a solid is a metal, a semiconductor, or an insulator. If an isolated atom is in its state of lowest energy, the energy levels are discrete and quantized, and a finite amount of energy, δE, is needed to raise it to an excited state. The same is true of small molecules. In a solid, the energy separations may be finite or may be so small that δE is negligible. The

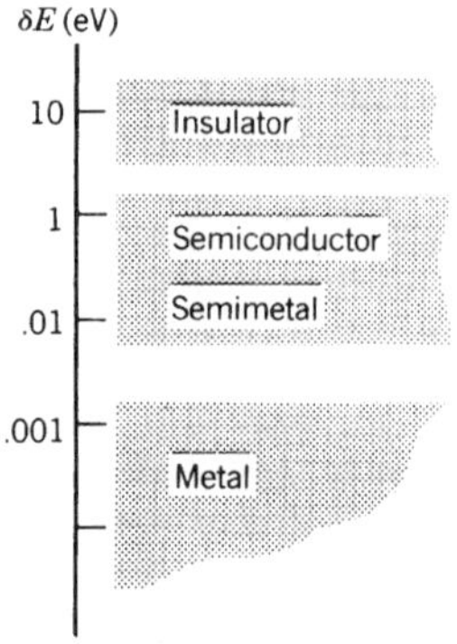

Figure 1.2 The energy needed to raise a solid to an electronically excited state, showing ranges of values corresponding to metals, semiconductors, and insulators at normal temperatures.

several regimes are shown in Figure 1.2, which does not depend on the chemical nature or physical state (e.g., crystalline versus amorphous) of the solid. δE for materials free of defects and impurities is known as the band gap and is represented as E_g.

We can correlate the electronic properties of the solid and the electronic properties of its component atoms in an illuminating way, again irrespective of whether the solid is crystalline or not. Just as in a simple diatomic molecule the atomic levels split into bonding and antibonding states as the component atoms approach, so the atomic levels split as the atoms are brought together to form a solid. In the solid the sheer number of atoms leads to a continuum of levels. As the atoms begin to overlap, the top and bottom of each continuum of states shows the same bonding and antibonding character found for molecules.

One convenient description for simple ionic solids is to regard the valence band as dominated by the anions and the conduction band by the cations. This is valid for NaCl and MgO, for instance, although not necessarily when transition metals are involved as in AgCl, FeO, and UO_2. The separation of the centers of the valence and conduction bands is determined by interactions between unlike species. Thus, in a binary compound AB it is often the B–B and A–A interactions that determine band widths ΔE_v and ΔE_c of the valence and conduction bands and the A–B interactions that fix the separation Δ of the band centers. The band gap is a combination of all three energies, that is,

$$E_g = \Delta - \tfrac{1}{2}\Delta E_v - \tfrac{1}{2}\Delta E_c \qquad (1.1)$$

because it is determined by the highest valence and lowest conduction states. The excitation of an electron from the valence band to the conduction band often involves charge transfer, in which an electron is transferred from species A to species B.

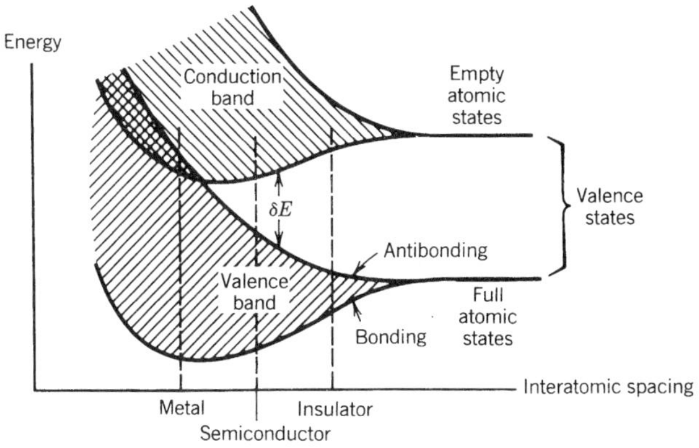

Figure 1.3 Electronic energies for a solid as a function of interatomic spacing. At large spacings, all solids show an atomic limit and at small spacings all show a metallic limit (see the text for some qualifications).

The excitation energy δE (Figure 1.3) decides whether the solid behaves like a metal ($\delta E \rightarrow 0$) or an insulator, and the trend of δE with atomic spacing is important. As a solid is compressed, δE eventually falls to zero (Figure 1.3). There may be intermediate structural changes, but all solids eventually become metallic if sufficiently high pressure is applied. The opposite limit of widely separated component species should correspond to free atoms. Here Figure 1.3 is an oversimplification in one important respect. The atomic-like states at large spacings are only the straightforward free atomic states for a few cases like simple metals or rare gases. In ionic crystals such as KCl the separated states are those of the ions K^+ and Cl^-. In molecular crystals like solid methane the separated states are the component molecules. In covalent crystals the widely separated species are in valence states (Coulson 1961), chosen to optimize covalent bonding (see Section 8.4), and these valence states are not simple eigenstates of the constituent atoms.

Band gaps of binary compounds and elements show a number of broad trends depending on the species involved. These can be seen in Figure 1.4, which shows a tendency to higher E_g with increasing ionicity and with decreasing atomic number and lattice spacing. Phillips (1970) has discussed the implications of trends of this sort.

In Section 1.1 we outline techniques used for the calculation of electronic structure. We discuss electronic transport in Section 1.2, referring to the pinning of the Fermi level by impurities. The subject of electron–lattice interaction is also outlined in this section, including the

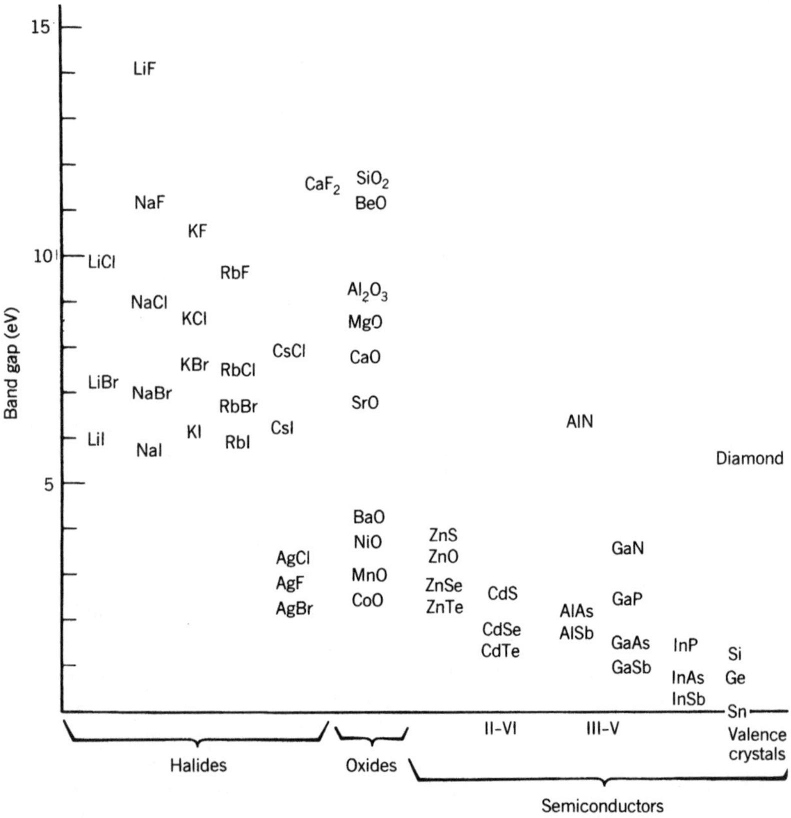

Figure 1.4 Band gaps, E_g, in nonmetals. The data available are often approximate, so the sizes of the printed chemical formulas in the diagram (e.g., NaCl) are often consistent with known accuracy. Note the trends with anion position in the periodic table, with the gap falling systematically as anions come from the more central columns of the periodic table or as anion sizes increase.

polaron phenomenon. Finally, in Section 1.3 we describe the optical properties of excitons and their condensates.

1.1 ELECTRONIC STATES OF SOLIDS

We have just suggested that an atomic picture, or a simple generalization of it, is useful at large interatomic spacings. As the spacings get smaller, Figure 1.3 implies large deviations from a free-atom model, evident in the large bandwidths. A complementary picture, the band model, becomes useful when there are large overlaps between close atoms. At its simplest, band theory suggests that one should think of states of the whole crystal, rather than of those of its component units, and that these states group themselves into well-defined continuous energy bands like those in Figure 1.3. These ideas are very general and still hold for disordered systems such as amorphous solids and liquids; the atomic and band limits are largely independent of the degree of disorder.

Practically all work on band structure has involved two restrictions. The first is the one-electron approximation, in which each electron is assumed to move in an average field of the other electrons and the nuclei. The second is the assumption of lattice translational periodicity. Taking the two assumptions together, the results of band theory give one-electron energies, $E_i(\mathbf{k})$, and wavefunctions for an electron with wavevector $\mathbf{k}$ in band i. From these energies and the corresponding wavefunctions, predictions can be made of results of experiments, including optical absorption, photoelectron emission, charge densities obtained using X-rays, and effective masses of electrons. A detailed comparison of these predictions with experiment lies outside the scope of this text. Suffice it to say that ground-state properties are often predicted very well indeed. Excited-state properties, especially when correlation (see Section 1.1.3) is important, are often poorly described even for simple systems [for a discussion of diamond, see Stoneham (1979); for transition-metal oxides, a notorious problem of band theory, see Kunz (1978)].

We shall concentrate on two special limits here, the so-called nearly-free electron models, which lead naturally to pseudopotential theory, and the so-called tight-binding models, which lead us to reexamine the relationship between atomic (or ionic) and band pictures.

1.1.1 Nearly Free Electron Models, Pseudopotentials, and Tight-Binding Models

Suppose an electron is moving in a region where the potential $V(r)$ is constant. Such electrons can be characterized by a wavevector $\mathbf{k}$. Their wavefunction is simply $e^{i\mathbf{k}\cdot\mathbf{r}}$, omitting a normalization factor, and they have a kinetic energy $E_k = \hbar^2 k^2/2m$ and a momentum $\hbar\mathbf{k}$. The density of states, $\rho(\mathbf{k})$, is proportional to $k^2 dk$ or, expressed as a function of energy,

$$\rho(E) = \frac{1}{2\pi^2}\left(\frac{2m}{\hbar^2}\right)^{3/2}\sqrt{E}\,dE \tag{1.2}$$

for a nondegenerate band. In this extreme, the band has infinite width, that is, $\rho(E)$ does not fall to zero at some higher energy.

Suppose that a weak periodic potential is applied with a characteristic repeat distance a. This corresponds most closely to the extreme left of Figure 1.3, where the bands overlap greatly and where the kinetic energy needed to localize an electron within a region of dimension a (of order $\hbar^2/2ma^2$) is much larger than the atomic energy splittings. One now finds, perhaps surprisingly, that many of the free-electron features remain to a useful extent.

The weak periodic potential changes the wavefunction from $e^{i\mathbf{k}\cdot\mathbf{r}}$ to $e^{i\mathbf{k}\cdot\mathbf{r}}u(\mathbf{k},\mathbf{r})$, where the Bloch function $u(\mathbf{k},\mathbf{r})$ has the periodicity of the lattice, and this extra factor can be looked at in two main ways. One can consider the change as a diffraction effect, in which the plane waves of the free electrons are diffracted by the perturbing lattice atoms. Alternatively, one can think of the effect as one in which the lattice potential mixes different electronic waves with wavevectors $\mathbf{k}$, $\mathbf{k}+\mathbf{k}_L$, $\mathbf{k}+\mathbf{k}_{L'}$, etc., where the $\mathbf{k}_L, \mathbf{k}_{L'}, \ldots,$ are determined by the important wavelengths in the periodic properties of the perturbing potential. The diffraction picture is useful because it allows one to think of the band gaps δE as stop bands [see Brillouin (1953), Slater (1958)] in which attenuation-free propagation of electronic plane waves is not possible. The mixing picture is also useful because it suggests, quite correctly, that $\hbar\mathbf{k}$ and the electron momentum may not be identical in some circumstances. This does not cause confusion very often, since, in practice, the selection rules for conservation of momentum and the conditions imposed by translational symmetry are complementary. In both the diffraction and mixing pictures one expects the kinetic energy to be modified. Near the bottom of the conduction band, for example, the energy of an electron becomes $\hbar^2k^2/2m^*$, where m^* is now an effective mass. The density of states is altered correspondingly, being multiplied by a factor $(m^*/m)^{3/2}$.

Physically, one is naturally suspicious of believing that lattice ions could provide merely a weak perturbation of free-electron states. It is more natural to imagine that the scattering of electrons would be very strong. However, there are two factors that give the weak-perturbation model fairly general validity. The first is that in any periodic system the one-electron states satisfy Bloch's theorem. Properly-chosen periodic eigenstates are, by definition, not scattered. The problem is thus not why scattering is weak, but whether the real and observable properties can be adequately represented by a simple generalization of free-electron behavior.

The second factor is seen most easily from the idea of a pseudopotential. The wavefunction of any valence electron must be orthogonal to the core orbitals, introducing the so-called orthogonality constraint. There are two ways of handling the constraint. One is to solve the one-electron Schrödinger equation

$$\left(-\frac{\hbar^2}{2m}+V\right)\psi=E\psi \tag{1.3}$$

using wavefunctions that are chosen to be orthogonal to the cores. Such functions are usually complicated and unwieldy. An alternative is to put in the orthogonality through an extra term in the Hamiltonian, taking care that the same energy will result:

$$\left(-\frac{\hbar^2}{2m}+V+h\right)\phi=E\phi \tag{1.4}$$

In this strategy, which introduces no approximation at this stage, the combination $(V+h)$ is the pseudopotential and the function ϕ is the pseudowavefunction. This is the preferable approach whenever $(V+h)$ is more easily understood and handled than V alone. The extra term in the pseudopotential that represents the orthogonality constraint has the effect of excluding the electrons of interest from the core region where, inside the inner-shell electrons, they might otherwise gain energy from the unscreened attractive Coulomb interaction with the nucleus. It transpires that these attractive and repulsive terms cancel rather completely (Cohen and Heine 1961), leaving only a residual weak interaction.

The idea of pseudopotential is a very useful one, and pseudopotential methods prove extremely powerful tools. They are introduced in two stages. One involves merely a rewriting of the Schrödinger equation in the one-electron approximation, that is, moving from Equation (1.3) to Equation (1.4). We emphasize again that no new approximations are introduced at this stage; all that happens is that the orthogonality constraint is represented in a new way. No assumption about the strength or weakness of the potential arises. The second stage does involve approximations, and the expressions that result can be broken into large and small components so that physically based assumptions can be introduced. In ionic crystals the assumptions are usually ones like "closed shells of neighboring ions do not overlap" rather than assumptions of weak scattering. There is no inconsistency in using a nearly-free electron model for the outer valence electrons and a tight-binding picture (see below) for the inner-core electrons.

Several important points now emerge. The first is that sometimes there is difficulty in making a clear-cut distinction between a valence orbital and a core orbital. If we have a single electron outside closed shells, as in the F center (Section 5.1.1), there is no problem. But if we consider a trapped-hole center such as the V^- center (Section 5.1.2), or a simple acceptor (Section 5.3.1), it is not clear whether the other electrons in the valence band count as core or valence electrons. This is also the case for the d electrons in transition-metal ions (Section 5.2). In practice, it is usually clear when the pseudopotential approach is applicable, and generally one can exploit the cancellation between V and h mentioned above. In many cases $(V+h)$ for an ion can be written as an interaction with a point charge plus a small correction. If the ion is replaced by a neutral atom, as in a semiconductor like Si, only the small correction appears, and acts as a rather weak perturbation.

Another point is that the pseudowavefunction ϕ and the real wavefunction ψ are not identical. ψ can be generated from ϕ by orthogonalizing to the core orbitals $|c\rangle_i$,

$$|\psi\rangle = \left(|\phi\rangle - \sum_i |c_i\rangle\langle c_i|\phi\rangle\right) \Big/ \left[1 - \sum_i |\langle c_i|\phi\rangle|^2\right]^{1/2} \tag{1.5}$$

However, $|\phi\rangle$ cannot be generated from $|\psi\rangle$ uniquely. Any linear combination of core orbitals can be added to $|\phi\rangle$ without affecting $|\psi\rangle$. This non-uniqueness in $|\phi\rangle$ means h is not unique either, and this can be exploited by adding a useful constraint. Thus, if ϕ is chosen to vary as smoothly in space as possible, the Cohen–Heine form of pseudopotential results: another choice that can be simpler to apply in some cases is the Phillips–Kleinman form of pseudopotential. The choice is at the discretion of the user. These options are discussed for solid-state defects by Stoneham (1975, Section 6.2.4). We shall give an example of the application of the pseudopotential technique to point defects in Section 5.1.

The free-electron models discussed above invoke electrons contained in a structureless box. The other extreme, the tight-binding model, stresses that solids are built of atoms or ions and corresponds to the right-hand limit of Figure 1.2. It adopts the ideas of the linear combination of atomic orbitals (LCAO) used in molecular physics (Coulson 1961). The states of the crystal in simple band theory correspond exactly to the molecular orbitals of molecules, and this is exploited in tight-binding approaches.

In the simplest tight-binding cases the Bloch functions $u(\mathbf{k}, \mathbf{r})$ reduce to linear combinations of atomic orbitals. If there is no overlap of wavefunctions and if we consider interactions with nearest neighbors only, we obtain an effective mass

$$m^* = \frac{\hbar^2}{JNa^2} \tag{1.6a}$$

and a bandwidth $2B$

$$2B = 2NJ = 2\hbar^2/m^* a^2 \tag{1.6b}$$

for a nondegenerate band with N nearest neighbors at distance a and with interactions J between nearest neighbors. The density of states near the band extremum contains the factor $(m^*/m)^{3/2}$ mentioned earlier for the free-electron limit. Thus, the density of states in this tight-binding limit is proportional to $(JNa^2)^{-3/2}$ or to $(Ba^2)^{-3/2}$. Several general features, therefore, show a strong resemblance at the two extremes of nearly-free and tightly-bound electrons (see also Section 1.1.2). This resemblance can extend even to results of detailed calculations; provided that the calculation uses a basis that is flexible enough, one can usually get equally good band structures whether one starts from a tight-binding model or from the plane waves of a nearly-free electron model.

The potential V which appears in Equations (1.3) and (1.4) comes partly from the interactions of the valence electrons with each other. It is often necessary to treat this self-consistently, that is to make some assumption for V, to solve the Schrödinger equation to obtain charge densities, to recalculate V, and to continue iteratively to self-consistency. The Hartree–Fock approach to the calculation of electronic properties (Slater 1963) includes exchange interactions, and the total wavefunction is correctly antisymmetrized. Unrestricted Hartree–Fock theory goes further in that electrons in given orbitals but with opposite spins are allowed to have different wavefunctions. If a Hartree–Fock calculation were done exactly, the results would be (a) one-electron wavefunctions $\phi_i(\mathbf{r})$, (b) one-electron energies ϵ_i that go with the $\phi_i(\mathbf{r})$, (c) a total wavefunction that can be written as a determinant whose elements are the $\phi_i(\mathbf{r}_i)$, and (d) a total energy E for a given occupancy of the one-electron orbitals. It is important to realize that the total energy E is not the sum of the one-electron energies $\sum \epsilon_i$; in essence $\sum \epsilon_i$ double counts the electron–electron interactions and misses nucleus–nucleus interactions. In the calculation of defect energies, for example, it is wrong to use $\sum \epsilon_i$ to predict the geometry that a defect will adopt.

Molecular-orbital methods lend themselves well to systematic approximations (Pople and Beveridge 1970). Since methods and their variations abound, it should be stressed that it is always best to use an approach that has been verified for the type of problem to be investigated. Approaches that have a good record for predictions of molecular geometry should work well for defect geometries; those that estimate electron paramagnetic resonance (epr) parameters accurately are likely to be good for other parameters depending on wavefunctions. More generally, when one turns to the approximate molecular-orbital methods, one finds that the usefulness of specific approximations depends on whether one wants to optimize the ϕ_i, the ϵ_i, or the total energy E. It is usually easiest to get good ϵ_i and acceptable ϕ_i; good total energies are harder to calculate.

Semiempirical methods are very useful, the word empirical here indicating explicit use of matrix elements containing parameters that describe the electronegativity, size, and extent of bond formation, with values of the parameters partly based on experimental information. Such theories have been used extensively by chemists, who have introduced their own approximations and nomenclatures. For the simplest calculations, extended Hückel theory is convenient (Hoffman 1963). If total energies are needed, or if there is charge transfer, a self-consistent theory is required; the so-called CNDO method has a good record in defect studies, and its generalizations, the so-called INDO and MINDO methods, may be useful (Pople and Beveridge 1963).

In comparing theory with experiment it is useful to take a specific example and we shall choose diamond. This is one of the best-studied systems, with its high symmetry and low atomic number, so that a useful comparison can be made. Predictions fall into two broad classes (Table 1.1). The first class describes only the ground state of the perfect crystal, with a full valence band and an empty conduction band. The calculation of the distribution of charge density $\rho(\mathbf{r})$ in the crystal is a good initial test of theory. Simply superimposing atomic charge densities does quite well for many of the Fourier components, $\rho(\mathbf{k}) = \int d^3\mathbf{r}\rho(\mathbf{r}) \exp(i\mathbf{k}\cdot\mathbf{r})$, measured in X-ray scattering (Table 1.2). The most important test comes from the $\langle 222 \rangle$ reflection, for which spherical atomic charge densities

Table 1.1 Predictions from Band Structure Calculations

Class I	Class II
Only the valence band states are important for the following crystal ground-state properties:	Both valence band states and conduction band states are important for the following properties:
Charge density	Optical dielectric constant (refractive index)
Compton profile	Optical absorption spectrum
Compressibility	Electron effective mass
Cohesive energy	
Lattice parameter	
X-ray emission spectra	
Hole effective mass	

Table 1.2 Fourier Components $\rho(\mathbf{k})$ of the Charge Density in Diamond Obtained from Experiment and from Theory[a,b]

Wavevector $\mathbf{k}$	Experiment	Densities of Spherical Atoms Superimposed	OPW Theory	Local Pseudopotential Theory	Tight-Binding Theory	Self-Consistent Theory
⟨111⟩	0.99	0.814	1.011	0.88	0.93	0.976
⟨220⟩	0.18	0.203	0.221	0.01	0.16	0.15
⟨311⟩	−0.04	+0.045	−0.037	−0.14	−0.02	−0.05
⟨222⟩	±0.15	0.0	−0.105	−0.15	−0.08	−0.12
⟨400⟩	−0.14	−0.013	−0.105	−0.13	−0.06	−0.10

[a] See Stoneham (1979) for full references.
[b] Units of wavevector **k** are (2π/lattice parameter) and of charge density are e per atomic volume.

Table 1.3 Comparison of Measured and Calculated Values of the Lattice Constant and Bulk Modulus of Diamond

	Lattice Constant (nm)	Bulk Modulus ($N\,m^{-2}$)
Experiment	0.3567	4.43×10^{11}
Theory	0.3545[a]	4.38×10^{11} [a]
	0.3561[b]	9.4×10^{11} [b]
	0.3602[c]	4×10^{11} [c]

[a] After Euwema et al. (1973).
[b] After Thomas (1975).
[c] Yin and Cohen (1981).

give no contribution associated with charge enhancement between atoms. The bonding contribution shows up well in most models, indicating that this feature is well understood. The Compton line profile (Cooper and Leake 1967), which reflects the electron momentum distribution rather than electron charge distribution, is less well studied, but again one finds that experiment and theory agree adequately. Lattice parameter, cohesive energy, and bulk modulus (Table 1.3) are more stringent tests. The bulk modulus, especially, is sensitive to details of the calculation, for example, to the number of terms included in a variational calculation.

The second class of tests of predictions involves the conduction-band states, either alone or in combination with valence-band states. Examples are given in Table 1.1. The band structure itself, E versus $\mathbf{k}$, is not measured directly. One obtains instead composite quantities such as bandwidths, optical absorption or emission spectra, and dielectric constants.

We show in Figure 1.5 a typical calculated band structure for diamond, and compare theory and experiment in Figure 1.6. Clearly the width and density of states of the valence band (which are ground-state properties only) are given satisfactorily. The optical absorption, however, is predicted adequately rather than accurately. The optical dielectric constant (identical with the static dielectric constant only in nonpolar crystals like diamond) is observed to be $\epsilon_0 = 5.7$ and predicted values are close. The calculations also confirm that the wavevector-dependent dielectric constant at zero frequency deviates from ϵ_0 only for wavelengths comparable with the interatomic spacing. Thus, in the

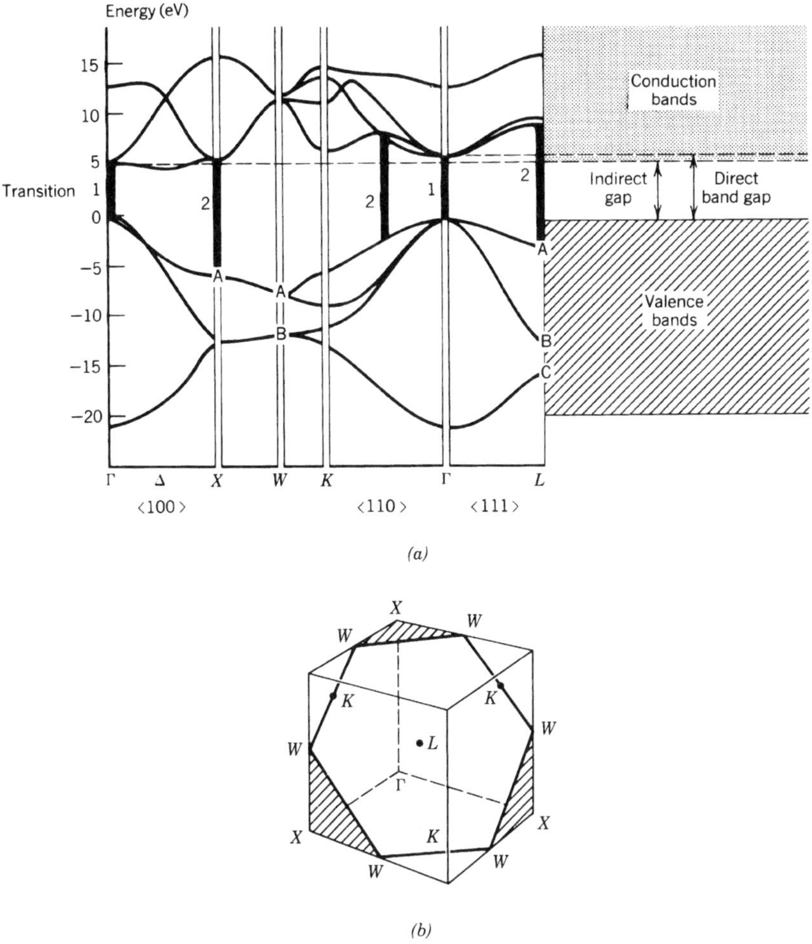

Figure 1.5 (*a*) Band structure of diamond [after Painter et al. (1971)]. The labels 1 and 2 and the labels *A, B, C* identify features that are compared with experiment in Figure 1.6. (*b*) The symmetry directions in the Brillouin zone.

screening of a point charge ($e^2 r \rightarrow e^2/\epsilon r$) one may take ϵ as ϵ_0 (ϵ_0 and ϵ_∞ are equal for diamond) down to about a nearest-neighbor distance, reducing ϵ to 1 as distances become shorter still. There are no accurate estimates or verified measurements of effective masses for conduction band electrons in diamond, although many calculations do locate the conduction band minima in the correct region of the Brillouin zone (see Figure 1.5).

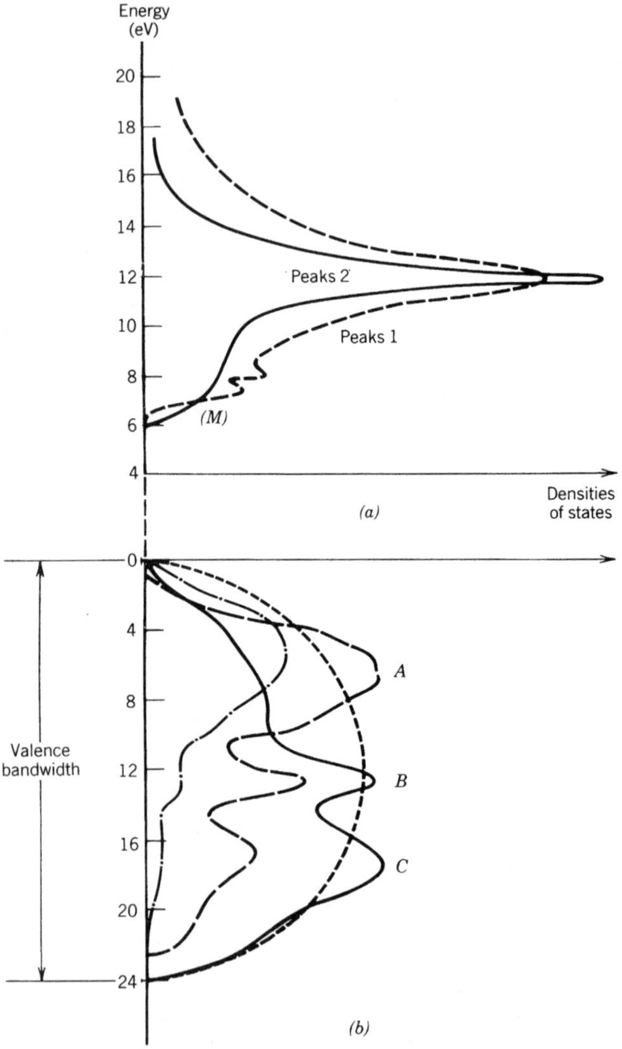

Figure 1.6 Comparison of experiment with the band structure calculations of Figure 1.5. (*a*) The valence-band density of states is obtained experimentally by X-ray photoemission [—— after McFeely et al. (1974)] and *K* X-ray emission [–·– after Wiech and Zöpf (1969)] and is compared (*b*) with the calculated valence-band density of states (–––, peaks *A*, *B*, *C* corresponding to regions of high densities of states indicated on Figure 1.5*a*) and a simple tight-binding model (- - -). The measured optical absorption (–·–), associated with peaks 1 and 2 of Figure 1.5 [after Roberts and Walker (1967)], is compared with the calculated combined density of states (——) for the initial valence and final conduction bands involved.

1.1.2 Calculations of Electronic Structure

We can make useful links between Bloch functions and atomic functions, which carry over into calculations of defect wavefunctions, discussed later in this section. In the tight-binding description of a monatomic solid there are localized atomic functions $c(\mathbf{r}, \mathbf{R}_i)$ centred on sites $\mathbf{R}_i$. The wavefunctions of the one-electron eigenstates take the standard form

$$\psi(\mathbf{k}, \mathbf{r}) = \frac{1}{\sqrt{N}} \sum_i \exp(i\mathbf{k} \cdot \mathbf{R}_i) c(\mathbf{r}, \mathbf{R}_i) \tag{1.7}$$

These are delocalized functions, in a form easily related to the Bloch form $\exp(i\mathbf{k} \cdot \mathbf{r}) u(\mathbf{k}, \mathbf{r})$, in which the Bloch function $u(\mathbf{k}, \mathbf{r})$ is itself periodic. Our expression (1.7) assumes that the atomic functions $c(\mathbf{r}, \mathbf{R})$ are orthogonal; the factor $\sqrt{N}$ normalizes ψ.

If we know the Bloch functions, $\psi(\mathbf{k}, \mathbf{r})$, the reverse transform is possible quite generally, that is, even when the nearly free electron model is appropriate. This transform defines Wannier functions $a(\mathbf{r}, \mathbf{R}_i)$,

$$a(\mathbf{r}, \mathbf{R}_i) = \frac{1}{\sqrt{N}} \sum_k \exp(-i\mathbf{k} \cdot \mathbf{R}_i) \psi(k, r) \tag{1.8}$$

where the sum is over the first Brillouin zone. If a tight-binding model is appropriate, the Wannier functions are just the atomic functions to lowest approximation. To the second approximation the Wannier functions contain small terms to make sure that the $a(\mathbf{r}, \mathbf{R}_i)$ with different $\mathbf{R}_i$ are orthogonal. If the free-electron approximation were valid, the form of the Wannier functions would change. For example, if the Brillouin zone were spherical with radius k_{max}, $a(\mathbf{r}, \mathbf{R}_i = 0)$ would reduce to the form

$$[\sin(k_{\text{max}} r) - (k_{\text{max}} r) \cos(k_{\text{max}} r)]/2\pi^2 r^3 \tag{1.9}$$

When there is more than one atom in the unit cell, careful generalization is needed, for the Wannier function (1.8) relates to a *cell* rather than to a site. One useful result is that the expectation value of the Hamiltonian between Wannier functions gives the average band energy,

$$\langle a(\mathbf{r}, \mathbf{R}) | \mathcal{H} | a(\mathbf{r}, \mathbf{R}) \rangle = \sum_{\mathbf{k}} E(\mathbf{k}) \Big/ \sum_{\mathbf{k}} 1 \tag{1.10}$$

where the sums are over all wavevectors $\mathbf{k}$ in the Brillouin zone and $\mathbf{R}$ labels the cell.

Table 1.4 Comparison of the Relative Merits of Different Theoretical Methods in Defect Studies[a]

Method	Examples of Use	Simplifications or Empirical Factors	Geometric Factors	Merits	Problems
Pseudopotential ion size corrections [Bartram et al. (1968)]	*F* center (Section 5.1.1)	Pseudopotential parameters may be approximated	No real restrictions	Good for systems with one electron outside closed shells	Total energies hard to obtain accurately
Tight-binding					
(a) Non-self-consistent cluster [e.g., extended Hückel theory, Hoffman (1963) and Messmer and Watkins (1972)]	Organic solids (see Section 8.4)	Simplified matrix elements	No real restrictions	Useful as a first step for a self-consistent calculation	One-electron energies and wavefunctions only
(b) Self-consistent cluster (e.g., CNDO). [Mainwood et al. (1978) and Itoh et al. (1977)]	Radiation-damage mechanism in alkali halides (Section 6.2.2); Semiconductor interstitials (Section 6.4.6)	Simplified matrix elements	No real restrictions beyond cluster size and shape	Good for total energies and wavefunctions	Convergence to self-consistency in excited states may be difficult

(c) Self-consistent cluster [Muffin-tin, $X\alpha$, e.g., Hemstreet (1980)]	Transition-metal impurities (Section 5.2)	Local approximation to exchange; muffin-tin approximation	Muffin tins should not overlap; cluster size and shape are important	Good for intratomic properties	One-electron features only; total energy not available without major extension
(d) Continued fraction methods [e.g., Haydock et al. (1972)]	Amorphous systems (Chapter 8)	Simplified matrix elements; not self-consistent	No real restrictions	Good for low symmetries when self-consistency is not needed	Only the electron density $\rho(\mathbf{r}, E)$ associated with a chosen site is given
Local density functional [e.g., Baraff et al. (1980)]	Vacancies and impurities in semiconductors (Sections 5.4 and 5.5)	Local approximation to exchange and correlation	High symmetry desirable	High accuracy for ground-state energies	Information on excited states may be limited or impossible to obtain; only the density, not the wavefunction, is found

[a]Only a few of the methods are discussed in the text, but fuller discussions applied to defects can be found in the references cited.

When we consider an electron in an otherwise empty band, we have two apparently alien ways of describing its wavefunction $\psi(\mathbf{r})$. One is to use atomic orbitals so that $\psi(\mathbf{r}) = \sum_i \alpha_i c(\mathbf{r}, \mathbf{R}_i)$. The other is to use Bloch functions, that is, $\psi(\mathbf{r}) = \sum_{\mathbf{k}} \beta(\mathbf{k}) u(\mathbf{k}, \mathbf{r})$. But the differences are partly superficial. If we write the Bloch functions in terms of atomic functions, or vice versa, the two expressions become identical to the extent that the atomic functions are a good representation of the Wannier functions. In particular, pseudopotential theory and effective mass theory (Section 5.3) have more in common than is clear at first sight. In pseudopotential theory, where one uses Equation (1.4), $|\phi\rangle$ is orthogonal to the lower-energy occupied states in a way that would be ensured automatically by writing the one-electron wavefunctions in the form $\sum \alpha_i c(\mathbf{r}, \mathbf{R}_i)$. However, in effective mass theory one starts from the opposite extreme, $\sum_{\mathbf{k}} \beta(\mathbf{k}) u(\mathbf{k}, \mathbf{r})$, but the Bloch function is that of the higher band, and modulation by a slowly varying function $f(r)$ replaces the sum over wavevectors.

Any detailed survey of defect theory would be out of place here. Nevertheless, it is useful to have some idea of the advantages and disadvantages of popular methods, and we shall mention just a few of the options. The main points are summarized in Table 1.4. The methods differ in two respects, namely, in *geometric* and *electronic* factors. By geometric factors we mean the number of atoms or ions in a crystal that are treated in detail, where they are positioned, and what assumptions are made for those atoms or ions that are treated in less detail. By electronic factors we mean the accuracy of any assumptions made in solving the Schrödinger equation and in deriving useful quantities like total energies. The total energy, E_{total}, is related to the one-electron energies, ϵ_i, by

$$E_{\text{total}} = \sum_i \epsilon_i - E_{ee} + E_N \tag{1.11}$$

where E_N is the nuclear–nuclear interaction and E_{ee} are the electron–electron interactions, which are double counted in $\sum_i \epsilon_i$. A fuller discussion of the details, applications, and merits of the theories is given by Stoneham (1975). Here we comment mainly on the factors by which one should assess defect theories and their status, dealing with electronic factors first and geometric factors second:

1. Electronic Factors. The theories vary in the form of the solutions they give. Ignoring for the moment all questions of accuracy, the solutions take these main forms:

(a) Essentially complete solutions giving total energies, one-electron energies, and wavefunctions.

(b) One-electron energies and wavefunctions alone, with total energies difficult to calculate.
(c) One-electron density as a function of energy, $\rho(\mathbf{r}, E)$. Thus wavefunctions are not obtained in this class of calculations, and usually total energies cannot be obtained either.

The choice of method in practice depends on three features, namely, the effort one is prepared to invest, the accuracy one requires, and the property one wants to predict. One might forego the extra information from calculations in classes (a) and (b) if the accuracy and convenience of (c) were of sufficient advantage. An example is self-consistency, which is sometimes easier for class (c). Self-consistent solutions are important if one anticipates substantial redistribution of charge from the initial assumed distribution or from the initial to final states in an electronic transition.

Another aspect of electronic factors concerns the specific assumptions made. The two major types of assumptions involve simplifications and the use of empirical factors. Examples of simplifications are:

(a) Use of a muffin-tin approximation in which the potential is assumed constant outside a spherical region around each atom.
(b) Dropping of certain terms in the Hamiltonian or replacing them by simpler expressions.
(c) Neglect (or simplification) of core orbitals.
(d) Use of a limited variational trial function.
(e) Neglect of self-consistency.

Empirical factors take many forms such as:

(a) Fitting of pseudopotentials or of matrix elements to observed structures.
(b) Fitting of matrix elements to observed cohesive energies, that is, to total energy factors rather than one-electron terms.
(c) Fitting of local exchange and correlation models (see Section 1.1.3) to atomic data.

Empirical factors are relatively benign provided their use is recognized explicitly. In many cases they mean that a theory is being used to extrapolate from some well-established results (e.g., cohesive energy) to related ones (e.g., the total energy of the system as the atomic positions are changed) that are not accessible to direct experiment. Very few theories are truly *a priori*, and properly used empirical terms are an important part of defect theory. Simplifications must be used with even more care for they affect different predictions differently. Muffin-tin approximations, for instance, have modest effects on intra-atomic features of complex atoms but major effects on changes in total energy as relative motions of atoms occur.

2. Geometric Factors. We can distinguish several distinct levels of calculation:

(a) Presumed geometry, for example, using atoms at their sites in the undistorted lattice or at chosen random atomic positions for an amorphous or disordered system.
(b) Restricted relaxation, for example, only radial movement of just the nearest neighbors of a defect.
(c) Longer-range relaxation, including polarization due to any net charge associated with a defect (see Section 2.3.3).
(d) Cluster calculations, which concentrate on a microcrystal of modest size. Cluster calculations carry their own limitation, since one must test carefully for dependence on cluster size and shape.

1.1.3 Correlation and its Consequences

In studies of electronic properties, effects of electron–electron interaction can be divided into two components. In one, each electron moves individually in an average field of the others. The other component, involving the motions of the other electrons, gives correlation corrections which we now discuss. In studies of simple molecules two extreme approaches are followed, namely, molecular-orbital methods and valence-bond methods (Coulson 1961). In the molecular-orbital method each electron moves in an *average* field of the nuclei and other electrons, and to this extent correlation effects are ignored. For the hydrogen molecule in its ground state, for example, the two electrons might be assigned wavefunctions $\phi(\mathbf{r})\alpha$ and $\phi(\mathbf{r})\beta$, with spin functions α and β and with the spatial wavefunction given approximately as $\phi(\mathbf{r}) \simeq [a(\mathbf{r}) + b(\mathbf{r})]/\sqrt{2}$. Here $a(\mathbf{r})$ is an atomic orbital on atom A and $b(\mathbf{r})$ is a similar orbital on atom B. The total wavefunction for the two electrons is given by a Slater determinant to ensure that the Pauli exclusion principle is obeyed, that is,

$$\Phi_{MO} = \det \begin{Vmatrix} \phi(\mathbf{r}_1)\alpha_1 & \phi(\mathbf{r}_1)\beta_1 \\ \phi(\mathbf{r}_2)\alpha_2 & \phi(\mathbf{r}_2)\beta_2 \end{Vmatrix} \tag{1.12}$$

If this is expanded there will be some terms like $a(\mathbf{r}_1)b(\mathbf{r}_2)$ and $a(\mathbf{r}_2)b(\mathbf{r}_1)$ (omitting the spin part for clarity) that have the two electrons on opposite sites. There will also be ionic terms, like $a(\mathbf{r}_1)a(\mathbf{r}_2)$ and $b(\mathbf{r}_1)b(\mathbf{r}_2)$, which put two electrons on the same site. However, even in the molecular-orbital model there is some limited correlation, for the Pauli exclusion principle ensures that the ionic terms only involve electrons with opposite spins. Hence terms of the type $a(\mathbf{r}_1)a(\mathbf{r}_2)\alpha_1\beta_2$ appear but not $a(\mathbf{r}_1)a(\mathbf{r}_2)\alpha_1\alpha_2$.

The valence-bond model (Coulson 1961) takes correlation further for it excludes all ionic terms and ensures that there are never two electrons on one site. Whenever electron 1 is close to site A, electron 2 must be close to site B. One possible ground state wavefunction is $a(\mathbf{r}_1)b(\mathbf{r}_2)\alpha_1\beta_2$. Since the electrons are indistinguishable, the spatial part of the ground state wavefunction has the (unnormalized) form $a(\mathbf{r}_1)b(\mathbf{r}_2)+a(\mathbf{r}_2)b(\mathbf{r}_1)$ in the valence-bond approximation.

In the valence-bond approach even electrons with opposite spins avoid each other: only the terms with electrons on distinct sites contribute. However, the avoidance is achieved by a description which associates the electrons in pairs. When there are only two electrons, this is straightforward; when there are more than two electrons, the procedure can be very clumsy.

The two extreme limits of molecular-orbital and valence-bond models represent, respectively, an underestimate and an overestimate of correlation. This can be seen by writing a more general form (Coulson 1961 p. 154) which contains these two approximations as limiting cases. If we consider only the spatial part of the wavefunctions, and if we abbreviate $a(\mathbf{r}_1)$ as a_1, etc., then the unnormalized molecular orbital form is $(a_1+b_1)(a_2+b_2)$ and the valence bond form $(a_1b_2+a_2b_1)$. A weighted sum of the two $\{(a_1+b_1)(a_2+b_2)+\Lambda(a_1b_2+a_2b_1)\}$ is a generalization which, if Λ is chosen variationally, allows one to adjust the extent of correlation by varying the proportions of correlated parts $(1+\Lambda)(a_1b_2+a_2b_1)$ and ionic components $(a_1a_2+b_1b_2)$. The degree of correlation is enhanced as Λ is increased.

Various alternative ways of interpolation between the molecular-orbital and valence-bond schemes exist. For molecules configuration admixture is especially popular; in effect one uses a weighted linear combination of the two extremes. The example of the last paragraph can be considered as a special case in which another molecular orbital configuration $(a_1-b_1)(a_2-b_2)$ is admixed into the original $(a_1+b_1)(a_2+b_2)$. In solids the Hubbard model (Hubbard 1963) is especially useful. This is an exactly soluble model in which the correlation is introduced as an energy penalty, U, when two electrons with opposite spins occupy the same site. If the bandwidth is $2B$, then $U/2B$ is a measure of the correlation. Vanishing U leaves the molecular-orbital description of conventional band theory, whereas large $U/2B$ represents strong correlation.

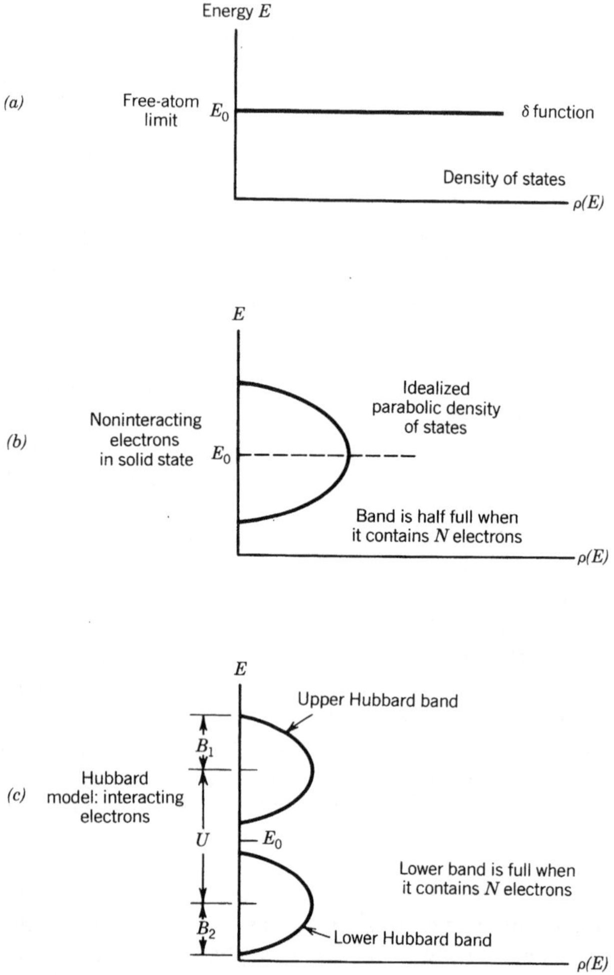

Figure 1.7 (*a*) and (*b*) represent the development of the band structure of noninteracting electrons on going from free atoms to a solid; (*c*) represents effects of electron correlation in band structure, giving rise to Hubbard bands separated by the correlation energy *U*. For *N* hydrogen atoms the neutral solid of *N* protons and *N* electrons would have a half-filled band for noninteracting electrons [case (*b*)]. For interacting electrons [case (*c*)] one could have the following situations:

1. Less than *N* electrons: partly-occupied lower Hubbard band (hole states).
2. *N* electrons: full lower Hubbard band and empty upper Hubbard band.
3. More than *N* electrons: partly occupied upper Hubbard band (electron states).

Qualitatively, the effects of correlation can be seen from two examples. The first is the metal–insulator transition (see Section 8.2). For a system with paired electrons, like a rare-gas solid or a simple ionic crystal, one can have an insulator at large interatomic spacings and a metal for close atomic separations (Figure 1.3). When there is only one electron per atom, as for an array of widely spaced hydrogen atoms, the picture is different. Simple band theory gives bands of one-electron states, as before, with the valence-band half full. Since there are always empty states in half-filled bands which can be reached for negligible δE, all such systems should be metals, contrary to observation. The anomaly can be resolved when correlation is considered. The point is that conduction would proceed by transferring an electron from one site ($H^0 \rightarrow H^+$) to another ($H^0 \rightarrow H^-$). Because of correlation δE must be of order U, and this energy inhibits metallic conduction at large spacings. As the interatomic distances decrease, a transition to the metallic state eventually occurs. This insulator–metal transition is known as the Mott–Hubbard transition (Section 8.2.1). When the number of electrons is less than the number of sites, conduction can occur without the inhibiting activation energy U.

The second example concerns carrier transport when there is either an excess of H^+ or of H^-. The energies of the extra electrons (H^-) and holes (H^+) are separated by the correlation energy discussed above. In effect, the original single band without correlation terms has been split into two bands, the so-called upper and lower Hubbard bands separated by the correlation energy U (Fig. 1.7). In the upper Hubbard band the extra electrons from the H^- ions provide the transport; in the lower Hubbard band the holes corresponding to H^+ are the mobile carriers. This picture is useful in understanding such systems as magnetic oxides and heavily doped semiconductors (Section 8.2.2).

1.2 ELECTRONIC TRANSPORT AND ELECTRON–LATTICE INTERACTION

1.2.1 Effects of Impurities on Conductivity

In nonmetals electronic currents are generally modest because of the significant energy δE needed to excite electrons. High conductivity is associated with small δE, especially with the partly filled bands characteristic of metals, but also possibly with heavily doped solids, or with solids highly excited by ionizing radiation. Figure 1.8 shows the conductivities of a wide variety of metals and nonmetals.

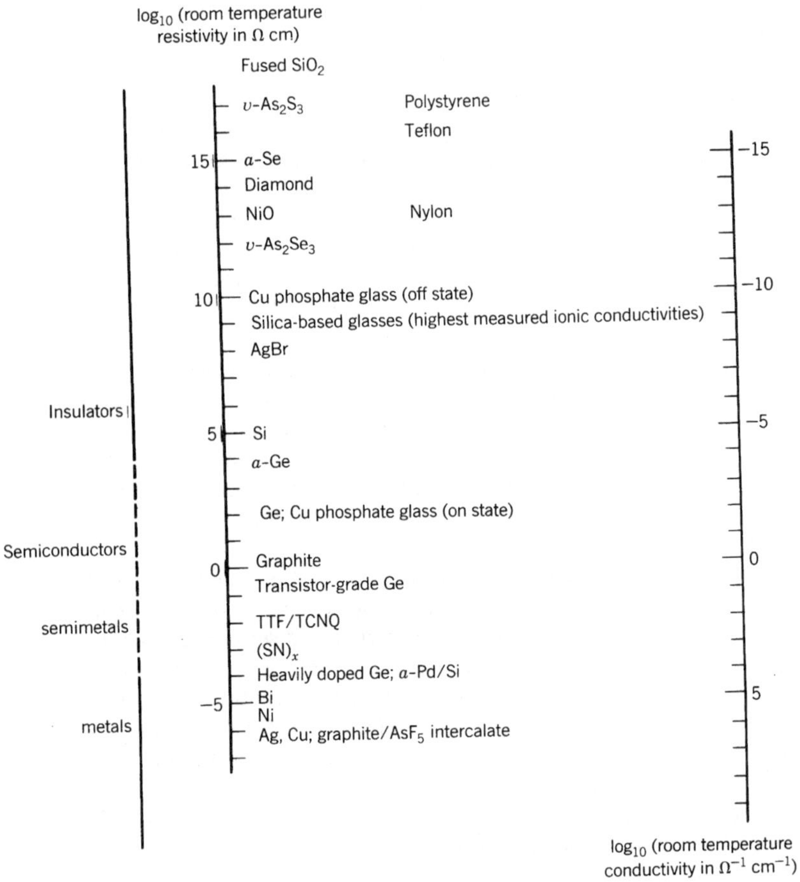

Figure 1.8 Resistivities at room temperature for a variety of solids. Unless stated, the solids are nominally pure. Prefixes are c- (crystalline), a- (amorphous), and v- (vitreous). Note that cases of switching are shown, for example, where Cu phosphate glass can occur in on and off states.

When there are very few electrons in a conduction band, the conductivity σ is given by

$$\sigma = n_e e \mu \tag{1.13}$$

where n_e is the number of electrons per unit volume and μ is the mobility. The number of carriers depends on the defects present through two factors, namely, the ratio of the defect concentration to the density of states in the conduction band and the ratio of the defect ionization energy to the thermal energy k_BT. The mobility μ may also be controlled by defects, either by scatter or by repeated trapping and release of carriers. An important point to note is that the effects of defects on carrier concentration and mobility are opposite, tending to compensate each other. Heavily doped crystals are thus often of intermediate conductivity (say 0.01–100 Ω^{-1} cm^{-1}), since there are too many carriers for them to be highly insulating and there are too many scatterers for high conductivity.

Electrons are the only carriers of current in the absence of ionic motion. When there are only a few electrons in the conduction band, they dominate in the transport of charge. However, when a band is almost full, it is simpler to concentrate on the missing electrons, referred to as holes. This idea of holes enters into conduction in discussing acceptor centers like Al in Si (Section 5.3.1) and in treatments of the incomplete *d* shells of transition-metal ions (Section 5.2). If the valence band of a semiconductor contains *N* electrons when full, the transport and optical properties when it contains only $N-n$ electrons are those expected from *n* carriers (holes) with the following properties:

1. Their charge is $|e|$. This charge describes consistently the interactions of holes with external electric fields, with other electrons, and with other holes. Thus holes repel each other and move toward the cathode.
2. We can define an effective mass tensor by $m^*_{ij} = \hbar^2/[\partial^2 E(\mathbf{k})/\partial k_i \partial k_j]$ for holes as well as for electrons. For free charges m^*_{ij} becomes $m_0\delta_{ij}$, giving a scalar mass. Many crystals have degenerate valence bands, often derived from atomic *p* states and often with significant spin–orbit coupling; in such circumstances the tensor form is required for hole masses.
3. The spin–orbit coupling has the opposite sign from that of electrons and this may be seen in the epr *g* factors for hole centers (Section 4.5).

If electrons are the dominant carriers, as in *n*-type material, most defects will be in neutral or negatively charged states, depending on the precise position of the Fermi level. If holes are dominant, as in *p*-type material, defects are principally in positively charged states.

Electrons are governed by Fermi–Dirac statistics. The Fermi function

$$f(E, E_F, T) = \frac{1}{\exp[(E-E_F)/k_BT]+1} \qquad (1.14)$$

describes the occupancy of both band and defect states. The form of $f(E, E_F, T)$ means that at $T = 0$ all states below the Fermi level E_F are occupied and all those above are empty; at finite temperatures there is a relatively narrow energy range of about $k_B T$ at E_F with partial occupancy. The Fermi level is fixed by noting that the total number of carriers N_c must be given by

$$N_c = \int_{-\infty}^{E_F} dE\, \rho(E) f(E, E_F, T) \tag{1.15}$$

where $\rho(E)$ is the density of states, including spin and other degeneracies. This equation is readily solved in simple cases. For undoped material E_F is near the center of the band gap. Doping by donors, which contribute electrons to the previously empty conduction band, raises the Fermi level. Acceptors, which contribute holes to the previously full valence band, lower the Fermi level.

We shall often be concerned with defects that can exist in more than one charge state. In thermal equilibrium, the charge state principally present may be controlled by doping. For example, consider a small concentration of a transition metal M, which can exist in charge states M^+, M^{2+}, and M^{3+} in a semiconductor. If the system is initially p type, with many acceptors, the M^{3+} state will predominate. As acceptors are removed and donors are added, the Fermi level will rise until M^{2+} states become numerous. The energy $E(3+/2+)$ defines the position of the Fermi level at which there is a precise balance between M^{3+} and M^{2+}, that is, $M^{3+} + e(E_F) \rightleftharpoons M^{2+}$ neither absorbs nor emits energy. Further increases in donor concentration and decreases in acceptors lead to formation of M^+. In many oxides, for example, MgO, thermal equilibrium is established very slowly, so quite different populations of charge states may be seen from sample to sample.

With a small concentration of dopants, M^{N+}, the position of the Fermi level can be controlled by independent donor or acceptor doping. However, reasonable concentrations of M can cause Fermi level pinning, that is, lead to only a very small shift of E_F despite substantial changes in donor and acceptor concentrations. This happens readily if introduction of M causes a local peak in the density of states [this is easily seen by differentiation of Equation (1.15) with respect to E_F; pinning corresponds to large values of dN_c/dE_F]. One specific mechanism corresponds simply to changing the charge state of a dopant M^{N+}. If electrons are added when there are many M^{3+} ions present; for example, E_F will be pinned at $E(3+/2+)$ until practically all M^{3+} have been converted to M^{2+}.

1.2.2 Electron–Lattice Interaction

Charged defects and localized carriers in solids produce electric fields that polarize the surrounding host lattice. In the case of simple ionic solids this polarization involves two main components. One is electronic polarization, proportional to ϵ_∞^{-1}, and hence characterized by the optical dielectric constant ϵ_∞. The other is the ionic polarization, in which there is relative motion of anions and cations. It is this second component, characterized by ϵ_0 (or strictly by $\epsilon_\infty^{-1} - \epsilon_0^{-1}$), that leads to the most important polarization effects. This ionic polarization contributes both to static and to dynamic properties of charged carriers and defects. We now consider these phenomena.

If a static point charge is placed in a solid, the atoms around it are both polarized and displaced (Figure 1.9). The polarization energy appears in many static properties such as defect formation energies, as well as in charge-transfer optical transitions, and in other defect optical and thermal energies. The polarization also affects carrier dynamics. When

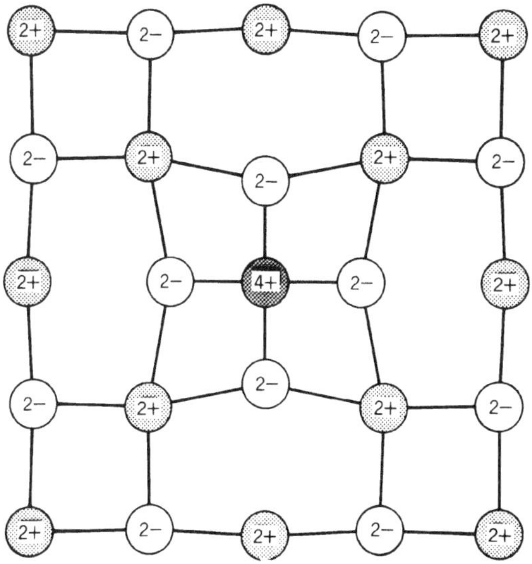

Figure 1.9 Effects of a localized charge. Here MgO has been modeled, with a 4+ ion replacing the central Mg^{2+}. A calculation by P. W. Tasker (private communication), using the Harwell HADES code, shows clearly the inward motion of the nearby anions, the outward motion of the cations, and the rapid fall-off of displacements with distance r from the defect. At large distances from the defect, theory (Sections 2.3.3 and 3.4) predicts displacements proportional to $1/r^2$.

the point charge moves, the electronic polarization and ion displacement adjust in an attempt to follow the charge. The acceleration per unit applied electric field will be altered and hence the apparent effective mass will be changed. If the carrier moves steadily, phonons will be generated as energy is given to lattice vibrations, and these energy losses ultimately limit the mobility. The word *polaron* is used for an electronic carrier with its associated distortion and polarization field. It is the polaron's properties, rather than those of the bare carrier, that determine observed transport properties such as mobility, cyclotron resonance, and thermoelectric behavior. In special cases one can regard the polaron as a carrier with altered mass m_p^* rather than the bare band mass m_b^*.

There are three main mechanisms by which a free carrier interacts with the lattice:

(a) *Fröhlich coupling*. Here the carrier charge couples to the electric field produced by the longitudinal optic phonons. This occurs only in polar crystals such as KCl and GaAs.
(b) *Piezoelectric coupling*. Here the carrier charge couples to the electric field produced by the acoustic phonons. This only occurs in crystals of appropriately low symmetry, such as quartz or zinc oxide.
(c) *Deformation potential coupling*. Here the carrier energy changes with the strain (mainly the dilatation) produced by acoustic modes and occurs in all crystals.

The three mechanisms are not all of equal importance. The piezoelectric contribution, absent by symmetry in many crystals, is readily screened by free carriers. All three mechanisms can be written in forms that apply to continuum solids, although they are readily written in atomistic forms. Likewise, all three interactions can be important for bound carriers (see, for example, Section 5.1.3).

The Fröhlich mechanism is the most important. For a static charge, the ionic screening associated with this coupling changes the potential at a distance r from the value with only electronic screening ($\epsilon = \epsilon_\infty$), to the value with ionic screening ($\epsilon = \epsilon_0$):

$$e^2/\epsilon_\infty r \rightarrow e^2/\epsilon_0 r \equiv (e^2/\epsilon_\infty r) - (e^2/r)(\epsilon_\infty^{-1} - \epsilon_0^{-1}) \qquad (1.16)$$

Clearly $(\epsilon_\infty^{-1} - \epsilon_0^{-1})$ is one factor in the electron–lattice coupling. In fact it proves convenient to use instead the dimensionless combination

$$\alpha = \tfrac{1}{2}(\epsilon_\infty^{-1} - \epsilon_0^{-1})[(e^2/r_p)/\hbar\omega_{\mathrm{LO}}] \qquad (1.17)$$

as a measure of polaron coupling. Similar expressions can be given for other coupling mechanisms (Duke and Mahan 1965). The polaron radius

Table 1.5 Typical Values of Polaron Coupling Constants α [Equation (1.17)] and Polaron Radii r_p [Equation (1.18)]

Type of Crystal	Typical Value of α	Typical Value of r_p (Å)
Alkali halide	2–4	10–15
Silver halide	1.5–2	13–15
II–VI Semiconductor	0.3–0.9	10–13
III–V Semiconductor	0.02–0.15	10–12

r_p, defined by

$$r_p^2 \equiv (\hbar^2/2m_p^*)(\hbar\omega_{LO}) \tag{1.18}$$

is essentially the Compton wavelength of a carrier with kinetic energy equal to the longitudinal optic phonon energy $\hbar\omega_{LO}$. The polaron radius nearly always lies in the range 10–15 Å and has no direct dependence on α. Typical values of α and r_p are given in Table 1.5 for a variety of materials.

Many cases of lattices distorted by carriers or defects can be understood by their relationship to a model calculation in which a harmonic oscillator, representing the host, is perturbed by a constant force that represents the defect or carrier. This simple, exactly soluble example occurs so often we shall discuss it here. As a preliminary we consider the effect of a carrier, trapped or free, in a harmonic lattice. Since we know that the harmonic lattice can be treated dynamically in terms of its independent normal modes, we concentrate on just a single one of these normal modes. When there is no carrier present, the Hamiltonian has the form

$$\mathcal{H}_0 = \tfrac{1}{2}M\dot{Q}^2 + \tfrac{1}{2}KQ^2 \tag{1.19}$$

where M is an effective ionic mass and K is a force constant; the mode Q has angular frequency $(K/M)^{1/2}$. The carrier introduces four main types of terms, (i) a constant term E_0, which has no dynamic effects although it appears in the energy change in moving the carrier in and out of the solid; (ii) a linear term $-FQ$, where F is the force the carrier exerts on the mode; (iii) higher-order terms in Q alone, which alter the effective force constant, that is, $K \rightarrow K'$; and (iv) higher-order cross terms in Q and Q', which mix the original normal modes.

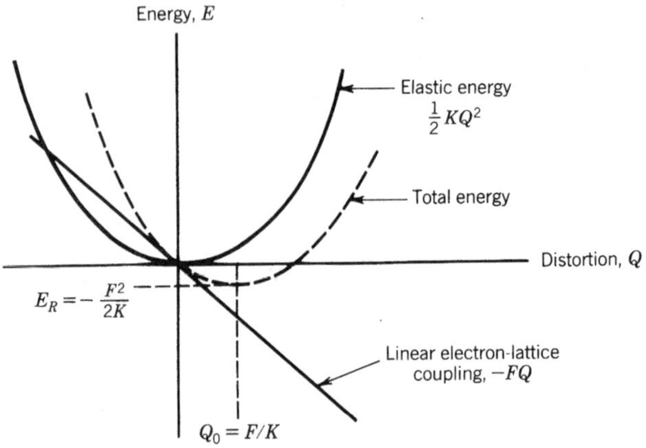

Figure 1.10 Energies associated with linear coupling to a lattice distortion Q. The two main terms in the energy are shown, namely, linear coupling $-FQ$ and elastic strain energy $+\frac{1}{2}KQ^2$. The total energy shows a minimum corresponding to a relaxation energy $E_R \equiv -F^2/2K$ at an equilibrium distortion $Q_0 = F/K$.

An exact solution is easily found if we ignore the constant term and the higher-order corrections, leaving only the term $-FQ$. The new Hamiltonian is

$$\mathcal{H} = \mathcal{H}_0 - FQ \tag{1.20}$$

which can be simply rewritten as

$$\mathcal{H} = \tfrac{1}{2}M\dot{\tilde{Q}}^2 + \tfrac{1}{2}K\tilde{Q}^2 - \tfrac{1}{2}F^2/K \tag{1.21}$$

by setting $\tilde{Q} = Q - F/K$. Physically, there is merely a displacement of the mode by $Q_0 = F/K$, linear in the force, as the oscillator vibrates about a new average position (Figure 1.10). This is accompanied by a relaxation energy $E_R = F^2/2K$, quadratic in the force F. The relaxation energy E_R consists of a reduction of $-2E_R$ in the coupling term $-FQ$ and an increase of $+E_R$ in the strain energy $\frac{1}{2}KQ^2$. In the polaron problem we anticipate that F is proportional to $\sqrt{\alpha}$ and K is proportional to ω_{LO}^2 so that the displacement F/K varies as $\sqrt{\alpha}/\omega_{LO}^2$ and the energy term $F^2/2K$ as α/ω_{LO}^2.

We also note the important technical point that we may combine arbitrarily any modes with the same frequency. Thus two such normal modes Q' and Q'' may be combined, keeping them orthogonal, to obtain

$\tilde{Q}'$ and $\tilde{Q}''$:

$$\begin{aligned}\tilde{Q}' &= \cos\theta\, Q' + \sin\theta\, Q'' \\ \tilde{Q}'' &= -\sin\theta\, Q' + \cos\theta\, Q''\end{aligned} \tag{1.22}$$

This leaves the perfect-crystal Hamiltonian unaltered for whatever value of θ is chosen. However, θ can be chosen so as to simplify the linear coupling terms and to eliminate one of the coordinates entirely from the coupling. This can be seen explicitly:

$$\begin{aligned}\tilde{F}'\tilde{Q}' + \tilde{F}''\tilde{Q}'' \rightarrow & (\tilde{F}''\cos\theta - \tilde{F}''\sin\theta)Q' \\ & + (\tilde{F}'\sin\theta + \tilde{F}''\cos\theta)Q''\end{aligned} \tag{1.23}$$

which can be reduced to a single term if $\tan\theta$ is taken as $\tilde{F}'/\tilde{F}''$:

$$\tilde{F}'\tilde{Q}' + \tilde{F}''\tilde{Q}'' \rightarrow \left[\tilde{F}'(\tilde{F}' + \tilde{F}'')/\sqrt{(\tilde{F}')^2 + (\tilde{F}'')^2}\right] Q'' \tag{1.24}$$

This simplification is important in calculating optical absorption line shapes (Section 4.2).

A second useful and general result gives the interaction between two defects or two carriers. Suppose the first defect applies forces F_1 to the lattice and the second defect applies forces F_2. Then the interaction energy between the defects or carriers is the difference between the separate and the total relaxation energies:

$$-(F_1 + F_2)^2/2K - [-(F_1^2/2K) - (F_2^2/2K)] = F_1F_2/K \tag{1.25}$$

that is, the interaction is simply the work done producing displacements F_2/K against forces F_1. This result may also be used to discuss superconductivity and also spin–spin interactions in some magnetic systems.

1.2.3 Properties of Polarons

Suppose an electron or hole has effective band mass m_b^* in a rigid lattice. If linear coupling to any lattice mode occurs, there will be two obvious measures of the effect. One measure is the relaxation energy (see Figure 1.10). From the modes of energy $\hbar\omega_\alpha$ there is a relaxation energy contribution of $N_\alpha\hbar\omega_\alpha$, where we define N_α as the number of phonons in the cloud accompanying the carrier; in all there are $\sum_\alpha N_\alpha$ phonons of all energies. The other measure is the polaron effective mass m_p^*, which is defined by the kinetic energy $\hbar^2k^2/2m_p^*$ (see Section 1.2.2).

In the case of weak coupling to optic modes by the Fröhlich interaction, the results are especially simple (see Harper et al. 1973):

$$\text{Number of phonons in cloud } N_\alpha = \alpha/2 \tag{1.26a}$$

$$\text{Polaron effective mass } m_p^* = m_b^*(1 + \alpha/6) \tag{1.26b}$$

As α increases, these expressions are replaced by more complex ones, the most useful expression for the polaron mass being an interpolation

$$m_p^* = \frac{1 - 0.08\alpha^2}{1 - \alpha/6 + 0.0034\alpha^2} m_b^* \tag{1.26c}$$

valid for $\alpha \lesssim 5$; this regime covers almost all practical cases (Table 1.5).

There are two distinct difficulties in checking experimentally expressions like (1.26b) and (1.26c) in ionic solids. First, although one can measure m_p^* from the cyclotron resonance frequency eH/m_p^*c, we have no way of measuring m_b^*, the band mass free from polaron corrections. Second, there is the practical problem that intrinsic carrier densities in pure ionic crystals are very small. With band gaps varying from 2.7 eV in AgBr to 8.5 eV in KCl, thermal carrier generation is negligible at room temperature. However, free carriers in both conduction and valence bands can be created by ultraviolet excitation in AgBr. When the band gap is higher, as in KCl, electrons can be produced in the conduction band by creating F centers and by photoionization by optical excitation on the high-energy side of the F absorption band (Section 5.1.1).

Optically produced carriers may have high kinetic energies at first. However, in ionic solids, they lose this energy rapidly, in about 10^{-13} sec, to longitudinal optic phonons by Fröhlich coupling. Once carrier energies fall below $\hbar\omega_{LO}$ further relaxation occurs by slower processes involving coupling to acoustic modes. In principle, thermalization should occur in about 10^{-11} sec with the distribution of carrier energies determined by lattice temperature. This equilibration occurs readily in highly pure semiconductors like Si and Ge. In highly purified crystals of some silver and alkali halides, the kinetic energy of the carriers can approach to within 2 K of a lattice temperature of 1.2 K (Harper et al. 1973). If, however, the crystal is impure or contains many intrinsic defects, trapping may remove carriers before thermalization is approached.

The most important transport property is the carrier mobility μ [Equation (1.13)]. One common technique for measuring mobility exploits the Hall effect, which uses an applied magnetic field to distinguish between electron and hole motion, through the Lorentz force. The mobility depends on temperature as well as on the Fröhlich coupling instant α. At the very lowest temperatures, below 15 K in the purest alkali and silver halides, mobility is largely independent of temperature.

At liquid-helium temperatures the mobility may be large, 2.5×10^4 cm^2/V sec in ordinary KBr and 1.3×10^5 cm^2/V sec in zone-refined KBr. As the temperature rises, the mobility falls, and in lowest-order perturbation theory one finds

$$\mu = \frac{e}{2m_b^* \alpha \omega_{LO}} [\exp(\theta_{LO}/T) - 1] \tag{1.27}$$

with θ_{LO} given by $\hbar\omega_{LO}/k_B$; the mobility is simply inversely proportional to the Bose–Einstein occupancy of the longitudinal optic modes.

There are aspects of carrier mobility that become important at very low temperatures. Clearly a carrier with kinetic energy less than $\hbar\omega_{LO}$ (typically 0.05 eV) cannot slow down by producing optic phonons. In the same way, electrons with a velocity below the sound velocity (typically electrons with energy below 10^{-4} eV) cannot slow down by generating acoustic phonons.

We now present an argument which suggests that localizing a carrier stabilizes the carrier, because of gain in relaxation energy. Suppose the carrier is spread equally over N sites (for simplicity we shall ignore interference effects since full calculations also confirm the point we are making). The defect forces at each site reduce to F/N, since Coulomb forces due to localized charges obviously fall in proportion to the fraction of charge at a particular site. However, there are N contributions with their own relaxation energies $(F/N)^2/2K$ [see Equation (1.21)] so that the total relaxation energy becomes $N[F/N)^2/2K]$ or $(1/N)(F^2/2K)$. There is therefore possible competition between the energy [essentially half the bandwidth; see Equation (1.9)] encouraging delocalization and the energy favoring carrier localization, to gain up to $F^2/2K$ in relaxation energy. The two limits give at one extreme the large polaron, essentially the case we have treated so far, and, at the other extreme, the small polaron. The simplest example of a small polaron might be considered as a lattice ion in an unusual charge state, for example, U^{5+} in UO_2 or Mn^{3+} in MnO, although the best-established examples are molecular ions, for example, Cl_2^- replacing $2Cl^-$ in KCl (Sections 5.1.3). For all polarons the inertia associated with lattice polarization and distortion affects the polaron dynamics. In the case of the small polaron, motion is by hopping, and the carrier may be totally immobilized at low temperatures.

This competition between different energy terms for a free carrier can be expressed more formally (Emin and Holstein 1976). Suppose the carrier is localized in a dielectric continuum in a region of dimension L [precisely how this localization is defined does not matter so long as there is scaling such that $\psi(\mathbf{r})$ goes to $(L/L_0)^{-3/2}\psi(\mathbf{r}/L)$ as L is changed from some initial value L_0]. Then simple scaling arguments show that:

(i) Kinetic energy $\sim +L^{-2}$

(ii) Coulomb terms $\sim -L^{-1}$ from any defect present and from deformation-induced electric polarization.

(iii) Short-range terms $\sim -L^{-3}$ from, for example, deformation potential coupling or molecular interactions.

(iv) Cross terms $\sim -L^{-2}$.

For various values of the proportionality constants one finds quite distinct regimes (Figure 1.11):

(a) Only Coulomb and kinetic-energy terms (Case I). A single minimum in total energy is found. The $1/L^2$ kinetic term dominates at small L, preventing collapse to a point charge.

(b) Only short-range and kinetic-energy terms (Case II). There are two minima, at $L = 0$ and $L = \infty$. The $L = \infty$ solution is simply the free

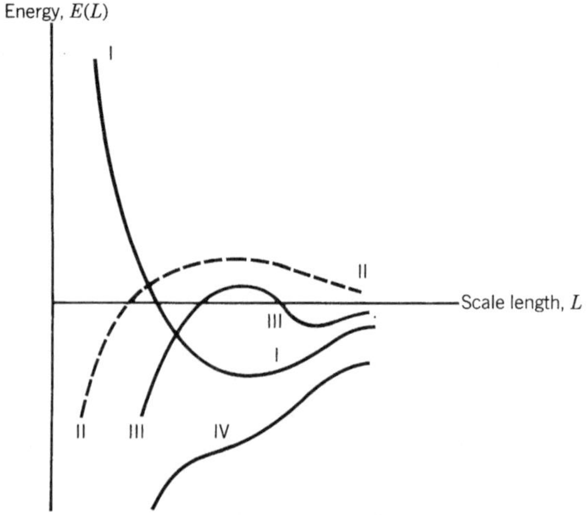

Figure 1.11 Regimes of behavior for an electron coupled to lattice distortion. The total energy is given as a function of scale length L characterizing a localized, normalized electron wavefunction $(L/L_0)^{-3/2}\psi(\mathbf{r}/L)$. I: Only a large polaron forms. No short-range deformation-potential interaction. II: Pure short-range coupling to dilatation. Note that there are only small-polaron ($L \rightarrow 0$) and nonpolaronic ($L = \infty$) minima. III, IV: General cases showing only small polarons (IV) or both small and large polarons separated by a barrier (III, the coexistence case). After Emin and Holstein (1976).

carrier, and the $L = 0$ solution (remember we have ignored atomic structure) corresponds to a highly localized small polaron. Note that there is no intermediate-range polaron with finiate L.

(c) When both long-range and short-range terms occur one always finds a small-polaron state; one may find the free ($L = \infty$) form converted into a large polaron (Case III) or the free form may be simply unstable (Case IV).

All these results are strictly for a continuum. The lattice structure dictates that L cannot fall significantly below the lattice spacing. This can have the qualitative effect of eliminating the small-polaron limit (e.g., Case III). Likewise, all the calculations assume a single band; there may exist no large polaron in one band but large-polaron states associated with another band (cf. Case IV).

From these considerations one can anticipate several unusual features when there are two distinct minima (Case III, Figure 1.11) corresponding to alternative forms of polarons. First, the nature of the ground state may change abruptly, although the ground-state energy will be continuous, at least in the model described. Second, if one has a short-lived carrier in a solid, for example, a positron e^+, a muon μ^+ (Section 4.5), or an exciton (Section 1.3.1), it may never reach the ground state associated with the lowest minimum. For kinetic reasons, that is, because some of the key transitions are relatively slow, the carrier may reach states associated with the upper minimum and decay before transfer. Third, if the carrier does end in the upper minimum, there is an energy barrier opposing transfer to the lower minimum. This may be discerned in the temperature dependence of the transfer, and in suitable cases both types of state may be seen. In alkali iodides, for example, the band-edge luminescence from conventional excitons is clearly distinguished experimentally from lower-energy luminescence of localized excitons; these are referred to as self-trapped excitons or excitonic polarons (Section 5.1.3). Observed intensities monitor populations in the upper minimum and lower minimum, respectively, and show a temperature dependence reflecting a barrier of some few meV against conversion. This conversion process is much slower (of the order of nanoseconds) than any direct relaxation to a self-trapped state, which would presumably take only a few lattice vibration periods (less than a picosecond).

Whether or not a small polaron forms depends on the relative magnitudes of two energies, the lattice relaxation energy (or stabilization energy) for the localized carrier, E_R (Figure 1.10), and the transfer energy J [as in Equation (1.5)], which determines the uncoupled carrier bandwidth and thus the energy gain from delocalization. When we look beyond polaron behavior to more general situations, we see that there is a third energy of importance, the on-site Coulomb repulsion U between two carriers (Section 1.1.3). There are three extreme cases of interest, which include ideas we have already outlined:

(a) Competition between correlation and band formation (see Section 1.1.3):
$U \gg J$ Valence-bond limit for H_2 and analogous cases
$J \gg U$ Molecular-orbital limit for H_2 and analogous cases

(b) Competition between lattice relaxation and band formation (see above):
$E_R \gg J$ Small polaron
$J \gg E_R$ Large polaron

(c) Competition between lattice relaxation and correlation:
$E_R \gg U$ Electrons prefer to share sites to gain lattice relaxation energy (Section 5.4)
$U \gg E_R$ Electrons prefer to remain on separated sites

The relation between these cases and a discussion of intermediate forms is given by Toyozawa (1981), who defines $S \equiv E_R/\hbar\omega$ and $T \equiv J$ in his (S, T, U) notation.

1.3 OPTICAL PROPERTIES

We have seen in the previous section that when a crystal is excited with photons of energy $h\nu$ larger than E_g electrons and holes are created with excess kinetic energy of order $(h\nu - E_g)$. The photoexcited carriers lose this excess energy and relax to band extrema primarily by interaction with phonons. In materials such as GaAs and AgCl the dominant electron–phonon interaction mechanism is polar optical phonon scattering, except at the lowest temperatures. Thermalized electrons and holes attract each other by the Coulomb potential $U(\mathbf{r}) = -e^2/\epsilon r$ to form excitons, stable bound states of the two particles, where ϵ is an appropriate dielectric constant. The exciton is an electrically neutral particle that can move through the crystal, transporting energy. Excitons can interact with each other to form molecular excitons, and if created with a sufficiently high density, can form an electron–hole plasma (Reynolds and Collins 1981).

1.3.1 Excitons and the Band Edge

In exciton formation the electron–hole correlation can take many forms and an indication of the variety of excitons possible is given in Table 1.6. In one limit, referred to as the Wannier case, both electrons and holes move in extended orbits around each other. In this case, for spherical band extrema that are nondegenerate, the energy level system is similar to that of the hydrogen atom. The optical absorption energies are given by a Rydberg-type series

$$E_n = E_g - \frac{\mu e^4}{2\hbar^2 \epsilon^2 n^2} + \frac{\hbar^2 k^2}{2(m^*_{pe} + m^*_{ph})} \tag{1.28}$$

where n is the principal quantum number, and $\mu^{-1} = (m^*_{pe})^{-1} + (m^*_{ph})^{-1}$ defines μ as the reduced polaron mass of the exciton. The second term in (1.28) is simply $(\mu/m_0)/(\epsilon^2 n^2)$ rydbergs. The third term in (1.28) is the kinetic energy associated with center-of-mass motion of the exciton with momentum (see below) $\mathbf{k} = \mathbf{k}_e + \mathbf{k}_h$; the exciton states are in fact band states characteristic of the crystal as a whole.

In principle it is possible to observe a Rydberg-type series of optical transitions from the valence-band edge to the conduction band, converging on the conduction-band edge for $n = \infty$, and such series have been observed in Cu_2O (Reynolds and Collins 1981). Generally, however, the exciton absorption at energies less than E_g is less structured; GaAs is a fairly typical case (Figure 1.12). Table 1.7 gives measured ionization energies, E_x, of excitons ($E_x = \mu e^4/2\hbar^2\epsilon^2$) if expression (1.28) applies; these are effective rydbergs for the hydrogenlike excitons.

In the more ionic solids the exciton wavefunctions are more highly localized and are referred to as Frenkel excitons (Table 1.6). In alkali halides, which are typical examples, the excitons of lowest energy have the hole localized in the valence p orbitals of the halogen and the electron extends to the nearest shell of alkali ions. These excitons have a tendency to self-trap at low temperatures (Section 5.1.3). The valence band is split into $j = \frac{1}{2}$ and $\frac{3}{2}$ states by the spin–orbit coupling of the halogen, and this structure is partly resolved in exciton absorption by chlorides and clearly resolved in bromides (Figure 1.13).

The main selection rule affecting optical absorption by crystalline solids is concerned with the wavevector, sometimes misleadingly referred to as crystal momentum. Wavevector selection rules simply ensure that, when light of wavevector $\mathbf{k}$ causes transitions between states ψ_i and ψ_f, the integral $\int d^3r \psi_i e^{i\mathbf{k}\cdot\mathbf{r}} \psi_f$ does not vanish (or nearly vanish) because of oscillations in the integrand. For light of wavevector $\mathbf{k}$ to cause an

Table 1.6 A Glossary of the Main Species of Excitons

Exciton	In essence, an electron and hole moving with a correlated motion as an electron–hole pair.
Wannier exciton	Electron and hole both move in extended orbits. Energy levels related to hydrogen atom levels by scaling [Equation (1.28)] using effective masses and dielectric constant. Occurs in covalent solids such as silicon.
Frenkel exciton	Electron and hole both move in compact orbits, usually essentially localized on adjacent ions. Seen in ionic solids, such as KCl, in absorption.
Self-trapped exciton	One or both carriers localized by the lattice distortion they cause (see Sections 1.2.3 and 5.1.3). Observed in ionic solids, such as KCl, in emission.
Bound exciton	Only a useful idea when a defect merely prevents translational motion of an exciton and does not otherwise cause significant perturbation.
Core exciton	Lowest-energy electronic excitation from a core state, leaving an unoccupied core orbital (e.g., the 1*s* level of a heavy atom) and an electron in the conduction band whose motion is correlated with that of the core hole.
Excitonic molecule	Complex involving two holes and two electrons (see Section 1.3.2).
Multiple bound excitons	Complex of many holes and a similar number of electrons, apparently localized near impurities. Some controversy exists, but up to six pairs of localized carriers have been suggested.
Exciton gas	High concentration of electrons and holes in which each electron remains strongly associated with one of the holes (an insulating phase) (see Section 1.3.3).
Electron–hole drops	High concentration of electrons and holes in which the motions are plasma-like (a metallic phase), not strongly correlated as in excitons (see Section 1.3.3) and included here only for comparison.

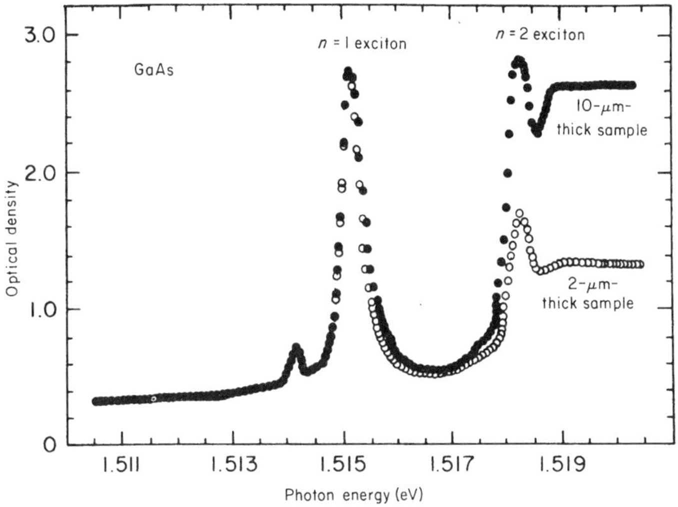

Figure 1.12 Optical density of GaAs at 2 K for sample thicknesses of 10 μm (●) and 2 μm (○). The peak at 1.5152 eV has been assigned to the $n = 1$ exciton transition and the peak at 1.5183 eV to the $n = 2$ transition [see Equation (1.28), text]. The band gap of this direct-gap material is close to 1.519 eV [after Sell (1972)].

allowed transition between electronic states characterized by $\mathbf{k}_i$ and $\mathbf{k}_f$, the relationship

$$\mathbf{k} = \mathbf{k}_i - \mathbf{k}_f \tag{1.29}$$

must hold. Since $\mathbf{k} \simeq 0$, this means that $\mathbf{k}_i \simeq \mathbf{k}_f$ and only vertical transitions on an $(E, \mathbf{k})$ diagram can occur (Figure 1.14). More general application of wavevector selection rules leads to the concept of indirect as well as direct transitions and allowed and forbidden transitions. A

Table 1.7 Binding Energies of Excitons[a]

Diamond	70	GaP	19.5	Cubic ZnS	36	CuCl	190
Cubic SiC	27	InP	5.12	ZnSe	19		
Si	14.3	GaAs	4.2	ZnTe	13.2		
Ge	4.15	GaSb	1.6				
		InSb	0.52				

[a]This table lists the energy needed (in meV) for exciton dissociation, that is, for the process $[eh] \rightarrow e + h$.

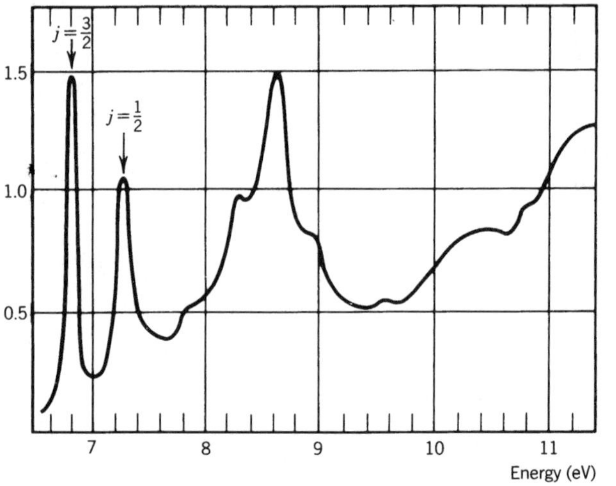

Figure 1.13 Optical absorption in arbitrary units of a thin film of KBr at 80 K. The arrows indicate the two spin–orbit components of an exciton transition, associated with the $j=\frac{1}{2}$ and $\frac{3}{2}$ components of the core hole with configuration $4p^5$, located on bromine and separated by 0.5 eV [after Eby et al. (1959)].

direct optical transition may be allowed or forbidden, depending on the symmetry of the crystal and perhaps on the electronic spin of the wavefunctions. An indirect transition is characterized by a change in wavevector between the electronic states (Figure 1.14) and wavevector conservation may be achieved by simultaneous excitation of a phonon of appropriate symmetry and wavevector $\mathbf{k}_{ph}$ such that

$$\mathbf{k}=\mathbf{k}_i-\mathbf{k}_f\pm\mathbf{k}_{ph} \tag{1.30}$$

In an optical absorption measurement the exciton absorption overlaps the region of band-to-band absorption which begins at E_g (Figure 1.14). The shape of the absorption edge immediately above E_g is calculated to vary as $(h\nu-E_g)^{1/2}$ for allowed direct transitions and as $(h\nu-E_g)^{3/2}$ for forbidden direct transitions (Elliott 1962). For indirect transitions the calculated shape is more complex, involving the temperature dependence of phonon populations.

Measured shapes of band-edge absorption do not always fall into the categories mentioned above. Absorption coefficients in a wide range of materials are found to have an exponential shape of the form

$$\alpha(h\nu)=\alpha_0\exp[-\gamma(h\nu-E_0)/k_BT^*] \tag{1.31}$$

referred to as Urbach's rule (Urbach 1953). Here γ is a constant of order

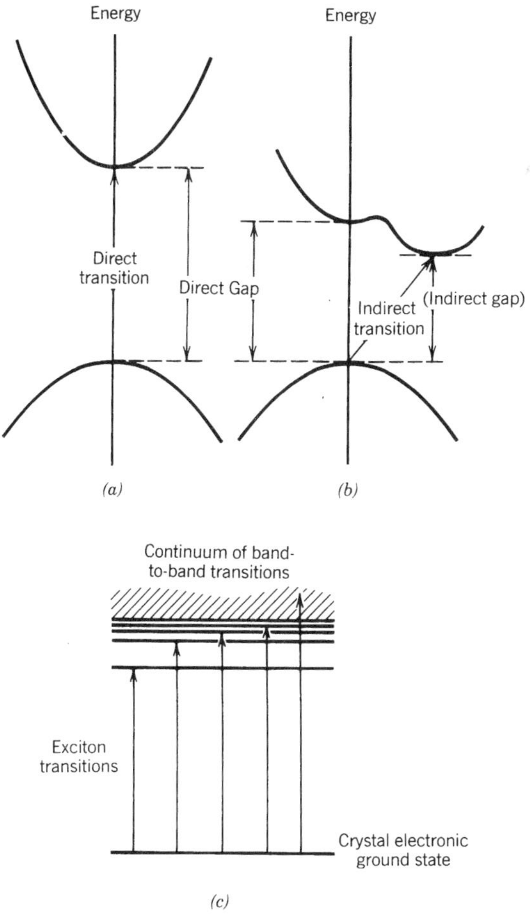

Figure 1.14 Schematic representation of (*a*) direct and (*b*) indirect band-to-band transitions for nondegenerate valence and conduction bands. Transitions (*b*) require participation of a phonon for wavevector conservation. (*c*) Exciton transitions converging on the band-to-band transitions of (*a*) and (*b*). Since the exciton involves correlated motion of electron and hole, it cannot be represented correctly on the one-electron level schemes of (*a*) and (*b*).

unity and α_0 and E_0 are parameters determined by the material; T^* replaces T because the low-temperature behavior is better represented by using $T^* = (\hbar\omega'/2k_B)\coth(\hbar\omega'/2k_BT)$ with $\hbar\omega'$ comparable in magnitude to phonon energies. In amorphous semiconductors the temperature dependence may be less marked.

The origin of Urbach's rule continues to be controversial. Thus Sumi and Toyozawa (1971) conclude that exciton–phonon interaction is responsible for the observed exponential shape. In their theory the exciton is treated as being partly mobile and partly localized, due to momentary trapping by lattice deformations. Indeed, their theory embraces the possibility of self-trapping of excitons. A different approach was adopted by Dow and Redfield (1972), who noted that Urbach's rule is quite often obeyed by damaged or disordered solids. They proposed a unified theory for the exponential shape, invoking electric field ionization of excitons; the electric field is assigned either to defects or to LO phonons (or to acoustic phonons in the case of piezoelectric materials). The dominant mechanism may, of course, change from case to case.

We now mention briefly the electronic states is layered crystalline structures in which the layers are so thin (typically 10 and 20 monolayers) that they are virtually two dimensional. These structures are made using techniques such as molecular beam epitaxy (MBE), in which a layer of one semiconductor is deposited on another under ultra-high-vacuum conditions [see, for example, Ploog and Döhler (1983)]. The lattice structure of the faces in contact in these epitaxial layers should be closely similar, to avoid strain. Because of the slow growth rate the thickness of the layers can be controlled to about one atomic layer (~0.3 nm resolution) and a relatively low growth temperature reduces diffusion, giving abrupt interfaces.

The MBE technique has been applied, for example, to the growth of thin films of GaAs, between 5 and 50 nm thick, between thick slabs of $Ga_{1-x}Al_xAs$ where $x \sim 0.2$. The heterojunctions formed by the different energy gaps of these two materials produce potential wells of different depths in GaAs for electrons and for holes. We bear in mind that a particle of mass m confined by an infinite potential barrier to a layer of thickness d_z has bound states given by

$$E_n = \frac{\hbar^2 \pi^2}{2m}\left(\frac{n}{d_z}\right)^2 + \frac{\hbar^2}{2m}(k_x^2 + k_y^2) \tag{1.32}$$

corresponding to quantized motion normal to the layer (z direction; $n = 1, 2, 3, \ldots$) and bandlike motion within each (x, y) plane. Optical absorption studies show peaks above the band edge of the GaAs layers which arise from transitions between the quantized electron and hole levels (referred to as subband structure); the positions of the peaks in these so-called quantum-well structures depend on the layer thickness (Figure 1.15), as suggested by Equation (1.32). Exciton energies are affected too whenever the effective orbital radius is comparable with the layer thickness.

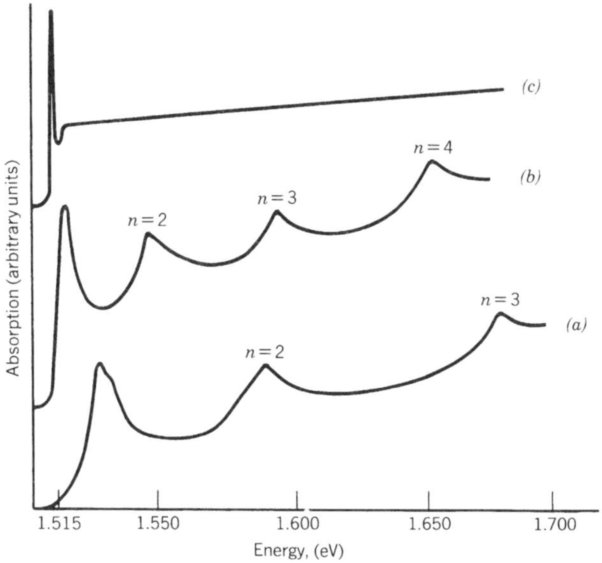

Figure 1.15 Optical absorption at 2 K of GaAs layers with thickness of (*a*) 14 (*b*) 21, and (*c*) 400 nm sandwiched between thicker layers of $Ga_{1-x}Al_xAs$ ($x \sim 0.2$). Electron and hole states with the same quantum number *n* [Equation (1.32), text] have maximum overlap and dominate exciton absorption. Trace (*c*) for the largest thickness, is characteristic of bulk GaAs (see Figure 1.12); note that the exciton peak characteristic of the bulk moves smoothly to higher energy with decreasing layer thickness [after Dingle et al. (1974)].

An extension of the heterojunction concept previously described involves the laying down of a periodic structure of thin layers of two semiconductors, for example, $GaAs/Ga_{1-x}Al_xAs$ or $CdTe/Hg_xCd_{1-x}Te$. The period in thickness lies in the range one to tens of nanometers, and the superlattice thus created can be regarded as a synthesized semiconductor not present in nature. The superlattice potential modifies the band structure of the component semiconductors creating minizones in wave-vector space and subbands in energy. For example, if the superlattice periodicity is L, the Brillouin zones of the component layers are split in the z direction into a series of subzones of width $2\pi/L$, altering transport properties in the z direction. This is an example of zone folding, which also affects phonons and which is discussed further in Sections 2.2.2 and 8.3.

A major advantage of layered structures is that it is possible to choose a combination of layers with properties that are free from some of the constraints of normal bulk material. An example is high carrier mobility, where values as high as 10^6 cm^2/V sec have been reported, very much higher than in common bulk systems. One limiting factor in bulk systems is that the carriers, for example, electrons, are present because donors are ionized, and the ionized donor has a Coulomb field that scatters the carriers and reduces their mobility. This feature can be virtually eliminated in layered structures by a process known as modulation doping. Suppose alternate layers are GaAs and $Ga_{1-x}Al_xAs$. The conduction-band minimum in the alloy occurs at a higher energy than in pure GaAs. If only the $Ga_{1-x}Al_xAs$ is doped with donors, the electrons released by ionization will move to the potential well formed by the GaAs. These carriers now have a high mobility, since they are only scattered weakly by the ionized impurities in the other layers. This approach illustrates one of several special features, namely, that one can separately optimize both the choice of defect for doping and the band structure features for high mobility. Another related example involves structures chosen so that electrons and holes are localized in different, separately optimized, layers. In such cases the spatial separation of electrons and holes means that recombination will be slow. These structures are now finding important device applications (Ploog and Döhler 1983; Kelly 1983).

1.3.2 Molecular Excitons

We saw in Section 1.3.1 that when free carriers are created in semiconductors with energy greater than E_g they relax to band extrema in times $\lesssim 10^{-12}$ sec by emission of phonons, and that thermalized electrons and holes can bind to form excitons. The excitons eventually decay to their ground state either by nonradiative mechanisms or by emission of radiation. However, radiative recombination in many semiconductors is a slow process ($\sim 10^{-5}$ sec). This is especially so in indirect-gap semiconductors, for example, Si, since optical emission requires participation of a phonon for wavevector conservation (Section 1.3.1). Using intense excitation sources, for example lasers, it is possible therefore to create high densities of excitons in many materials at low temperatures and to study their interactions. Two distinct phases of interacting excitons have been identified, namely, molecular excitons and the electron–hole liquid. We shall deal with the former in this section and the latter in the following section.

The theory of molecular excitons has a long history [see, for example, Lampert (1958)]. It has been explored in recent years by Brinkman et al. (1973), for example, who investigated the effect of the mass ratio $\sigma = m_e/m_h$ on E_{x2}, the exciton–exciton binding energy, for isotropic electron and hole bands. They find that the binding energy decreases monotonically with increasing σ, being only a few percent of the single-exciton binding energy E_x in the limit $\sigma = 1$. We can obtain a qualitative understanding of this variation by recognizing that in the limit $\sigma \to 0$ the exciton molecule is analogous to the hydrogen molecule (dissociation energy 4.7 eV) and the exciton is analogous to the hydrogen atom (ionization energy 13.53 eV). Hence we expect (Hopfield 1964) for $m_e \ll m_h$

$$\frac{E_{x2}}{E_x} \sim \frac{4.7}{13.53} = 0.35 \tag{1.33}$$

In the limit $\sigma \to 1$ the excitonic molecule is analogous to the positronium molecule, whose binding energy is close to 0.13 eV, while the ionization energy of the positronium atom is 6.8 eV, half of that of the hydrogen atom. Hence we expect

$$\frac{E_{x2}}{E_x} \sim \frac{0.13}{6.8} = 0.02 \tag{1.34}$$

for $m_e = m_h$. There is evidence for the existence of excitonic molecules in a variety of crystals such as CuCl, CuBr, ZnSe, and CdS [Haken and Nikitine (1975) and Ueta and Nishina (1976)]. The most detailed investigations have been carried out on CuCl, and we shall confine our discussion to this crystal.

CuCl has the zincblende crystal structure below 680 K and is generally assumed to have a direct band gap. It has a small value of σ (~ 0.02) and a relatively large binding energy ($E_{x2} \sim 34$ meV) for the molecular exciton. The spin–orbit splitting of the valence band is $\lambda = -63$ meV giving an upper Γ_7 ($j = \frac{1}{2}$) doublet and lower Γ_8 ($j = \frac{3}{2}$) quartet. The absorption spectrum of the single exciton, which has a large binding energy (Table 1.7), has been reported, for example, by Cardona (1963). It is generally assumed that in the first excited state of the molecular exciton the two holes are in the Γ_7 valence band with a resultant $J_h = 0$. The two electrons are associated with the s-like Γ_6 conduction band with $J_e = 0$. Hence the first excited state has $J = J_h + J_e = 0$ (Figure 1.16).

The first excited state of the exciton is split into two levels, Γ_5 ($J = 1$) and Γ_2 ($J = 0$), by the electron–hole exchange interaction. In contrast to single-exciton luminescence, which terminates on the crystal ground state, the luminescence of the molecular exciton terminates in excited states of the single exciton. This luminescence is expected to arise from an allowed electric dipole transition between the $J = 0$ state of the molecular exciton to the $J = 1$ state of the single exciton (Figure 1.16). If the molecular exciton has a kinetic momentum $\hbar\mathbf{K}$, the final exciton may

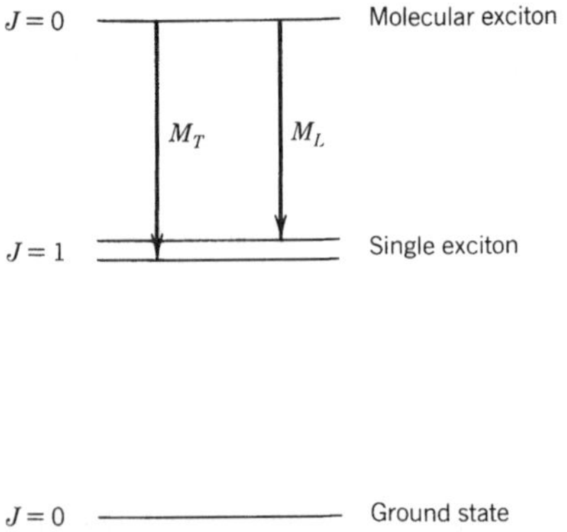

Figure 1.16 Schematic representation of the energy levels of a single exciton and a molecular exciton in CuCl involved in the M_T and M_L transitions (see text for additional details).

have either a longitudinal or a transverse character. In the case of CuCl this leads to the appearance of two characteristic bands near 3.170 eV, labeled M_L (longitudinal) and M_T (transverse), separated by 5.4 meV (Figure 1.17). The shape of the bands yields the momentum distribution function of the molecular exciton, and this is fairly close to Maxwell–Boltzmann at relatively low densities. Under these conditions the expected intensity ratio of M_T to M_L should be about 2:1, because of the twofold degeneracy of the transverse exciton, in rough agreement with observation (Figure 1.17).

Evidence for the existence of larger molecular complexes of excitons is still controversial (but see Table 1.6). We may also mention speculation about the possibility of Bose condensation of excitons. This is a

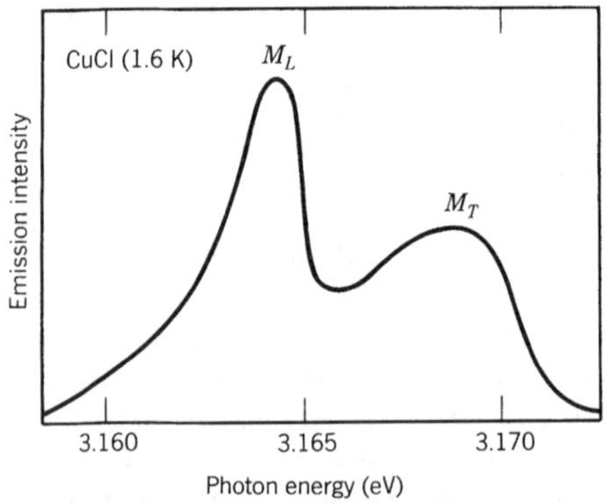

Figure 1.17 Emission bands of the molecular exciton ([*eheh*]→[*eh*]) in CuCl at 1.6 K (see Figure 1.16) following excitation in the band-to-band region with a nitrogen laser.

phase transition for particles with integral spin in which those particles exceeding a critical density $n_c(T)$ condense into a single quantum state at low temperatures, allowing the possibilities of observing quantum effects on a macroscopic scale, for example, superfluid helium and Cooper pairs in superconductors. Excitons are especially interesting in this context because the small effective mass of the particles suggests a high transition temperature. Since excitons decay by emission of radiation, study of a condensed state should be possible by optical means. A number of optical studies of excitons have been made giving results that have been interpreted as possibly arising from Bose condensation (Peyghambarian et al. 1983).

1.3.3 Electron–Hole Drops

We saw in Section 1.3.2 that excitons created at high densities may interact to form excitonic molecules. In some materials a more intimate interaction between excitons is possible, leading to the formation of an electron–hole plasma. The possibility of observing this metallic phase was first discussed by Keldysh (1968), who drew an analogy between monovalent atoms and excitons. He pointed out that the atoms can form a molecular phase, as with molecular hydrogen, or a metallic phase, as with sodium. Since the molecular state of excitons is only weakly bound (Section 1.3.2), the analogy to sodium is in some cases more appropriate.

It is perhaps not surprising that when excitons reach high densities the spatial correlation between a given electron and a given hole should become transitory. In this limit electrons and holes may give up their exclusive association and behave as two interpenetrating Fermi fluids. Since this so-called electron–hole (e-h) liquid is made up of independent electrons and holes, it is metallic, in contrast to the insulating exciton gas in equilibrium with it. We shall see below that the e-h liquid forms drops which in Ge are about 4 μm in size, with a density of about $2{\cdot}4 \times 10^{17}$ electrons and holes per cm^3. In Ge there are about 5×10^7 particles in a drop, corresponding to about 3×10^4 lattice atoms for each electron and hole, so that the carrier density of the e-h liquid is very much smaller than that of a normal metal. The highly excited states of Si and Ge are by far the best understood, although other systems, such as GaP and AgBr, have been investigated.

A matter of primary importance for understanding the e-h liquid is the dependence of the ground-state energy per e-h pair, $E_p(n)$ (the sum of kinetic, exchange, and correlation energies), on the pair density n [see Rice (1977)]. This has a minimum value E_p at a density n_0 (Figure 1.18). If this collective state is to be stable against emission of excitons at

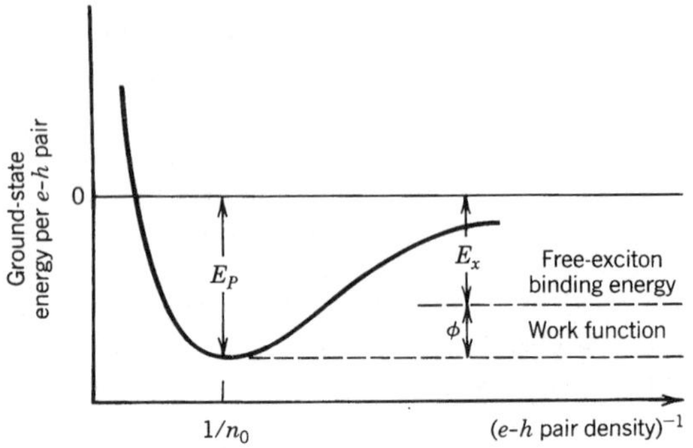

Figure 1.18 Schematic representation of the dependence of the ground-state energy of an *e-h* pair, E_p, on the reciprocal of the *e-h* pair density.

low temperatures, then E_p must be below the exciton ionization energy E_x (see Section 1.3.1). The difference

$$\phi(T) = E_p - E_x \qquad (1.35)$$

is known as the binding energy, or work function, of the condensate; it is temperature dependent.

A phase diagram for Ge is shown in Figure 1.19. At low n there is

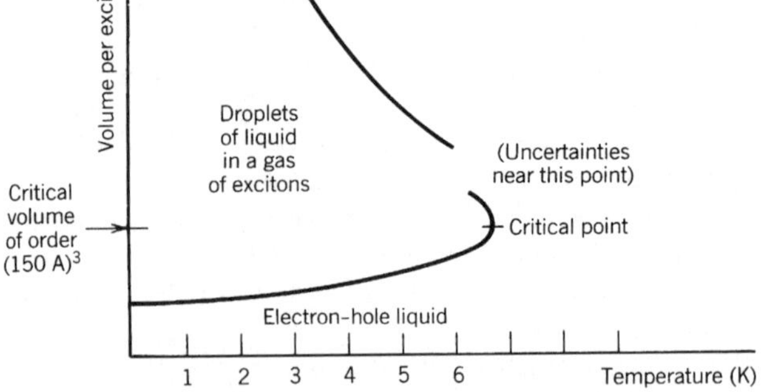

Figure 1.19 Phase diagram for a gas of excitons in equilibrium with the *e-h* liquid in germanium. The phase boundaries are determined from luminescence lineshape fits (see Figure 1.20); the incomplete region of the phase boundary is not well understood. The critical temperature is 6.5 K and the critical density is $n_c = 3\times10^{17}$ cm^{-3}.

Table 1.8 Some Properties of Electron–Hole Drops in Si and Ge

	Si	Ge
Droplet radius R	$<1\ \mu m$	$\sim 4\ \mu m$
Binding energy, $\phi(0)$, that is, energy to remove one exciton at $T=0$	8.2 ± 0.1 meV	1.8 ± 0.2 meV
Ground-state energy per e-h pair, E_p	22.8 ± 0.5 meV	6.0 ± 0.2 meV
Density of e-h pairs, n_0	$3.33 \pm 0.05 \times 10^{18}$ per cm^3	$2.38 \pm 0.05 \times 10^{17}$ per cm
Critical density of e-h pairs, n_c, at $T=0$	$1.2 \pm 0.5 \times 10^{18}$ per cm^3	$0.8 \pm 0.2 \times 10^{7}$ per cm^3
Critical temperature, T_c	25 ± 5 K	6.5 ± 0.1 K
Principal luminescence band for e-h droplet	1.082 eV	0.709 eV
Principal luminescence band for single exciton	1.097 eV	0.713 eV

only one phase (the excitons); at high n there is also only one phase (e-h liquid). For intermediate values of n the two phases coexist. In the two-phase region the amount of e-h liquid relative to exciton gas increases with n at a fixed T but the density of each phase is constant.

Electron–hole drops can scatter electromagnetic radiation without change of wavelength because there is a mismatch between their refractive index and that of the crystal. Germanium is transparent at 4 K from 1.66 μm to longer wavelengths and the $\lambda = 3.39\ \mu$m He–Ne laser line and the $\lambda = 10.6\ \mu$m CO_2 laser line can be used for scattering studies [see Hensel et al. (1977) for a general discussion]. It is possible to determine the drop radius R from the angular dependence of the intensity of scattered light for $R > \lambda$. The value of R for Si is smaller than that for Ge (Table 1.8). In crystals of Ge subject to nonuniform stress unusually large values of R are found, up to 0.25 mm, and this is still a subject for investigation.

The questions of the nucleation of drops and drop sizes are complex. The e-h drops continually absorb and emit excitons and also emit electromagnetic radiation at an energy below the crystal band gap, arising from recombination of electrons and holes. After the exciting radiation is turned off, drops in Ge have a lifetime of about 4×10^{-5} sec

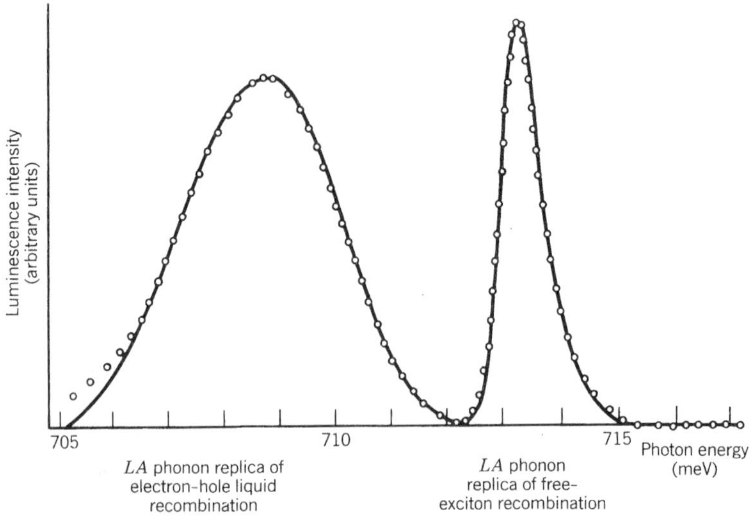

Figure 1.20 Luminescence spectrum of Ge at 3.5 K (circles) showing the *LA* phonon replicas of the free exciton and the *e-h* liquid. The solid curve is a calculated fit to the lineshape (see the text).

at 1 K, comparable to that of the excitons themselves. The maximum radius of the *e-h* drop under a given set of experimental conditions corresponds to a stable equilibrium associated with balance between internal radiative decay and surface inflow of *e-h* pairs. Since the radiative decay increases as R^3 and the surface inflow as R^2, there must be a maximum R at equilibrium.

The richest information about *e-h* drops is provided by the recombination luminescence spectrum, shown for Ge in Figure 1.20; this has an energy smaller than the crystal band gap. As with the process of photogeneration, this emission is an indirect process, requiring the participation of a phonon.* It is also accompanied by phonon-assisted emission of free excitons. The most intense *e-h* transition in Ge involves a single longitudinal-acoustic (*LA*) phonon, whereas in Si a nearly degenerate transverse-optic (*TO*) plus longitudinal optic (*LO*) pair are involved. Much weaker two-phonon replicas also occur. For both Ge and Si the phonons accompanying *e–h* droplet luminescence and free exciton

*For a discussion of phonons see Section 2.2 and of phonon-assisted electronic transitions see Section 4.2. Such transitions are sometimes referred to as phonon replicas and sometimes as vibronic transitions.

luminescence are the same. The energies of the principal luminescence transitions for e-h droplets and for free excitons in Si and Ge are given in Table 1.8. The e-h peak is broader and more asymmetric than the free exciton peak.

The excitons obey Maxwell–Boltzmann statistics, and the narrow peak shape is determined by statistics and the density of states. The electrons and holes in the liquid obey Fermi–Dirac statistics, and there are two nearly identical energy distributions, one for electrons and one for holes. The shape of the spectrum is deduced by assuming that any electron is equally likely to fall into any hole. Recombinations in which both particles have very high or very low energy will be rare and the most common energy difference will have some median value. From analysis of the recombination luminescence spectrum it is possible to determine the ground-state energy per pair, E_p, the work function, $\phi(T)$, and the equilibrium concentration, n_0, of electrons and holes in a drop [see, e.g. Hensel et al. (1977)].

To establish the phase diagram (Figure 1.19) describing equilibrium between the exciton gas and the e-h liquid, it is necessary to measure the exciton density $n_{ex}(T)$ at the onset of e-h drop formation. Estimates of this number may be obtained, for example, from a knowledge of the incident laser power and the known optical properties of the crystal. Such methods must be treated with caution, however, because of the possibility of supersaturation of the exciton gas and because the nucleation kinetics of the drops are not fully understood.

The liquid side of the phase diagram can be determined from the temperature dependence of the width of the luminescence bands, as described earlier. The distinction between the exciton and e-h drop luminescence spectrum disappears with increasing temperature leading to the identification of a critical temperature T_c and density n_c (Table 1.8). That part of the phase diagram of Ge between 5 and 6 K associated with the high-density exciton gas is not represented in Figure 1.19 because of uncertainty about its nature.

For $T < 4$ K the exciton gas in equilibrium with the drops is weakly ionized. Near T_c, however, the exciton gas becomes a metallic plasma. At intermediate temperatures it is possible that the exciton gas may dissociate to form a gas of electrons and holes before condensing to form e-h drops. Such a gas would differ from the e-h drop only in its density. Evidence for the existence of such a phase must be regarded as tentative at present.

CHAPTER TWO

Interatomic Forces and Atomic Motions

Crystal cohesion is clearly a consequence of interatomic bonding. As noted in Chapter 1, crystal structures themselves indicate important aspects of the bonding: close packing arises from short-range forces, anions are correlated with cations as a result of Coulomb forces, and open structures tend to be associated with directed covalent bonds. A more complete picture of the forces emerges from studies of other phenomena, such as elastic constants, dielectric constants, and lattice vibrations (see Table 2.1). The elastic constants are a measure of the energy change when a solid is strained uniformly and are related to lattice vibrations of very long wavelength. The dielectric constants reflect the response of the solid to an electric field; the optical constant ϵ_∞ gives the response of the electrons alone, whereas the static constant

Table 2.1 Monitors of Interatomic Forces

Physical Property	Physical Source
Elastic constants	Long-wavelength acoustic modes associated with relative motion of unit cells
Dielectric constants	ϵ_∞: electronic polarization
	ϵ_0: electronic and lattice polarization. The lattice polarization depends on relative motions within each unit cell, that is, on long-wavelength optic modes
Infrared absorption	Long-wavelength modes with an electric dipole moment: ω_{TO}: the imaginary (out-of-phase) component of $\epsilon(\omega)$ has a zero at this frequency ω_{LO}: the real (in-phase) component of $\epsilon(\omega)$ has a pole at this frequency These working definitions need not give the same values as neutron-scattering data in complex crystal structures
Raman scattering	Long-wavelength modes that change the polarizibility
Neutron scattering	Phonons throughout the Brillouin zone; gives both frequency and wavevector of phonons and polarization in special cases

ϵ_0 involves a balance between the direct force of the applied field, the interatomic forces, and the induced polarization. The lattice vibrations are investigated using neutron scattering, Raman scattering, infrared absorption, and specific-heat techniques. The vibrations of perfect crystals are described by dispersion curves, $\omega(\mathbf{k})$, which define phonon energies for a given wavevector $\mathbf{k}$. We shall consider the different contributions to the interatomic forces in Section 2.1 and the most useful models for these forces in Section 2.2. We pay particular attention to long-wavelength phenomena, such as elastic constants, in Section 2.3.

The interatomic forces control many defect phenomena. When a defect is created or an impurity is introduced, the energy change depends strongly on the polarization and distortion of the surrounding host. The entropy factors associated with the defect also depend on any frequency changes produced by the defect. When diffusion occurs, the energy to overcome the barriers between adjacent sites depends on how the nearby atoms respond to motion of the diffusing particle.

We discuss vibrational excitations associated with point defects in crystals in Section 2.4 and the more complicated case of vibrations of mixed crystals in Section 2.5. Finally, in Section 2.6, we discuss vibrations of amorphous solids.

2.1 TYPES OF INTERATOMIC FORCES

In all solids there is a repulsion between atoms or ions at short distances. The cohesive forces which bind and stabilize the solid are more varied. They can have quite a different character from one system to another. The factors that determine crystal structure (Section 1.1) and those that determine restoring forces when the structure is distorted are closely related. One aim is to divide the energy of distorted crystals into convenient components. Since this cannot be done uniquely, one attempts in practice to ensure physical sense, simplicity, and transferability (so that if atoms A and B are present in different crystals, the form of the AB interactions is always closely the same). For many purposes it is useful to have a simple phenomenological description of each contribution to the energy, and it is such parametrizations that we discuss now.

The repulsive forces vary far more rapidly than most other terms determining the crystal energy. They are usually described by a radial potential of the form

$$V_{R1} = A_1 r^{-N} \tag{2.1}$$

or of the form

$$V_{R2} = A_2 \exp(-r/\rho) \tag{2.2}$$

There is no special advantage in either form. Tradition favors V_{R1} in molecules and rare-gas solids, with N typically in the range 9–12. For other systems V_{R2} is usually chosen, and we shall generally adopt this choice here. Values of ρ do not vary much from case to case; there is a slight increase of ρ with atomic number, but $\rho \sim 0.3$ Å is typical.

Two useful general observations can be made about forces in solids. The first is a combining rule, which states that if the repulsive interaction between atoms A and B is V_R^{AB}, one finds to a good approximation that $V_R^{AB} = (V_R^{AA} V_R^{BB})^{1/2}$. Second, the repulsive forces are very important as forces but rather unimportant as a contribution to the energy. This is easily seen from a simple example. Suppose that in equilibrium the repulsive forces from V_{R1} exactly balance those from an attractive interaction $-Br^{-M}$. For the forces to balance, the derivative of the energy per atom, $(\partial/\partial r)(V_{R1} - Br^{-M})$, vanishes, so that

$$NA_1 r_0^{-(N+1)} = MBr_0^{-(M+1)} \tag{2.3}$$

and the attractive energy at equilibrium, Br_0^M, is bigger than the repulsive energy $A_1 r_0^{-N}$ by a factor N/M, where $N \gg M$ typically. The slowly varying attractive contribution dominates in the energy at equilibrium. For Coulomb attraction ($M = 1$) and short-range repulsion varying as r^{-9}, for example, 90% of the cohesive energy of the solid is Coulombic.

The Coulomb interaction between ions is the commonest and probably the most important interaction in solids. It takes the standard form

$$V_c = Z_1 Z_2 e^2/r \tag{2.4}$$

between charges $Z|e|$. Normally the dielectric screening of charges is accounted for separately by direct calculation of polarization dipole moments, so it is not included explicitly here. The energy of any specific ion involves a sum of Coulomb interactions over all other ions in the crystal, and this is expressed in terms of a Madelung constant α_{Mi},

$$E_i = \alpha_{Mi} Z_i e^2/a \tag{2.5}$$

where a is the nearest-neighbor distance. Caution is required in using published tables of Madelung constants, since there are several conventions depending on whether E_i is the energy per ion or per unit cell and on whether a is the lattice parameter or an interatomic spacing.

The charges Z in Equation (2.5) are generally taken as ± 1 in alkali halides and ± 2 in alkaline-earth oxides, with obvious extensions to other highly ionic systems. To a first approximation, these values are the ones suggested by observed phonon dispersion curves or cohesive energies. Indeed, because some of the smaller energies are hard to identify (and hence to parametrize), it is usual to assume an integral Z and to make

Table 2.2a Calculated and Observed Crystal Properties for Al_2O_3[a]

Properties	Calculated	Observed
Lattice formation energy (eV)	−160.21	−160.4
Elastic constants		
c_{11} (10^{11} dyne cm^2)	42.96	49.69
c_{12}	15.48	16.36
c_{13}	12.72	11.09
c_{33}	50.23	49.8
c_{14}	−2.99	−2.35
c_{44}	16.66	14.74
c_{66}	13.70	16.67
Dielectric constants		
$\epsilon_{0\parallel}$[b]	9.38	9.34
$\epsilon_{0\perp}$	11.52	11.54
$\epsilon_{\infty\parallel}$	2.08	3.1
$\epsilon_{\infty\perp}$	2.02	3.1

[a] After Catlow and Mackrodt (1982), who give references to experimental data.
[b] $\parallel$, $\perp$ refer to the c axis of the crystal.

phenomenological changes in the other, smaller, terms to fit experiment. Thus, for UO_2 a fit to phonon data gives $Z(U) \sim 3.97$ as optimal (Cochran 1971), and it is normal to take $Z(U) = 4$, with compensating changes in other terms. In systems where there is some covalent bonding, like SiO_2 or GaAs, there is no single definition of ionic charge Z. For the purposes of calculating interatomic forces, the correct definition is determined by the dipole moment per unit displacement of the ion in question. In this context Z is not the charge within some arbitrarily chosen atomic sphere, so it is not predicted by standard band-structure calculations, nor is it simply related to bulk-crystal properties in general. Table 2.2 gives examples of how well relatively simple ionic models can describe some bulk-crystal properties. A general discussion of ionicity and covalency is given by Catlow and Stoneham (1983).

In covalent crystals one must recognize that there is a charge build-up between atoms, the so-called bond charge. In many cases, the same framework used for ionic crystals works quite well, that is, ignoring bond charges explicitly, but adjusting other parameters for short-range interactions. This approach effectively describes a restricted range of behavior. More general methods need an explicit covalent contribution.

Table 2.2b Calculated and Observed Crystal Properties for CaF_2[a]

Properties		Experimental Value	Calculated Value
r_0: lattice constant (angstroms)	r_0	2.722	(2.722)
Second-order elastic constants (10^{11} dyne cm^{-2}): c_{11}, c_{12}, c_{44}	c_{11}	17.124	(16.9)
	c_{12}	4.675	(4.80)
	c_{44}	3.624	3.23
Dielectric constants: ϵ_0, ϵ_∞	ϵ_0	6.47	(6.42)
	ϵ_∞	2.05	(2.01)
ω_{TO}, ω_R: long-wavelength transverse optic and Raman frequencies (cm^{-1})	ω_{TO}	270.0	(259.2)
	ω_R	330.5	310.7
Third-order elastic constants (10^{11} dyne cm^{-2}), which may be defined as strain derivatives of second-order elastic constants: c_{111}, c_{112}, c_{123}, c_{144}, c_{166}, c_{456}	c_{111}	−124.6	−107.8
	c_{112}	−40.0	−33.8
	c_{123}	−25.4	−17.5
	c_{144}	−12.4	−9.3
	c_{166}	−21.4	−23.2
	c_{456}	−7.5	−7.8
H_L: lattice formation energy (eV)	H_L	−26.76	−28.06

[a] After Catlow and Norgett (1973) who give references to experimental data. Values in parentheses were used in fitting the shell model parameters.

One approach is to write the build-up of interatomic charge as a bond charge within a Coulomb interaction model. An alternative is to add a valence-force potential, V_{VF}, in which the main effects of covalency are included phenomenologically. The covalent terms in the potential favor specific bond angles and lengths. One therefore constructs a potential in terms of the degree of bond bending, that is, deviations $\Delta\theta$ of the actual angles between bonds from the undistorted crystal values, and bond stretching, that is, changes of bond lengths Δr from the undistorted crystal values, for example,

$$V_{VF} = \tfrac{1}{2} V_{\theta\theta}(\Delta\theta)^2 + \tfrac{1}{2} V_{rr}(\Delta r)^2 + V_{r\theta}\Delta r\Delta\theta \qquad (2.6)$$

Finally, there is the van der Waals interaction. This is dominant in rare-gas solids, and is occasionally important in ionic systems where it may provide attraction between second-neighbor ions. The van der

Table 2.3 Summary of Forces Active in Nonmetallic Solids

Type of Material	Type of Force				
	Short-Range Repulsion	Coulomb	Valence (Angle-Dependent Bonding)	Dispersion (van der Waals)	Free-Carrier Screening
Ionic (e.g., KCl)	Normally included	Normally included; dominates the energy	Negligible	Weak	Negligible
Polar semiconductors (e.g., ZnSe)	Normally included	Normally included	Often included	Probably weak	Extrinsic carriers can screen some interactions
Valence systems					
(a) Crystals, (e.g., Si, diamond, and III–V compounds	Not important; covalent bonds give an equilibrium interatomic spacing too large for core–core repulsion	Not too important though present in III–V compounds	Important since bonding dominates	Usually omitted	Extrinsic carriers can screen some interactions
(b) Amorphous semiconductors	Bonding factors dominate	Not too important apparently	Important since bonding dominates	Usually omitted	Rebonding minimizes densities of carriers
(c) Molecular crystals, including organics	Bonding factors dominate	Not too important	Bonding dominates intramolecular terms	Can give important intermolecular terms	Negligible
(d) Layer compounds (e.g., MoS_2 or graphite-based systems)	Normally included between planes; within planes, bonding dominates	Included where appropriate	Only within planes	Mainly important between planes	Carriers are only mobile within layers and cannot easily screen interactions by motion perpendicular to layers
Rare-gas solids (e.g., Ar)	Normally included	Negligible	Negligible	Dominant binding term	Negligible

Waals interaction is always assumed to have its longest-range form

$$V_{VDW} = -C/r^6 \tag{2.7}$$

There is little doubt that this is inappropriate at the short distances for which it is most important. If one were modelling a pure van der Waals term, without incorporating many unidentified small terms, it would be more rigorous to use a saturating (i.e., non-diverging) form, for example,

$$V_{VDW} = -C/[c^6 + r^6] \tag{2.8}$$

where c is the distance at which saturation becomes important, typically comparable with an atomic radius; this would have the added advantage of avoiding instabilities in some calculations.

Table 2.3 contains a summary of the types of forces active in non-metallic solids and refers to materials in which they are important.

2.2 LATTICE VIBRATIONS

There are three central ideas that underly theories of lattice vibrations. The first is the *adiabatic approximation*, which is the idea that the electrons move so much more rapidly than the heavy atomic nuclei that we can assume they provide a unique potential energy for each geometry of the nucleus, and that we can ignore the electrons' detailed dynamics. The second idea is the *harmonic approximation*. In classical mechanics the theory of small vibrations relies on the fact that, if displacements $\mathbf{x}$ from the equilibrium of even a complicated system are small, the total energy can be expanded in the form

$$\begin{aligned} E &= \text{kinetic energy} + \text{potential energy} \\ &= \sum_i \tfrac{1}{2} M\dot{x}_i^2 + \sum_{ij} \tfrac{1}{2}\mathbf{x}_i \cdot \mathbf{A}_{ij} \cdot \mathbf{x}_j + \text{higher-order terms} \end{aligned} \tag{2.9}$$

Terms linear in $\mathbf{x}$ are missing because the system is in static equilibrium at $\mathbf{x} = 0$. The harmonic approximation ignores the higher-order terms in Equation (2.9) and, to this extent, there always exists a linear transformation among the coordinates $\mathbf{x} \leftrightarrow \mathbf{Q}$, which casts Equation (2.9) into a form corresponding to independent harmonic oscillators:

$$E = \sum_j \tfrac{1}{2}\tilde{M}\dot{Q}_j^2 + \sum_j \tfrac{1}{2}\tilde{A}_j Q_j^2 \tag{2.10}$$

In quantum mechanics precisely the same transformation works.

In both classical and quantum mechanics the importance of the harmonic approximation is that the harmonic-oscillator problem can be solved exactly. Thus, the behavior of any harmonic lattice can be related to the soluble problem of a set of independent oscillators, and this leads to some important and general results. One immediate conclusion is that, in thermal equilibrium, the probability that there is a displacement x in a mode from the mean position $x = 0$ is given by a normal distribution

$$P(x) = P(0) \exp[-x^2/2\overline{x^2}] \tag{2.11}$$

where the mean square displacement, $\overline{x^2}$, is related to mass M, frequency ω, and temperature T by

$$\overline{x^2} = \frac{1}{2}\frac{\hbar}{M\omega}\coth\left(\frac{\hbar\omega}{2k_BT}\right) \tag{2.12}$$

For practical purposes the harmonic approximation only breaks down at high temperatures, near phase transitions, or in restricted regions near defects.

The third approximation, referred to as the *dipole approximation*, is important when one begins to construct specific models of interatomic interactions. It assumes that in calculating the longer-range Coulomb interactions between the components of a distorted or polarized lattice one may limit multipole expansions to the dipole level. The dipole approximation is central to the Mott–Littleton method for calculating static polarizations near static defects (Section 2.3.3). Quadrupole terms are ignored except in special cases.

2.2.1 Vibrations of Molecules

It is useful to make a diversion to molecular vibrations before we treat the solid state. This has direct bonuses, as well as being a convenient introduction to the case of the crystalline solid, for one useful way of analyzing the vibrations of amorphous or defective solids is to consider a piece of the whole solid as if this portion were a large molecule.

In principle, it is relatively simple to calculate the normal modes of a molecule once the interatomic forces are defined. If there are n atoms, one can write down $3n$ equations of motion for the atomic displacements, $\mathbf{u}(i)$, $i = 1 \ldots n$, that ignoring the special features of rotations and translations, have the form

$$M_i \frac{d^2u_x(i)}{dt^2} = -F_x(i) \tag{2.13}$$

The molecule will have some equilibrium geometry $\mathbf{u} = 0$ for which the net forces $F_x(i)$ vanish. If Φ is the total potential energy, then $\partial\Phi/\partial u_x(i)|_{\mathbf{u}=0}$ is zero. This condition can be used to limit the number of independent constants needed to model the forces. The forces will be dominated by the terms linear in the displacements in most cases, so that we can write

$$F_x(i) = \sum_{j,\beta} A_{x\beta}(i, j) u_\beta(j) \tag{2.14}$$

The force-constant matrix **A** is just the second derivative of the total energy with respect to the displacements, that is,

$$A_{\alpha\beta}(i, j) = \frac{\partial^2 \Phi}{\partial u_\alpha(i) \partial u_\beta(j)} \tag{2.15}$$

We recognize Equations (2.13) and (2.14) as being those of coupled harmonic oscillators. Instead of the simple equation

$$m \frac{d^2 u}{dt^2} = -Au \tag{2.16a}$$

we have a matrix equation

$$\mathbf{M} \cdot \frac{d^2 \mathbf{u}}{dt^2} = -\mathbf{A} \cdot \mathbf{u} \tag{2.16b}$$

for the molecule where $M_{ij} = M_i \delta_{ij}$ and where there are $3n$ coordinates for the cartesian components of the displacements. Some of these displacements are linearly related because we exclude rigid rotations and translations, but this relationship can be handled without difficulty.

Since (2.16a, b) are linear equations, there always exist linear combinations of the displacements,

$$e(k, t) = \sum_{i,\beta} B_\beta(k, i) u_\beta(i) \exp[i\omega(k)t] \tag{2.17}$$

that are dynamically independent, that is, for which each $e(k)$ has a well-defined frequency $\omega(k)$, and which can be regarded as an oscillator decoupled from the others. Here k is simply a label; it will have greater significance in the crystal case where it becomes a wavevector, reflecting translational symmetry (Section 2.2.2). Equation (2.17) is just the standard result of the classical mechanics of small vibrations. The frequencies $\omega(k)$ are easily found. If we seek displacements $\mathbf{u}$ varying as $\exp(i\omega t)$, Equation (2.16b) has consistent solutions only if the secular equation is satisfied:

$$\mathrm{Det}\|A_{\alpha\beta}(i, j) - M_i \omega^2 \delta_{ij}\| = 0 \tag{2.18}$$

For a diatomic molecule, this is merely a (1×1) determinant; the $6n$

coordinates are reduced by three translations and the two rotations about axes normal to the molecular axis, leaving just the relative motion of the two atoms. For a large three-dimensional molecule the secular determinant is of size $(3n-6)\times(3n-6)$.

The normal modes $e(k)$ [Equation (2.17)] can be obtained by matrix diagonalization. As a practical point it is convenient to work with the equations

$$(\mathbf{M}^{-1/2}\cdot\mathbf{A}\cdot\mathbf{M}^{-1/2}-\omega^2\mathbf{1})\mathbf{u}=0 \tag{2.19}$$

where $\mathbf{1}$ is the unit matrix. Furthermore, it is usually helpful to use symmetry as much as possible. If the molecule has inversion symmetry (e.g., SF_6 or C_2H_6), the odd-parity and even-parity modes can be treated separately. Also, Equation (2.19) determines $\mathbf{u}$ [and hence $e(k)$] apart from a possible multiplicative factor. This factor is fixed by noting that the mean square amplitude in normal mode $e(k)$ is given by Equation (2.12).

2.2.2 Vibrations of Crystalline Solids

The description of molecular vibrations in the previous section provides a convenient starting point for the study of crystal vibrations. Clearly, one does not want to solve simultaneous equations for $n\sim 10^{23}$ atoms in a crystal. Fortunately this is not necessary since techniques exist for obtaining results for crystals parallel to those for molecules. One example is the continuum approach of elasticity theory (Section 2.3). Another exploits crystal translational symmetry. This leads us to the important conclusion that in a crystalline solid, with n atoms per unit cell, the phonon frequencies and modes for a chosen wavevector $\mathbf{k}$ can be obtained by equations exactly like (2.18) and (2.19) for a molecule of n atoms. The parallels go further because there are also important equilibrium conditions for the crystal.

For the crystalline solid, ionic positions are defined by the unit cell (I) and the position within the cell (i). All the equations of Section 2.2.1 go through as before, subject only to the replacement of $u_{\alpha 0}(i)$ by $u_{\alpha 0}(iI)$ and of $A_{\alpha\beta}(i, j)$ by $A_{\alpha\beta}(iI, jJ)$. The $\sim 10^{23}$ equations resulting are unmanageable as they stand. However, we know that the translational symmetry of the crystal will be reflected in the normal modes. Consider the modes with a chosen wavevector $\mathbf{k}$. We anticipate that the displacements can be factored into components that represent relative displacements within the unit cell, $\mathbf{u}_0(i, \mathbf{k})$, and a phase, $\exp(i\mathbf{k}\cdot\mathbf{R}_I)$, that depends on which unit cell is considered; here $\mathbf{R}_I$ is a convenient origin in cell I. Hence we shall seek normal modes $\mathbf{e}(\mathbf{k}, t)$ that are linear combinations of terms like

$$\mathbf{u}_0(i, \mathbf{k}) \exp(i\mathbf{k} \cdot \mathbf{R}_I) \exp[i\omega(\mathbf{k})t] \tag{2.20}$$

in which a plane wave is modulated by a factor depending only on the relative motion of atoms within the unit cell.

This development exploits translational symmetry. Instead of $3nN$ equations for the potentially infinite number N of unit cells of n atoms, we shall have N separate groups of $3n$ equations for each of the N values of $\mathbf{k}$. We shall be able to choose $\mathbf{k}$, and solve a $3n \times 3n$ secular equation like (2.19) for molecules. The full demonstration that this is so relies partly on simple mathematical properties of Fourier transforms discussed in detail by Ziman (1960), Cochran (1973), and Stoneham (1975). For present purposes, it is most important to note that the force-constant matrix $A_{\alpha\beta}(iI, jJ)$ depends only on the relative ionic positions, and not independently on the two ionic positions iI and jJ. We may choose $\mathbf{R}_I = 0$ without loss of generality and define a dynamical matrix by

$$D_{\alpha\beta}(i, j; k) = \sum_J A_{\alpha\beta}(iI, jJ) \exp(i\mathbf{k} \cdot \mathbf{R}_J) \tag{2.21}$$

The frequencies $\omega(\mathbf{k})$ are the roots of a secular equation:

$$\mathrm{Det}\|D_{\alpha\beta}(i, j; \mathbf{k}) - M_i \delta_{ij} \omega^2(\mathbf{k})\| = 0 \tag{2.22}$$

We can now see the pattern of crystal-lattice vibrations. For each value of $\mathbf{k}$ there are $3n$ modes, that is, one for each vibrational degree of freedom in the unit cell. When there is only one atom per cell, the three modes must reduce to the continuum elastic waves at small $\mathbf{k}$ (large wavelength). When there are more atoms per cell, different modes occur, with relative motion of ions within the cell. These have frequencies that are shown on dispersion curves (cf. Figure 2.1). The relative motions can be classified further by symmetry in many cases. The relative motions of ions with different charges change the electrical dipole moment of the cell, giving infrared-active modes with $\mathbf{k} \simeq 0$. The relative motions of ions with identical charges can still change the polarizability, giving Raman-active modes with $\mathbf{k} \simeq 0$.

It is helpful to look at phonon dispersion in one dimension. When there is only one atom per unit cell, the dispersion curve, giving the variation of frequency ω with wave vector $\mathbf{k}$, is monotonic. We can choose, quite arbitrarily, the unit cell to be twice the size. This is equivalent to choosing a Brillouin zone of one-half the original size and requires that we must fold back the dispersion curve (Figure 2.1). This now appears to have an optic branch joining the acoustic branch at the zone edge. Any perturbation that causes the two atoms in the cell to be

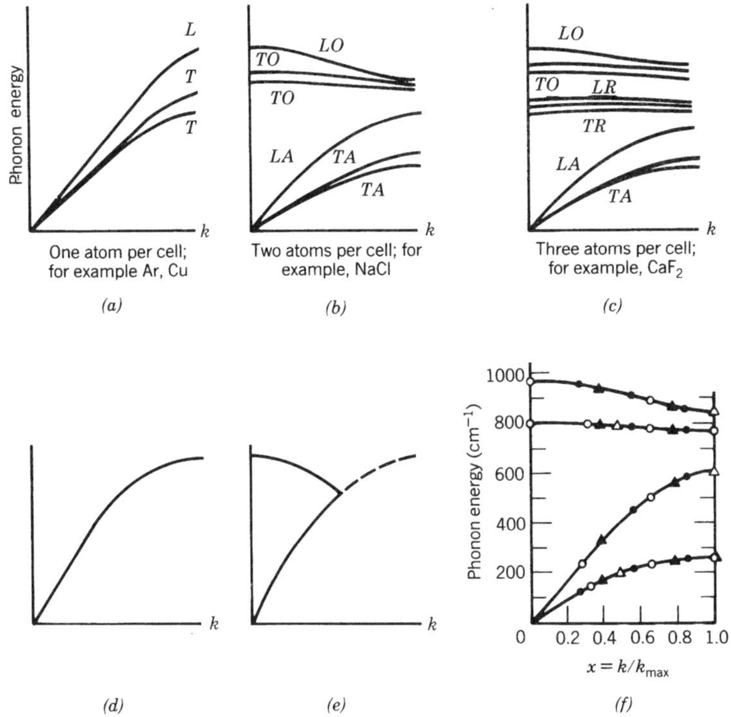

Figure 2.1 Examples of phonon dispersion curves. *L* indicates (mainly) longitudinal modes, *T* (mainly) transverse modes. *A, O, R* correspond to acoustic, optic and Raman modes. The upper curves—(*a*), (*b*), (*c*)—correspond to one, two, and three atoms per unit cell. The lower curves—(*d*) and (*e*)—show how these are related using the monatomic linear chain with two different sizes of Brillouin zone. The final set of curves—(*f*)— shows the same ideas for polytypes of SiC, where there are several cell sizes but very similar interatomic forces [after Feldman et al. (1968)]. Measurements of zone-center phonons for different polytypes are used to build up these phonon dispersion curves for the polytype with the smallest unit cell.

physically inequivalent will raise the zone-edge degeneracy, giving two separate branches in the usual way. We have already encountered zone folding associated with electronic excitations in Section 1.3.1.

The so-called folding-back procedure is useful in other ways. Some properties, like entropy and specific heat, involve sums of the sort $\sum_{\mathbf{k}} f(\omega(\mathbf{k}))$ with $\mathbf{k}$ covering the whole Brillouin zone. If a suitably shaped large unit cell is chosen, it may suffice to sum only the $\mathbf{k}=0$ points (including those folded back) with a weighting factor, since the folding leads to a sampling of all the zone. This is very useful computationally.

There are also more direct applications of folding back. One is to the so-called staging compounds of graphite in which regularly spaced layers of interstitial metal atoms are intercalcated between the planes of carbon atoms (Section 8.3). Another is to the polytypes of SiC or ZnS, where the stacking order of atomic planes along the trigonal axis can have any one of a number of ordered sequences (see Figure 3.29). In such systems as these the lattice vibrations of the different cases should be related by folding back provided only that the interatomic forces are sensibly constant from case to case. This is illustrated for SiC in Figure 2.1.

2.2.3 Atomic Models of Lattice Dynamics: The Shell Model

In principle, the models of interatomic forces are sufficient to calculate the frequencies $\omega(\mathbf{k})$ for all wavevectors $\mathbf{k}$ and branches α of crystal vibration. The same model that explains crystal cohesion should explain lattice vibration, although the importance of some terms will differ. One such term concerns ionic polarizability, since ions in their perfect-crystal positions are often at sites where the electric field (and hence dipole moment) is zero by symmetry. The so-called shell model, which generalizes the simplest models, has proved a success in virtually all types of nonmetals, and is the basis for almost all modern work on lattice dynamics. It provides a convenient scheme for parametrizing observations and also a suitable target for basic theoretical studies.

The essence of the shell model (Dick and Overhauser 1958; Cochran 1971) is that the outer electrons of each ion constitute a shell of charge $Y|e|$, essentially massless, and the nucleus plus inner electrons provide a massive core of charge $X|e|$; clearly the total ionic charge is $Z|e| \equiv (X+Y)|e|$. The core and shell are assumed coupled by harmonic springs of force constant K (Figure 2.2). Since there are still harmonic, short-range-repulsive, interatomic forces (mainly, and often exclusively, between shells of nearest neighbors), the shell model is basically harmonic. Since the shells are massless, they respond instantly and adiabatically to changes in core positions. Electrostatically, core and shell are both treated as point charges, so relative displacement gives a dipole associated with each ion, but no higher multipoles.

We may see more clearly the procedure for applying the shell model by considering a diatomic molecule AB. Suppose ion A is an unpolarizable point charge $Z_A e$, and that ion B is polarizable with net charge $Z_B e$. The shell model pictures B as two particles, a shell of zero mass and charge $Y_B e$ and a core of mass M_B and charge $X_B e$. We can write down

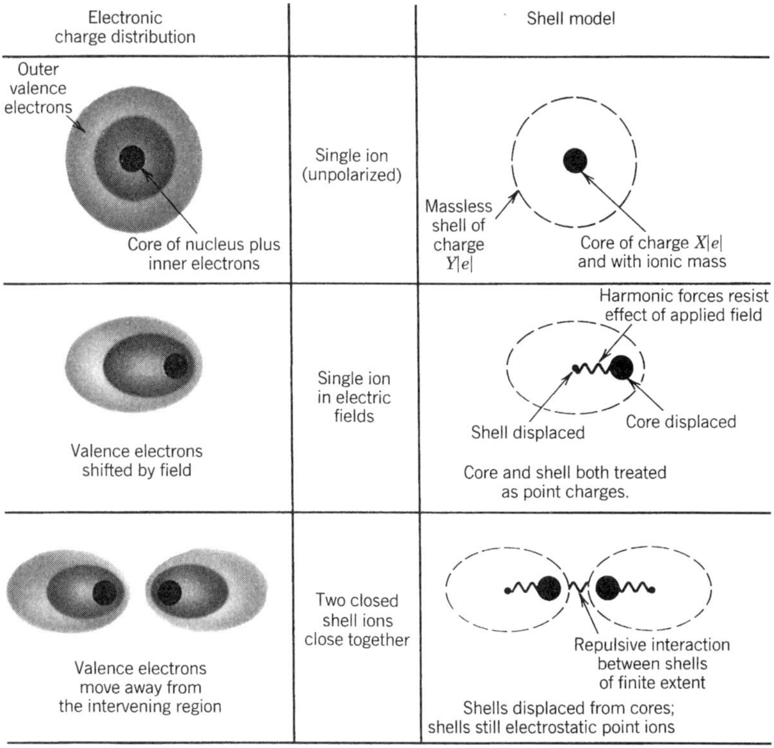

Figure 2.2 The shell model. This figure compares electron densities with their shell-model description.

the terms in the energy and, for simplicity, we do this in one dimension. The two distances we need (Figure 2.2) are then R, the B shell distance from A, and W, the shell-core spacing of ion B. Firstly, there is an electrostatic interaction

$$\phi_{es}(R, W) = Z_A Y_B e^2/R - X_B Z_A e^2/(R + W) \tag{2.23}$$

In the short-range repulsion the shell has, in effect, a finite size and a typical form (cf. Section 2.1) might be

$$\phi_{int}(R) = B \exp(-bR) - C/R^6 \tag{2.24}$$

Finally, there is a polarization energy,

$$\phi_{pol}(W) = \tfrac{1}{2}KW^2 \tag{2.25}$$

assuming harmonic coupling of shell and core with force constant K. The total potential energy for the diatomic molecule is thus

$$\Phi(R, W) = \phi_{es}(R, W) + \phi_{int}(R) + \phi_{pol}(W) \tag{2.26}$$

If we make the adiabatic assumption that the massless shell follows the motion of the heavy ionic cores instantly, we may obtain the potential energy used to calculate molecular vibration frequencies from

$$\frac{\partial \Phi}{\partial W} = 0 \tag{2.27}$$

This enables us to eliminate the shell position explicitly and to express Φ solely as a function of the nuclear spacing $R + W$. As Sangster (1974) has shown, the same shell model describes both ionic crystals and ionic molecules well.

A second result concerns polarizability. For the free B ion the applied field E displaces the shell. The equilibrium displacement balances electrical and shell-core harmonic forces,

$$\frac{\partial}{\partial W}(EY_B eW + \tfrac{1}{2}KW^2) = 0 \tag{2.28}$$

so that $W = (EY_B e/K)$ and the dipole moment is $WY_B e$. The dipole moment per unit field for the free ion, that is, the polarizability, is thus

$$\alpha = Y_B^2 e^2/K \tag{2.29}$$

The electrical forces on shells and cores in the molecule are balanced by the several terms in $\Phi(R, W)$ and the effective ionic polarizability now depends on the environment of the ion.

As described, the shell model is more than a means of introducing extra parameters. It goes beyond the simple point-charge models, which account quite well for crystal cohesion (Born and Huang 1954) in two main respects. First, the ions are individually polarizable. Second, the polarizability depends on the neighbors of a given ion since the repulsive forces affect the relative core-shell motion under an electric field. This is an important physical feature since it avoids some of the difficulties of more approximate point, polarizable-ion models.

Problems can occur when one tries to identify the shell too closely with specific electrons, for example, the outer shell. Partly, the problems concern overlap, since the regions most susceptible to overlap of neighboring ions are those one would usually associate with shells. A consequence is that the *effective* shell charge may sometimes be positive (i.e., opposite to that of the electrons which constitute the shell). A related

Table 2.4 Experimental Techniques for Measurement of Core and Shell Positions

Core positions	Shell positions
Neutron scatter	X-ray scatter
Low-energy electron diffraction (LEED)	Low-energy rare-gas scatter
Extended structure in the X-ray absorption edge (EXAFS)	

difficulty occurs when one asks which coordinate is monitored experimentally. The position is outlined roughly (for the situation is not simple) in Table 2.4.

Since the shell model is phenomenological, there need be no unique way of fitting shell and core charges. Indeed, experimental data are basically incomplete since even full phonon dispersion curves, without eigenvectors, are not sufficient (Leigh et al. 1971). There are two common ways of obtaining shell-model parameters. The commonest, and least satisfactory, is to do a best fit to phonon data in which a computer is allowed to run freely over parameters to reduce the mean-square differences between model and experiment. Here the problem is that a physically irrelevant fit may be marginally better than one which is more consistent with a broader understanding of the material. The alternative approach, therefore, is to constrain either certain parameters or certain physical quantities. One might require the lattice parameter, or the dielectric constants, to be fitted exactly. Alternatively, one might argue that certain shell parameters remain the same along a series of crystals containing the same ion. Although uniqueness cannot be achieved, the results obtained in this way are likely to be more useful and generally satisfactory.

2.3 LONG-WAVELENGTH LIMIT: ELASTICITY AND DIELECTRIC CONTINUA

Many important aspects of lattice dynamics can be understood by considering distortions on a scale much greater than atomic dimensions. The atomic structure then ceases to be of major importance, and often it does not matter if the solid is crystalline or amorphous, or has some other microscopic inhomogeneity. Continuum theory of elasticity and of electrostatics describes most of the long-wavelength properties (see below for exceptions) and provides a limit to which any atomic model must reduce. Important cases for which continuum methods are available are:

1. Elasticity, in which the most important result is Hooke's law, which states that the stress tensor $\boldsymbol{\sigma}$ and the strain tensor $\boldsymbol{\epsilon}$ are linearly related. One can write this in the general form $\boldsymbol{\sigma} = \mathbf{c}\boldsymbol{\epsilon}$, in terms of elastic constants $\mathbf{c}$, or as $\boldsymbol{\epsilon} = \mathbf{s}\boldsymbol{\sigma}$, in terms of elastic compliances $\mathbf{s}$. We shall give specific examples later.

2. Macroscopic electric fields, which can occur for two reasons:

   (i) For longitudinal optical modes the motion of positive and negative ions conspires to produce a macroscopic electric field that splits the longitudinal and transverse optic branches at long wavelengths. This splitting can be observed in several ways and is a measure of ionicity; it can be expressed in terms of the Lyddane–Sachs–Teller relation, the standard form being $\omega_{LO}^2/\omega_{TO}^2 = \epsilon_0/\epsilon_\infty$. Here ω_{LO} and ω_{TO} are the zone-center longitudinal and transverse optic mode frequencies, and ϵ_0 and ϵ_∞ are the static and optic dielectric constants. Generalizations are discussed by Cowley and Cochran (1962).

   (ii) In piezoelectric crystals, a uniform strain of the right symmetry will produce a macroscopic electric field. This means that elastic constants need an extra boundary condition in their definition, depending on whether or not the end faces of the stressed crystal are electrically connected. Piezoelectric crystals are especially important as transducers, which convert oscillating electromagnetic fields into acoustic energy within the solid and vice versa. Such transducers have an immense range of applications, including ultrasonic methods of nondestructive testing, the measurement of very small displacements, and the wealth of uses of surface wave devices. Other major applications of piezoelectric crystals (Pointon 1982) include spark generation in ignitors.

3. Magnetoelastic coupling in magnetic solids, in which changes of magnetic field (among other possibilities) can be used to control velocities of sound. This effect has a variety of applications, including delay lines.

4. Effects of surfaces on lattice vibrations, which we shall discuss separately in Chapter 7.

One important qualification is needed. Although all atomic models have a unique continuum limit, each continuum model can have many atomic models consistent with it. If we only know that a solid has cubic symmetry in macroscopic elasticity, we do not know if its structure is simple cubic, body-centered cubic, face-centered cubic, or something much more complicated. In the same way, if one knows the strain $\boldsymbol{\epsilon}$, one needs more information to know where all the atoms are in the unit cell. The extra parameters are the so-called *internal strain* parameters (Cousins 1982). If we consider the diamond structure compressed along a $\langle 111 \rangle$ axis we conclude that there are two possible extremes. The strain may be taken up in bond length changes, that is, mainly in the $\langle 111 \rangle$ bond lengths, or it may be taken up in changing the angles between the $\langle 111 \rangle$ bond and the basal $\langle 1\bar{1}\bar{1} \rangle$, $\langle \bar{1}1\bar{1} \rangle$, and $\langle \bar{1}\bar{1}1 \rangle$ bonds. Thus one needs an extra parameter to decide whether the strain is mainly taken up between the widely spaced or the closely spaced atomic planes. It also happens that the internal strain parameter for nonpolar crystals, such as diamond, is closely related to the piezoelectric constant for the polar crystals of the same underlying structure, such as GaP.

2.3.1 Elastic Constants

For our purposes a discussion of the elasticity of cubic crystals will be both illustrative and useful. If we define strain components e_{ij} as derivatives of displacements $\mathbf{u}$ with respect to position $\mathbf{r}$,

$$e_{ii} = \frac{\partial u_i}{\partial x_i}, \qquad e_{ij} = \frac{\partial u_i}{\partial x_j} + \frac{\partial u_j}{\partial x_i} \tag{2.30}$$

and if the stress components σ_{ij} correspond to forces directed along r_i acting on areas normal to axis j, then the stresses and strains of a cubic crystal are related by only three independent elastic constants (c_{11}, c_{12}, c_{44}) or elastic compliances (s_{11}, s_{12}, s_{44}). These are in turn related by

$$(c_{11}+2c_{12})(s_{11}+2s_{12}) = (c_{11}-c_{12})(s_{11}-s_{12}) = c_{44}s_{44} = 1 \tag{2.31}$$

The combinations $c_{11}-c_{12}$ and c_{44} correspond to shear distortions, without volume change, and the bulk modulus (the reciprocal of the compressibility, determining volume change solely) is $\frac{1}{3}(c_{11}+2c_{12})$. For many systems there is an approximate equality, $c_{12}=c_{44}$. This Cauchy relation would be an exact equality if there were only pairwise forces between atoms at sites with inversion symmetry.

We now state without proof some results that are often useful. First, the elastic energy can be written:

$$U = \tfrac{1}{2}c_{11}(e_{xx}^2 + e_{yy}^2 + e_{zz}^2) + c_{12}(e_{yy}e_{zz} + e_{zz}e_{xx} + e_{xx}e_{yy}) + \tfrac{1}{2}c_{44}(e_{xy}^2 + e_{yz}^2 + e_{zx}^2) \tag{2.32}$$

a form that shows clearly the cubic symmetry. Second, suppose we take a cubic crystal with cuboidal shape, but with the faces cut with axes having arbitrary orientation to the atomic crystal axes. If a compressive pressure P is applied to faces with direction cosines (l_1, l_2, l_3) relative to the atomic axes, and if we write $P_{ij} = Pl_il_j$, then the resulting strains can also be written in a symmetrized form:

$$\begin{aligned} e_{xx} + e_{yy} + e_{zz} &= -(s_{11} + 2s_{12})(P_{11} + P_{22} + P_{33}) \\ 2e_{zz} - e_{xx} - e_{yy} &= -(s_{11} - s_{12})(2P_{33} - P_{11} - P_{22}) \\ e_{xx} - e_{yy} &= -(s_{11} - s_{12})(P_{11} - P_{22}) \\ e_{xy} &= -s_{44}P_{xy} \\ e_{yz} &= -s_{44}P_{yz} \\ e_{zx} &= -s_{44}P_{zx} \end{aligned} \tag{2.33}$$

For an isotropic solid, for example, a glass, the two independent shear constants $\frac{1}{2}(c_{11} - c_{12})$ and c_{44} are equal, so

$$c_{11} = c_{12} + 2c_{44} \tag{2.34}$$

In such an isotropic case it is usual to quote Lamé constants λ and μ. The relation to the cubic case is defined by

$$\begin{aligned} c_{11} &\to \lambda + 2\mu \\ c_{12} &\to \lambda \\ c_{44} &\to \mu \end{aligned} \tag{2.35}$$

Poisson's ratio is now uniquely defined as $\nu = \lambda/2(\lambda + \mu)$ and Young's modulus as $\mu(2\mu + 3\lambda)/(\lambda + \mu)$.

2.3.2 Elastic Waves

The time-dependent equations of motion lead to the Green–Christoffel equations for elastic waves (Hudson 1980) relating wavevector $\mathbf{k}$ to frequency ω:

$$\det\|\Lambda(\mathbf{k}) - \omega^2\mathbf{1}\| = 0 \tag{2.36}$$

where $\mathbf{1}$ is the unit matrix and where the Green–Christoffel tensor, Λ, is defined in terms of density, elastic constants c, and wavevector $\mathbf{k}$ by $\Lambda_{\alpha\beta} = \rho^{-1}\sum_{\gamma,\delta} c_{\alpha\gamma\beta\delta}k_\gamma k_\delta$. For an isotropic solid the longitudinal waves move with velocity v_l,

$$v_l = \sqrt{(\lambda + 2\mu)/\rho} \tag{2.37}$$

and the two degenerate branches of shear waves with velocity v_t,

$$v_t = \sqrt{\mu/\rho} \tag{2.38}$$

These correspond to the acoustic branches mentioned earlier. For solids in general, the acoustic modes have pure longitudinal character (displacements parallel to **k**) or pure transverse character (displacements perpendicular to **k**) only along special high-symmetry directions.

Long-wavelength elastic waves are important in three main cases; (1) when they are generated by macroscopic stress fields (as in ultrasonic nondestructive testing), (2) when they are generated thermally at low temperatures, whether as a pulse (Wolfe, 1980) or in equilibrium, and (3) when they are generated by the nonradiative decay of a defect with very close energy levels, as in spin–lattice relaxation (Section 4.3.3).

2.3.3 Electronic and Ionic Polarization

In simple ionic crystals like NaCl, the long-wavelength modes are of two main types. The acoustic modes, in which both Na^+ and Cl^- move in phase, can be handled by continuum elasticity. The optic modes, in which Na^+ and Cl^- move out of phase, can be handled using Maxwell's equations, that is, continuum electromagnetism, slightly generalized. These points are clearly discussed by Cochran (1973) and Kittel (1966). For our purposes, we need to know a number of specific points. First, both the electronic and ionic polarization determine the lattice dynamics. Second, there is a qualitative difference between the longitudinal modes (for which the relative ionic motion generates a macroscopic electric field and raises the frequency) and the transverse modes (for which no such field is generated). Third, whenever there is an applied electric field, whether due to a defect or due to an external field, one must consider the direct effect of this field at any given site and the effect of the other induced electric dipoles. This means that *effective field corrections* are needed whenever one calculates spectroscopic transition rates induced by applied electromagnetic fields (see Section 4.1), or when one calculates dielectric constants or optic mode frequencies explicitly. These aspects have been discussed extensively elsewhere for perfect crystals (Cochran 1971; Mott and Gurney 1948). The continuum result for defect-induced polarization is less well known, and is conveniently treated here.

When we discussed interatomic forces in Section 2.1 there was an implicit assumption that, with exceptions like covalent systems, we were dealing with two-body interactions, unchanged in the presence of lattice defects. Obviously there are other cases for which this assumption is not valid. If the solid is grossly defective, for instance if it is so highly doped that carriers screen the electrostatic interactions, then the forces will be changed. There is, however, one pervasive and universal exception to the assumption of two-body interactions. When there are charged defects present, or in almost any circumstances in which atoms are at sites of low symmetry, each atom or ion in the crystal will have an induced electric dipole moment. The interactions between these dipoles are of long range and are an important part of the energy of the defective crystal. Moreover, the magnitudes of the dipoles must be calculated consistently, recognizing the importance of all of them. Thus if a defect at site 1 induces a dipole at site 2, this dipole affects the value of the dipole induced at site 3 by the defect at site 1.

One important approximate approach for the calculation of the dipoles and their interactions, due to Mott and Littleton (1938), exploits known continuum limits. The Mott–Littleton method is a self-consistent, iterative procedure for incorporating the interactions between defect-induced dipoles. Suppose there are defect charges that give an electric displacement $\mathbf{D}$ (for a point charge Ze at $\mathbf{r}=0$, $\mathbf{D}$ is simply $Ze\mathbf{r}/r^3$). The macroscopic polarization $\mathbf{P}$ is given in terms of $\mathbf{D}$ by

$$\mathbf{P}=\frac{1}{4\pi}(\mathbf{D}-\mathbf{E})=\frac{1}{4\pi}\left[1-\left(\frac{1}{\epsilon_0}\right)\right]\mathbf{D} \tag{2.39}$$

This polarization is the sum of an electronic polarization,

$$\mathbf{P}_{\text{el}}=\frac{1}{4\pi}\left[1-\left(\frac{1}{\epsilon_\infty}\right)\right]\mathbf{D} \tag{2.40}$$

and an ionic polarization arising from the relative displacement of the ions in each unit cell:

$$\mathbf{P}_{\text{ion}}=\frac{1}{4\pi}\left(\frac{1}{\epsilon_\infty}-\frac{1}{\epsilon_0}\right)\mathbf{D} \tag{2.41}$$

We have already seen that this ionic polarization is important in polaron behavior (Section 1.2.3). The interactions between the dipoles are included empirically through the macroscopic dielectric constants.

We now wish to calculate the individual electric dipole moments. First, we must assign a dipole moment $\mathbf{M}_i$ to each ion i in each unit cell. Suppose for simplicity that there are only two ions per cell ($i=+$ or $-$). Then

$$\mathbf{M}_+ + \mathbf{M}_- = 2\Omega\mathbf{P} \tag{2.42}$$

where Ω is the volume of the unit cell. We may argue that each moment $\mathbf{M}_i$ is proportional to a polarizability. In free space these electronic polarizabilities would be α_+ and α_-. In a crystal there is an additional displacement contribution from ionic polarization, which we define by

$$2\alpha_{\text{disp}}/(\alpha_+ + \alpha_-) = |\mathbf{P}_{\text{ion}}|/|\mathbf{P}_{\text{el}}| \tag{2.43}$$

If we assign the displacement polarizability symmetrically, then

$$\mathbf{M}_+/\mathbf{M}_- = (\alpha_+ + \alpha_{\text{disp}})/(\alpha_- + \alpha_{\text{disp}}) \tag{2.44}$$

Collecting the various expressions together, we can deduce the moment of each ion when the electric displacement is $\mathbf{D}$:

$$\mathbf{M}_\pm = \frac{\Omega}{2\pi} \frac{\alpha_\pm + \alpha_{\text{disp}}}{\alpha_+ + \alpha_- + 2\alpha_{\text{disp}}} \left(1 - \frac{1}{\epsilon_0}\right) \mathbf{D} \tag{2.45}$$

where

$$\alpha_{\text{disp}} = \tfrac{1}{2}(\alpha_+ + \alpha_-) \frac{\epsilon_0 - \epsilon_\infty}{\epsilon_0(\epsilon_\infty - 1)} \tag{2.46}$$

Thus the moments and the displacement polarizability are defined by the dielectric constants, by $\mathbf{D}$, and by individual ionic polarizabilities.

These continuum results are usually applied in regions distant from a defect under consideration. Close to the defect, displacements and dipole moments are adjusted explicitly to minimize energy, with more distant ions treated by approximations like (2.45). These ideas are used in the theoretical methods described in Section 3.4 for the calculation of defect properties, such as formation energies.

2.4 VIBRATIONS OF CRYSTALS CONTAINING DEFECTS

So far we have emphasized the lattice vibrations of perfect crystals. These vibrations have special features that are important when we consider defective solids. First, perfect-crystal modes can be identified by a wavevector $\mathbf{k}$. Second, the modes are delocalized, a related consequence of translational symmetry. If the amplitude of a mode is known in one region, then the amplitude many wavelengths away is determined, and is neither damped nor amplified. Of course localized behavior will be found for any solid, perfect or imperfect, if a localized force is applied, driving the solid at a frequency that is not one of the normal mode frequencies. The vibrational motion in phase with the applied force will be damped with distance from the point of application of the force, just as an electromagnetic wave decays on a transmission line at frequencies outside the pass band.

When we consider a crystal containing point defects, we have two distinct descriptions. One is to regard the defects as scatterers of the original perfect-lattice modes. The defects thus scatter phonons from one mode $\mathbf{k}$ to another $\mathbf{k}'$. This is useful in discussing effects of defects on transport properties like thermal conductivity (Flynn 1972, p. 137; Challis et al. 1976). The second description notes that the imperfect lattice is still as good a harmonic system as the perfect lattice (though we shall qualify this later), so that it has a perfectly valid description in terms of its own normal modes and phonons. Although we cannot use a wavevector $\mathbf{k}$ as a label, nor can we exploit translational symmetry, we can still relate the behavior to that of dynamically independent harmonic oscillators.

2.4.1 Lattice Modes Associated with Point Defects

Suppose a perfect crystal contains a single point defect. We can envisage three distinct types of vibrational mode. First, there may be some lattice modes completely unaffected by the defect; if the defect only involves a change of mass, modes with a node at the defect site will not be affected, since they do not involve motion of the altered mass. Second, most modes will have amplitude changes, either increases or decreases, near the defect. If the motion is strongly enhanced near the defect, but tends to a small though finite value far away, there is a *resonance*, with frequency ω_R; antiresonances, when the motion is correspondingly diminished, are of less interest. Third, the motion may only be local, dying away to zero far from the defect. This is called a *local mode*, with frequency ω_L, which lies outside those of the perfect-lattice modes. In an extreme case, the *Einstein oscillator* limit, only the defect atom moves.

Suppose the effective defect mass is M' and the effective harmonic force constant is K'. If the defect mass is low enough, or the force constant at the defect is increased enough, a local mode will appear with high frequency. Thus, for a local mode we have $\omega_L \equiv \sqrt{K'/M'} > \omega_{\max}$, where $\omega_{\max}$ is the highest frequency of the unperturbed lattice. Exactly the opposite conditions lead to low-frequency resonances, where modes with low frequencies centered on $\omega \simeq \omega_R$ have enhanced local amplitude. Qualitatively, we expect the picture shown in Fig. 2.3.

A further general point concerns degree of localization. The root-mean-square amplitude of a given atom in a given mode depends on how localized the mode is. Suppose for simplicity a mode involves the motion of N atoms, all with equal amplitudes u. The mean potential energy $\langle V\rangle$ will be $\frac{1}{2}NKu^2$, where K is a force constant. But, for a harmonic oscillator the total energy E (including kinetic energy) is just $2\langle V\rangle$, and is given by $\frac{1}{2}\hbar\omega\coth(\hbar\omega/2k_BT)$; this is determined entirely by ω and T, independent of N. Clearly, if $\frac{1}{2}NKu^2$ is independent of N, we expect the

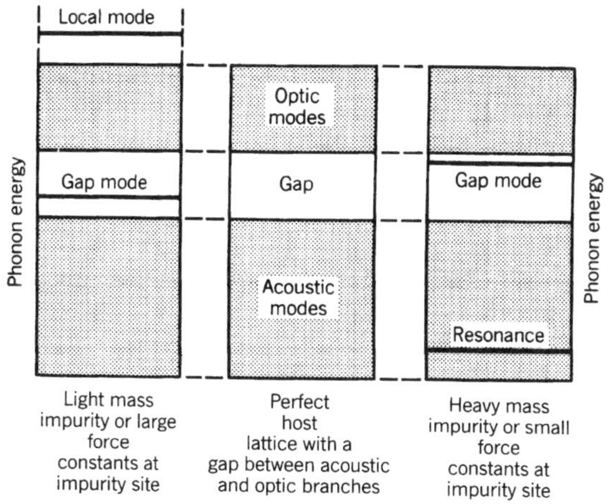

Figure 2.3 The various types of localized vibrations for impurities with altered masses and force constants.

root-mean-square amplitude $\langle u^2 \rangle^{1/2}$ to vary with localization as $1/\sqrt{N}$. If the defect affects amplitudes strongly over a localized region of N_L atoms in a host of N_∞ atoms, we expect the root-mean-square amplitude, $\langle u^2 \rangle^{1/2}$, relative to a single atomic oscillator of the same frequency and mass to be:

	Close to the Defect	Far from the Defect
Perfect-crystal mode	$O(N_\infty^{-1/2})$	$O(N_\infty^{-1/2})$
Resonance	$O(N_L^{-1/2})$	$O(N_\infty^{-1/2})$
Local mode	$O(N_L^{-1/2})$	0

If vibrational amplitudes are enhanced near the defect, as in a local mode, effects of anharmonicity can be increased.

Although local modes were predicted by Lord Rayleigh as early as the 1880s, Schäfer (1960) was the first to give firm experimental identification. Gap modes were seen soon after in hosts with a gap between acoustic and optic branches. Such gaps are favored by large differences in mass between lattice ions, as in KI. Gap modes are analogous to localized modes to the extent that they occur in a region in which there are no vibrational states of the host crystal.

Defect-induced vibrations within the range of host-lattice frequencies need further classification, for there are several distinct effects. The first class of behavior comes merely from changes in selection rules. Wavevector conservation, related to translational symmetry of perfect crystals, means that only optic phonons near $\mathbf{k} \sim 0$ are infrared active. Defects break this symmetry, so band modes throughout the Brillouin zone become infrared active (odd-parity case) or Raman active (even-parity case). The defect nature does not matter greatly; the defect-induced spectra, extensively studied for ionic hosts like KCl and CaF_2 and semiconductors like GaP and Si, mainly reflect the density of states of the perfect crystal (Newman 1974).

The second class includes resonances in the sense described earlier. These are especially apparent when the resonance frequency ω_R is in a region where the host density of states is low, principally at very low frequencies. The width of the resonance depends on the extent to which the vibrational motion is defect-like, favoring a sharp resonance, or band-like. Internal vibrations of molecular defects (e.g., CH_4) are very sharp, whereas resonances arising from a modest mass increase (e.g., K in NaCl) are much less so.

The third class of interest covers resonances where the harmonic approximation is insufficient, notably tunneling states. Examples include molecules that can reorient (O_2^- in KCl; HCl in Ar), off-center substitutional ions (Li^+ in KCl; F^- in NaBr), and some interstitial systems, like O in Si (see Section 2.4.2).

So far we have considered isolated defects in single crystals. Typical concentrations would be 1000 ppm or less. In such cases, as at higher concentrations, infrared and Raman techniques provide a wealth of detailed spectroscopic information (Barker and Sievers 1975). If the defect concentration is increased, the first consequence is the formation of defect pairs, followed by increasing numbers of larger complexes. However, even the concept of clusters loses its meaning in due course and mixed crystals are formed, needing special treatment (Section 2.5).

Suppose we have a single isolated oscillator, driven by an applied periodic force:

$$m\ddot{x} + m\omega_0^2 x = Fe^{i\omega t} \tag{2.47}$$

The motion in phase is $x \equiv x_0 \exp(i\omega t)$, where

$$x_0 = F[m(\omega_0^2 - \omega^2)]^{-1} \tag{2.48}$$

This shows (a) a *linear response*, with the amplitude x_0 proportional to the applied force F and (b) a *resonance*, so that the amplitude is enormously enhanced as ω approaches ω_0. In any real system there will be damping and the apparent divergence will be avoided.

In defect-vibration problems the force comes either from an applied external field or from the effect of an impurity atom on a host, usually assumed perfect otherwise. Although linear response is still appropriate, we must note that the force is applied to a single ion and not to a single mode. Equation (2.48) may now be written with a *response function R* (equivalently a Green's function in the standard sense of differential equations) relating amplitude component $x_{\alpha i}$ at site i to force component $F_{\beta j}$:

$$x_{\alpha i} = R_{\alpha\beta,ij} F_{\beta j} \tag{2.49}$$

For the static lattice this simply expresses Hooke's law; when the force has frequency ω, we have both in-phase and out-of-phase displacements to determine.

We can show that an altered mass ($m \rightarrow m + \delta m$) in Equation (2.47) leads to the expression

$$x_0 = F[m(\omega_0^2 - \omega^2) - \omega^2 \delta m]^{-1} \tag{2.50}$$

This form persists in the more general solid-state cases, with the response function R_{00}^{-1} replacing the factor in square brackets; the 00 subscripts refer to the diagonal element ($\alpha = \beta$) at the site at which the mass changes. The resonance condition for this mass-defect case is

$$R_{00}(\omega)\omega^2 \delta m + 1 = 0 \tag{2.51}$$

the simplest solution of importance in defect lattice dynamics.

Figure 2.4*a* shows the real part of $R_{00}(\omega)$ for a crystal with properties similar to silicon. The intersections with the mass-defect curve, $(\omega^2 \delta m)$, for a light impurity correspond to a local mode (energy > 600 cm^{-1}), an antiresonance (500 cm^{-1}), and a resonance (near 150 cm^{-1}). The two main peaks in the real part of $R_{00}(\omega)$ correspond to the principal peaks in the phonon density of states, proportional to the imaginary part of $R_{00}(\omega)$ (Figure 2.4*b*).

More complicated solutions involve changes in force constants or in defect charges, or defects in more complex lattices. Formal calculations are straightforward. Frequently, however, they merely parametrize the altered force constants, rather than analyze microscopic interatomic forces. This happens because local lattice distortion changes the approach distances of neighboring ions (and hence effective force constants, related to $\partial^2 V/\partial r_\alpha \partial r_\beta$) over quite large distances, and these changes cannot be determined uniquely from experiment.

Calculations of defect lattice dynamics are used in various ways, for

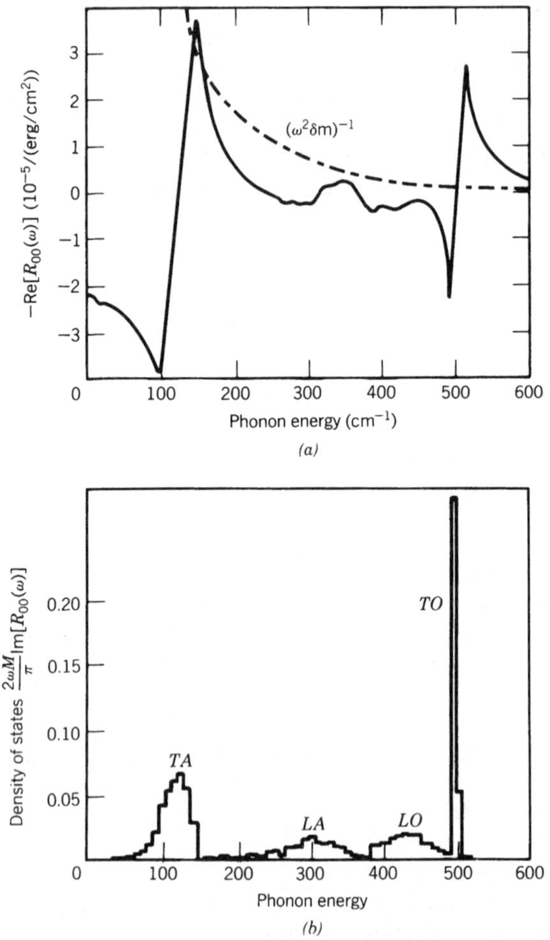

Figure 2.4 (*a*) The energy dependence of Re($R_{00}(\omega)$) (solid line), and also $1/\omega^2\delta m$, as needed in local mode calculations for Si. (*b*) Density of states for Si lattice vibrations, proportional to Im($R_{00}(\omega)$) [after Elliott and Pfeuty (1967)]. Note the Debye-like *TA* modes and Einstein-like *TO* modes.

example, to determine mean square amplitudes $\langle u^2 \rangle$, which occur in defect Debye–Waller factors, and the mean square momentum $\langle p^2 \rangle$, which appears in the impurity Mössbauer effect (Wertheim 1971; Sawicka 1980; Stanek et al. 1981). There exist well-defined and tested techniques of calculation of defect lattice dynamics within reasonably realistic models (Maradudin et al. 1971; Stevenson 1966; Stoneham 1975, Chapter 11).

2.4.2 Examples of Local Modes, Resonance Modes, Gap Modes, and Off-Center Ions

A striking example of a local-mode system is provided by the hydride ion in simple ionic solids such as CaF_2 and KCl. Hydrogen dissolves substitutionally in these materials as H_s^- ions, replacing lattice anions. Because of its very light mass, it gives rise to localized vibrational modes at frequencies higher by factors of 2–3 than the highest band modes. One may thus use an adiabatic approximation to describe the local oscillator, and assume that the H_s^- ion vibrates in an effective static potential having the symmetry of the host lattice at the impurity site. In CaF_2, for example, the fluorine sites have tetrahedral (T_d) point symmetry (Figure 2.5*a*). The vibrational Hamiltonian for the local oscillator has the form

$$\begin{aligned}\mathscr{H} &= Ar^2 + Bxyz + C_1(x^4 + y^4 + z^4) \\ &\quad + C_2(x^2y^2 + y^2z^2 + z^2x^2) \qquad (2.52)\end{aligned}$$

obtained from a Taylor expansion of the energy to fourth order in the displacement of the H_s^- ion. In the case of alkali halides the anions are at centers of inversion (Figure 2.5*b*), with point symmetry O_H, and the $Bxyz$ term in Equation (2.52) is absent. Equation (2.52) can be used to predict the energy level structure of an anharmonic oscillator with either T_d or O_H symmetry, as shown in Figure 2.5. Comparison with the observed transitions makes it possible to determine the constants A, B, C_1, *and* C_2 of Equation (2.52), which characterize the effective potential.

The static form of Equation (2.52) neglects two important effects arising from the motion of lattice ions. One is that the anharmonic coupling of the local mode to band modes gives rise to a pronounced temperature dependence of the local-mode linewidths and peak positions as well as to two-phonon transitions involving simultaneous excitation of a local phonon and a lattice phonon (Figure 2.6). The lattice sidebands associated with the latter type of transition are related to the single-phonon density of states of the host crystal. The other effect is that the correct mean square displacement of the hydrogen, as needed in Debye–Waller factors, is not given by Equation (2.52) alone. The local mode is determined with reasonable accuracy by considering only relative hydrogen–host motion. However, the motion of the cage of neighbors around the hydrogen, carrying the vibrating hydrogen with them, makes an important contribution to the hydrogen Debye–Waller factor [for a discussion of Debye–Waller factors, see, e.g., Kittel (1966), p. 70].

Oxygen in silicon is another important impurity that can be described by a local potential. Isolated oxygen forms an electrically neutral defect

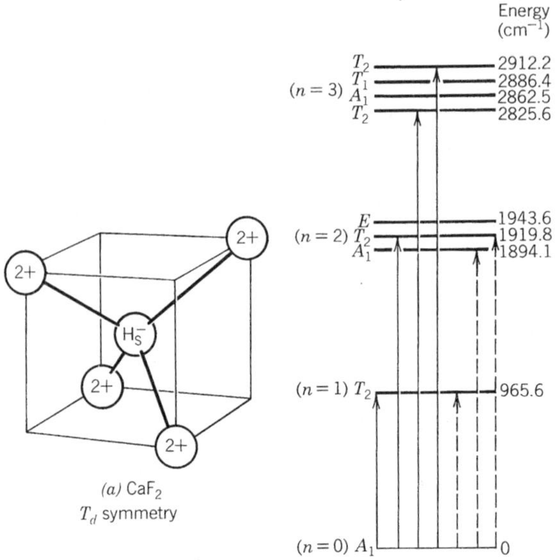

(a)

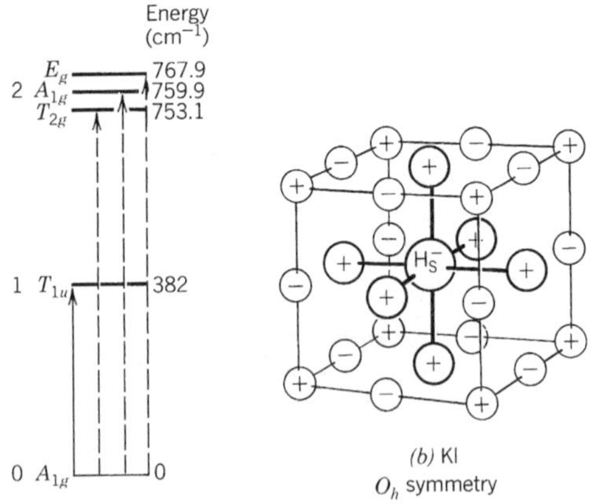

(b)

Figure 2.5 Substitutional hydrogen (H_s^-) in ionic crystals. The local lattice structure and schematic vibrational energy level diagrams for (a) CaF_2 and (b) KI hosts are shown. Solid lines indicate observed infrared transitions and dashed lines observed Raman transitions. Note the different energy scales for (a) and (b).

center whose structure has been established through analysis of its vibrational spectrum. It consists of an oxygen atom in an interstitial position covalently bonded to two nearest silicon atoms (Figure 2.7), and to a reasonable approximation it can be treated as an Si_2O molecule embedded in the crystal. The molecule has three normal modes of vibration (Figure 2.8), and these are infrared active. The ω_2-type vibration involves two degrees of freedom for the oxygen in the plane perpendicular to the Si–Si axis. The energy levels of a two-dimensional anharmonic oscillator may be obtained using the Hamiltonian

$$\mathcal{H} = \frac{1}{2M} P^2 + \tfrac{1}{2} M\omega^2 R^2 + A \exp[-(bR)^2] \qquad (2.53)$$

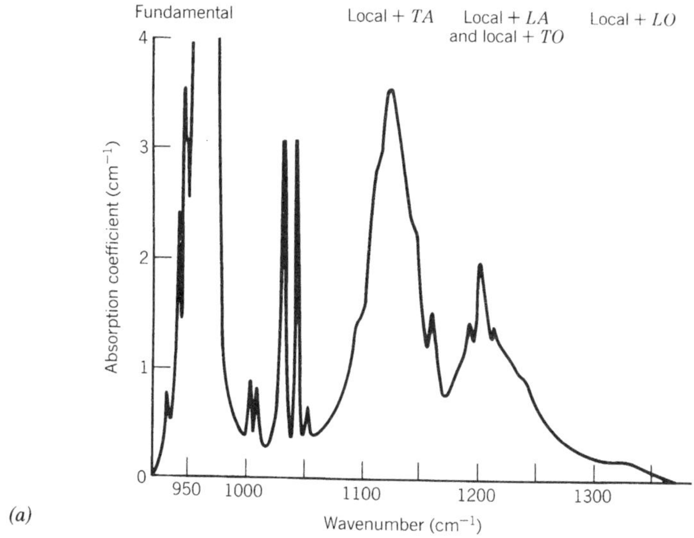

Figure 2.6 (*a*) Structure at 20 K associated with the fundamental vibrational line of H_s^- in CaF_2, at 966 cm^{-1}. The sharp lines between 925 and 1050 cm^{-1} are assigned to H_s^- lines associated with other point defects inadvertently present in the crystal. The broad bands between 1050 and 1400 cm^{-1} are due to simultaneous excitation of band phonons and the fundamental vibration of H_s^-; the sharp structure in this region is due to singularities in the single-phonon density of states of CaF_2. (*b*) Measured variation with temperature of the peak position and width of the second harmonic vibrational line of H_s^- in CaF_2 (fundamental shown in Figure 2.6*a*) [after Elliott et al. (1965)].

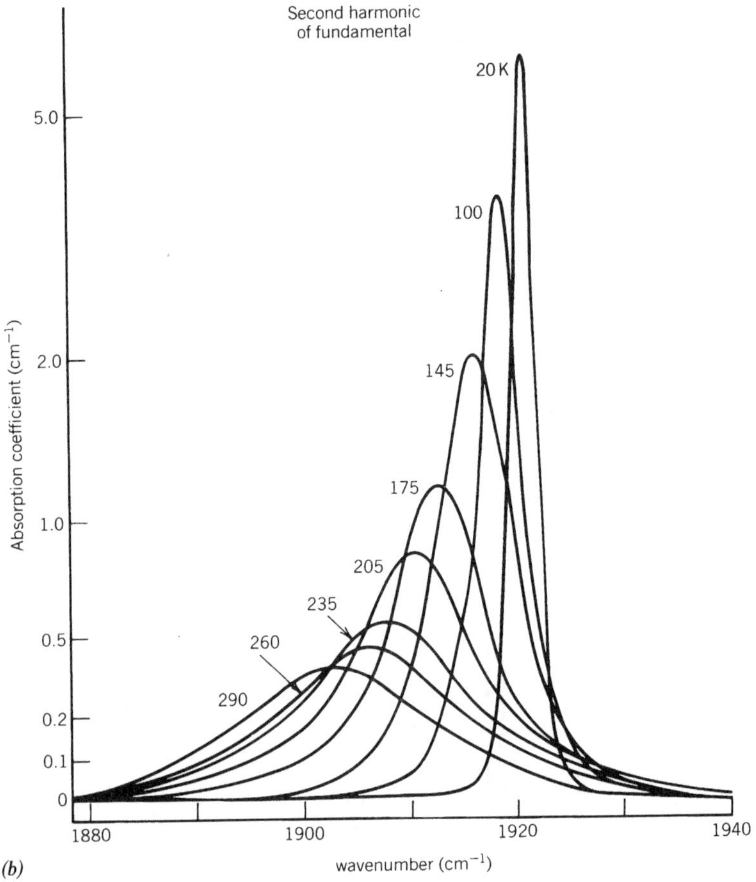

(b)

Figure 2.6 ***(Continued)***

where R is the radial distance from the Si–Si axis and P is the corresponding momentum. Equation (2.53) implies rotation about the Si–Si axis. The anharmonic term $A\exp[-(bR)^2]$, with constants A and b to be determined by experiment, leads to a splitting and mixing of the ω_2 harmonic oscillator levels. The levels are labeled $|u, m\rangle$, where u is the vibrational quantum number and m is the angular momentum quantum number corresponding to azimuthal rotation. Observed infrared transitions corresponding to ω_2-, ω_3-, and $(\omega_3 + \omega_2)$-type vibrations are shown in Figure 2.9.

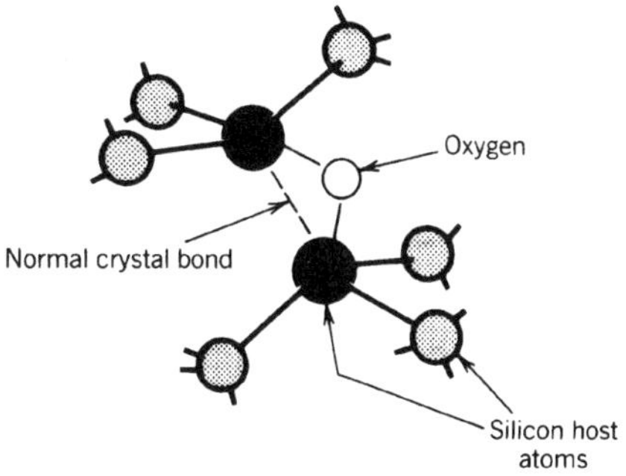

Figure 2.7 Neutral oxygen interstitial in Si, forming, in effect, an Si_2O molecule.

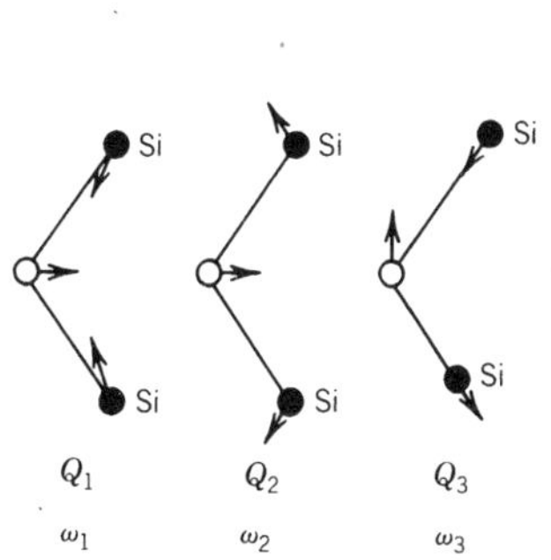

Figure 2.8 Classification of normal vibrational modes of the neutral oxygen interstitial in silicon, regarded as an Si–O–Si molecule.

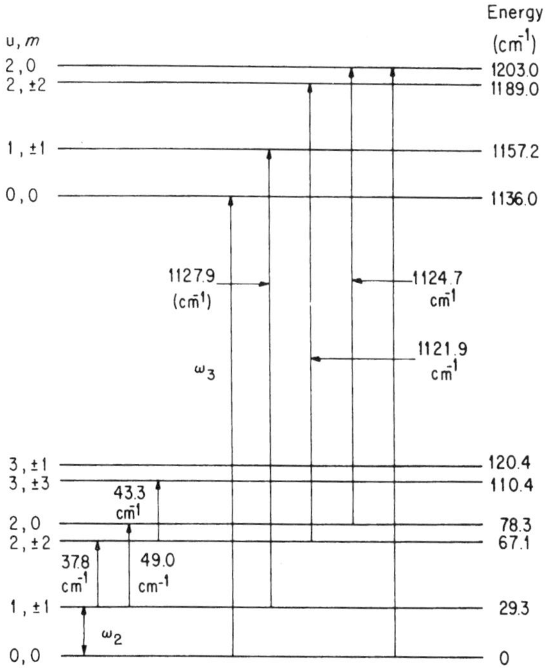

Figure 2.9 The ω_2 and ω_3 vibrational energy levels of interstitial oxygen in silicon. Observed infrared transitions are indicated by arrows [after Bosomworth et al. (1970)]. The ω_1 mode (see Figure 2.8) has not been clearly identified (see text).

An infrared transition observed at 517 cm^{-1} does not fit into the scheme of Figure 2.9. This line is slightly below the highest band mode of silicon, at 523 cm^{-1}, and near a peak in the phonon density of states, at 500 cm^{-1}. Hence it may be more appropriate to describe the 517 cm^{-1} peak as an oxygen-induced resonance mode rather than the ω_1 normal mode of Si_2O.

Gap modes have been studied most extensively in alkali halides. Five of these materials—LiCl, NaBr, NaI, KBr, and KI—have gaps between the acoustic and optic modes, and gap modes have been observed in all except the first. The gap in KI extends from 70 to 96 cm^{-1} and a gap mode of H_s^- occurs at 93 cm^{-1}. *F* centers in KI give rise to an infrared-active gap mode at 82 cm^{-1} and a Raman-active gap mode at 78.4 cm^{-1}

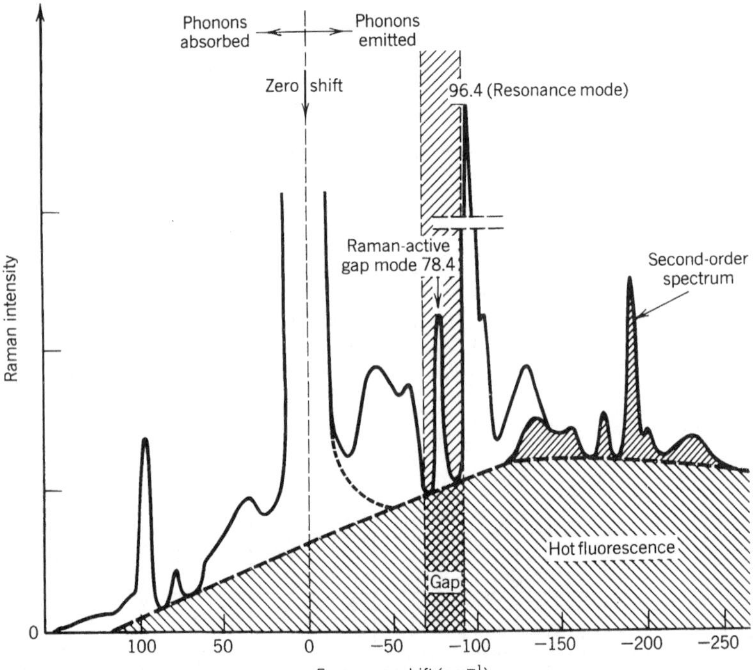

Figure 2.10 Raman spectrum for KI containing *F* centers. Note the Raman-active gap mode (78.4 cm^{-1}), the resonance mode (96.4 cm^{-1}), the second-order spectrum (shaded ///) and the background of hot fluorescence (shaded \\\\\\), probably due to emission from vibrationally excited states of the *F*-center excited state [after Buisson et al. (1975)].

(Figure 2.10) [the *F* center consists of an electron replacing an anion (Section 5.1.1)]. The positions of these modes can be explained by assuming a 70% reduction in the central force constant acting between the electron and the nearest cations, compared with the corresponding cation–anion force constant.

The reduction of force constant for *F* centers is high compared with the percentage change generally required for chemical impurities. This large reduction also explains the existence of a resonance mode for *F* centers in KI at 96.4 cm^{-1} (Figure 2.10). However, a 10% reduction suffices to explain resonance modes observed for the Tl^+ ion replacing the K^+ ion in KCl. A similar reduction of force constant explains the resonance observed for Ag^+ in KI (Figure 2.11). An unusual situation arises when Li^+ replaces K^+ in KBr. This light ion gives rise to a sharp resonance mode at low energy (16.3 cm^{-1} for ^7Li and 17.9 cm^{-1} for ^6Li) (Figure 2.12), apparently requiring that the Li–Br coupling constant be only about 0.6% of the K–Br coupling constant.

The fact that the Li–Br force constant is so small suggests that a negative force constant might be possible, and this, in effect, occurs for

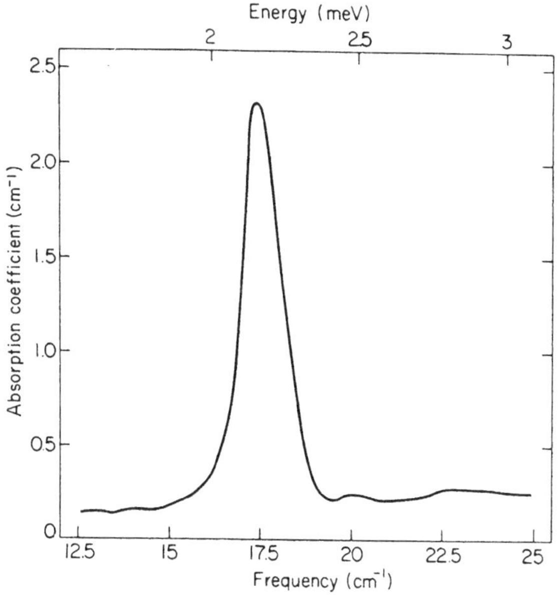

Figure 2.11 Far-infrared absorption at 2 K due to Ag^+ ions in KI at a concentration of 10^{18} cm^{-3} [after Sievers (1964)].

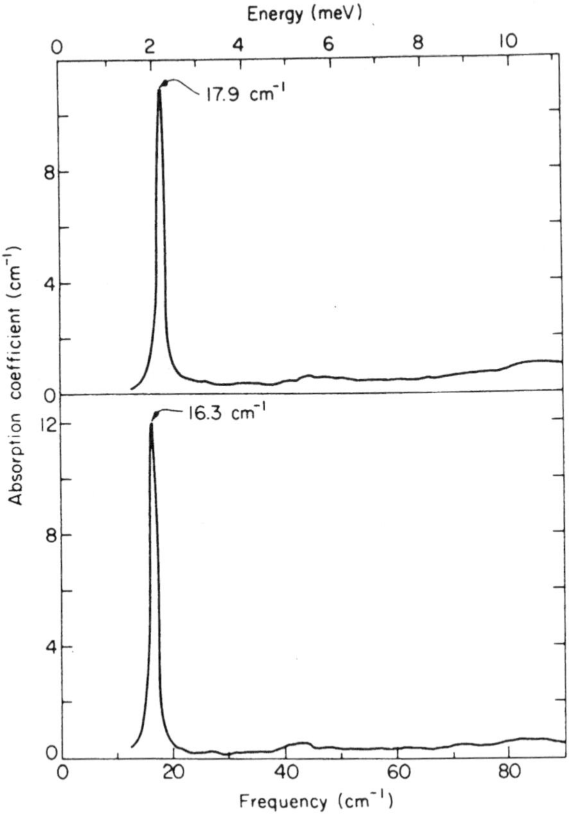

Figure 2.12 Far infrared absorption at 1.5 K of KBr containing (*a*) 1.5×10^{18} ^{6}Li ions cm^{-3} and (*b*) 1.3×10^{18} $^7Li^+$ ions cm^{-3} [after Sievers and Takeno (1965)].

Li^+ replacing K^+ in KCl. The K^+ lattice site is not a stable equilibrium position for the small Li^+. This ion can gain energy by polarizing the neighbors as it moves off-center along a $\langle 111 \rangle$ direction until a new equilibrium position is reached. In this case there are eight equivalent equilibrium positions for the off-center ion and, if there is a high-energy barrier between them, the Li^+ ions act as independent oscillators with energy $\hbar\omega$ at each of the eight off-center sites (Figure 2.13). To the extent that tunneling between the barriers occurs the eightfold spatial degeneracy is raised giving rise to four levels—two singlets and two triplets—with roughly equal level separation δ (Figure 2.13). For Li^+ in

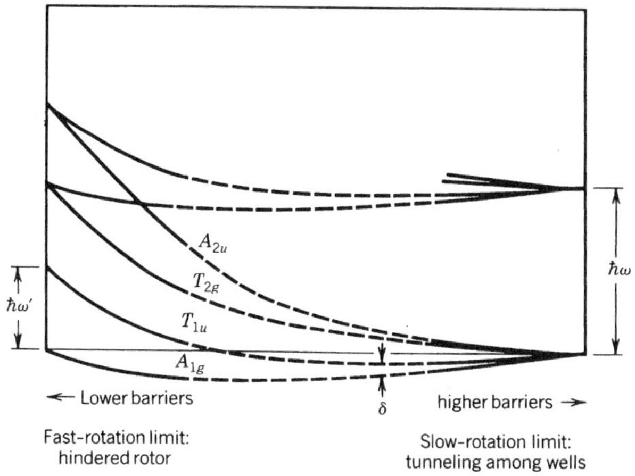

Figure 2.13 Energy levels for a system that can reorient, for example, KCl:Li. The left-hand side corresponds to low barriers. The limit of a zero barrier gives the free rotor case, with equal spacings $\hbar\omega'$. The right-hand side corresponds to slow tunneling among wells. The degeneracy (eight) of the lowest state is equal to the number of wells (i.e., the eight ⟨111⟩ directions in cubic symmetry) [after Brown et al. (1966)].

KCl the observed spacings can be fitted by $\hbar\omega = 50\ \text{cm}^{-1}$ and $\delta = 0.8\ \text{cm}^{-1}$, although the fit is not unique. If the energy barrier were zero, the Li^+ ion should not be off center, and we should expect equally spaced harmonic oscillator energy levels of energy $\hbar\omega'$ (Figure 2.13).

2.4.3 The Configuration Coordinate Diagram: Reaction Coordinates

For many defect properties the interactions with lattice vibrations can be successfully mimicked by considering interactions with a single effective mode. When there is a real local mode, as in cases we have just described, this will usually be the effective mode. In other cases the effective motion is referred to as a configuration coordinate. We shall see in Chapter 4 that the concept of a configuration coordinate has wide application to defect optical spectra and nonradiative transitions. The value of the configuration coordinate model stems from the fact that there are a number of important exact results arising from those cases in which we concentrate on a single effective frequency for the lattice vibrations of interest. Such cases include the defect modes just considered, although, as we shall now see, the concept of a single frequency has much wider usefulness.

Although a defect in a solid may be influenced directly or indirectly by any of the large number of possible lattice distortions, this complexity is partly deceptive. First, most of the consequences of defect–lattice coupling can be understood conveniently by starting from a picture in which only one distortion matters, and this can be generalized when necessary. Second, it is often true that a very specific distortion (perhaps the outward motion of nearest neighbors) is all that matters, and this alone is represented by a single coordinate. Third, the simplification of using a single coordinate has support from a basic technical result. Suppose we know the mean square displacement $\langle x^2 \rangle$ or mean square momentum $\langle p^2 \rangle$ in a particular coordinate at a particular temperature. Then there are very tight limits on these values at other temperatures, provided only that the lattice is harmonic. Thus we will be able to predict the temperature dependence of $\langle x^2 \rangle$ or $\langle p^2 \rangle$ rather well, even using a single-frequency model which simplifies the distortions and dynamics considerably.

Suppose that the host solid contains a defect of mass M_g in electronic state g. In a one-coordinate model the total energy of the defect for a harmonic lattice can be written

$$E_g = E_{g0} + \tfrac{1}{2}K_g(Q_g - Q_{g0})^2 \tag{2.54}$$

Here K_g is a force constant [so there is an effective frequency $\omega_g \equiv (K_g/M_g)^{1/2}$] and there is an equilibrium distortion Q_{g0} in coordinate Q (Figure 2.14). Also shown are the different levels of vibrational excita-

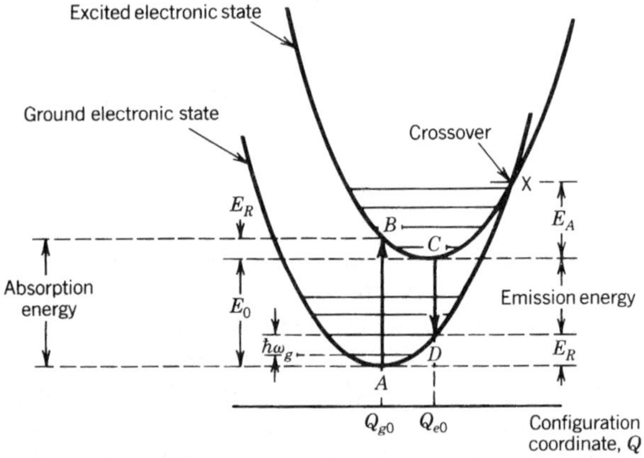

Figure 2.14 Configuration coordinate diagram. Note the vibrational levels associated with each electronic state. E_0 is the zero-phonon energy, equal to the thermal excitation energy. No distinction is made between the relaxation energy, E_R, in ground and excited states in Equation 2.56 (text). Particular vibrational levels are ignored in drawing the transitions *AB* and *CD*.

tion, with energies $(n_g + \frac{1}{2})\hbar\omega_g$. In each of these vibronic (vibrational/electronic) levels (see Section 4.2) the vibrational motion takes place principally between the two classical turning points where the line for each n_g meets the parabola. Indeed, even quantitatively, in any state but the few lowest, the probability distribution of Q is dominated by the regions near the parabola.*

If the defect is excited to state $|e\rangle$, the energy becomes

$$E_e = E_{e0} + \tfrac{1}{2}K_e(Q_e - Q_{e0})^2 \tag{2.55}$$

We can now define several important transitions. The Franck–Condon transitions ($A \to B$, $C \to D$) take place at constant Q (Figure 2.14). The zero-phonon transitions correspond to $(n_g = 0) \to (n_e = 0)$, with a generalization when excited vibrational states are involved. In discussing the relation between the various energies one usually assumes identical force constants in the two states, that is, $K_g = K_e$. The important energies are the relaxation energy E_R; the activation energy E_A, which measures the crossover position† above the excited state minimum; and the zero-phonon energy E_0 (see Figure 2.14):

$$\begin{aligned} E_A &= (E_0 - E_R)^2/4E_R \\ E_R &= S_0\hbar\omega \\ E_0 &= p\hbar\omega \end{aligned} \tag{2.56}$$

The two dimensionless factors S_0 and p appear often in discussions of defect–lattice interactions; S_0 is called the Huang–Rhys factor [see Equation (4.13)].

Where are the limitations of this model? First, Q is not usually a normal mode. We have already seen in Section 2.2 that in both classical and quantum physics harmonic systems can be treated exactly as a collection of dynamically-independent oscillators, the normal modes. However, the coordinate Q is, in general, a combination of normal modes determined by the defect's electronic structure; it is a single normal mode only in very special cases, for example, when it is a local mode. A coherent motion in Q will die away as the component normal modes get out of phase. Second, the best choice of Q may depend on which property interests us. There may be different choices depending on whether we are interested in optical or thermodynamic properties.

*See Schiff (1955, p. 66) where this point is illustrated. Note that for thermal equilibrium the distribution over the vibrational levels is gaussian, centered on Q_{g0}.

†The crossover is drawn by extrapolation here; in reality there may be good reasons for an *avoided crossing*.

Third, we may need different choices of Q (not just of K_g or K_e) in ground and excited states.

The configuration coordinate is a special case of a *reaction coordinate*, an idea that covers diffusion, chemical reaction theory, and many other areas (Stoneham 1981). In the wider fields of reaction coordinates there is an additional qualification. Suppose a diffusing particle has mass M_D. When it moves from site to site, clearly the particle's own coordinate is a good first choice of reaction coordinate. But this would give deceptive results for isotope effects, since other atoms in the solid also move with an effective mass, M_S; a reaction coordinate should have an effective mass $(\alpha M_D + \beta M_S)$ to reflect the isotopic changes properly.

2.5 VIBRATIONS OF MIXED CRYSTALS

In the cases where crystals AC and BC form a mixed crystal, represented by $A_yB_{1-y}C$ over all compositions from $y = 0$ to $y = 1$, one may see a localized vibrational mode of A in BC change frequency and broaden into a lattice band mode in a continuous way, ultimately showing the modes of isolated B atoms in AC. This type of study can be done quite readily for y less than 1% using Raman and infrared techniques, and neutron scattering methods can also be used for values of y and $1-y$ greater than a few percent. For simple mixed crystals two distinct types of behavior are commonly observed:

1. *Two-mode systems.* In these systems, for finite values of y, there are two quite separate TO modes and two separate LO modes, and the crystals show two separate restrahl bands (Figure 2.15). Mixed crystals represent a level of theoretical complexity intermediate between that of isolated impurities in crystals (Section 2.4) and that of amorphous solids (Section 2.6). In general, the theory of vibrational excitations in mixed crystals presents formidable difficulties, and phenomenological models are sometimes used. The so-called isodisplacement model is of this genre, and generates two-mode behavior. To provide a basis for this model we use the two-atom crystal (Figure 2.16) and note that for a perfect GaP crystal, for example, the long-wavelength modes can be written

$$\begin{aligned} M_g\ddot{x}_g &= -K(x_g - x_p) + Z_gE \\ M_p\ddot{x}_p &= -K(x_p - x_g) + Z_pE \end{aligned} \tag{2.57}$$

where E is the macroscopic electric field, and displacement amplitudes of gallium and phosphorus are x_g and x_p. The Ga ions have the same

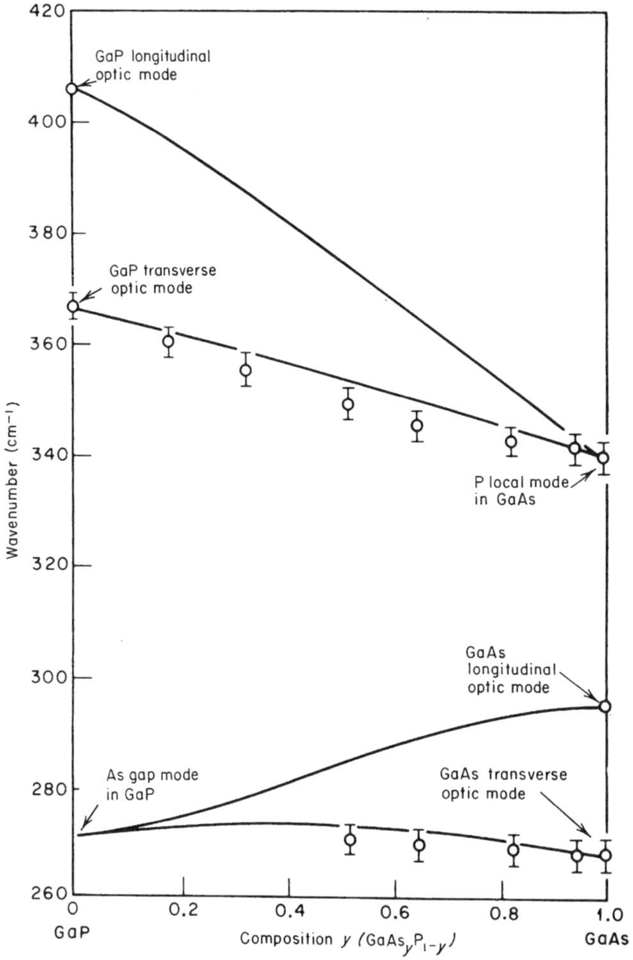

Figure 2.15 Observed positions of restrahl peaks (ǂ) for $GaAs_yP_{1-y}$. For $y \sim 1$ the uppermost peak becomes a local mode of P in GaAs. The lines represent calculations by Chang and Mitra (1968) based on an isodisplacement model (see text).

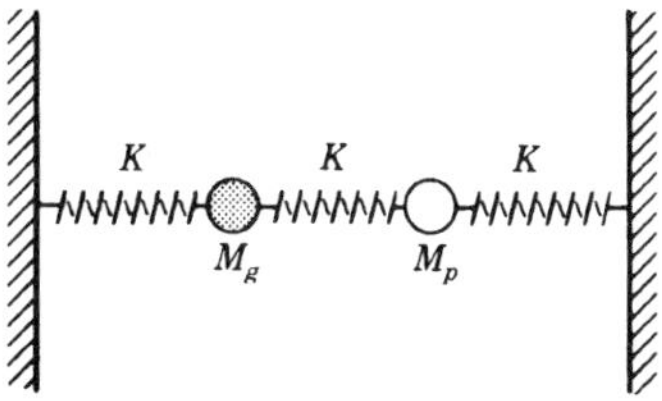

Figure 2.16 Model of a crystal composed of two atoms M_g and M_p and with a single force constant K [see Equation (2.57), text].

displacement x_g in every unit cell, corresponding to $k=0$. If this assumption of displacements identical in each cell is made for the mixed crystals $GaAs_yP_{1-y}$ too, we have the isodisplacement model. For this model equations (2.57) become

$$\begin{aligned} M_g\ddot{x}_g &= -yK_a(x_g - x_a) - (1-y)K_p(x_g - x_p) + Z_gE \\ M_a\ddot{x}_a &= -K_a(x_a - x_g) - (1-y)K_2(x_a - x_p) + Z_aE \\ M_p\ddot{x}_p &= -K_p(x_p - x_g) - yK_2(x_p - x_a) + Z_pE \end{aligned} \tag{2.58}$$

where K_2 is a second-neighbor (anion–anion) force constant.

Figure 2.15 shows a calculation of the restrahl modes for $GaAs_yP_{1-y}$ using equations of the type (2.58). The weak high-frequency mode, which occurs at 350 cm^{-1} for $y \to 1$, is the phosphorus local mode in GaAs. Similarly the 225 cm^{-1} mode found for $y \to 0$ is a highly localized As gap mode in GaP.

The isodisplacement model may be made more realistic by making force constants depend on y. It should be emphasized that, in fitting procedures using such models, force constants and effective charges are sometimes used which are difficult to reconcile with the microscopic properties of real crystals; the models may be adequate for interpolation, but do not bear detailed physical scrutiny. The model may also be elaborated by considering the possibility in the case of $GaAs_yP_{1-y}$, for example, of five different isodisplacement sublattices with a Ga ion surrounded by four, three, two, one, or zero As ions. This approach generates fine structure in calculated restrahl spectra corresponding to that sometimes observed experimentally.

2. *Single-mode behavior*. Many mixed crystals, for example, $Ca_ySr_{1-y}F_2$, exhibit a single restrahl peak, perhaps with some structure, over the complete range of mixture, and the phenomenological virtual-ion crystal model may be used to represent this type of behavior. In this case one replaces the mixed ions by a single average ion producing, in effect, a pure-crystal problem. In the case of $Ca_{0.3}Sr_{0.7}F_2$, for example, we replace Ca and Sr by an average mass of $0.3(40)+0.7(88)=73$ amu. For detailed fitting of restrahl spectra using this model, it is also necessary to average force constants and effective charges.

There has been some discussion in the literature of the crystal properties that determine whether one-mode behavior should occur. One important rule is that, if a gap mode and a local mode are formed at opposite ends of the range of mixture, we expect to get two-mode behavior. Finally, it should be emphasized that in crystals with more complex chemical composition than those discussed above (e.g., ZnSe and GaP form a completely miscible quaternary alloy) more complex mode behavior occurs.

2.6 VIBRATIONS OF AMORPHOUS SOLIDS

The well-ordered structure of crystals can be described systematically by making use of their known translational symmetry. Amorphous solids inevitably have a statistical component in their structure, and their underlying features can be both varied [see, for example, Ziman (1979)] and hard to isolate. This complexity has given rise to three important models: (1) microcrystals, separated by boundary regions of ill-defined structure; (2) continuous random networks in which each atom always has the same neighbors (e.g., Si might always have four O neighbors in SiO_2) so that there is no coordination disorder, but merely variations in the angles and bond lengths that determine precise geometry; (3) hard sphere systems, disordered as in a liquid. (See Table 2.5.)

Most amorphous solids are macroscopically isotropic. Structural information comes principally from X-ray or neutron scattering, combined with physical or chemical intuition and knowledge of related systems. The pair-distribution function, $g(R)$, is one of the most important observed quantities, since it determines many thermodynamic and structural features. Suppose atom 1 in a monatomic solid is at site $\mathbf{R}_1$. Then, ignoring any time dependence, the probability that there is an atom 2 at

Table 2.5 Structural Models for Amorphous Materials

Model	Example[a]
Continuous random network	a-Si; a-SiO_2
Continuous random network plus interstitials	Most conventional glasses, for example, borosilicate glasses with a (B + Si + O) framework and Na present interstitially
Microcrystal	"Amorphous" ionic crystals where the Coulomb interactions force crystalline ordering, but only over a short range
Hard-sphere structures	"Amorphous" rare-gas or molecular solids where the strong short-range repulsions and weak cohesive forces lead to structures like those modeled with ball bearings.

[a]In several cases these specific examples remain controversial, although most would believe that they are described correctly here as a first approximation.

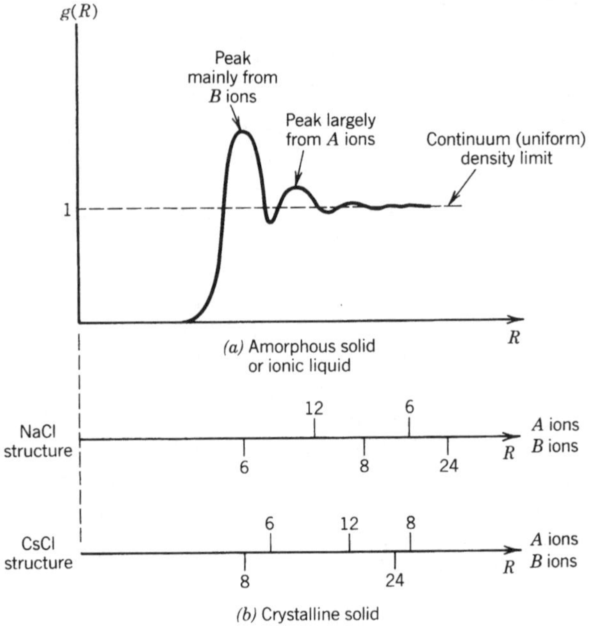

Figure 2.17 Schematic representation of the pair distribution function $g(R)$ for an amorphous solid AB. The atom at $R=0$ in (a) is assumed to be of type A. In (b) the positions of the neighbors are shown for the NaCl and CsCl structures, with spacings normalized to have closest neighbors at the peak of (a).

distance $R=|\mathbf{R}_2-\mathbf{R}_1|$ is $g(R)$. For a crystal, $g(R)$ is finite only at shells of known spacing, so that the pair-distribution function has a series of spikes; if shell i contains Z_i atoms, $g(R)=\sum_i Z_i\delta(R-R_i)$. The lack of long-range order in amorphous solids broadens the peaks, although the structure of the first coordination shell is clear (Figure 2.17). At large values of R, $g(R)$ tends to 1, so that the probability of a given spacing is determined by the average density used to normalize $g(R)$.

For systems involving several species, more complex descriptions are needed, and more complicated experimental techniques are necessary, such as isotopic substitution in neutron diffraction (Enderby and Nielson 1981).

In amorphous solids it is clear that we cannot exploit translational symmetry (except perhaps at very long wavelengths). There will be a well-defined density of states $\rho(\omega)$, but no wavevector to characterize modes. If $\rho(\omega)$ has well-defined, reproducible, features, these will usually

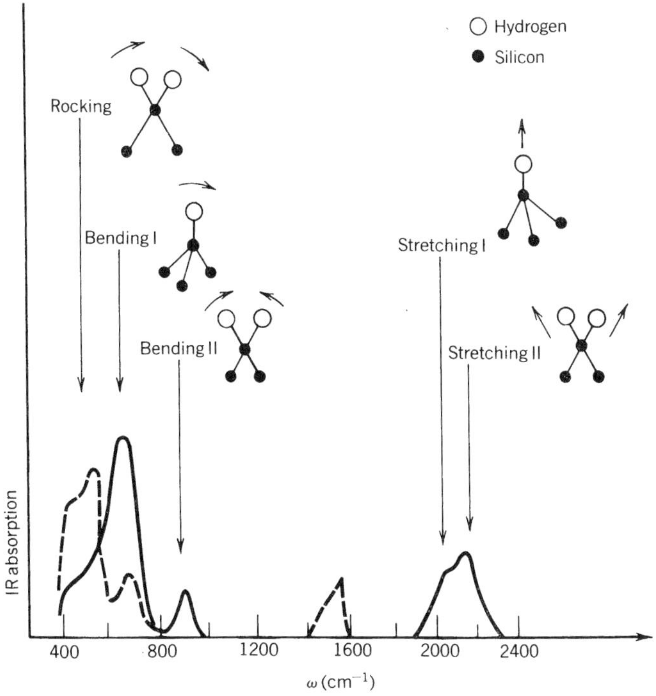

Figure 2.18 Infrared absorption for hydrogenated amorphous Si. The various distinct vibrational modes for Si_3–H and S_2–H_2 are illustrated schematically. Spectra for both hydrogen (——) and deuterium (- - -) are shown.

have a structural interpretation. This is especially so when there are highly localized modes associated with impurities; Figure 2.18 shows examples for amorphous silicon formed from silane (SiH_4) and hence containing much hydrogen (the so-called *a*-Si:H; see also Section 8.1).

The precise identification of features of $\rho(\omega)$ with types of atomic motion is conveniently done by computer experiment. For amorphous silicon (*a*-Si), for example, one starts with two assumptions. One concerns interatomic forces (there are simple models that invoke both bond-stretching and bond-bending forces to describe both amorphous and crystalline forms (see Section 2.1)). The other assumption is structural. Many covalent crystal structures (e.g., Si or SiO_2) can be looked at as connected rings. In the diamond structure, six-membered rings are easily seen (Figure 2.19). On going to an amorphous solid a large unit cell is constructed, maintaining fourfold coordination, but also

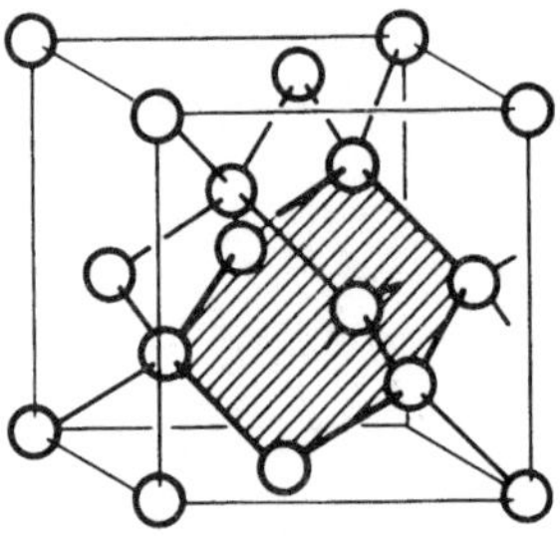

Figure 2.19 Structure of the diamond lattice showing a six-membered ring, which is the smallest closed circuit for this crystalline lattice. For corresponding structures in amorphous material see Figure 8.1.

including other rings of bonds, for example, five-membered rings. Just as in the calculations on crystals with point defects, one can calculate the density of states and the anticipated infrared and Raman spectra. An example is given in Figure 2.20 (Alben et al. 1975). While the densities of states for both amorphous and crystalline solids are similar, the spectra show some significant differences. In essence, the spectra follow $\rho(\omega)$ more closely for a-Si, and lack some of the sharp structure seen in crystalline silicon (c-Si). This is because the selection rules are much less strong for the amorphous system. Indeed, the cancellations that occur in perfect crystals are violated in disordered cases sufficiently that some modes are both infrared and Raman active (Hayes and Loudon 1978). Analysis of infrared data (Klug and Whalley 1984) suggests that in a-Si the Si atoms have a distribution of apparent charges with a root-mean-square value of $0.24|e|$.

The lattice specific heat of simple dielectric solids such as KCl follows a T^3 law at very low temperatures (Kittel (1953), Chap. 5). This behavior is predicted by the Debye model of specific heats. At first glance the continuum approximation underlying the Debye model, which predicts this dependence, might be expected to apply to amorphous as well as to crystalline solids. However, at very low T the measured specific heat of amorphous (glass) and crystalline (α-quartz) phases of SiO_2 have a different temperature dependence (Figure 2.21). The specific heat of the glass decreases much more slowly than the T^3 prediction (the hump for both materials at ~50 K arises from dispersion of acoustic vibrations) and fits the expression

$$C = C_1 T + C_3 T^3, \quad 0.1\,\mathrm{K} < T < 1\,\mathrm{K} \tag{2.59}$$

This type of behavior occurs in a wide range of glassy materials, for example, amorphous As_2Se_3. It is generally assumed that the specific heat anomaly is an intrinsic property of the amorphous state, although it is difficult to prove that extrinsic effects, for example, impurities, are not responsible to some extent.

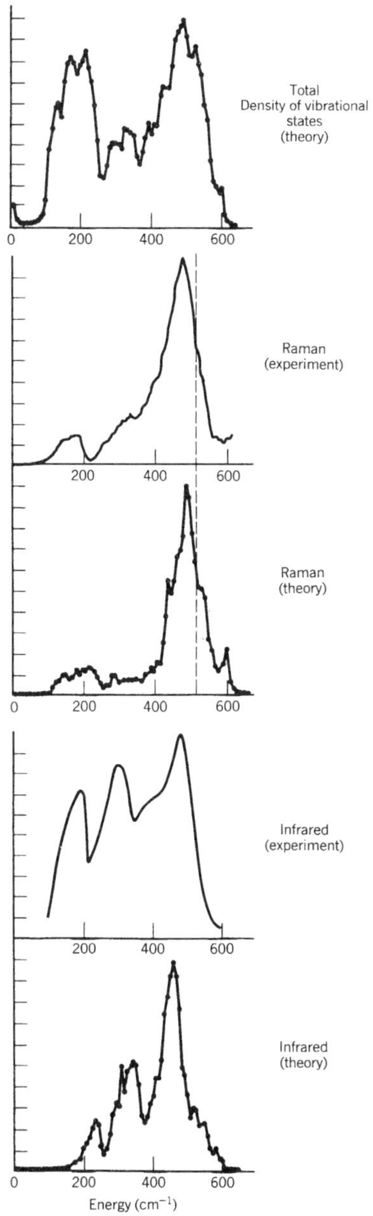

Figure 2.20 Calculations of the vibrational density of states of amorphous Si, using a valence-force potential, compared with measured Raman and infrared spectra [after Alben et al. (1975)].

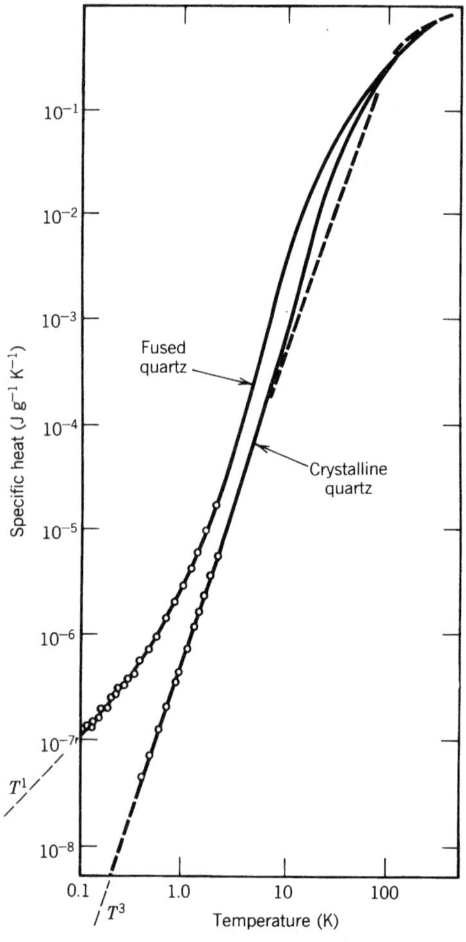

Figure 2.21 Temperature dependence of the specific heat of SiO_2 in both the glass (fused quartz) and crystal (α-quartz) phases. The dashed line gives the T^3 prediction. The specific heat of the glass varies almost as T at the lowest temperatures [after Zeller and Pohl (1971)].

Amorphous materials showing a specific heat anomaly also show an anomaly in thermal conductivity κ at low T. Heat transport in insulating crystals is due to phonons and is given by the expression (Kittel 1953, Chap. 5)

$$\kappa = \frac{1}{3}\sum_i \int_0^{\omega_D} C_i(\omega) v_i(\omega) l_i(\omega)\, d\omega \qquad (2.60)$$

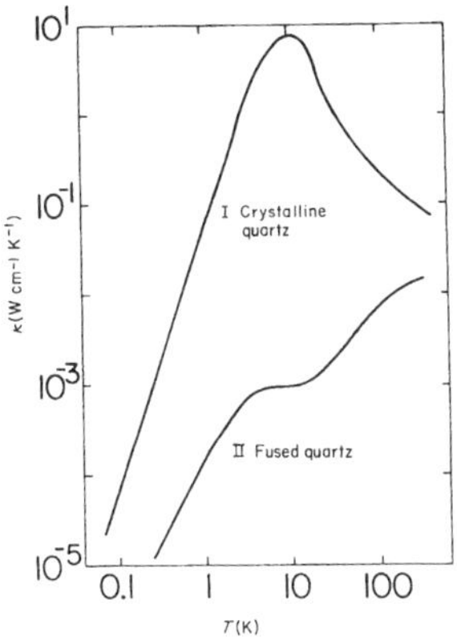

Figure 2.22 Temperature dependence of the thermal conductivity of crystal quartz (curve I) and fused quartz (curve II). Curve I varies as approximately T^3 below 10 K and curve II varies as roughly T^2 at the lowest temperatures. For a detailed discussion of the shape of curve II see Anderson (1981).

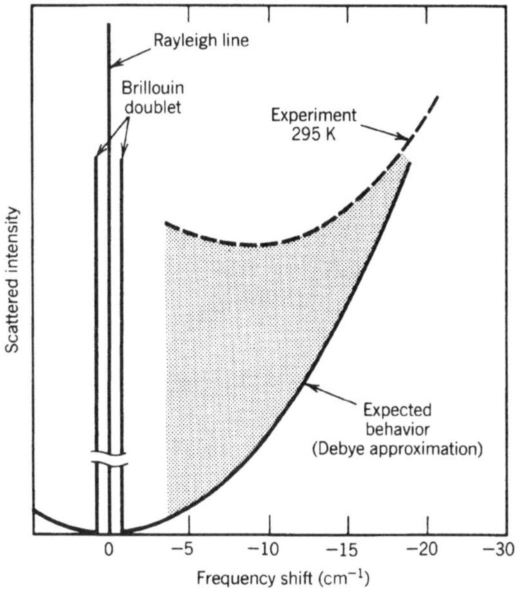

Figure 2.23 The energy dependence of the intensity of light scattered by fused quartz at low energies. The dashed curve gives the experimental results and the solid curve the prediction of a Debye model; the shaded area is the excess scattering. The inner three lines are the Rayleigh line and a Brillouin doublet [after Winterling (1975)].

for an isotropic material. $C(\omega)$ is the phonon specific heat, dependent on frequency, $v(\omega)$ is the phonon velocity, $l(\omega)$ is the phonon mean free path, and the summation is over one longitudinal and two transverse modes. In the Debye limit of no phonon dispersion $\kappa \propto T^3$, a behavior shown by crystalline quartz at low T (Figure 2.22). However, κ for fused quartz varies as T^2 at low T (Figure 2.22), and this behavior is typical of amorphous materials below ~ 10 K.

To explain the low-temperature anomalies in C and κ of amorphous solids, the vibrational density of states at low energies needs to be higher than in crystalline solids. This might come from a local resonance, for example. Many workers have suggested a phenomenological tunneling model assuming a spread in both symmetry and height of potential barriers between neighboring sites of some unspecified ion; quantum mechanical tunneling between sites can then give rise to the fairly constant density of states at low ω required to give a linear term in C and a quadratic term in κ. Thus the model supposes delocalized Debye continuum modes, with $\rho(\omega) \sim \omega^2$, and localized resonances, probably involving tunneling, with constant $\rho(\omega)$ over some range of frequencies. Additional evidence for the existence of low-energy resonances is provided by measurement of ultrasonic absorption, which is much stronger in glasses than in crystals, and which shows the phenomenon of saturation (Hunklinger and Schickfus 1981)]. Still more direct evidence results from the study of quasielastic light scattering by vitreous silica (Figure 2.23). However, because macroscopic defects like surfaces give unwanted scattering, it is difficult to obtain a complete picture to as low a frequency as desired. Nevertheless, an upturn of the scattered intensity is observed in the range 10 to 5 cm^{-1}. In Figure 2.23 the extrapolated scattering from higher energies is drawn assuming a Debye model for the vibrational density of states. It is suggested that the additional scattering in the region 10 to 5 cm^{-1} is due to tunneling levels, assuming that the dielectric susceptibility of a defect is different in the configurations between which it tunnels. The technique of quasielastic light scattering has also been used with success on other amorphous materials using more sophisticated experimental techniques (Jäckle 1981).

The quantitative application of the tunneling model is not without its problems, however, For example, attempts to fit the measured quasielastic light scattering for glasses over a range of temperatures have not been very successful (Jäckle 1981). Moreover, detailed work by Black and Halperin (1978) shows that tunneling densities of states predicted from specific heat measurements are substantially larger than those derived from ultrasonic measurements.

CHAPTER THREE

Lattice Defects

In a solid at the absolute zero of temperature atoms only move about their perfect lattice sites because of their zero-point vibrations. As the temperature is raised, a variety of changes may occur. The thermal vibrations increase in amplitude, the vibrational anharmonicity leads to change in lattice spacing, lattice defects are produced, and electronic excitations may be generated. There is also the possibility of a change of phase, which may be to another solid phase, to the melt, or to vapor. In this chapter we concentrate largely on thermally induced lattice defects together with consequent effects on electrical conductivity, diffusion, and lattice spacing.

We shall also occasionally come across thermally induced electronic defects. In many semiconductors like Si and Ge the electrons and holes produced by thermal ionization are delocalized. In some ionic systems like UO_2, the thermal generation of electrons and holes is believed to correspond to charge transfer, producing U^{3+} (from the electron) and U^{5+} (from the hole) in addition to the normal U^{4+} ions. The most fully documented case of this type of small-polaron behavior is $La_{1-x}Sr_xCrO_3$ (Karim and Aldred 1970), although small polarons are likely in most oxides in which the cations can change valence easily (Table 3.1a).

There are two principal types of lattice point defect in crystals (Table 3.1b):

1. *Frenkel Defects*. These are vacancies and interstitials of the same species in equilibrium. This type of disorder occurs in AgCl (Section 6.3.3) where we find Ag^+ vacancies and Ag^+ interstitials in equal numbers, and also in CaF_2 where we find F^- vacancies (F_v^-) and F^- interstitials (F_i^-).
2. *Schottky Defects*. Only vacancies occur in the body of the crystal, with extra atoms at the crystal surface or in some other internal defect sink (see below). This type of disorder is characteristic of materials with close-packed structures where the energy required to insert an interstitial into the structure is high (a few electron volts). In monatomic materials such as rare-gas solids and metals, only one type of vacancy occurs. Schottky defects are also characteristic of alkali halides, but here both cation and anion vacancies occur in equal numbers for charge compensation (Section 3.1).

The descriptions Frenkel and Schottky are not exclusive, and one should question whether the interstitital ion of a Frenkel pair has the same charge as the normal ion in the perfect crystal. Although these are the commonest forms of disorder, there are other important cases such as

Table 3.1 Types of Disorder in Crystalline Solids

Type of Disorder	Crystalline Solid
(a) Electronic disorder	
Delocalized charges	Si, Ge, III-V, most II-VI
Localized charges	$M_{1-x}O$ (probably for M = Fe, Ni; perhaps for M = Co, Mn).
	UO_{2+x}
	$La_{1-x}Sr_xCrO_3$
(b) Ionic disorder	
Frenkel	
Anion Frenkel	CaF_2, SrF_2, BaF_2, UO_2, CeO_2, ThO_2
Cation Frenkel	AgCl, AgBr, $NaNO_3$, KNO_3
Schottky	Alkali halides
	Cesium halides
	BeO
	Alkaline-earth oxides, although these are extrinsic (i.e., impurity dominated) in practice
(c) Other forms of disorder	
Antisite defects	GaP, GaAs
	Complex oxides with several cation species, for example, spinels and garnets
Shear planes or similar large-scale disorder	TiO_2, Nb_2O_5

antisite defects (e.g., *B* on *A* sites in a crystal *AB*; see p. 163) and macroscopic defects like shear planes and dislocations.

The defects we shall mostly consider are in (or very close to) equilibrium in the bulk of the solid. Quite different populations may be present very close to dislocations or surfaces and may also be introduced by quenching from high temperatures or by irradiation (see Section 6.1).

We generally find for a given crystal that the energies of formation of Frenkel and Schottky defects are sufficiently different that one type is dominant. The lattice relaxation in the vicinity of a defect and the polarization by its net charge are critical in determining the formation energy (Section 3.4). These factors decide, for example, the form of the lattice configuration of an interstitial and its environment. An interstitial may share a lattice site with a second ion forming a so-called split interstitial (Figure 3.1*a*). If the axis of the pair of ions lies along a close-packed direction, the split interstitial is sometimes known as a crowdion (Figure 3.1*b*).

In addition to the intrinsic defects mentioned above there may be impurity ions present in crystals, giving so-called extrinsic defects. The impurities may be present inadvertently, in trace concentrations, or they may be deliberately added. Trace impurities commonly found are H, O (from water vapor), and C, together with species chemically similar to the host ion replaced (K in NaCl, Si in Ge). If an impurity ion in an ionic

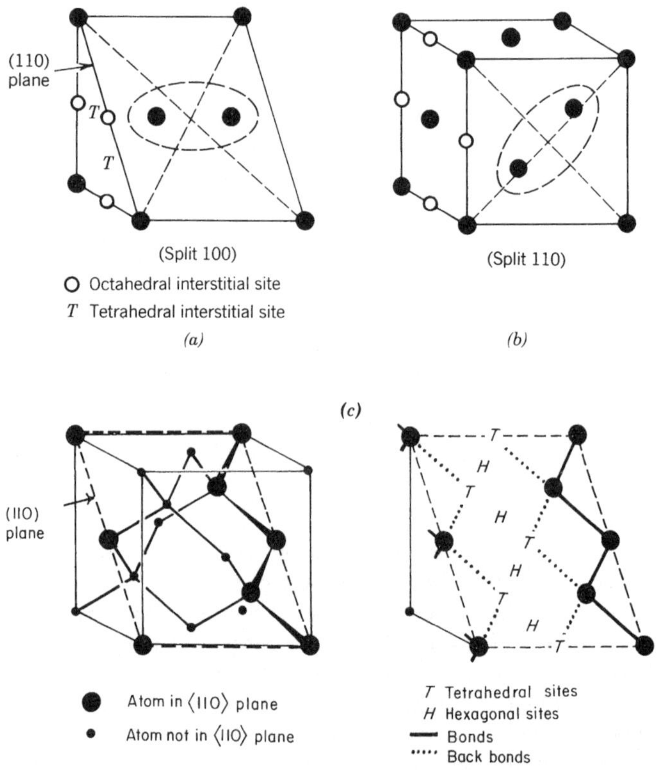

Figure 3.1 Two examples of split interstitial configurations, with (*a*) a ⟨100⟩ axis in a bcc host, corresponding to the cation sublattice in the cesium halide structure and (*b*) a ⟨110⟩ axis in an fcc host, corresponding to the cation sublattice in the NaCl structure. Split interstitials occur in the diamond structure also; (*c*) shows the tetrahedral (*T*) and hexagonal (*H*) interstitial sites for diamond but does not show the split form explicitly.

crystal has a net charge different from that of the host ion it replaces (aliovalent impurity), other defects must be introduced for charge compensation. Thus, if Ca^{2+} in CaF_2 is replaced by Y^{3+} or a trivalent rare-earth ion (Re^{3+}), the extra positive charge may be balanced by a fluorine interstitial (F_i^-). Since the impurity and its compensating defect have opposite effective charge in the crystal, they tend to associate in pairs with a binding energy of order $q_1q_2/\epsilon r^2 \sim 0.5$ eV. Similarly, a Ca^{2-} ion replacing K^+ in KCl may be charge compensated by a cation vacancy, and these also tend to form pairs with a binding energy of $\sim$0.4 eV. Hence these defects are almost entirely paired at room temperature if equilibrium is established. We shall see later (Section 3.2) that such paired defects make a measurable contribution to the dielectric relaxation of the crystal.

Our remarks so far refer to defect concentrations of less than $\sim$1 mole %. At higher defect concentrations defect interactions may lead to reduction in defect formation energy and to defect clustering (Sections 3.3 and 3.5). Clustering is especially important in nonstoichiometric oxides such as iron oxide ($Fe_{1-x}O$) and uranium oxide (UO_{2+x}), and on an altogether grander scale one finds ordered shear planes in certain high-dielectric-constant oxides like TiO_{2-x} (Section 3.3.2).

These varieties of behavior reflect differences in production and interaction energies, as well as in the configurational entropy factors, which also determine equilibria. The application of classical statistical mechanics to dilute defect concentrations will be outlined in Section 3.1 and in Section 3.2 we provide experimental background. In Section 3.3, we deal with some aspects of nonstoichiometry, emphasizing the role of lattice defects in this phenomenon. Techniques for the calculation of energies of point defects are outlined in Section 3.4. Materials with unusually high ionic conductivity are discussed in Section 3.5. These materials are referred to as superionics or fast-ion conductors and generally have one component sublattice that is highly disordered. Defects in more covalent crystals (e.g., Si and GaP) are touched on in Section 3.6 and, finally, in Section 3.7 we provide an overview of macroscopic defects (dislocations, grain boundaries) not normally found in thermal equilibrium with the lattice.

3.1 DEFECTS IN THERMAL EQUILIBRIUM

Any departure from a perfectly ordered lattice structure may be classified as a lattice defect. Defects may be intrinsic or due to impurities and they may be localized or extended. The introduction of a defect into a crystal always increases the lattice energy. However, this effect is counterbalanced by entropy changes arising from the introduction of defects, so that a crystal in thermodynamic equilibrium at a finite temperature always contains a finite concentration of defects.

We may define the work necessary to create a single defect (e.g., a vacancy plus an interstitial for Frenkel disorder) in a variety of ways, depending on which thermodynamic variables (pressure P, volume V, temperature T, or entropy S) are kept constant. Most experiments are done at constant temperature and pressure, so it is the Gibbs free energy g_p (where the subscript means constant pressure) that is relevant. That is often split into two components,

$$g_p = h_p - Ts_p \tag{3.1}$$

involving the enthalpy h_p and entropy s_p. Only the nonconfigurational entropy is included in Equation (3.1), and this is principally due to changes in lattice vibrational frequencies. Theorists often calculate u_v, the internal energy appropriate for constant volume (strictly for constant lattice parameter) and zero temperature. The several alternative forms are related to each other, and the main results (Catlow et al. 1981) are summarized in Table 3.2. In practice the orders of magnitude are 0.5–5 eV for the enthalpy h and a few k_B, that is, a few times 10^{-4} eV/K for the entropy, so that the entropy term Ts_p will only matter at high temperatures and is often ignored in approximate treatments. (There will be other effects also at high temperature arising from thermal expansion.)

Table 3.2 Relations among Parameters in Defect Thermodynamics[a]

If β_p is the volume thermal coefficient of expansion, V is the crystal volume, and K_T is the isothermal compressibility, then we have the relations (Catlow et al. 1981):

$$g_p = f_v = h_p - Ts_p$$
$$h_p = u_v + (T\beta_p/K_T)v_p$$
$$s_p = s_v + (\beta_p/K_T)v_p$$
$$v_p = -K_T(\partial f/\partial \ln V)_T$$

where, for defect formation, for instance,

g_p = Gibbs free energy of formation (constant T, P)
f_v = Helmholz free energy of formation (constant V, T)
h_p = Enthalpy of formation (constant P, S)
u_v = Internal energy of formation (constant V, S)
s_p, s_v = Entropy of formation (constant P, V)
v_p = Formation volume (constant P, T)

[a]These expressions are correct to first order only in the defect concentrations.

Likewise, terms like Pv_p (Table 3.2) only become important when the pressure reaches a few kilobars (10^8 N/m^2).

In equilibrium at constant temperature and pressure, the Gibbs free energy of the whole system (i.e., including all defects and configurational entropy) is extremal under variations of the defect concentration, subject to some constraints. We shall give examples below of defect energetics, starting with simpler cases, and working to more complex ones. The arguments we use are principally those of standard statistical mechanics.

3.1.1 Schottky and Frenkel Defects

Consider first a crystal of N atomic sites that, for simplicity, we may imagine to be a rare-gas crystal. We suppose there are no complications due to electronic or orientational degeneracy. Let g be the work necessary to create a single vacancy at constant temperature and pressure. We shall assume the vacancy concentration to be so low that defect–defect interactions can be ignored and hence that g is independent of the vacancy content. The free energy of the defective crystal contains three principal terms:

(i) A term G_0 corresponding to the perfect crystal.
(ii) A term $G_1 = ng$ coming from the work done in creating n vacancies.
(iii) A term $G_2 = -TS_{\text{config}}$ from the configurational entropy.

The concentration of vacancies at equilibrium is determined by the extremal condition

$$\frac{\partial G}{\partial n} = \frac{\partial}{\partial n}(G_0 + G_1 + G_2) = 0 \tag{3.2}$$

In our case G_0 is independent of n by definition, and $(\partial G_1/\partial n)$ is simply g. Equation (3.2) reduces to

$$g - T\frac{\partial}{\partial n}(S_{\text{config}}) = 0 \tag{3.3}$$

The configurational entropy term is obtained using a standard combinatorial method. The vacant lattice points can be arranged in P_L ways,

$$P_L = N!/[(N-n)!] \tag{3.4}$$

The configurational entropy is given by $k_B \ln P_L$, so that we have

$$S_{\text{config}} = k_B \ln\{N!/[(N-n)!n!]\} \tag{3.5}$$

We need $T\partial S_{\text{config}}/\partial n$, and this is found using Stirling's theorem that $\ln(x!)$ tends to $x \ln x$ at large x. If N, $(N-n)$ and n are all large, we find

$$T\frac{\partial S_{\text{config}}}{\partial n} = k_B T \ln\left(\frac{N-n}{n}\right) \tag{3.6}$$

If the defect concentration $f \equiv n/N$ is small, the right-hand side of Equation (3.6) reduces to $-k_B T \ln f$. Collecting together the terms in $\partial G/\partial n$ we find

$$g + k_B T \ln f = 0, \tag{3.7}$$

giving the fraction f of vacant sites as

$$f = \exp(-g/k_B T) \tag{3.8}$$

The fractional vacancy concentration depends exponentially on the Schottky formation energy g, and increases exponentially with increasing temperature.

Even the simple example discussed above raises some general points. The first concerns the reference energy state. Although for the simple Schottky defect this is the perfect crystal, there may be some choice in other cases. As an example we consider the energy of solution of an interstitial impurity, for example, oxygen, in a crystal. The reference state will be different when one considers the equilibrium of the crystal with a gas phase X_2 and when one considers the equilibrium of the same crystal with a liquid compound MX. The two cases can be related, one to the other, but there is opportunity for confusion in analyzing published data on energies of solution.

A second point that can cause confusion arises in connection with Schottky disorder. The number of defects present in equilibrium will depend to a large extent on the energy required to create them. If one creates a vacancy by moving a bulk atom to a surface site, the energy will depend on the precise surface site, and hence the equilibrium would at first sight seem to depend on details such as the crystal habit. The resolution of this problem can be given in several forms. One way is to define Schottky disorder for an infinite crystal with no free surfaces or sinks. The vacancy is introduced by rearranging N atoms on $N+1$ sites, with a volume change if the process is to occur at constant pressure. It is the formation energy so defined that determines the equilibrium concentration of defects, in principle.

In many simple estimates of formation energies the experimental cohesive energy or the sublimation energy is required. It is here that problems could arise if the experiment measured some special property, like the energy of removal of atoms from special surface sites. Fortunately, experiments generally measure the correct bulk-crystal cohesive energy, and it is not necessary to know the precise atomic mechanism involved.

Special difficulties can occur in polar crystals where, for instance, a surface develops a net charge by the preferential production of defects of one sign of charge (see Section 7.5). Here the surface acts as a sink for, say, more cations than anions. Since there is no universal way to divide the cohesive energy into anion and cation contributions, one has to know the detailed site to which each defect component goes.

A third general point concerns internal degrees of freedom. These might be electronic, when an ion can be in one of several possible spin states, or in a low-lying excited electronic state. Degrees of freedom may also be orientational, as in the case of a divacancy (Section 3.2) or a molecular impurity ion such as O_2^-. The contribution of internal degrees of freedom to the free energy takes the form

$$\tilde{g}_d = k_B T \ln(Z_0/Z_d) \tag{3.9}$$

per defect where Z_0 is the internal partition function in the reference state and Z_d is that in the defect state. The partition function has the general form of a sum over the states $\sum_i \exp(-\epsilon_i/k_B T)$; it is temperature dependent in general, and there is a higher effective degeneracy at high temperature since more states are populated. How one includes $\tilde{g}_d$ in equations like (3.2) is partly a matter of personal choice. One can build $\tilde{g}_d$ into the main part of the free energy of formation g, or one may put it into an extended entropy term G_3. The effect is simplest for Schottky defects when one regards $\tilde{g}_d$ as an extra term in the free energy of formation g, giving a total free energy g_t. It is then easily seen from Equations (3.7), (3.8), and (3.9) that the defect fraction becomes

$$f = (Z_d/Z_o)\exp(-g_t/k_B T) \tag{3.10}$$

Table 3.3 contains examples, some of which will be encountered later in the book, for which Equation (3.10) is useful.

The extended form of the extra term G_3 is useful for cases such as simple Frenkel defects, where the vacancy and interstitial are independent to a certain extent. If the interstitial has access to N' sites, the configurational entropy for n interstitials contains an extra term like (3.5), but with N' replacing N. Equation (3.6) becomes

$$T\frac{\partial S^F_{\text{config}}}{\partial n} = k_B T \ln\left[\left(\frac{N-n}{n}\cdot\frac{N'-n}{n}\right)\right] \tag{3.11}$$

Table 3.3 Examples of Defect Systems where Equation (3.10) Applies[a]

Host Lattice	Defect	Type of Degeneracy	Z_d/Z_0
NaCl	F center (neutral anion vacancy)	Spin $S=\frac{1}{2}$	2
bcc rare-gas solid	Divacancy oriented along $\langle 111\rangle$ and equivalent directions	Orientational	4 [note (111) and $(\bar{1}\bar{1}\bar{1})$ are equivalent]
Diamond	Neutral vacancy	Vibronic, lowest state with degeneracy 2; higher state with degeneracy 1 at energy Δ	$(2+e^{-\Delta/k_BT})$
NaCl	M center (neutral anion divacancy) oriented along $\langle 110\rangle$ and equivalent directions	(i) Spin, with $S=0$ in lowest state and $S=1$ in excited state at energy Δ; (ii) orientation (sixfold)	$(1+3e^{-\Delta/k_BT})\times 6$

[a] In these we assume a perfect-crystal reference state, and that Z_0 is unity. The final-state value $Z_d \equiv \sum_i \exp(-\epsilon_i/k_BT)$ is determined by any degeneracies and excited states, although only a few of the excited states (those with ϵ_i/k_B less than the melting temperature) really matter.

for Frenkel defects, and this reduces to $-k_BT\ln(f^2N'/N)$ at low defect concentrations. The appearance of f^2 is important, resulting in

$$f=(N/N')^{1/2}\exp(-g_F/2k_BT), \tag{3.12}$$

where g_F is the free energy of formation of a Frenkel pair. Note that $g_F/2$ rather than g_F appears in the exponential.

3.1.2 Law of Mass Action

Our previous examples refer to a single defect species, but often a whole series of interacting defects are involved, with reactions between them which lead to equilibrium ultimately. A typical reaction might be

$$n_1X_1 + n_2X_2 + \cdots \rightleftarrows \nu_1 Y_1 + \nu_2 Y_2 + \cdots \tag{3.13}$$

where n_i, ν_i are simply the numbers of the species X_i, Y_i in the reactant and product components. The Law of Mass Action asserts that, in equilibrium, the concentrations $[X_i]$, $[Y_i]$ are related by

$$\frac{[X_1]^{n_1}[X_2]^{n_2}\cdots}{[Y_1]^{\nu_1}[Y_2]^{\nu_2}\cdots} = \varphi(T) \tag{3.14}$$

where the reaction constant φ depends on temperature and pressure, but not on the concentrations. In general $\varphi(T) \propto \exp(-\Delta E/k_BT)$, where ΔE is the free energy absorbed in the forward reaction.

An important example is the effect of oxygen gas pressure on the conductivity of a nonstoichiometric oxide, for example, UO_{2+x} (see Section 3.3.1). Oxygen enters interstitially and the first reaction is

$$(O_2^0)_{gas} \rightarrow 2O_i^0 \tag{3.15}$$

followed by trapping of electrons by the neutral interstitial,

$$O_i^0 \rightarrow O_i^{2-} + 2h \tag{3.16}$$

as two valence-band electrons are captured (holes created). For the first reaction we have

$$|O_2^0|/|O_i^0|^2 = \varphi_1 \tag{3.17}$$

or, since $[O_2^0]$ is proportional to the gas pressure, p,

$$[O_i^0]^2 \propto p/\varphi_1 \tag{3.18}$$

For the second reaction,

$$[O_i^0]/[O_i^{2-}][h]^2 = \varphi_2 \tag{3.19}$$

If we rearrange (3.19) and use (3.18) we can get

$$[h] = \{[O_i^0]/\varphi_2[O_i^{2-}]\}^{1/2} \propto p^{1/4} \tag{3.20}$$

showing that the hole concentration varies as $p^{1/4}$. This result is typical of p-type conduction in oxides; as the external oxygen pressure increases, there is a slow increase in carrier concentration and also conductivity.

Other, simpler, results can also be cast in this form. Suppose an undoped intrinsic semiconductor contains thermally generated electrons and holes. Then mass action indicates $np = \varphi$, where n is the electron concentration and p is the hole concentration. Here $\varphi \propto \exp(-E_g/k_BT)$ is determined by the band gap E_g. Since every electron is generated with an accompanying hole, that is, $n = p$, we expect

$$n = p \propto \exp[-(\tfrac{1}{2}E_g)/k_BT] \tag{3.21}$$

The factor $\frac{1}{2}$ occurs here for exactly the same reason as in Equation (3.12) for Frenkel defect production.

3.2 EXPERIMENTAL INVESTIGATIONS OF LATTICE DEFECTS

Four important types of measurement are used to determine concentrations of defects and defect motion. One type of experiment measures the diffusion and ionic conductivity of crystals, related to the long-range motion of defects. A second type of experiment measures dielectric or elastic relaxation, that is, the rate at which polarization or strain follow a changing field. If there is an oscillating electric field, for example, there is a dielectric loss from the polarization that is out of phase with the applied field. Relaxation experiments can be used to monitor reorientation of defects like impurity–vacancy pairs, which may reorient without any long-range motion. The third type of experiment is static; defect concentrations can be obtained from measurements of X-ray lattice parameters, or density changes, or the two in combination. Finally, there are resonance methods which monitor the time dependent local fields experienced by moving particles; these include nmr (see Section 3.5) and epr and μsr (see Section 4.5).

Reviews of the energetics of point-defect formation in crystals and of defect dynamics are contained in Crawford and Slifkin (1972) and in Lidiard (1974).

3.2.1 Diffusion and Ionic Conductivity

The most commonly used techniques to obtain values for the free energy of formation and of motion of point defects in nonmetallic crystals involve measurement of the temperature dependence of the ionic conductivity of undoped and of doped crystals. The technique of tracer diffusion is also used (Flynn 1972, p. 390). The ionic conductivity of a cubic crystal may be written

$$\begin{aligned}\sigma &= \sum_i q_i c_i \mu_i \\ &= N \sum_i q_i x_i \mu_i \end{aligned} \tag{3.22}$$

where q_i is the effective ionic charge and μ_i is the mobility (drift velocity per unit applied field) of the defect species i. The concentration c_i can be written as the product Nx_i of the molar fraction x_i of defects and N, the number of formula units per unit volume.

If species i has only one important class of diffusion jump, the mobility takes the form

$$\mu_i = \frac{q_i R_i^2}{k_B T} d_i \nu_{\text{eff}} \exp(-h_i^M / k_B T) \tag{3.23}$$

where R_i is the displacement distance of the net charge q_i, h_i^M is the enthalpy of motion, and d_i is a degeneracy factor, determined by the number of possible equivalent jump directions. The effective frequency ν_{eff} contains several independent factors [see Flynn (1972, Chapter 7) for a full discussion], and is often approximated by the Debye frequency in default of better information. The major component of ν_{eff} is an approach frequency, that is, the number of times per second the thermal motion of the defect carries it toward the barrier to be surmounted. An extra factor in ν_{eff} [see Equation (3.1)] is $\exp(s^M/k_B)$, from the entropy of motion. A further factor is that of mobility drag, arising from Coulomb interactions between the various defect species.

Those processes occurring in ionic conductivity are also important in the phenomenon of diffusion. Diffusion is governed by Fick's two laws (Shewmon 1963). The first law notes that the flux $\mathbf{j}$ of diffusing atoms is proportional to the concentration gradient, that is,

$$J = -D\nabla c \tag{3.24}$$

Here c is the concentration of diffusing particles per unit volume and D is the diffusion coefficient (cm^2/sec). This law must be generalized when there are applied fields or when several species diffuse (Howard and Lidiard 1964). The second law adds the equation of continuity, $\partial c/\partial t = -\boldsymbol{\nabla} \cdot \mathbf{J}$, giving

$$\frac{\partial c}{\partial t} = \frac{\partial}{\partial x}\left(D\frac{\partial c}{\partial x}\right) \tag{3.25}$$

The mean square displacement of diffusing particles in a time t is given by

$$\overline{r^2(t)} = 6Dt \tag{3.26a}$$

and in three dimensions it follows that

$$D = \tfrac{1}{6}\Gamma a^2 \tag{3.26b}$$

where Γ is the jump rate for diffusion (i.e., the total jump probability per unit time from an initial site to all other sites) and a is the average jump distance. Equation (3.26b) ignores correlation effects between successive jumps (see e.g., Howard and Lidiard 1964). If ionic conductivity and diffusion occur by the same mechanism, it is straightforward to show that μ and σ [Equation (3.22)] are related to D [Equation (3.24)] by the so-called Nernst–Einstein relationships,

$$\mu = \left(\frac{q}{k_B T}\right) D \tag{3.27a}$$

or, alternatively,

$$\sigma = \left(\frac{Nq^2}{k_B T}\right) D \tag{3.27b}$$

Expressions (3.27) follow by considering the steady state set up in an applied field, when the diffusion fluxes driven by the concentration gradient just balance the gradient enforced by the applied field [see, for example. Mott and Gurney (1948, page 63)]. Again, correlation effects of order unity are omitted from these expressions. Relations among μ, D, and σ are given in practical units in Table 3.4.

Care is required in the use of Equations (3.27). In alkali halides, for example, paired cation and anion vacancies do not contribute to σ but do contribute to D. Diffusion may not take place primarily in the bulk of the crystal; it can occur readily along grain boundaries (Section 3.7) or, although it is less well documented, along dislocations (so-called pipe diffusion). Indeed such short-circuit processes generally dominate below a few hundred degrees centigrade, when thermally generated defects are relatively rare.

In a crystal such as NaCl, with $E_g \sim 9$ eV (Figure 1.4), the contribution of thermally generated electron carriers at 1000 K is expected to give an electrical conductivity of only $\sim 10^{-4}\ \Omega^{-1}\ \text{cm}^{-1}$. If there were no ionic

Table 3.4 Relations among Ionic Conductivity, Mobility, and Diffusion[a]

$$\frac{\sigma\ (\Omega^{-1}\ \text{cm}^{-1})}{\mu\ (\text{cm}^2/\text{V sec})} = 1.602 \times 10^{-19} Nq$$

$$\frac{\mu\ (\text{cm}^2/\text{V sec})}{D\ (\text{cm}^2/\text{sec})} = 1.14 \times 10^4 T^{-1} q \quad \text{(Nernst–Einstein relation)}$$

$$\frac{\sigma\ (\Omega^{-1}\ \text{cm}^{-1})}{D\ (\text{cm}^2/\text{sec})} = 1.8 \times 10^{-15} Nq^2 T^{-1}$$

[a] N is the number of carriers of charge $q|e|$ per cubic centimeter; T is in degrees kelvin. Note that the SI unit of conductivity, S (the siemen), is identical with Ω^{-1} (or mho), so that $1\ \Omega^{-1}\ \text{cm}^{-1} = 1\ \text{S cm}^{-1} = 100\ \text{S m}^{-1}$.

defects in the crystal, the contribution of ionic motion to the electrical conductivity would be negligible. However, the measured electrical conductivity of NaCl at 1000 K is $\sim 0.5\,\Omega^{-1}\,\text{cm}^{-1}$, due largely to the motion of cation vacancies. An isolated cation vacancy will diffuse over $\sim 10^3$ lattice spacings per second at this temperature.

For a simple vacancy transport mechanism both σT and D obey relationships of the form

$$\sigma T = A \exp(-Q/k_B T) \tag{3.28}$$

$$D = D_0 \exp(-Q/k_B T) \tag{3.29}$$

where A and D_0 are constants and where $Q = h^F + h^M$; here h^F is the activation enthalpy for creation of a vacancy and h^M is the activation enthalpy for motion of the vacancy. Since both h^F and h^M are of order 1 eV for ionic solids, the exponential factor in Equations (3.28) and (3.29) varies rapidly with temperature. Indeed, for $Q \sim 1$ eV, the exponential factor varies by $\sim 10^{11}$ between 300 and 1000 K and diffusion via point defects is a very slow process at room temperature. As we remark below, the situation in real crystals is more complicated.

It is standard procedure in the study of ionic conductivity to plot $\ln(\sigma T)$ against T^{-1} [Equation (3.28)]. Details of such a plot for undoped NaCl are shown in Figure 3.2. Clearly this plot is curved over its entire length and fitting sections with straight lines is an approximation. Analysis of the data of Figure 3.2 requires a model of the imperfect crystal, which includes Schottky defects, divalent impurity ions present as trace impurities, divalent impurity–vacancy complexes, and vacancy pairs. Four stages are identified, with various levels of confidence. Stage I of Figure 3.2 is predominantly intrinsic conductivity, dominated by cation vacancies. At higher temperatures (stage I′) it seems that thermally induced anion disorder makes a noticeable contribution. The conductivity in stage II is extrinsic, dominated by cation vacancies arising from the presence of residual divalent cation impurities. In stage III association between the divalent impurities and the cation vacancies becomes important, the increased slope arising from the energy of association.

It is clear from Figure 3.2 that the apparent value of Q [Equation (3.28)] varies in a complicated way with temperature. It is not possible to separate Q into a motion component h^M and a formation component h^F from the temperature dependence alone, even in simple cases. One must control defect concentrations by other means, notably by selective doping. From the study of $\sigma(T)$ in both undoped and doped alkali halides, it is possible to derive values for the enthalpy of formation of Schottky pairs (h^F), the enthalpy of motion of cation vacancies (h_c^M) and

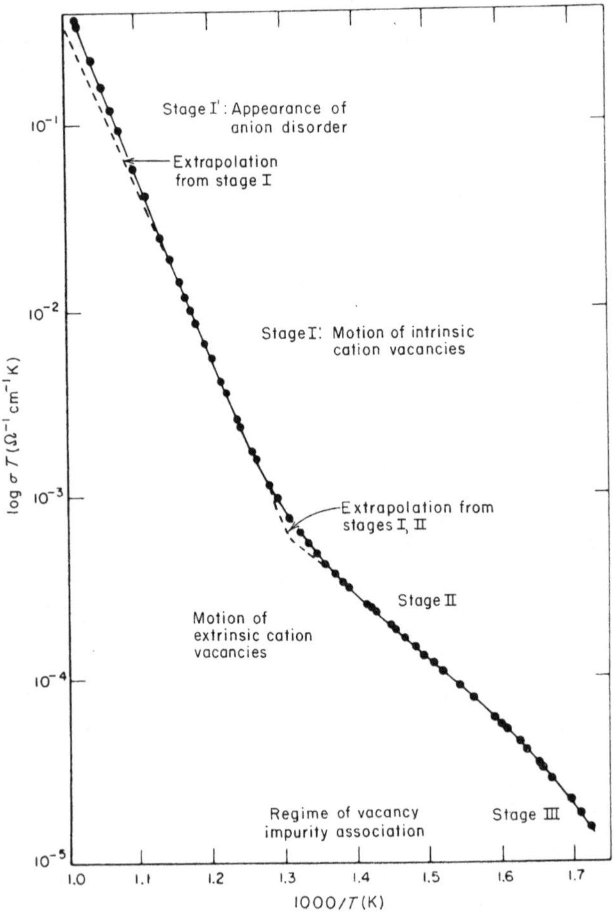

Figure 3.2 Temperature dependence of the ionic conductivity of NaCl. The dotted line shows straight-line fits to stages I and II [after Kirk and Pratt (1967)].

of anion vacancies (h_a^M), and the enthalpy of association of divalent cation impurities and cation vacancies (h^A). Some representative values are given in Table 3.5 and Figure 3.3.

The study of defect energetics in crystals with the fluorite structure is not as well established as in alkali halides. Mass transport studies have shown that in CaF_2 motion of Ca^{2+} is very much slower than that of F^-. It is generally agreed that at temperatures below the transition to the

Table 3.5 Enthalpies of Formation and Motion and Entropy of Formation of Point Defects in Ionic Crystals[a]

	Schottky Disorder					Anion Frenkel Disorder					
	LiF	NaCl	KCl	KBr	CsI	CaF_2	SrF_2	BaF_2	PbF_2	$SrCl_2$	
Enthalpy of Schottky pair formation (eV)	2.5	2.3	2.3	2.4	1.9	2.7	2.3	1.9	1.1 (obtained from theory; see section 3.5)	1.7	Enthalpy of Frenkel pair formation (eV)
Entropy of Schottky pair formation (k_B)	9.6	6	6.5	8.6	—	5.4					Entropy of Frenkel pair formation (k_B)
Cation vacancy motion enthalpy (eV)	0.7	0.7	0.7	0.6	0.6	~1.0	0.8	0.7			Anion interstitial motion enthalpy (eV)
Anion vacancy motion enthalpy (eV)	0.7	1.0	1.0	0.9	0.3	0.6	0.9	0.6			Anion vacancy motion enthalpy (eV)

[a] Values are taken from Flynn (1972) and Lidiard (1974).

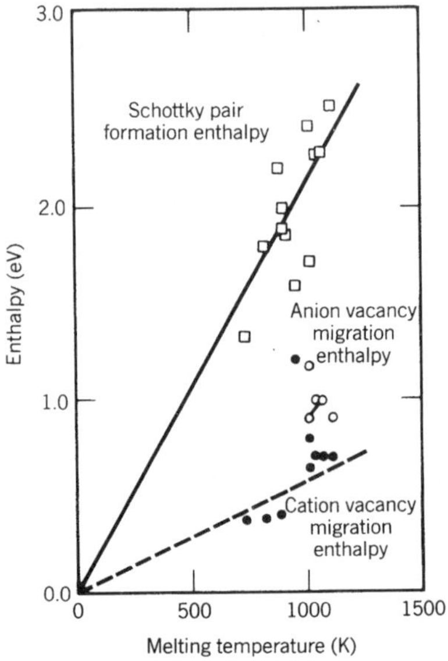

Figure 3.3 Phenomenology of defect enthalpies in ionic crystals. Schottky pair formation enthalpies and cation and anion migration enthalpies are shown as a function of host melting point [data from Süptitz and Teltow (1967) and from Barr and Lidiard (1970)]. (□, Schottky pair formation enthalpy; ○, anion vacancy migration enthalpy; ●, cation vacancy migration enthalpy.)

superionic state (Section 3.5) anion Frenkel defects are the dominant intrinsic defect. The defect content can be readily altered by doping with impurities, for example, oxygen, readily dissolves in F^- sites in CaF_2 as O^{2-} and can be charge compensated by a fluorine vacancy. Also, Ca^{2+} can be replaced by Na^+, charge compensated by a fluorine vacancy, or by Y^{3+}, charge compensated by a fluorine interstitial.

In alkaline-earth fluorides it is generally accepted that the fluorine interstitial moves through the crystal by an ion replacement mechanism, also referred to as an interstitialcy mechanism (Figure 3.4). In this case the interstitial displaces a nearest lattice anion into another interstitial site. However, lattice anions move even more readily through the crystal by interchange with anion vacancies. The energies of migration of vacancies and interstitials in alkaline-earth fluorides, typically about 1 eV, are reviewed by Corish et al. (1977). Values of the enthalpy of formation of anion Frenkel pairs for a variety of crystals with the fluorite structure are given in Table 3.5.

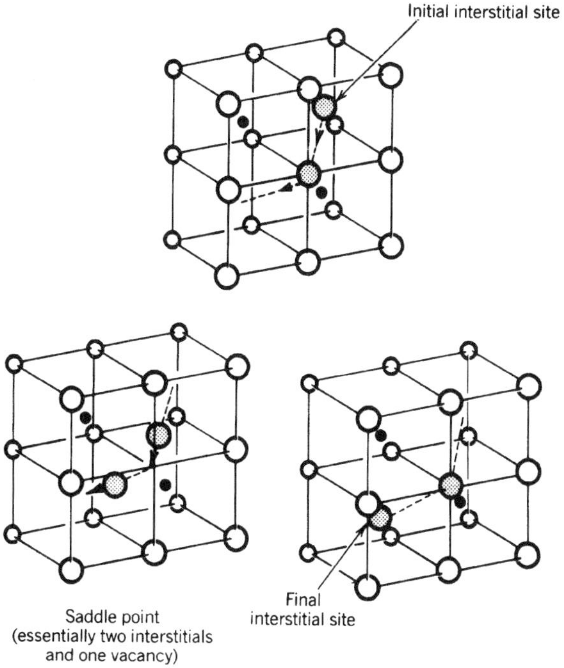

Figure 3.4 Interstitialcy mechanism of F^- motion in CaF_2. Note that the interstitial ion at the final stage is not the initial interstitial ion. The diffusion path is the broken line. Ca^{2+} ions are shown as solid circles; F^- ions are open circles, except those participating directly in the interstitialcy replacement sequence, which are shown as cross-hatched circles.

Although there are no experimental values for Schottky defects in fluorites, calculation suggests (see Section 3.4) that the energy required to produce a Schottky pair in CaF_2 is greater than 5 eV and to produce a cation Frenkel pair the energy is 6 eV or even larger. These large values suggest that such defects are not significant.

Little is known experimentally at the present time about the energetics of point-defect formation in stoichiometric oxides such as MgO and CaO (for a discussion of nonstoichiometric oxides see Section 3.3). It seems that the minimum energy of defect formation in such crystals is $\gtrsim 4$ eV and hence the concentration of intrinsic defects even at temperatures near the melting point is very much less than the impurity concentration (~ 100 ppm). In such circumstances the experimental study of defect energetics is extremely difficult.

3.2.2 Dielectric and Anelastic Relaxation of Defects

We have already pointed out that defect pairs are often formed from charged impurities and the intrinsic defects which compensate them. Examples such as (Y^{3+}–F_i^-) occur in CaF_2 doped with Y and (Ca^{2+}–cation vacancy) in KCl doped with Ca. Such pairs have a permanent electric dipole moment and their thermally induced reorientation can be biased by an applied electric field. This alignment contributes a term $P_d \sim n_d(\mu^2 E/k_B T)$ to the crystal polarization, where n_d is the concentration of defect pairs and μ is the electric dipole moment of the pair. Such defect pairs have elastic as well as electric dipole moments and can also be preferentially aligned by application of uniaxial stress to the crystal. The same is true of some impurity molecular ions, like $(CN)^-$ or $(OH)^-$. Other defects, for example, O_2^- substituting for Cl in KCl, have an elastic dipole moment but not an electric dipole moment.

If the polarizing electric field E is suddenly changed, the polarization evolves to a new equilibrium value with a relaxation time $\tau(T)$ determined by the frequencies of the various possible jumps reorienting the pair. A similar effect occurs with stress, but the relaxation time may be different. We can write this in terms of a susceptibility χ, which relates the response R of the polarization (electric or elastic) to the applied field F (electrical or stress). One part of the response is instantaneous, that is, $R_i = \chi_i F$. This corresponds to electronic polarization in an electric field or to the usual elastic strain under an applied stress. The rest of the response R_d relaxes toward a saturation value R_{ds}, that is, if $\delta R_d = R_d - R_{ds}$, we anticipate

$$\delta R_d = R_{ds} \exp(-t/\tau) \tag{3.30}$$

with a relaxation time $\tau(T)$. The saturation response, $R_{ds} = \chi_d F$, defines the other term in the susceptibility.

If we apply an oscillating electric (or stress) field, $F(\omega) = F_0 e^{i\omega t}$, the response $R(\omega)$ will lag behind the field, and this may be described in terms of a complex susceptibility $\chi(\omega) \equiv \chi'(\omega) - i\chi''(\omega)$. In terms of the time dependence of the response, we have

$$\frac{dR}{dt} = \frac{d}{dt}(R_i + R_d) = \frac{dR_i}{dt} - \frac{1}{\tau}(R_d - R_{ds}) \tag{3.31}$$

giving, for the frequency domain,

$$i\omega\chi = i\omega\chi_i - \frac{1}{\tau}(\chi - \chi_i - \chi_d) \tag{3.32}$$

Simplifying, we obtain the Debye relaxation expressions

$$\chi' = \chi_i + \frac{\chi_d}{1 + \omega^2\tau^2} \tag{3.33}$$

$$\chi'' = \frac{\chi_d \omega\tau}{1 + \omega^2\tau^2} \tag{3.34}$$

Investigation of this Debye-type response as a function of frequency allows a determination of $\tau(T)$. One convenient way of displaying and analyzing Equations (3.33) and (3.34) is to use the Cole–Cole plot of χ'' versus $\chi' - \chi_i$ for many frequencies. The equations are easily manipulated to show that the points should lie on a semicircle centered on ($\chi' = \frac{1}{2}\chi_d + \chi_i$, $\chi'' = 0$) and passing through the origin (Figure 3.5). Measurement of the temperature dependence of $\tau(T)$ gives a measure of the activation energy for relaxation E_R, through a relation of the form $\tau(T)^{-1} = \tau_0^{-1}\exp(-E_R/k_BT)$. Application of an oscillating stress field also gives rise to a Debye-type response, referred to as internal friction (Nowick and Heller 1963, 1965).

The energy absorption in the oscillating field is proportional to χ'', and is a maximum at frequency $\omega = 1/\tau$. This is called nonresonant absorption, that is, an absorption determined by a rate (e.g., a jump rate) rather than an energy difference, δ, which gives resonant absorption when $\hbar\omega = \delta$. Both cases can be seen in some systems; one can show that

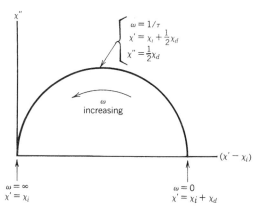

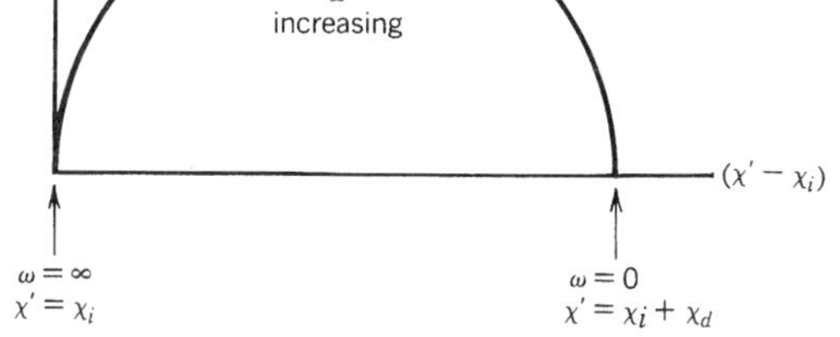

Figure 3.5 Cole–Cole plot showing the relation between the real ($\chi' - \chi_i$) and imaginary (χ'') parts of the susceptibility for a Debye-type system at various frequencies (see text).

the frequency for maximum absorption moves from $1/\tau$ to $\delta/\hbar$ as $\omega\tau$ increases (Fröhlich 1958; Stoneham 1975, page 504).

In general, the reorientation of a defect pair will involve a number of distinct relaxation modes with different relaxation times τ_i. These modes can be labeled by symmetry, rather like vibrational modes of molecules. One can make general statements based on symmetry arguments about how defects will move or reorient as a particular electric field or stress component is applied, and this defines the types of relaxation mode that occur.

The simplest case is a defect with two states only, parallel and antiparallel to the applied field. The rate equations describing the occupancies n_i of the two states are

$$\frac{dn_1}{dt} = -\omega_{12}n_1 + \omega_{21}n_2 \tag{3.35a}$$

$$\frac{dn_2}{dt} = \omega_{21}n_2 - \omega_{12}n_1 \tag{3.35b}$$

where the total number $(n_1 + n_2)$ is constant. Combining these equations we find

$$\frac{d}{dt}(n_1 - n_2) = -(\omega_{12} + \omega_{21})(n_1 - n_2) \tag{3.36}$$

with a single relaxation time $(\omega_{12} + \omega_{21})$.

Suppose we now have a three-state system, for example, when the dipole also has a state normal to the field. There will then be two relaxation constants in general, one electrically active and one elastically active. In simple cases [e.g., Lewis and Stoneham (1967)] these two constants may describe, for example, the recovery of the population difference of the parallel and antiparallel states alone $(n_1 - n_2)$, and the recovery of the population of the normal state relative to the parallel and antiparallel states $[n_0 - \frac{1}{2}(n_1 + n_2)]$.

Such analyses can be done systematically, and we summarize these for defects with trigonal, tetragonal, or orthorhombic symmetry in cubic crystals in Table 3.6. It is apparent that a tetragonal defect such as Y^{3+}-F_i^- in CaF_2 may have a triply degenerate T_{1u} relaxation mode that will couple to an electric field, a doubly degenerate E_g mode that will couple to uniaxial elastic stress applied in ⟨100⟩ or ⟨110⟩ directions (but not ⟨111⟩), and a nondegenerate A_{1g} mode that will couple to hydrostatic stress.

Table 3.6 Coupling of External Fields to Defects in Cubic Crystals[a]

		Possible for Defects of Symmetry			Symmetry Coordinates	
Mode Symmetry	Degeneracy	Tetragonal	Orthorhombic ⟨110⟩	Trigonal	Electric Field	Stress, σ
A_{1g}	1	Yes	Yes	Yes	None	$\sigma_{xx} + \sigma_{yy} + \sigma_{zz}$
E_g	2	Yes	Yes	No	None	$(2\sigma_{xx} - \sigma_{yy} - \sigma_{zz}, \sigma_{yy} - \sigma_{zz})$
T_{2g}	3	No	Yes	Yes	None	$(\sigma_{yz} + \sigma_{zy}, \sigma_{zx} + \sigma_{xz}, \sigma_{xy} + \sigma_{yx})$
T_{1u}	3	Yes	Yes	Yes	(E_x, E_y, E_z)	None
A_{2u}	1	No	No	Yes	None	None
T_{2u}	3	No	Yes	No	None	None

[a]The possible normal modes, their symmetry type, and the symmetrized combinations of field components to which they couple are given.

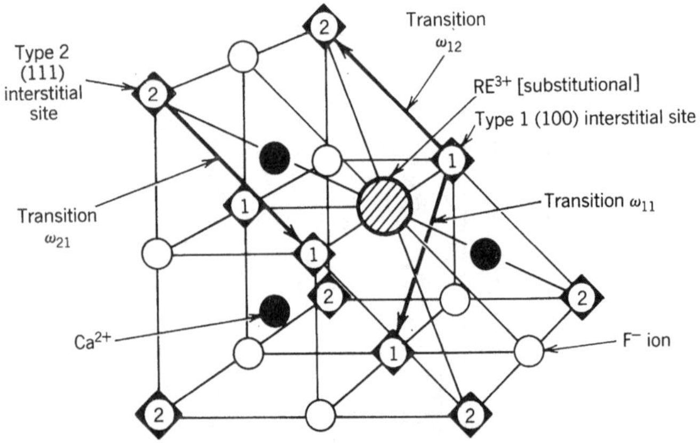

Figure 3.6 Jumps of an F^- interstitial bound to a substitutional trivalent rare-earth ion (cross-hatched circle) in CaF_2. Calciums are labeled with solid circles and lattice fluorines are labeled with open circles. Two types of fluorine interstitial sites are considered [labeled (1) (the nearest ⟨100⟩ site) and (2) (the nearest ⟨111⟩ site)]. The three transition rates ω_{11}, ω_{12}, ω_{21}, involving the two interstitial sites 1 and 2, are shown.

Pursuing the example of Y^{3+}-F_i^- in CaF_2, and assuming that the Y^{3+} does not move, we see (Figure 3.6) that there are six equivalent nearest-neighbor interstitial sites denoted by the subscript 1 and eight equivalent next-nearest-neighbor interstitial sites denoted by the subscript 2. The jump frequency of F_i^- from an interstitial site i to interstitial site j is denoted by ω_{ij}. The relaxation rates (τ^{-1}) of the various relaxation modes are given in Table 3.7. The A_1 mode, for example, readjusts the balance between the inner (1) and outer (2) shells of interstitial sites. The x component of the T_{1u} mode (see Table 3.6) readjusts the relative fractions of interstitials in the positive and negative x directions from the Y^{3+} ion, and so on.

Some specific assumptions are usually made about the rates. Here, for instance, it has been assumed that ω_{22} is too slow to contribute to τ^{-1}. When more than one mode contributes to either dielectric or anelastic relaxation, a detailed kinetic analysis is required to assign the relative strengths of these modes to the relaxation.

In practice, it is important to subtract extraneous contributions. The extra contributions from dislocations to mechanical relaxation and of carriers or diffusing point defects to dielectric loss usually give distinctive frequency dependences, so that a clear separation is possible.

Measurements of dielectric relaxation can also be made in ionic crystals by the method of ionic thermocurrents (ITC). Here the defects

Table 3.7 Relaxation Rates of Dipoles in the Fluorite Structure[a] for Different Modes of Relaxation

Symmetry of Relaxation Mode	Degeneracy of Relaxation Mode	Relaxation Rate τ^{-1} in Units of ω_{12} ($x \equiv \omega_{21}/\omega_{12}$)
A_{1g}	1	$4+3x$
A_{2u}	1	$3x$
E_g	2	$4+6x$
T_{1u}	3	$2+\frac{7}{2}x \pm \sqrt{2+\frac{1}{2}x}$
T_{2g}	3	$3x$

[a]See Figure 3.6.

are polarized at a high temperature and the polarization is frozen in by cooling to a temperature so low that reorientation cannot occur. The electric field is then removed and the crystal is slowly warmed, giving a current pulse when the dipoles reorient. This technique has the advantage that the conductivity background is negligible. Figure 3.7 shows ITC peaks measured in $CaF_2{:}Er^{3+}$ by (a) polarizing with an applied

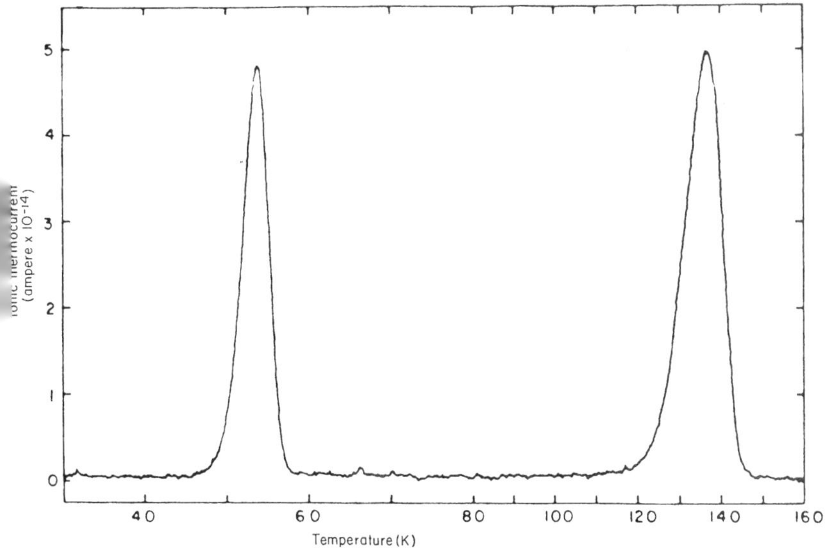

Figure 3.7 Ionic thermocurrent (ITC) spectrum of CaF_2:Er. The heating rate is 0.1 K/sec below 80 K and 0.2 K/sec above 80 K [after Stott and Crawford (1971)].

electric field of 500 V/cm for 5 min at 140 K, (b) cooling in the electric field to 12 K, and (c) warming at the rate of 0.105 K sec from 12 to 80 K and at the rate 0.203 K sec above 80 K. Two well-defined peaks are found at 53.6 and 137.9 K. The 137.9 K peak is assigned to reorientation of ($Er^{3+} - F_i^-$) pairs (cf. Figure 3.6), and detailed analysis of the lineshape gives $\tau_0 = 2 \pm 0.3 \times 10^{-13}$ sec and $E_R = 0.385 \pm 0.008$ eV. The origin of the 53.6 K peak is uncertain. Since the Er^{3-} ion is paramagnetic, the technique of paramagnetic resonance (Section 4.5) can also be used very effectively to study the structure of complexes of the type Er^{3+}-F_i^-.

The Debye model presumes that the response falls exponentially

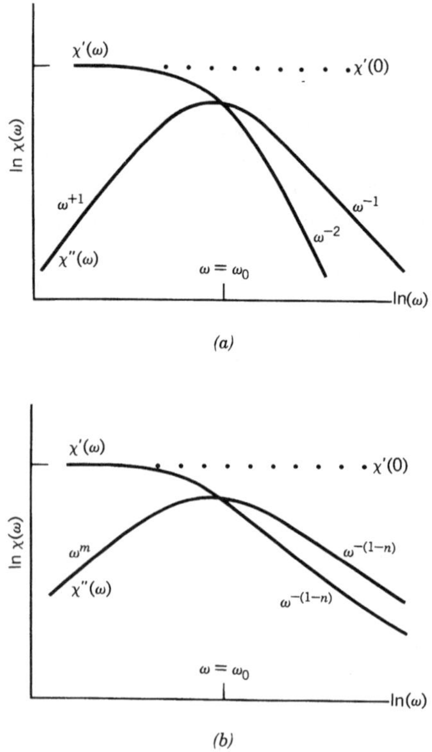

Figure 3.8 Universal dielectric behavior of $\chi(\omega)$. The Debye model (*a*) is compared here with (*b*) the more general behavior observed in many systems. The diagrams are drawn so that the maximum absorption occurs at the same frequency, ω_0, in both cases.

toward its saturation value. This leads to characteristic frequency dependencies for $\omega \to 0$ ($\chi' \sim \omega^0$, $\chi'' \sim \omega^1$) and $\omega \to \infty$ ($\chi' \sim \omega^{-2}$, $\chi'' \sim \omega^{-1}$). However, nature is inevitably more complicated. When accurate measurements are made over very wide ranges of frequency, somewhat different features are found in many systems (Jonscher 1977, 1981) (Figure 3.8). At low frequencies the Debye form must be generalized ($\chi' \sim \omega^0$, $\chi'' \sim \omega^m$, where $m = 1$ is the Debye case). At high frequencies, however, the different behaviors of χ' and χ'' predicted in the Debye model are not found; a better fit is $\chi' - \chi'' \sim \omega^{-(n-1)}$. The reasons for this very general behavior are not fully agreed, but are clearly of interest. The practical implications are significant, because measurements of conductivity versus temperature are often made at a single frequency, and it is important to know that this frequency is suitable over the entire temperature range.

3.2.3 Dimensional Changes Caused by Defects

When defects are created by heat or by irradiation, or are incorporated by doping, there are changes of density, of total volume, and of lattice parameter. Such changes are important practically because they can lead to modification of mechanical properties. This subject has been studied in detail for materials that are heavily irradiated, for example, nuclear fuels (like UO_2) or insulators for possible use in fusion reactors.

These dimensional changes can be measured and exploited scientifically in several ways (see e.g., Eshelby 1955). First, the density will show whether a dopant enters interstitially, as a replacement for a host ion, or whether it displaces a group of ions to the surface. Clearly an interstitial atom will increase the density, however light it is, whereas a substitutional atom may reduce the density if it replaces a heavier host atom. Second, if a Schottky defect is produced, we can see two terms in the volume change, ΔV_1, arising because an atom has been transferred to the surface, increasing the volume enclosed by the perimeter, and ΔV_2, because there are internal strains caused by the defect. Thus, if we measure the change in volume from the density change, we get $\Delta V = \Delta V_1 + \Delta V_2$. If we use X-ray diffraction to measure the altered lattice parameter, we obtain merely ΔV_2. The difference, ΔV_1, is basically the atomic volume Ω. From the difference (exploiting simple assumptions like statistical uniformity for the defect distribution) the number of Schottky defects is given by

$$n = \text{Number of vacancies per unit volume}$$

$$= \frac{\text{Volume change from density data} - \text{Volume change from X-ray data}}{\text{Atomic volume}}$$

This can be rearranged into a more familiar form by noting that the change in linear dimension, Δl, is related to the density change, $\Delta \rho$, by

$$\Delta \rho/\rho = -3\Delta l/l \tag{3.37}$$

Hence, if we write the fractional change in lattice parameter as $\Delta a/a$, we see that

$$n = 3(\Delta l/l - \Delta a/a) \tag{3.38}$$

Very precise measurements of both terms on the right-hand side of Equation (3.38) are needed to give reliable results, since both are of order 10^{-3} or less.

This approach is especially useful in rare-gas solids and metals, where there are no ionic charges, so ionic conductivity gives no information directly. From a study of the temperature dependence of n, an activation energy for vacancy formation is found, giving 0.06 eV for Ar, 0.077 eV for Kr, and 0.1 eV for Xe. However, it is not always possible to use this approach unless new lattice sites are created. As shown in Figure 3.9,

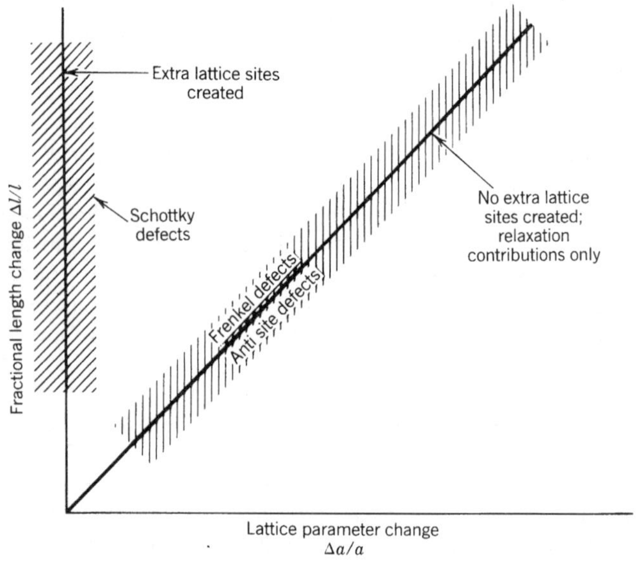

Figure 3.9 Relative values of overall length and lattice parameter change for different defect types. When no extra lattice sites are created (e.g., Frenkel defects; antisite defects), both lattice parameter change $\Delta a/a$ and length change $\Delta l/l = \frac{1}{3}\Delta V/V$ are comparable. When extra sites are created (e.g., Schottky defects), there is an extra expansion that dominates so that $\Delta l/l$ is enhanced above $\Delta a/a$.

Frenkel defects behave differently and do not contribute to $\Delta l/l - \Delta a/a$.

It is usual to characterize defect strain fields by an elastic dipole tensor P_{ij} which we write in terms of defect forces. Defect forces are those constant forces $\mathbf{F}(\mathbf{R}_I)$ which, if applied to the perfect lattice at site $\mathbf{R}_I$ would give the same ionic displacements as the defect itself. The elastic dipole tensor is then given by $P_{ij} = \sum_I F_i(R_I)R_{Ij}$. This tensor can be determined in several ways. One is from the lattice relaxation contribution to $\Delta l/l$ or $\Delta a/a$. Formally, the formation volume of a defect is the pressure derivative of the formation free energy dE_f/dp. For a cubic, perfectly harmonic, solid containing a defect with at least cubic symmetry, one finds a relaxation volume ΔV_{rel}, which is the change in volume when the ions are allowed to relax from perfect lattice sites:

$$\Delta V_{\text{rel}} = 3G/(c_{11} + 2c_{12}) \tag{3.39}$$

where $G = P_{xx} = P_{yy} = P_{zz}$ and c_{11} and c_{12} are elastic constants (Section 2.3.1). Results from Equation (3.39) can mislead unless they include an important term from the pressure dependence of the elastic and dielectric constants of the host. This may even change the sign of ΔV_{rel} [see, e.g., Gillan (1981)].

X-ray measurements of the defective crystal show both a shift of Bragg peaks, corresponding to $\Delta a/a$, and a diffuse scatter component between the peaks (see also Section 3.5.1). When there are several different defects i, the shift of the peak is proportional to $\sum_i n_i P_i$, where the diffuse scatter allows one to measure $\sum_i n_i P_i^2$, so that careful experiments enable one to distinguish and assess contributions of several different types of defect present simultaneously [see Peisl (1975)]. In fact, measurements of the strain broadening of sharp epr and nmr lines, or optical absorption and luminescence lines, allow one to monitor random internal strains, from which values of the dipole tensor may also be obtained (Stoneham 1969).

When the defect has lower symmetry than the local crystal symmetry, it can be aligned by stress. Measurements of the anisotropic changes in dimension caused by preferential alignment permit estimates of other components of the dipole tensor. Aligning tetragonal defects in a cubic host gives $2P_{zz} - P_{xx} - P_{yy}$, for example. These anisotropic components are the defect strengths that determine intensities in internal friction (Section 3.2.2).

3.3 NONSTOICHIOMETRY

Most of the materials we have dealt with so far have a well-defined phase with a chemical composition precisely reproducible, for example, KCl, and are referred to as stoichiometric. However, compounds exist that are characterized by a single phase over a range of chemical composition, and these are referred to as nonstoichiometric. Examples are wüstite, $Fe_{1-x}O$, where there is cation deficiency indicated by x, or uranium oxide, UO_{2+x}, where there is anion excess indicated by x. The important feature is that there is a continuous range of compositions without gross structural change. This is rationalized in terms of lattice defects, which remain disordered to some degree; there may be short-range order into well-defined structural units without a fully ordered new phase developing. The range of composition is limited at large defect concentrations by the formation of a separate phase (e.g., Fe_3O_4, corresponding to $x = 0.25$, or U_4O_9, corresponding to $x = 0.25$). At low defect levels, there is a mere qualitative and arbitrary distinction between nominally stoichiometric and nonstoichiometric crystals. Examples of typical defect concentrations are shown in Figure 3.10.

Experimentally, a range of techniques can be used to establish the

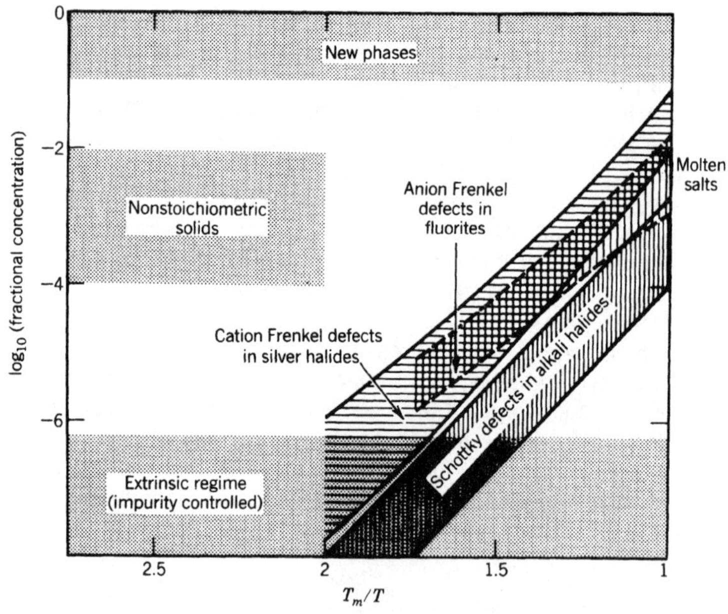

Figure 3.10 Fractional defect concentrations in thermal equilibrium. Typical values for ionic crystals are compared here with those for nonstoichiometric systems. T_m is the melting temperature.

structure and composition of nonstoichiometric crystals. Some, like thermodynamic properties (specific heat, vapor pressure) or magnetic properties, are simple but give little guidance on structure on an atomic scale. More direct information about actual defect distributions may be obtained from diffuse scattering studies, mentioned in Section 3.5 [see Fender (1973) for neutrons and Guinier (1963) for X-rays]. The electron microscope shows up contrast features reflecting local inhomogeneities in the crystal, such as dislocations. The electron microscope has the additional advantage that it can show the existence of two phases intergrown in a random way, a situation that would almost certainly lead to confusion in an X-ray or neutron-scattering study. In general, neutron and x-ray-scattering techniques are not sufficiently sensitive to readily study defect concentrations of $\lesssim 1$ mole%.

Our present brief account of nonstoichiometry must be based on a few examples [for further examples see, e.g., Greenwood (1968) and Sørensen (1981)]. Several general features are found. First, the low formation enthalpies needed to give high defect concentrations are often achieved by clustering. Defects may aggregate into point-defect clusters, often based on a simple building block, or they may evolve into planar or three-dimensional superstructures, which might be regarded as totally new phases. Second, many systems of interest require that there should be species present with variable valence (e.g., Fe^{2+} and Fe^{3+} readily occur) so that charge compensation is easily handled despite the asymmetry between two sublattices. Third, a high dielectric constant can help, partly by screening electrostatic interactions and partly because a large ionic polarizability (as in TiO_{2-x}) may indicate that there is an incipient phase-change instability that can contribute to the energy reduction. Of course, one must not forget the other constraints. In particular, the nonstoichiometric crystal, full of defects, should still be stable against disproportionation. Thus $Fe_{1-x}O$ should not simply form a mixture of FeO and Fe_3O_4. We now proceed to examples of the several forms.

3.3.1 Materials with the Fluorite Structure

Oxides with the fluorite structure show three important types of behavior. Some are anion deficient, that is, MO_{2-x}, suggesting that both M^{4+} and M^{3+} charge states are present. Others show an oxygen excess, that is, MO_{2+x}, suggesting that M^{4+} and M^{5+} (or O^{2-} and O^-) are present. A third class are the doped oxides, notably $M_{1-y}M'_yO_{2-x}$, where a modest fractional concentration y of a second cation with a smaller charge is specifically introduced.

Rare-earth ions generally occur in the tripositive valence state in compounds. Other valence states occur, however, generally with the especially stable $4f^0$ (Ce^{4+}), $4f^7$ (Eu^{2+}, Tb^{4+}), or $4f^{14}$ (Yb^{2+}) configurations. The oxides of those rare earths that readily form the +4 valence state, for example, Ce, Pr, and Tb, exhibit wide ranges of stoichiometry at high temperatures, associated with the presence of two valence states. For $T > 700°C$, cerium dioxide can be reduced to the nonstoichiometric anion-deficient form CeO_{2-x}. This has a defect fluorite structure down to $x = 0.28$. In this structure, one oxygen vacancy compensates two Ce^{4+} ions reduced to Ce^{3+}. Similarly, above 570°C there is a nonstoichiometric phase of praseodymium oxide (PrO_{2-x}) within the range $0 \lesssim x \lesssim 0.25$. Below 570°C ordering of oxygen vacancies develops, with the formation of a homologous series of phases Pr_nO_{2n-2} with $n = 7$, 9, 10, 11, 12 (effectively $x = 2/n$ here).

Solid solutions formed by dioxides of Zr, Hf, Ce, Th, and U with oxides of trivalent ions such as Y_2O_3 and oxides of divalent ions such as CaO also become anion deficient. At room temperature ZrO_2 has a monoclinic crystal structure, which changes to a tetragonal form above ~1200°C and to the cubic fluorite structure above ~2300°C. However, addition of ~15 mole% of CaO to ZrO_2 stabilizes the fluorite structure at lower temperatures, each calcium ion being charge compensated by an oxygen vacancy. The high vacancy concentration that results leads to a high oxygen mobility, which may be exploited in gas sensors and other ceramic devices (see Section 3.5.2). The phase diagram for ZrO_2:CaO has been studied by many authors; results conflict, but it appears to be generally accepted that the cubic phase is stable only above ~900°C.

Anion excess compounds with the fluorite structure also occur. However, only UO_{2+x} shows this behavior among the binary oxides. The phase diagram of the U–O system has been studied in detail. Above 375°C UO_2 can accommodate further oxygen atoms to a limit of $UO_{2.20}$ ($x = 0.20$) at 1400°C. Neutron-diffraction studies by Willis (1963) on $UO_{2.12}$ at 800°C suggested that the extra oxygen atoms tend to interact forming the so-called 2:2:2 clusters (Figure 3.11). In this complex two of the extra oxygens are in interstitial sites displaced from the center of a cube edge in $\langle 110 \rangle$ directions and, in addition, two oxygen atoms on normal lattice sites relax in $\langle 111 \rangle$ directions, creating two oxygen vacancies. It is possible that the excess oxygen is charge compensated by U^{5+} ions, although there is no definite proof of this. There is also evidence for a phase with composition U_4O_9, which appears to be an ordered derivative of UO_{2+x}. One puzzling aspect of the postulated 2:2:2 model is the short distance (2.1 Å) between the centers of the $\langle 110 \rangle$ displaced interstitials compared with the diameter of O^{2-} (2.8 Å). The UO_2 system has been investigated theoretically by Catlow (1977) with considerable success.

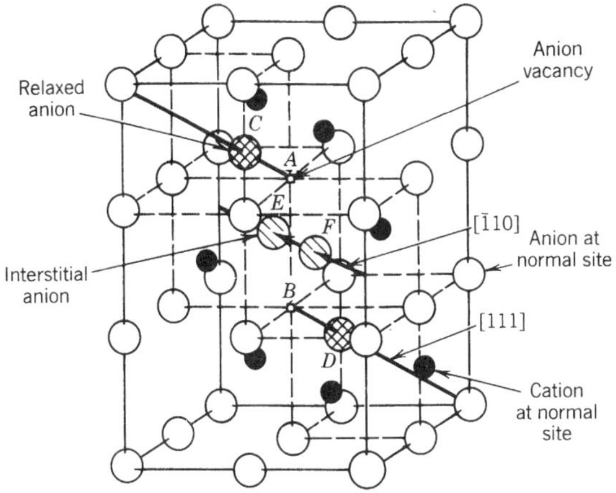

Figure 3.11 2:2:2 cluster in the fluorite structure, proposed for short-range order of anion defects. The notation indicates (number of true anion interstitials *E, F*):(number of anions with substantial displacements *C, D*):(number of anion vacancies *A, B*). The large open circles are the normal anions and the small solid circles are the cations on normal sites.

The fluorites CaF_2, SrF_2, and BaF_2 also form anion-excess phases with trivalent rare-earth fluorides and YF_3. The resulting phase may be presented as MF_{2+x}, where the additional fluorine ions provide charge recompensation for RE^{3+} ions substituting for M^{2+}. When x is small, the extra F^- ions are present in the crystal close to the center of empty fluorine cubes. However, when $x \sim 0.2$, the excess fluorine ions appear to form clusters of the 2:2:2 type, and even larger clusters.

3.3.2 Materials with the Sodium Chloride Structure

Nonstoichiometric materials with the sodium chloride structure tend to be metallic, for example, carbides, nitrides, and monoxides of Ti, Zr, Hf, V, Nb, and Ta. Only a few nonmetallic materials have been identified, the best known being $Fe_{1-x}O$, and the analogous oxides of Mn, Co, and Ni. $Fe_{1-x}O$ is stable in equilibrium conditions only above 550°C; at lower temperatures it disproportionates into iron and Fe_3O_4. Neutron-diffraction studies indicated that the number of iron vacancies in the metal lattice was greater than expected from chemical composition, and that the atoms removed to create the extra vacancies were located in tetrahedral interstitial sites. It is generally agreed (Koch and Cohen 1969;

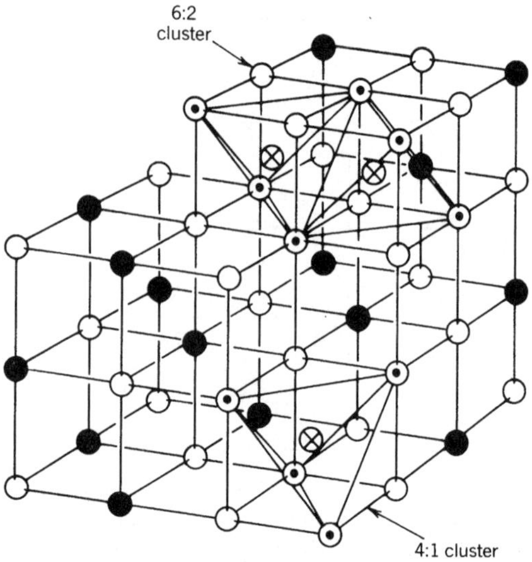

Figure 3.12 4:1 and 6:2 clusters in oxides, $M_{1-x}O$, with the NaCl structure. Here ⊗ represents M^{3+}, ○ represents M^{2+}, ⊙ is an M^{2+} vacancy, and ○ represents O^{2-}. The notation indicates (number of M^{2+} vacancies):(number of M^{3+} interstitials). Note that the 6:2 cluster corresponds to two edge-sharing 4:1 clusters.

Catlow and Fender 1975) that the iron interstitials, assumed to be in the form Fe^{3+}, and the iron vacancies form clusters based on the so-called 4:1 cluster of four vacancies and one interstitial (Figure 3.12). Clearly, if 4:1 clusters aggregate, for example, to give 6:2 or 8:3 clusters by edge sharing, the ratio of vacancies $(x + y)$ to interstitials (y) gives a measure of cluster size. Both experimental and computational studies suggest that edge-sharing tetrahedra are the most likely configuration. One can predict (Catlow and Stoneham 1981) the proportions of vacancies and clusters as a function of temperature and nonstoichiometry (Figure 3.13).

3.3.3 Shear Structures

It is possible in some materials to eliminate point defects almost entirely, forming planar defects and, in effect, generating an intermediate phase with a structure related to that of the parent phase. This situation is commonly found in transition-metal oxides with the rhenium oxide (ReO_3; see Figure 8.25) and rutile (TiO_2) structures, in which MO_6 octahedra share corners and edges, respectively. As oxygen is removed,

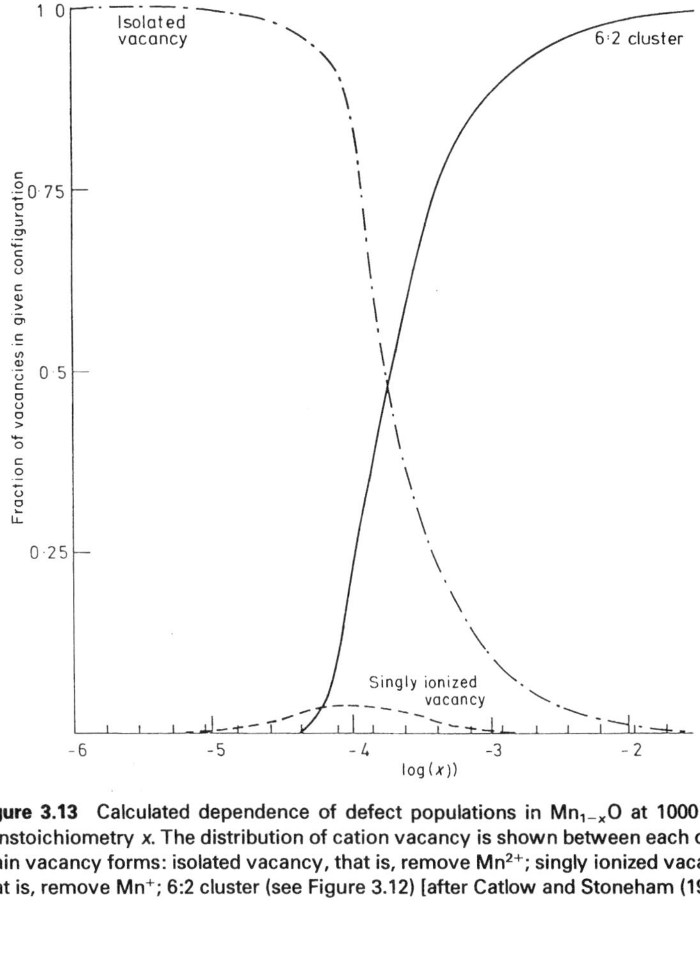

Figure 3.13 Calculated dependence of defect populations in $Mn_{1-x}O$ at 1000 K on nonstoichiometry *x*. The distribution of cation vacancy is shown between each of the main vacancy forms: isolated vacancy, that is, remove Mn^{2+}; singly ionized vacancy, that is, remove Mn^{+}; 6:2 cluster (see Figure 3.12) [after Catlow and Stoneham (1981)].

the structure can adjust by changing the linking of one plane of octahedra to another; the basic MO_6 octahedral units, with their strong Coulomb bonding, remain intact. A simple example (not found naturally) is shown in Figure 3.14. Many cases are known, notably nonmetallic oxides of Ti, V, Nb, Ta, Cr, Mo, and W.

Molybdenum oxide, MoO_{3-x}, based on the rhenium oxide structure, has a series of intermediate compounds with the general formula Mo_nO_{3n-m}. These structures contain slabs of the perfect ReO_3-type structure, separated by crystallographic shear planes of MoO_6 octahedra-

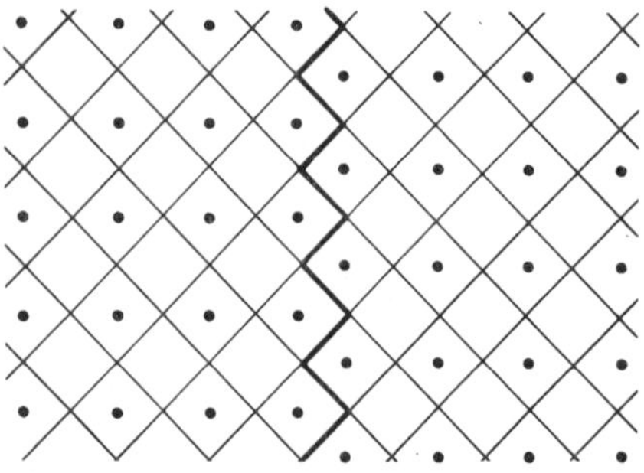

Figure 3.14 Idealized shear plane in the ReO_3 structure. The shear plane, idealized by the thicker line, runs normal to the page. The oxygens lie at the vertices of the squares, and the cations are shown as solid circles.

sharing edges (Figure 3.15). In this way the oxygen vacancies are eliminated, and the repetition of the shear planes at regular intervals gives rise to ordered intermediate structures.

Rutile (TiO_{2-x}) also shows nonstoichiometry by forming shear planes. In this case there are two extra interesting features (Bursill and Hyde 1971). One is that the shear planes tend to order at specific spacings. The other is that, as x increases, the shear (in fact, it is a more complex reconstruction that just shear) changes to a different crystallographic plane. Both these features result from interactions, principally elastic, between the planes (Stoneham and Durham 1973; Catlow 1982). This explains both the ordering and the observation that, once the interaction energies are comparable with formation energies, there is a switch of plane from one orientation to another.

If the ordering were always complete, giving uniform plane–plane spacings (though dependent on x), we should have well-defined phases at each stage rather than nonstoichiometry. The formation of irregularly spaced crystallographic shear planes provides a mechanism by which compounds may deviate from any ideal composition, so that shear structures are best regarded as a form of nonstoichiometry [see Anderson (1970a, b) for a discussion of this and related points].

We note finally that several systems (ZnS and SiC being the best known) show both disorder and various degrees of order in the stacking sequence of the crystal planes. This does not involve point defects, nor an excess of one or other species, and the formation of these polytypes is quite distinct from shear planes (see Figure 3.29).

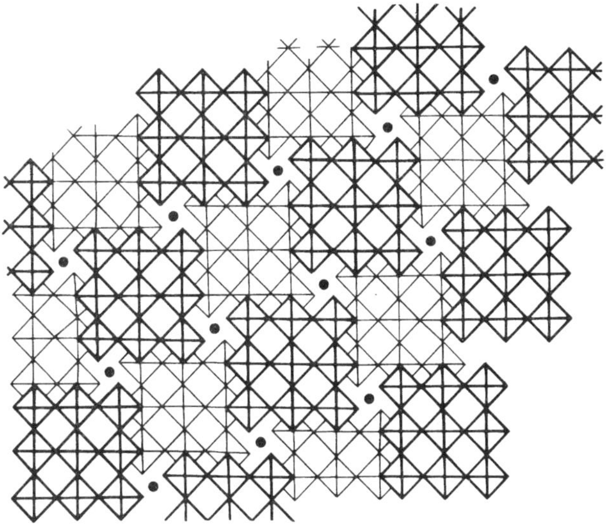

Figure 3.15 Block structure in a nonstoichiometric oxide. The figure shows the idealized structure of PNb_9O_{25} [after Cheetham and Taylor (1977)]. The diamond shapes are NbO_6 octahedra; those with thicker outlines lie above those with thinner outlines. Solid circles represent tetrahedral tunnel sites.

3.4 CALCULATION OF DEFECT ENERGIES IN IONIC SOLIDS

In ionic crystals there is a long history of successful quantitative calculation of defect energies. As a result, theory and experiment have often combined in interpretation, in analysis of complex situations, and in extrapolation to extreme conditions. We discuss here examples of the way in which this is done.

Most calculations involve three ingradients. First, an ionic model is assumed, usually the shell model developed for lattice dynamics (Section 2.2), and most of the systems we consider are ionic, either halides or oxides. The shell model avoids some of the problems, notably unphysical instabilities, found with simpler models of polarizable point ions. Suitably generalized, this picture can be extended to covalent systems, although this raises several complications (Stoneham and Harding 1982; Catlow and Stoneham 1983). The second ingredient is an interatomic potential, also discussed in Section 2.2. While a shell model describes the harmonic properties of a crystal at small displacements, more is needed to describe the larger displacements near defects and the interactions of defects with impurities. Third, an efficient computer code is required to use these models in predicting observable defect energies and entropies. In this section we discuss the third component, and its degree of success [for general discussion see Lidiard and Norgett (1972) and Catlow and Mackrodt (1982)]. We shall only consider closed-shell systems here, or at least cases where the electronic structure can be built into an interatomic potential, so that one can avoid explicit solution of the Schrödinger equation.

There are two main types of calculation of defect properties, with complementary strengths and weaknesses:

1. Static methods, in which the total potential energy of the defect in its host is minimized. This approach corresponds to finding the internal energy for chosen defect states and is especially useful for low-temperature properties. The energies which are measured experimentally then correspond to differences between calculated energies. Formation energies or energies of solution, for example, are obtained by subtracting the energy with a defect present from that in a defect-free reference state. Activation energies are found by subtracting the energy in its usual site from that in the saddle point, that is, the point of the lowest-energy diffusion-jump trajectory where the energy is a maximum. Static methods are often supplemented by two other types of calculation:

(i) *Perturbations*. Thus, to obtain a defect volume, for example, the change of energy with lattice parameter is needed. One evaluates the defect formation energy, E_F, for several lattice spacings and obtains the defect formation volume, ΔV_F, from

$$\Delta V_F = -K_T(\partial E_F/\partial \ln V)_T \equiv \delta E/\delta p \tag{3.40}$$

where K_T is the isothermal compressibility.

(ii) *Harmonic Vibrations*. The difference between the defect vibrational entropy of two different defect states, for example, paired and separated impurities, is given by

$$S_1 - S_2 = -k \ln\left(\prod_i \omega_{1i} \Big/ \prod_j \omega_{2j}\right) \tag{3.41}$$

Thus one has to consider harmonic vibrations about the equilibrium geometry in each of the states 1 and 2. The products in Equation (3.41) are over all vibrational modes of the defect lattice. This evaluation is straightforward in principle when one knows the force-constant matrix, giving the second derivative of the energy with respect to displacements from equilibrium. This matrix is closely related to the dynamical matrix of Section 2.2. Local mode frequencies (Section 2.4) can be evaluated from this force-constant matrix.

2. Dynamic methods, in which the time-dependent positions and momenta of the ions are obtained by solving Newton's equations of motion numerically. This technique of molecular dynamics involves the calculation of ion trajectories and the analysis of these trajectories to extract properties related to experiment. This approach is particularly useful for high-temperature properties.

Before we compare the strengths of these two methods, it is useful to describe some of the important technical ideas. The first general point concerns the handling of the host lattice atoms far from the defect, and three main methods are used:

1. Rigid boundary conditions are applied in which ions outside a given region are assumed immobile and unpolarizable. Occasionally one finds a variant which uses a finite crystallite with a free surface.
2. Periodic boundary conditions are applied in which one considers a regular array of defects. Using this convenient artifice we can consider the lattice as consisting of large supercells, each containing one defect. We may then exploit the periodicity which maintains that the displacement $\mathbf{u}(\mathbf{r}_i)$ of the ith ion in any supercell N is independent of N. Sometimes periodic boundary conditions are assumed explicitly, as is usual in molecular dynamics, or in large-unit-cell calculations of harmonic vibrations following static calculations. At other times the assumption of periodic boundary conditions is implicit, as in the Kanzaki method (Stoneham 1975, Chap. 8) in which displacements of selected wavelengths (or wavevectors) are sampled on a mesh in the Brillouin zone. This, it is readily verified, is exactly equivalent to a periodic boundary condition. Indeed, when one encounters real periodic defect arrays (e.g., shear planes or void lattices) the Kanzaki approach can be made exact.
3. Mott–Littleton boundary conditions are applied in which the central region containing the defect is matched beyond a certain point onto a dielectric elastic continuum. Technically, the solid is divided into an inner region I and an outer region II. Thus the total energy can be written as the sum of three terms:

$$E = E_{\mathrm{I}}(\mathbf{X}_{\mathrm{I}}) + E_X(\mathbf{X}_{\mathrm{I}}, \xi_{\mathrm{II}}) + E_{\mathrm{II}}(\xi_{\mathrm{II}}) \tag{3.42}$$

one for the inner and one for the outer region, and a cross term. The inner region I is treated in terms of discrete ions with shell or core displacements $\mathbf{X}_{\mathrm{I}}$. In the outer region, II, one wants to avoid considering all the ion–ion interactions explicitly. This is readily done by assuming that E_{11} is a quadratic (i.e., harmonic) function $\frac{1}{2}\tilde{\xi}\cdot\mathbf{A}\cdot\xi$ of the displacements ξ, with $\mathbf{A}$ the force-constant matrix of the perfect crystal, and also that region II is in equilibrium under the forces exerted by region I. The problem reduces to an explicit minimization of a function $E(\mathbf{X}_\mathrm{I})$ in which the displacements in the outer region are no longer explicit. It proves a useful strategy to divide region II into two regions. The outer, region IIB, is regarded as a dielectric continuum, with displacements calculated using Equation (2.45) and assuming the total net charge of the defect to be located at the origin. The inner, IIA, recognizes that the short-range repulsions depend strongly on ion positions, and evaluates their part of the interaction between regions, assuming discrete ions with coordinates given by the Mott–Littleton prescription. Long-range Coulomb sums must always be calculated properly in all regions.

Another general point concerns numerical methods. There are several approaches, mainly in three classes: (a) direct searches, in which $E(\mathbf{X}_\mathrm{I})$ is evaluated for many possible displacements $\mathbf{X}_\mathrm{I}$; (b) conjugate-gradient methods, which evaluate first derivatives $\partial E/\partial \mathbf{X}_\mathrm{i}$ and which, after each iteration, concentrate on those displacements which appear to offer the largest energy reduction at the next step; and (c) Newton–Raphson methods, in which both first and second derivatives are used. This third class is by far the most efficient as regards speed but makes heavy demand on storage. There are special cases where conjugate-gradient methods have advantages.

For crystals with pairwise interactions, $\partial^2 E/\partial \mathbf{X}_{\mathrm{I}i}\partial \mathbf{X}_{\mathrm{I}j}$ is particularly simple. Initial evaluation of the second-derivative matrix, $\mathbf{W}$, is not difficult, although it is too time consuming to attempt often. A useful strategy starts from a convenient configuration $\mathbf{X}_0$, for which one calculates first derivatives $\mathbf{g}_0$, second derivatives $\mathbf{W}_0$, and the inverse $\mathbf{H}_0 \equiv \mathbf{W}_0^{-1}$. At each iteration a better geometry is calculated, together with new forces and an updated approximation to the inverse force-constant matrix:

$$\mathbf{x}_N \rightarrow \mathbf{x}_{NH} = \mathbf{x}_N - \mathbf{H}_N \cdot \mathbf{g}_N \tag{3.43}$$

$$\mathbf{H}_N \rightarrow \mathbf{H}_{N+1} = \mathbf{H}_N + (\boldsymbol{\delta}\mathbf{x} \cdot \boldsymbol{\delta}\mathbf{x}^T)(\boldsymbol{\delta} g^T \cdot \boldsymbol{\delta} g) - (\mathbf{H}_N \boldsymbol{\delta} g \cdot \boldsymbol{\delta} g^T \mathbf{H}_N)/(\boldsymbol{\delta} g^T \cdot \mathbf{H}_N \boldsymbol{\delta} g) \tag{3.44}$$

where $\boldsymbol{\delta} x = \mathbf{x}_{N+1} - \mathbf{x}_N$, $\boldsymbol{\delta} g = \mathbf{g}_{N+1} - \mathbf{g}_N$ and the superscript T denotes the matrix transpose. Such methods are exploited in the HADES (Harwell Automatic Defect Evaluation System) code (Lidiard and Norgett 1972) which has dominated defect studies of this type.

In comparing static and dynamical methods we find ourselves in the fortunate situation that the two approaches are complementary, so that choice of approach is determined mainly by the precise question one needs to answer. The special advantages of static methods are these. First, they are flexible and economical. For a single defect the Mott–Littleton boundary condition can be exploited, and this boundary condition gives no significant limit on accuracy. Defect clusters (even aggregates of up to tens of defects), dislocations, or grain boundaries can be handled. There are no real restrictions on symmetry or charge. Second, they readily produce energies (motion, formation, binding, etc.) that can be compared directly with experiment. Third, with slight generalizations already mentioned, one can determine defect electrical and elastic dipole moments, volume changes, and entropies. Fourth, one can calculate some important electronic transition energies such as charge transfer energies, when polarization and distortion energies dominate, and also the energies of formation and motion of small polarons. Finally, it is perfectly practical to combine the present types of static structure calculations with electronic structure calculations in which the Schrödinger equation is solved for the defect.

However, static methods have their weaknesses. In particular, it is not trivial to predict free energies of defects. Static methods usually concentrate on internal energies, and generalization to give free energies is not always easy. Furthermore, one has to postulate a specific defect process for assessment, and although many processes can be tried, one always wonders if a critical process has been left out.

Molecular dynamics has its own strengths and weaknesses. It is not economical, usually exploiting periodic boundary conditions with a rather small repeat unit. Dislocations, complex defects, small polarons, and dipole moments have not been treated by dynamic methods so far. However, it does provide rates of processes (probabilities per unit time) more directly. It can treat the liquid state, or other highly defective state with transient defects, or nonequilibrium processes following a sudden perturbation, on exactly the same basis as the solid. Moreover, in principle it automatically finds for itself any process occurring with reasonable probability so the user does not need to find the right mechanism in advance. This is especially important in analyzing neutron- or optical-scattering data, where observed quantities depend on displacement ($\mathbf{u}_i$) or momentum ($\mathbf{p}_i$) correlation functions, for example, time averages like $\langle \mathbf{u}_i(t+\tau)\mathbf{u}_i(t)\rangle$. (These averages can be obtained relatively simply from molecular dynamics.)

Examples of the degree of success possible with careful calculations are shown by comparison with experiment in Tables 3.8 and 3.9. Clearly, formation energies and energies of motion are estimated successfully; indeed, accuracies of calculation are often comparable with the accuracy

Table 3.8 Comparison of Calculated and Observed Defect Formation Energies in Ionic Crystals[a]

Compound	Defect	Calculated Energy (eV)	Experimental Energy (eV)
NaCl	Schottky pair	2.32	2.30
CaF_2	Frenkel pair	2.75	2.7
MgO	Schottky pair	7.5	5.7
ZnO	Cation Frenkel pair	2.51	—
AgCl	Cation Frenkel pair	1.4	1.45

[a] After Catlow and Mackrodt (1982).

of the best experiments.

The techniques described above can also be used to investigate small polaron problems. Here, a first question concerns the mechanism of charge transport. One wishes to know if free holes move as large polarons, with a long mean free path, like the holes in GaAs, or if they move as small polarons, as in AgCl, randomly hopping from cell to cell. To answer these questions one needs to know the relative importance of the polarization and distortion produced by a localized charge and the competing reduction in kinetic energy achieved by delocalization (Section 1.2). In general, a large polaron is expected when the carrier's bandwidth exceeds a specific value. While there is no general consensus about the experimental results (Catlow et al. 1977), they are consistent

Table 3.9 Comparison of Calculated and Observed Defect Energies of Motion in Ionic Crystals[a]

Compound	Process	Experimental Energy (eV)	Calculated Energy (eV)
NaCl	Cation vacancy migration	0.67	0.65–0.75
CaF_2	Anion interstitial migration	0.91	1.0
MgO	Cation vacancy migration	2.1–2.3	2.2

[a] After Catlow and Mackrodt (1982).

Table 3.10 Polaron Energies (eV) in Transition-Metal Oxides[a]

		$Mn_{1-x}O$	$Fe_{1-x}O$	$Co_{1-x}O$	$Ni_{1-x}O$
Free Carrier	Estimated minimum bandwidth, $2B$, for the large polaron to be stable	3.06	2.18	1.88	1.6
	Type of polaron found experimentally	Probably small	Probably small	Probably small	Probably large
Carrier associated with alkali impurity	Binding energy of polaron, E_B				
	Theory	0.24	—	$0.52 - \Delta$ [b]	$0.54 + \Delta$ [b]
	Experiment	0.2	—	0.38	0.31–0.43
	Energy for hole, E_H, hopping around alkali				
	Theory	—	—	0.24	0.19
	Experiment	—	—	0.20, 0.25	Complex behavior 0.05–0.16

[a] After Catlow et al. (1977).
[b] Here Δ is the energy difference between large and small unbound polarons.

with predicted trends. Charge transport in MnO, for example, probably involves small polarons, associated with conversion of Mn^{2+} to Mn^{3+} and that in CoO and NiO is a close balance, marginally favoring large polarons (see Table 3.10).

A second question concerns the effect of impurities on charge transport. It is interesting to know how the motion of holes is affected when an oxide MO is doped with alkali atoms (lithium or sodium), producing $M_{1-x}O{:}A_{2x}$. One can study both the binding of a hole to the impurity and the hopping of the hole around the impurity. Experimental data are more complete for the impurity systems. They indicate that the trapped hole is localized on one of the host cations adjacent to the substitutional alkali impurity (Catlow et al. 1977). The fact that the predictions are in good general accord with experiment is a formidable test, for the energies are only a few tenths of an electron volt (see Table 3.10), as compared with Madelung energies (essentially the electrostatic energy of the ions in the crystal) and polarization energies, amounting to a few tens of electron volts.

3.5 FAST ION CONDUCTORS (SUPERIONICS)

3.5.1 Physical Properties of Fast Ion Conductors

Suppose we have a nonmetal in which electrical conduction is dominated by a single ion. We have already seen (Section 3.2.1) that we can characterize the motion of the ion by its diffusion constant D, by its mobility μ, or by its conductivity σ, and that these quantities are related generally by Equations (3.22) and (3.27). We mentioned in Section 3.2.1 the limitation of the Nernst–Einstein relation, Equation (3.27), that electrically neutral defects can contribute to tracer-diffusion coefficients, but not to the electric current. A second limitation arises from the possible correlation between successive jumps, for which the correlation factor f lies between 0 and 1 (1 corresponds to completely uncorrelated jumps; see Howard and Lidiard 1964). Correlation affects tracer measurements of D, but not values of D obtained from conductivity using Equation (3.27). If D_{tracer} is measured for a system in chemical equilibrium, then the Haven ratio ($D_{\text{tracer}}/D_{\text{cond}}$) is often a useful guide to the mechanism of conduction (Flynn 1972; see also Weber and Friauf 1969).

We can go further by using a typical upper bound to D. The jump rate, Γ, for diffusion [Equation (3.26b)] is not expected to exceed the number of lattice vibrational cycles per second, say 2×10^{13} Hz. With

$a^2/6 \simeq 10^{-16}$ cm^2, a typical value, we expect that D will be less than about 2×10^{-3} cm^2/sec. For $q = 1$ and $T = 1000$ K the mobility μ will be unlikely to exceed 2×10^{-2} cm^2/V sec and the highest ionic conductivity will be about 2 Ω^{-1} cm^{-1}, when all ions are mobile.

These arguments may be generalized in several ways. One is to allow a finite time for the jump, assuming typically a thermal velocity of motion ($\frac{1}{2}mv^2 \sim \frac{3}{2}k_BT$). Obviously it will be inappropriate to think of rapid jumps between thermalized states unless the time of flight (a/v) is short.

High values of D, typically 10^{-4} cm^2/sec, are found in molten salts, where all the ions are mobile. In solid ionics one usually finds much smaller values, even near the melting point, since the fraction of mobile ions rarely exceeds 10^{-2} in stoichiometric systems. For typical cases like LiF one finds jumps every few picoseconds just below the melting temperature, with a time of flight from site to site an order of magnitude less. At lower temperatures the mobility μ falls, roughly as $\exp(-h_m/k_BT)$, where h_m is the activation enthalpy of motion. The conductivity σ [see Equation (3.28)] may contain an additional formation enthalpy, h_f, if the mobile ions are thermally generated, so that $Q \equiv h_m + h_f$. Clearly, three features help if high ionic conductivity is needed, especially at moderate temperatures: (i) a large mobile ion concentration, (ii) a high mobility for the mobile species, and (iii) a low value of Q, generally associated with (i) and (ii). Materials possessing these features are referred to as fast ion conductors or superionics.

Apart from scientific curiosity, three major technical requirements demand materials with high ionic conduction (Hooper 1978). First, there are ion monitors in which the presence of a low concentration of some species is measured electrochemically as a voltage across a suitable electrolyte. The important features are discrimination (because one might wish to detect oxygen and not carbon in a liquid-metal coolant, for example) and low electronic conductivity, to avoid dissipating the potential generated in driving an electronic current. Indeed, for electrochemical monitors there is no need to draw a current at all. Second, high ionic conductivity can be valuable in electrolytes for batteries and fuel cells. These electrolytes need low electronic conductivity to avoid leakage losses. Third, for electrodes for batteries and fuel cells both high electronic conductivity and high ionic conductivity are desirable (see Table 3.11).

High power density, high energy density and reliability are essential if energy-storage systems involving fast ion conductors are to be competitive. In addition, one also hopes for structural integrity, for low weight, for stable conductivity not limited by degradation, and for immunity to environmental factors like oxygen or water vapor. It is

Table 3.11 Practical Needs for Fast Ion Conduction

Use	Ionic Conduction (Ω^{-1} cm^{-1})	Electronic Conduction (Ω^{-1} cm^{-1})	Special Needs	Examples
Ion monitor	$\gtrsim 1$	Low	High discrimination	ZrO_2/Y_2O_3 as an O_2 monitor
Fuel cell electrode	$\gtrsim 1$	$\gtrsim 100$	High-temperature stability	$PrCoO_3$ (note that the fuel gas is the other electrode)
Fuel cell electrolyte	$\gtrsim 1$	$\lesssim 10^{-8}$	Impermeable to fuel gas	ZrO_2/Y_2O_3 for H_2 combustion
Battery anode	$\gtrsim 1$	High	Electropositive material; no reactions or solubility in electrolyte	Usually metals
Battery cathode	$\gtrsim 1$	High	Electronegative Material	NiS; FeS_2; Li_xCoO_3
Battery electrolyte	$\gtrsim 1$	Low	Low leakage current; chemical inertness	β-alumina for the Na/S battery; Li_3N for Li-based batteries

especially useful if the device will operate at room temperature, but will survive heating on charging or discharging.

The general requirements mean that any ionic conductor worth serious consideration should have $\sigma_{\text{ionic}} \gtrsim 0.1\ \Omega^{-1}\ \text{cm}^{-1}$, that is, have a diffusion constant $D \gtrsim 10^{-6}\ \text{cm}^2/\text{sec}$, and an activation energy for ion conductivity of 0.1 eV or less. These ideals are rarely met, although there are candidates with properties close enough to justify large technical investments.

In addition to the usual transport methods used for the measurement of ionic conductivity, one finds that nuclear-magnetic-resonance (nmr) techniques are extremely useful, especially for the study of the dynamics of superionics. In a rigid lattice the magnetic dipole interaction between neighboring nuclei gives a Gaussian linewidth to a good approximation, and the width is of the order of the local field produced at the site of the spin by its neighbors, that is, in the range 10^{-4}–10^{-3} tesla. However, in liquids and gases, where fast relative motion of the ions occurs, linewidths are an order of magnitude smaller than in a rigid lattice and the shape is more nearly Lorentzian than Gaussian (for discussion of lineshapes, see p. 181). If the nuclear spins are in rapid relative motion, the local field seen by a spin will fluctuate rapidly in time. Only the average value taken over a time long compared with the fluctuations will be seen and this is smaller than the fluctuation value. The rate of fluctuation of the local field can be described by a fluctuation time τ_c, which is identical with the average jump time in diffusing systems. To get narrowing, the fluctuations should be fast compared with the Larmor precession in the instantaneous local field, that is,

$$\frac{1}{\tau_c} > \overline{(\Delta\omega_0^2)}^{1/2} \tag{3.45}$$

where $\overline{\Delta\omega_0^2}$ is the second moment of the lineshape in the absence of diffusion. Under these conditions the real linewidth is

$$\Delta\omega \sim (\Delta\omega_0^2)\tau_c \tag{3.46}$$

The temperature dependence of τ_c in a diffusing system can be studied by measuring the nuclear spin–lattice relaxation time T_1 and by applying the relationship (Bloembergen et al. 1948; see also Gordon and Strange 1978)

$$\frac{1}{T_1} \sim \frac{H_p^2\tau_c}{1+\omega^2\tau_c^2} \tag{3.47}$$

where H_p is the local magnetic field. This gives a maximum in $1/T_1$ when $\omega\tau_c \sim 1$. It also predicts that for $\omega\tau_c > 1$ (low temperature) $T_1 \propto \omega^2\tau_c$ and that for $\omega\tau_c < 1$ (high temperature) $T_1 \propto \tau_c^{-1}$, independent of ω.

Both light- and neutron-scattering methods are also used to obtain information about the dynamics of superionics. For both types of scattering in perfect crystals both energy and wavevector are conserved. The latter condition has the consequence that for light-scattering processes only excitations close to the center of the Brillouin zone can be observed. However, in neutron-scattering studies the entire range of the Brillouin zone is open to investigation. Light-scattering studies of lattice vibrations involve two different experimental approaches, one for the study of zone-center optical phonons (excitations with energy $\gtrsim 10\,\text{cm}^{-1}$), referred to as Raman scattering and one for the study of zone-center acoustic phonons (energy $\sim 1\,\text{cm}^{-1}$), referred to as Brillouin scattering (Hayes and Loudon 1978). We have already seen in Section 2.6 that the presence of disorder in a crystal breaks down the wavevector conservation rule and gives rise to additional scattering related to the single-phonon density of states.

The nature and extent of thermally induced disorder in a crystal can in principle be determined through precise measurement of the temperature dependence of the intensity of Bragg peaks using neutron-scattering methods. In addition, the defect hopping spectrum can be investigated through the quasielastic scattering of neutrons and laser light.

X-ray diffraction techniques have also been applied, with useful results, to the study of disorder in superionics, particularly the methods of diffuse X-ray scattering (Guinier 1963). The radiation scattered by an imperfect crystal is not as well localized as by a perfect crystal. There are still strong peaks corresponding to the Bragg law for the average lattice, but in other directions interference does not result in perfect cancellation. This off-peak scattering is called diffuse scattering and, in general, is weak compared with the average lattice diffraction. Thermal diffuse scattering is a particular example of off-peak scattering. In general, however, the interpretation of disorder-induced diffuse scattering is not straightforward.

The availability of intense continuous radiation in the X-ray region from synchrotron sources has given rise to an upsurge of interest in recent years in the study of extended X-ray absorption fine structure (EXAFS) in solids. This oscillatory structure is observed to an energy of about 1000 eV above an absorption edge (K, L, etc.) of a particular ion in a solid. It is due to structure-dependent interference effects that lead to modulation of the absorption coefficient by neighboring atoms and gives information about the environment of the absorbing atom. The method has been used to study the environment of disordered species in superionics (Boyce and Hayes 1979).

In addition to the study of transport and general spectroscopic properties the investigation of specific heat has also proved a useful tool in the study of superionics. The onset of disorder and of high electrical conductivity in many superionics is accompanied by a specific heat anomaly, and the entropy associated with the anomaly may be comparable with the entropy of melting.

Theoretical modeling has been of help in understanding fast ion conduction. Specific mechanisms can be reliably checked for the plausibility of the energies evolved by static methods (Section 3.4). Work on Li_3N and β-alumina provide good examples [see Perram (1983) and also Beniere and Catlow (1983)]. Molecular dynamics can often go further than static calculations, since mechanisms may show up that were not anticipated; the prediction that isolated cube-center fluorine interstitials are not present in fluorite fast ion conductors is a good example (Dixon and Gillan 1980).

In addition to the atomistic modeling just mentioned, phenomenological approaches have been pursued (Dieterich et al. 1980). These have been stimulating rather than directly useful.

In recent years, many fast-ion conductors have been investigated intensely (McGeehin and Hooper 1977; Salomon 1979; Hayes 1982). Some have large defect concentrations generated intrinsically, for example, in the fluorites, where anion Frenkel defect formation is relatively easy at high temperatures. Others, including β-alumina, have channels, planes, or more complex open paths allowing rapid ion motion. Finally, there are intriguing two-phase systems, like LiI/Al_2O_3, where two poor ion conductors combine to produce a highly conducting composite.

3.5.2 Examples of Fast Ion Conductors

(a) Silver and Copper-Based Compounds

Some examples are AgI, $RbAg_4I_5$, and CuI, in which the disorder occurs in the silver and copper sublattices. The earliest experiments on the abnormal conductivity of the classic example, AgI, were made by Tubandt and Lorenz (1914). The structure of AgI is hexagonal wurtzite (β form) at room temperature, and its conductivity is not unusual. At a critical temperature $T_c = 147°C$ there is a first-order structural phase change to a body-centered-cubic structure (α-AgI). The dc conductivity changes dramatically at T_c. The value of σ just below T_c is $\sim 3 \times 10^{-4}\ \Omega^{-1}\ \mathrm{cm}^{-1}$. Just above T_c σ increases to $\sim 1.3\ \Omega^{-1}\ \mathrm{cm}^{-1}$, an increase of some four orders of magnitude (Figure 3.16). Thereafter σ does not change appreciably with increasing T until the crystal melts at $T_M = 552°\mathrm{C}$, when σ actually falls by about 12%.

Structural studies show that in the fast ion α phase the iodine ions constitute a fairly rigid bcc lattice, and that the silver ions occupy principally the tetrahedrally coordinated (referred to as $12d$) interstitial

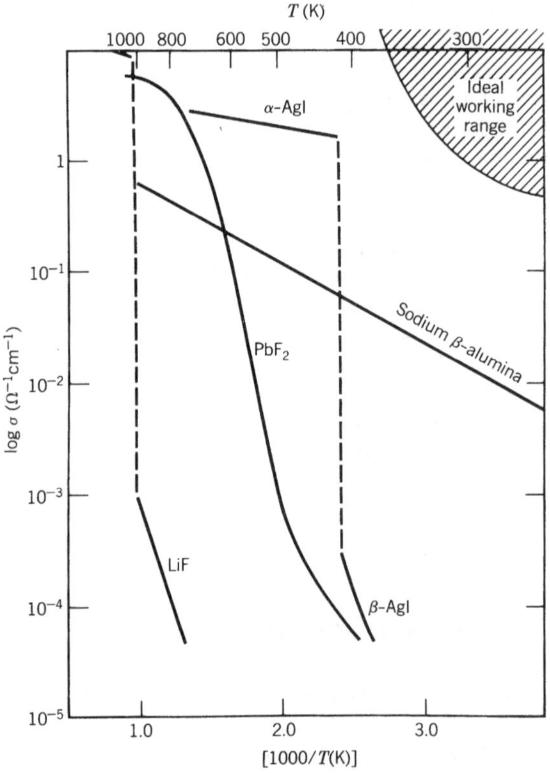

Figure 3.16 Conductivity of three fast ion conductors, contrasted with the normal behavior of LiF, where the conductivity only becomes high at the melting point (1140 K). The change in behavior of AgI above and below the $\beta \leftrightarrow \alpha$ transition at 420 K is discussed in the text. Practical applications need, ideally, conductivities in the shaded region, that is, $\sigma \gtrsim 1\ \Omega\ \mathrm{cm}^{-1}$ for $T \lesssim 100°\mathrm{C}$.

sites (Figure 3.17). There are 12 such sites in the unit cell. A variety of experiments suggests that motion of Ag^+ ions between neighboring tetrahedral sites is the basic step in migration. A quasielastic light-scattering study by Winterling et al. (1977) showed a central peak centered on $\omega = 0$ with 3.8 cm^{-1} full width at half-maximum (see Figure 2.22 for a related study). It was suggested that this peak was due to hopping of Ag^+ ions between neighboring $12d$ sites. The width is given by $\Gamma/2\pi$, with the jump rate Γ corresponding to a jump every 1.4 psec. Knowing the diffusion constant $D \simeq 2 \times 10^{-6}$ cm^2 sec^{-1}, one can deduce a jump distance of order 1.4 Å. This is encouragingly close to the actual separation of nearest-neighbor $12d$ sites (1.79 Å) although it would be imprudent to read too much into such agreement.

The frequency dependence of the electrical conductivity of α-AgI has been studied by several groups. Funke and Jost (1971) reported a pronounced and puzzling variation of $\sigma(\omega)$ in the region of 1 cm^{-1}, using microwave techniques. However, more recently Gebhardt et al. (1980), also using microwave techniques, find that $\sigma(\omega)$ is independent of frequency in the region 0.13–1.3 cm^{-1}, with a value of $\sim 1\ \Omega^{-1}\ cm^{-1}$, and conclude that the conductivity associated with diffusive motion of the Ag^+ ions has a Drude-type behavior, $\sigma(\omega) \propto (\omega_0^2 + \omega^2)^{-1}$, with $\hbar\omega_0$ corresponding to 30 cm^{-1}.

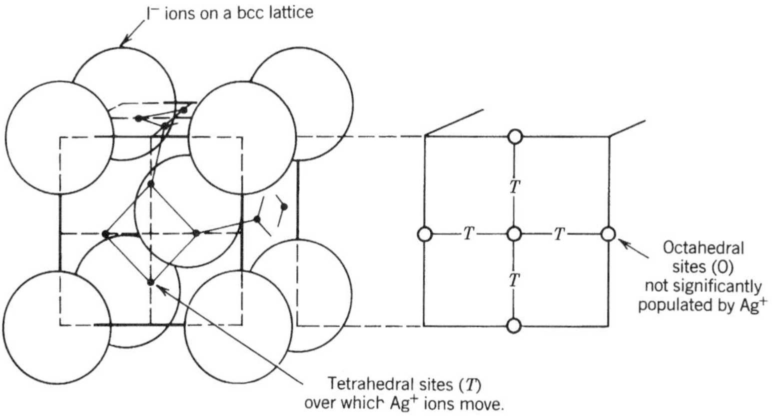

Figure 3.17 Ionic motion in superionic AgI. The I^- ions form an essentially rigid bcc framework, with Ag^+ ions moving among tetrahedral interstitial sites, T (i.e., interstitial with respect to the bcc I^- lattice).

(b) β-alumina and Related Structures

The prototype material here is sodium β-alumina, which, if available in stoichiometric form, would have the formula $Na_2O \cdot 11Al_2O_3$. However, crystals grown from the melt contain up to 30% excess sodium, and the formula is written more appropriately as $(1+x)Na_2O \cdot 11Al_2O_3$, with $x \simeq 0.3$. The crystals have hexagonal symmetry, consisting of blocks of aluminum oxide with a spinel structure, separated by mirror planes containing sodium and oxygen ions (Figure 3.18). The spinel blocks of aluminum oxide are 11.25 Å thick along the c axis, so that the length of the unit cell is 22.5 Å. The oxygen ions in the mirror plane [O(5) in

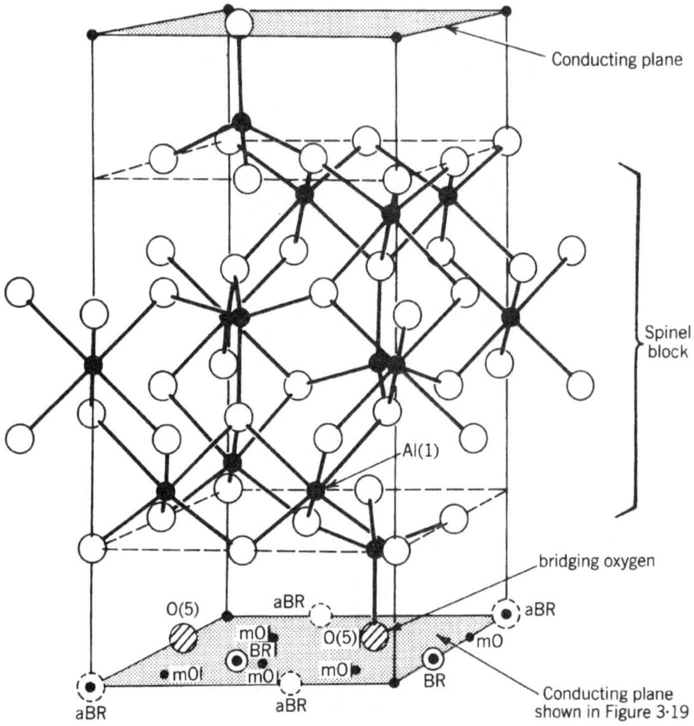

Figure 3.18 β-alumina lattice structure. Note the spinel blocks, separated by the conducting planes (for detail see Figure 3.19) containing the bridging oxygen (cross-hatch circles). The spinel oxygens are drawn as open circles and the aluminums as solid circles; the Al(1) site, referred to in the text, lies above a mid-oxygen site (see Figure 3.19) although the perspective does not show this clearly (see p. 157).

Figure 3.18] are coordinated by Al^{3+} ions in the adjoining oxide blocks, so that these bridging oxygens effectively link the blocks together. Each mirror plane can be regarded as a hexagonal network of O(5) ions, interspersed with cation sites referred to as Beevers–Ross (BR), as anti-Beevers–Ross (aBR), and as mid-oxygen (mO) sites (Figure 3.19).

Structural studies suggest that at room temperature ~66% of the sodium ions are near BR sites, ~30% are near mO sites, and a relatively small part, ~4%, are near aBR sites. It is generally accepted that the charge-compensation mechanism for the nonstoichiometric excess of sodium in the mirror plane occurs through the presence of interstitial oxygen ions (O_i^{2-}) in mO sites, bound by two aluminum ions [Al(1) in Figure 3.18] displaced from their normal positions toward the O_i^{2-}.

Sodium β-alumina is one of relatively few materials with high ionic conductivity at room temperature ($\sigma \simeq 0.02\ \Omega^{-1}\ cm^{-1}$ for the electric field **E** perpendicular to the c axis and negligibly small for **E** parallel to the c axis. The activation energy for ionic conductivity at room temperature is $Q = 0.15$ eV. It is possible to prepare more stoichiometric material with $x \simeq 0.07$, but, for this material at room temperature, σ is about an order of magnitude smaller and Q is about a factor of four higher. This suggests that the predominant mechanism of conductivity in the two materials is different, and that if it were possible to prepare perfectly stoichiometric sodium β-alumina, the conductivity would not be especially high. Although it is clear that the high ionic conductivity of

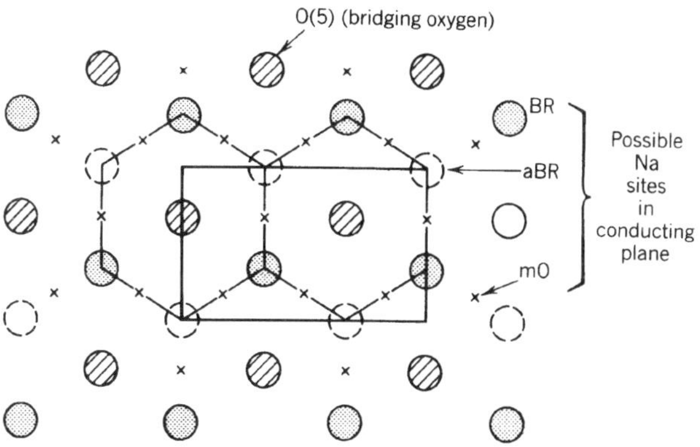

Figure 3.19 Conducting plane of β-alumina. The sites labeled are the bridging oxygen, (O(5)), the mid-oxygen sites, (mO), lying midway between bridging oxygens, and the Beevers–Ross (BR) and anti-Beevers–Ross (aBR) sites. The BR sites are the most stable sites for sodium ions. The rectangle is the part of the conduction plane shown in Figure 3.18.

sodium β-alumina is due to migration of Na^+ ions in the mirror plane, a satisfactory picture of the precise details of the conductivity mechanism has proved elusive.

Sodium β''-alumina has a structure closely related to that of sodium β-alumina and is the currently preferred material for use in the sodium–sulfur battery (Subbarao 1980). The ideal structure is represented by the formula unit $Na_2O{\cdot}MgO{\cdot}5Al_2O_3$, with the Mg^{2+} ions dissolved primarily in tetrahedrally coordinated aluminum sites in the aluminum oxide blocks. The crystal has rhombohedral symmetry with the aluminum oxide blocks arranged in a triple stacking sequence along the c axis, rather than the double stacking that exists in sodium β-alumina. The aluminum oxide blocks are separated by planar conducting zones containing sodium and oxygen ions, but the sites corresponding to the BR and aBR positions of Figure 3.19 are now equivalent. Again, as-grown crystals are not stoichiometric, containing sodium vacancies with a concentration of $\geq$15 mole%. It is generally agreed that the high ionic conductivity in sodium β'' at room temperature ($\sim$0.06 Ω^{-1} cm^{-1}) is due to migration of sodium vacancies.

(c) Crystals with the Fluorite Structure

Some crystals with the fluorite structure show a pronounced specific heat anomaly at a temperature T_c well below the melting temperature T_m (Table 3.12 and Figure 3.20). Above T_c the crystals are extensively disordered giving rise to a high ionic conductivity (see σ for PbF_2 in Figure 3.16). This disorder occurs in the anion sublattice, resulting in the generation of anion Frenkel pairs (Section 3.1) at concentrations of $\sim$10 mole%. The cation sublattice remains essentially intact. Well below

Table 3.12 Approximate Melting Temperatures T_m of Fluorite Crystals and Temperatures T_c for Transition to the Superionic State

Crystal	T_m (K)	T_c (K)
CaF_2	1633	1430
SrF_2	1723	1400
BaF_2	1550	1230
$SrCl_2$	1146	1000
PbF_2	1158	712

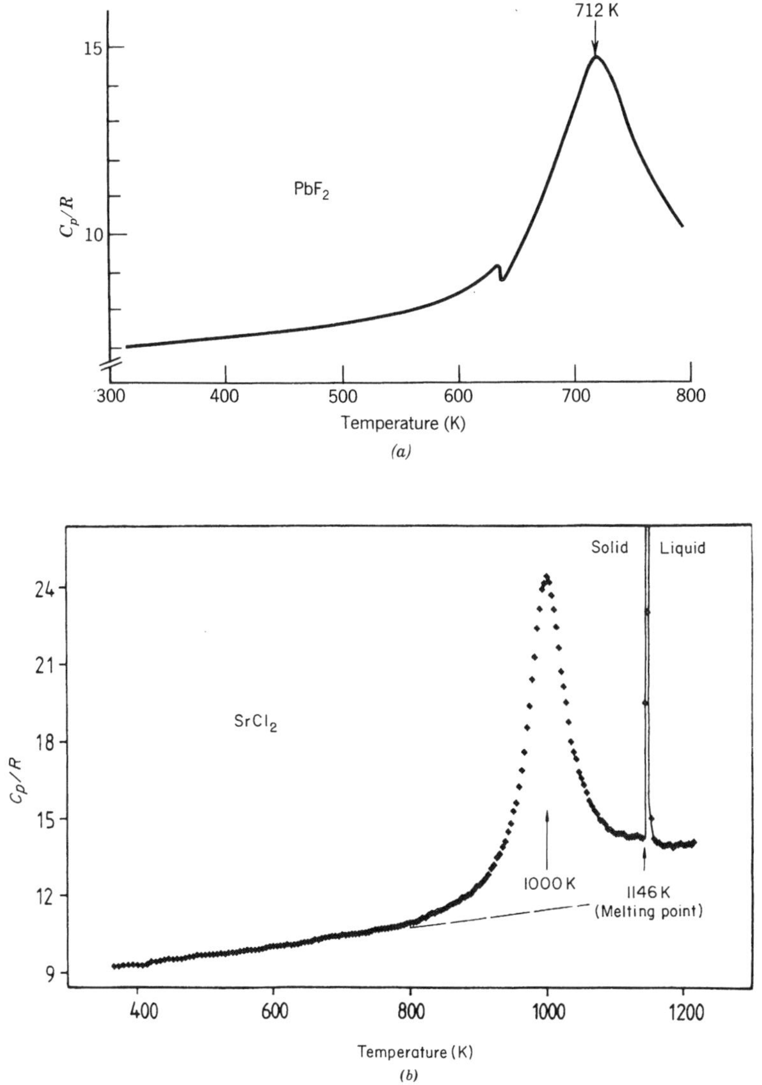

Figure 3.20 Molar heat capacity of (*a*) PbF_2 (cf. Figure 3.16) and (*b*) $SrCl_2$ at constant pressure, as a function of temperature [after (*a*) Nölting (unpublished) and (*b*) Schröter and Nölting (1979)].

T_c the concentration of anion Frenkel pairs is small, and anion interstitials inhabit the center of empty halide cubes (Section 3.1). However, there is now extensive experimental and computational evidence to indicate that above T_c the anion interstitials do not occupy the cube-center sites.

The heat capacity anomaly suggests the onset of an order–disorder transition, which is suppressed by higher-order effects before massive interstitial disorder develops. There has been extensive discussion of the cooperative interaction between Frenkel defects that gives rise to the anomaly, and it is generally agreed that this is primarily electrostatic in origin; a net Coulomb attraction between anion vacancies and anion interstitials reduces the defect formation energy when some defects are already present. The net attraction dominates up to a limiting concentration of defects determined by the build up of vacancy–vacancy and interstitial–interstitial repulsion.

Experimentally, the most useful structural information about the highly conducting state of fluorites has come from coherent neutron-scattering studies (Hutchings et al. (1984)). These studies suggest that anion vacancies and interstitials form transient clusters related to the 2:2:2-type cluster (Figure 3.11) at temperatures above T_c. Static energy calculations suggest that such clusters and even larger clusters are energetically stable and the stabilization energy may provide part of the driving energy required for the order–disorder transition at T_c. Although molecular dynamics simulations have not so far shown the existence of defect clusters the cell sizes used in such calculations have not been large enough to rule out the existence of clusters. It should be said that cluster models do not give a clear description of the diffusion process whereas dynamical models emphasize this aspect.

Finally, in this section we point out that superionics that conduct by oxygen-ion transport are useful for high-temperature fuel cells as well as for the measurement of thermodynamic and kinetic properties of systems containing oxygen (Subbarao 1980). Most interest has centered on ceramic oxides with the fluorite structure, principally calcia-stabilized zirconia (Section 3.3.1). This material is a reasonably good ionic conductor above ~700°C, due to motion of charge-compensating oxygen vacancies arising from the presence of aliovalent Ca^{2+} ions.

(d) Lithium Iodide / Alumina

Lithium iodide is a typical ionic conductor, with unacceptably low Li conductivity for practical purposes. Alumina has a completely negligible ionic conductivity. Yet LiI, grown so as to contain embedded alumina powder, is a reasonably good Li conductor ($\sigma = 10^{-1}\,\Omega^{-1}\,\mathrm{cm}^{-1}$ at 300°C). Detailed analysis of dependence on alumina particle size and density suggests that around each alumina grain a layer of higher conductivity exists, the conductivity falling off gradually rather than abruptly over a few hundred angstroms. Whether this layer is a space-charge layer, or another phase (possibly hydrated), or induced by misfit of the two crystal structures, is not clear. Nevertheless, conduction via these fast-conducting regions can be very rapid. These systems are examples of inhomogeneous conductors with regions of high and low conductivity. Theoretically, effective-medium theory (Landauer 1952) gives a good semiquantitative description of their behavior (Stoneham et al. 1979).

(e) Nasicon

An important type of fast ion conduction is typified by $Na_3Zr_2PSiO_2O_{12}$, sometimes referred to as Nasicon. This compound consists of a combination of corner-shared octahedra and tetrahedra with some sodium ions octahedrally coordinated to oxygen at the intersection points of continuous networks of three-dimensional tunnels. Conductivity at room temperature is comparable to that of the best sodium β-alumina, but with a somewhat larger activation energy ($Q \sim 0.3$ eV).

3.6 DEFECTS AND DIFFUSION IN SEMICONDUCTORS

Diffusion in semiconductors is important in several respects. First, controlled diffusion is of immense technological importance since doping by electrically active impurities is done principally by diffusion. This process involves both accurate spatial control and stability under subsequent treatment. Second, if ion-implantation doping is used (the alternative main technique; Section 6.4.4), it is essential to anneal any radiation damage, which would otherwise lower efficiency and perhaps provoke degradation. Third, there can be enormous differences in device lifetimes depending on whether they are operating or not. These differences are associated with recombination-enhanced ionic diffusion (Section 6.4.6), in which the presence of many free carriers (whether introduced by means of optical excitation, by irradiation with charged particles or simply by electrical bias methods) is critical.

3.6.1 Defects in Semiconductors

We shall be concerned with four main types of defect in semiconductors.

(a) Vacancies

Even in elemental semiconductors the vacancy can exist in many charge states. Thus, for Si, experimental data are interpreted in terms of five charge states, either seen directly by epr (V^+, V^-), or inferred indirectly from defect reactions, diffusion, or behavior under Fermi-level shifts (V^{2+}, V^0, V^{2-}). In diamond, V^0 (the so-called GR1 center) and V^- (the so-called ND1 center) are both observed optically.

In the polar semiconductors, the position is less clear. Only one isolated vacancy center has been identified in III-V compounds, namely, V^0 in GaP, although several have been tentatively assigned. In the tetrahedrally coordinated II-VI crystals, the V^- center (i.e., remove M_{II}^+) is seen in many hosts (e.g., ZnO, ZnS, ZnSe, ZnTe, BeO); the F^+ center (remove X_{VI}^-) and F^0 center (remove X_{VI}^0) are seen less frequently and are less easily identified. The vacancies observed in both fourfold- and six-fold-coordinated II-VI compounds are very similar. We shall discuss vacancy centers in more detail in Sections 5.1 and 5.4.

(b) Intrinsic Interstitials

These are elusive experimentally. Their existence can be inferred in Si from an indirect reaction

$$Si_i + X_s \rightarrow X_i \qquad (3.48)$$

postulated to explain displacement of substitutional impurities, X_s, to interstitial sites by transient silicon interstitials, Si_i, created in crystals subjected to electron irradiation at low temperatures (see Section 6.4.1). In electron-irradiated diamond there are tentative identifications of intrinsic interstitials from epr spectra. Explanation of the behavior of interstitials invokes the assumption that the interstitial is mobile even at the lowest temperatures (Watkins 1964; McKeighan and Koehler 1971), probably because of motion enhanced by recombination of electrons and holes produced by the same irradiation that created the interstitial. Theory supports this (Mainwood et al. 1978; Masri et al. 1983), arguing for a split (100) interstitial for diamond; in Si, theory suggests a charge-state dependence of the interstitial configuration, I^0 and I^- favoring the split (100) interstitial and I^+ the hexagonal interstitial site (see Fig. 3.1).

While more recent work on Si (Bar-Yam and Joannopolous 1984; Carr et al. 1984) suggests defect geometries differing in detail, their predictions support similar mechanisms of athermal diffusion.

(c) Substitutional Impurities

The staple dopants of devices are substitutional, for example, the donors (P, Sb, As, Bi) and the acceptors (B, Al, Ga, Te) in Si and Ge, and isoelectronic dopant species such as N in GaP.

(d) Interstitial Impurities

These can be notoriously mobile, key cases being Li and Cu in Si and Ge. Transition-metal impurities appear to be interstitial in Si and Ge, with only a small substitutional fraction likely (Weber 1983).

In addition to these main cases, there are a few other examples of defects in semiconductors that can be important. First, there are the antisite defects, for example, P on a Ga site in GaP (Van Vechten 1975; Table 3.1). These are certainly seen in III-V compounds, although there is still discussion of their precise importance in device behavior. Second, there are defects that can exist in both (immobile) substitutional and (mobile) interstitial form, the so-called dissociative defects. Examples proposed include Ag, Au, and perhaps Ni, in both Si and Ge. Third, there are defect aggregates, ranging from pairing of impurities with each other or with vacancies, to the formation of colloids, precipitates, and dislocation loops (see Section 3.7). Dislocations and grain boundaries are of special importance, since, as we have already pointed out (Section 3.2.1), they offer paths of rapid diffusion to many impurities. They also act as sinks, and may trap impurities either within the dislocation core or in an atmosphere of impurities in the region surrounding the dislocation.

3.6.2 Diffusion in Semiconductors

Despite the large number of diffusion mechanisms proposed for semiconductors in early studies, the number of mechanisms firmly identified is quite modest (Shaw 1973). Mechanisms are investigated in many ways, but the two central approaches exploit the (radioactive) isotope effect and the response to applied electric fields. Table 3.13 summarizes the position for mechanisms and gives typical activation energies of diffusion. Absolute diffusion rates are conveniently compared just below the melting point, and Table 3.14 does this for intrinsic diffusion in a variety of materials and/or impurity diffusion in Si and GaAs.

Table 3.13 Diffusion Mechanisms in Semiconductors

Mechanism	Typical Activation Energy for Si (eV)	Comment	Examples[a]
Interstitial diffusion	0.5–1.5	Included here are possible interstitialcy cases, where host and impurity interstitials exchange (only important under irradiation or perhaps near the melting point)	Interstitial diffusion of Li, Na, Cu, Ag, Au, and most transition metals in Ge or Si; interstitialcy diffusion of Si:Au (?)
Diffusion by dissociative defect (see Section 3.6.1(d))	1–3	Motion of interstitial fraction; significant substitutional fraction coexists	Si:Au, Ag, Ni? Ge:Au, Ag, Ni? Co? GaAs:Cd
Diffusion by vacancy mechanism	3.5–4.5	Substitutional impurity makes only jumps to adjacent vacancy; diffusion rate related to that of vacancy motion (self-diffusion)	Most donors and acceptors in Si and Ge (but not alkali metals)
Short-circuit diffusion	3–3.5	Diffusion along dislocation or grain boundary	CdS:Cu
Enhanced diffusion	Small	Either *radiation enhanced* because of vacancies and interstitials produced or *recombination enhanced* because of carrier densities involved	Summarized separately (Section 6.4)

[a]The fact that some impurities are shown under several mechanisms probably means that *all* th mechanisms can occur under suitable conditions. However, experimental identification of the mechanisr may be uncertain.

Most defects and impurities can exist in more than one charge state, and one expects defect charge states and diffusion rates to change with the Fermi level. This is seen with vacancies in Si where activation energies of 0.33 eV (V^0) and 0.18 eV (V^-) are reported. These are relatively small energies of motion; self-diffusion in Si is observed with the much higher activation energy of ~4 eV (Table 3.13) because of the additional formation energy required. Possibly related to this dependence of vacancy properties on charge state, we note that while both donors

Table 3.14 Diffusion Coefficients (cm^2/sec) in Solid Semiconductors Close to Their Melting points[a]

(a) Intrinsic Diffusion[b]

<table>
<tr><th>Material</th><th>Anion Diffusion Coefficient</th><th>Cation Diffusion Coefficient</th><th></th></tr>
<tr><td>Silicon</td><td colspan="2">9×10^{-12}</td><td rowspan="4">Relatively slow</td></tr>
<tr><td>Germanium</td><td colspan="2">6×10^{-12}</td></tr>
<tr><td>GaAs</td><td>2×10^{-12}</td><td>$\sim 10^{-12}$</td></tr>
<tr><td>InSb</td><td>1.3×10^{-14}</td><td>2.3×10^{-14}</td></tr>
<tr><td>CdTe</td><td>1×10^{-9}</td><td>1×10^{-7}</td><td rowspan="3">Relatively fast</td></tr>
<tr><td>ZnSe</td><td>$\sim 10^{-8}$</td><td>$\sim 10^{-7}$</td></tr>
<tr><td>AgI</td><td>Slow</td><td>$\sim 10^{-5}$</td></tr>
</table>

(b) Impurity Diffusion Coefficients in Si and GaAs

Diffusion Coefficient		Species in Si	Species in GaAs
Very fast	$>10^{-5}$	Cu, Li	Cu, Li
	10^{-6}–10^{-5}	Na, K, Fe, He, Au Co, Ni, Zn, Cr, Mn	Ag, O
	10^{-7}–10^{-6}	Ni, S	
	10^{-8}–10^{-7}	Ag	
	10^{-9}–10^{-8}	O, C	
Slow	10^{-11}–10^{-9}	Group-III acceptors and Group-V donors	

[a]See Shaw (1973).
[b]Note that for Frenkel disorder one component diffuses much faster than the other; for Schottky disorder, values are closer for the two components.

and acceptors diffuse by a vacancy mechanism in silicon, the acceptors move faster; this is not merely a size effect, for indeed opposite size effects are noted in Si and Ge. In other cases, for example, Fe interstitials in Si, no charge-state effect is evident in diffusion.

The quantitative analysis of diffusion involves complications that can be especially obtrusive in semiconductor studies. Diffusion is an atomic mechanism for removing concentration gradients (strictly, gradients in chemical potential), and, clearly, control of the concentration is itself a feature. At the simplest level, this raises the issue of what activation energy is observed. Suppose the diffusion flux associated with a single vacancy in a chosen charge state is $j_r \equiv j_{r0} \exp(-h_{Mv}/k_BT)$, where h_{Mv} is the enthalpy of motion. As the temperature changes, the number of vacancies $N_v \equiv N_{v0} \exp(-h_{Fv}/k_BT)$ will alter too, so that the total flux will show a temperature dependence characterized by the sum of the formation and motion enthalpies $(h_{Fv} + h_{Mv})$, which may be hard to unravel. When the impurities enter by diffusion from a distinct phase adjacent to the semiconductor surface (e.g., a doped melt, as in the liquid phase epitaxy), the segregation coefficient, that is, the ratio (concentration in the solid phase)/(concentration in liquid phase) at the interface, may depend on temperature and also on the facet. Second, it is usual to assume independent diffusion of the impurities. However, this is difficult to achieve since even very low concentrations of electrically active impurities can affect the Fermi level. The electrical activity raises the question of electroneutrality, that is, whether charge equilibrium of the carriers is achieved at all times, as well as the more obvious effects on defect charge states. In addition, device production is often carried out at such high defect concentrations that dislocations may be generated by the stress fields from defect misfit. The stress effects are probably the source of the so-called emitter dip and emitter pull effects in Si, in which B or Ga acceptor diffusion is influenced by the previous diffusion of P donors into an adjacent region.

3.6.3 Solubility and Distribution Coefficients of Dopants

In many cases one wants to achieve a particular concentration of dopant, so that the properties which matter are solubility as well as diffusion. Solubilities are, however, only defined relative to defined reference states of the component phases. In simple cases, as for the solubility of He in Si at room temperature, the reference state for He is the gas phase, whose state is defined once the pressure of the He gas is defined. The solubility of oxygen in silicon is less clear, since one could choose oxygen gas or an oxide (e.g., SiO_2) as the reference state for oxygen. Practically, most choices are acceptable provided equilibrium is achieved experimentally and provided the user knows which reference state was chosen. In simple cases one would expect Arrhenius behavior for the solubility, that is, a concentration varying as $\exp(-E_s/k_BT)$, where E_s is the energy of solution. However, E_s may have either sign. If work is done in removing an atom of the impurity species from the reference phase and placing it in the semiconductor (endothermic process), E_s is positive. If the process is exothermic, less impurity will remain in the semiconductor at higher temperatures.

The retrograde solubility seen in many cases is associated with both types of behavior ($E_s \gtrsim 0$). Examples include the silicon/transition-metal systems. At low temperatures, the reference system is the solid metal (we shall ignore for present purposes the probable role of metal silicides, notably M_2Si, in the equilibrium), and the energetics are such that $E_s > 0$, that is, more metal is dissolved at higher temperatures. Above the eutectic temperature,* T_{eu}, however, the reference phase is the melt formed at the Si/metal interface. T_{eu} may be several hundred degrees below the melting point T_m (1685 K for pure Si). We must now consider the equilibrium between a transition metal M in solid Si and an Si/M melt. The ratio

$$[\mathrm{M}]_{\mathrm{solid}}/[\mathrm{M}]_{\mathrm{liquid}} = k_d \tag{3.49}$$

defines k_d, the distribution coefficient. When $k_d < 1$, the concentration in the solid tends to zero at T_m. The solubility starts to diminish at some temperature between T_{eu} and T_m. As Thurmond and Struthers (1953) point out, the basic requirements for this retrograde solubility are a eutectic temperature well below the melting temperature and a rather low solubility in the solid state. The limiting equilibrium solubilities of transition metals in Si are indeed low, with maximum values (in atomic ppm) of 0.2 (Cr), 0.4 (Mn), 0.5 (Fe), 0.7 (Co), 10 (Ni), and 30 (Cu). These differences appear to correspond to two different classes of interstitial transition metal (Weber 1983). One class (Cu, Fe, Cr) has lower solubility and higher partial enthalpy of solution in solid Si, diffuses as M^0 and may be quenched (Fe and Cr at least) into tetrahedral interstitial sites. Impurities of the other class (Cu, Ni) are more soluble, perhaps diffusing as M^+, and vanish from interstitial sites on quenching.

Nonequilibrium maximum solubilities are discussed later under laser annealing (Section 6.4.5).

Distribution coefficients may depend on crystal facet, on the actual concentrations involved adjacent to the interface, and on temperature. For Ni in Si, for example, k_d is given roughly by $\exp\{-[0.2 + (1.25\ \mathrm{eV}/k_B T)]\}$ between 1000°C and the melting temperature. There is also a dependence on growth rate, that is, on how fast the solid–melt interface is moving. Values of k_d are given in Table 3.15 for several semiconductors. Various rules have been proposed to explain trends, usually combining atomic volume factors with simple chemical arguments.

*For a two-component system, T_{eu} is the lowest temperature at which liquid occurs (Guy 1971).

Table 3.15 Distribution Coefficient k_d at the Melting Point of Some Semiconductors for a Variety of Impurities[a]

Range of k_d	Impurities in Si	Impurities in Ge	Impurities in GaAs	Impurities in InSb
≥ 1 (more impurity in solid)		Si, B	K, Zr, Al, Si, P	As, Ca, Zn, Ca
0.1–1	B, O, Ge, P, As, C	Ga	Ca, S, O, Se, C	Cd, Si, Te, P, S
10^{-3}–10^{-1}	Sn, Al, Ga, Sb	Sn, Al, In, P, As, Sb		
$<10^{-3}$ (most impurity in liquid)	In, Tl, Bi, Zn, S Cu, Ag, Au	Pb, Tl, Bi, Zn, Cd, Hg, Te, Cu, Ag, Au	Cu, Cd, Te	Cu

[a] See Shaw (1973).

3.7 DEFECTS NOT IN THERMAL EQUILIBRIUM: LINE AND PLANAR DEFECTS

Most of the point defects encountered in solids represent some sort of thermal equilibrium, although one can make observations on nonequilibrium transients. When we come to line defects (dislocations) and planar defects (grain boundaries, surfaces, dislocation loops, stacking faults, shear planes), we are often dealing with defects whose concentration and configuration may be determined by factors other than thermal equilibrium. Dislocations, for example, are often introduced mechanically or during growth. Applied stresses may generate dislocations and cause them to move through the crystal. Screw dislocations offer a mechanism of rapid crystal growth from a liquid phase, because they ensure that a convenient ledge is always available for an atom to join the solid phase; the screw dislocation is then present because it has catalyzed growth. Both stress relief and growth may be involved in the generation of dislocations. When a layer of one crystal is grown on a substrate of another, and when the two lattice parameters do not quite match, dislocations may be grown in simply to provide stress relief.

The detailed understanding of linear and planar defects is at a lower level than that of point defects. This has several causes, notably that the concentrations (in effective numbers of atoms directly involved) are rather low (Table 3.16) and that there are many minor variants which make it hard to extract unique, well-identified features. Great progress has been made theoretically, using methods like those of Section 3.4, and this offers a way of obtaining reliable predictions of behavior inaccessible by experiment alone.

Table 3.16 Typical Densities of Dislocations and Typical Grain Sizes

Defect	Typical Values	Atoms Involved per cm^3 [a]
Dislocation	L = cm/cm^3 of dislocation;	
	$L = 10^5$ cm/cm^3 (typical of ionics)	3×10^{12} atoms/cm^3 on the dislocation line
	$L = 10^3$ cm/cm^3 (typical of semiconductors)	3×10^{10} atoms/cm^3 on the dislocation line
Grain boundary	Cubic grains of side l	
	$l = 10$ μm (relatively fine grain)	6×10^{18} atoms/cm^3 at the grain boundary
	$l = 600$ μm (relatively large grain)	10^{16} atoms/cm^3 at the grain boundary

[a] In a typical solid there are 5×10^{22} atoms/cm^3.

While there is only limited direct understanding of linear and planar defects, we shall have occasion to refer to them in quite a few places in this book. We shall limit ourselves here to descriptions and to the vocabulary of the cases of interest. Dislocations are especially important as traps of point defects (notably because of an elastic strain field which falls off only as $1/r$ with distance r from the dislocation line, giving a strong, long-range interaction with point defects) and of charge carriers and as modifiers of the properties of impurities in their vicinity. Because of their large strain field, dislocations have an energy of the order of 10 eV per interatomic spacing. Grain boundaries are especially important in matter transport, giving connected fast-diffusion paths. Both dislocations and grain boundaries can modify mechanical behavior, but a discussion of this extremely important aspect would take us beyond the scope of the book.

Dislocations are normally defined by two vectors, the axis **g** and the Burgers vector **b**. The axis is usually easy to recognize. The Burgers vector is a concept necessary because the topology of the lattice (i.e., the connectivity) is altered by the dislocation [for a discussion, and a definition of the Burgers vector see Kittel (1953), p. 327]. It emerges that for monatomic crystals there are two extreme types of dislocation, *edge* dislocations ($\mathbf{b} \perp \mathbf{g}$) and *screw* dislocations ($\mathbf{b} \| \mathbf{g}$). Simple examples are shown in Figure 3.21. It is very important to realize that these are not the only cases. In the diamond structure there are various forms defined by the angle between **b** and **g**, giving the 30°, 60°, and related dislocations. Also, in a real crystal the direction of the dislocation may change with **b** constant. If the axis forms a closed loop, one has a so-called *dislocation loop* that can be regarded as a region in which there are extra (or fewer) planes of atoms (Figure 3.22).

Dislocations can also move and take part in solid-state reactions. The important concepts here are *glide*, in which only modest motions of any individual atom are involved (the traditional analogy is with moving in a ruck in a carpet) and *climb*, which needs the capture or loss of point defects (Figure 3.23). Both kinks and jogs on dislocation lines (where the line has a sudden step) are important because the motion of such point defects on the dislocation provide a mechanism of climb (Figure 3.24). These kinks or jogs may capture carriers (see also Section 6.3.5), contributing to the dislocation charge. Furthermore, kink motion may be enhanced in the presence of ionization or excess carriers (see Section 6.4.6).

So far we have referred to the dislocations present in a monatomic crystal, for example, argon. Even in monatomic crystals, dislocations may be charged. In diatomic hosts (e.g., MgO or NaCl or GaP) there are additional possibilities; examples of edge and screw dislocations are

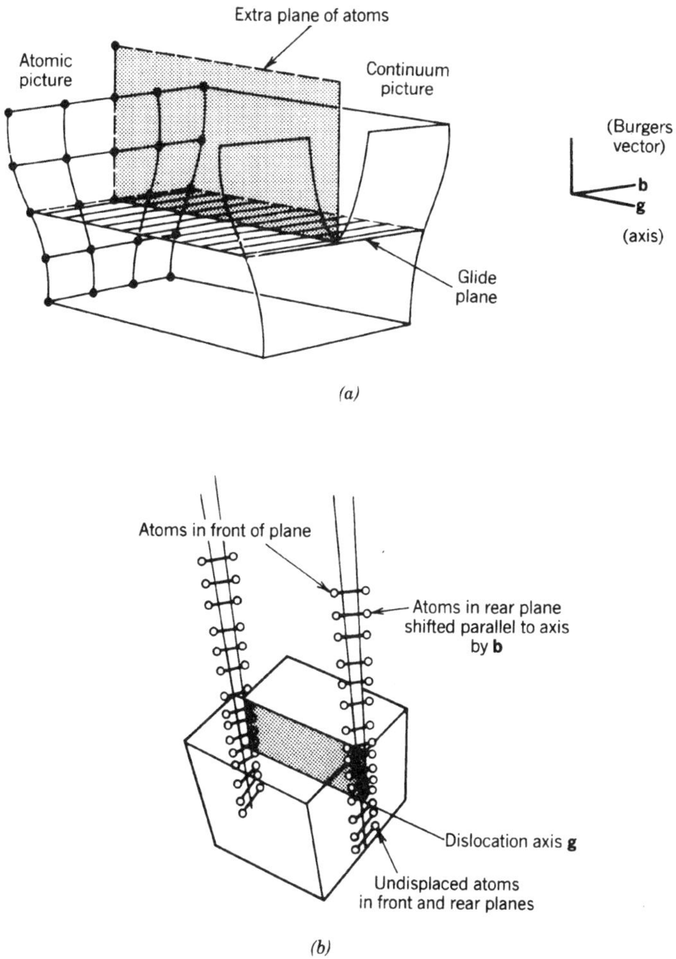

Figure 3.21 Simple models of (*a*) the edge dislocation and (*b*) the screw dislocation in monatomic lattices. Continuum limits are shown for (*a*). For the edge dislocation, Burgers vector **b** and axis **g** are perpendicular; for the screw dislocation, **b** and **g** are parallel.

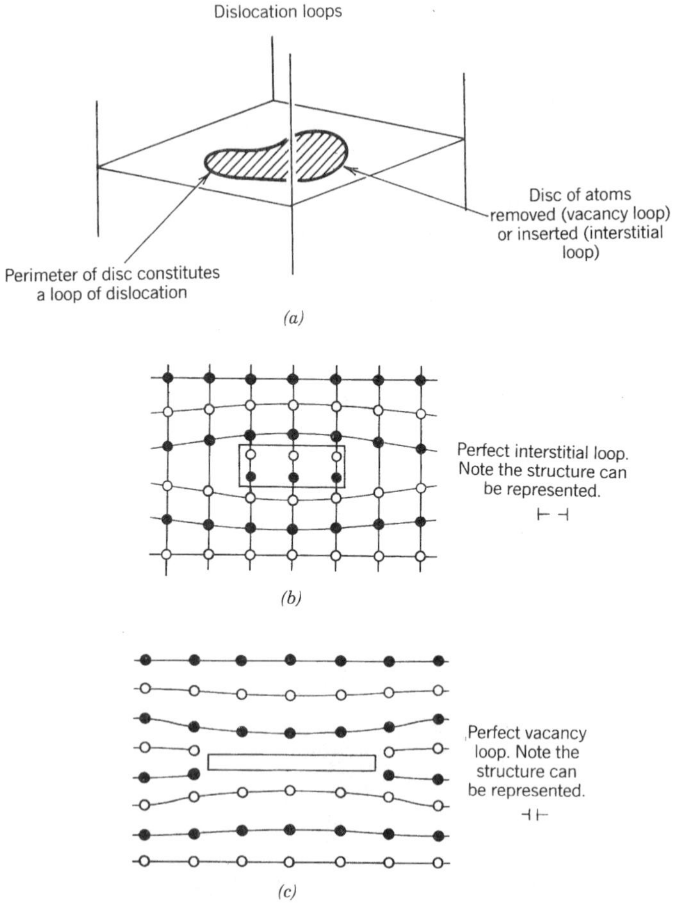

Figure 3.22 (*a*) shows schematically a disc of material added or removed whose perimeter can be regarded as a loop of dislocation line. Also shown are sections normal to perfect loops (i.e., free from stacking faults) of (*b*) interstitial and (*c*) vacancy form in a diatomic lattice. If dislocation energies are E_d per unit length, a large loop of radius R will have energy $2RE_d$, associated with R^2/a^2 atoms removed, that is, the energy per atom removed varies as $1/R$.

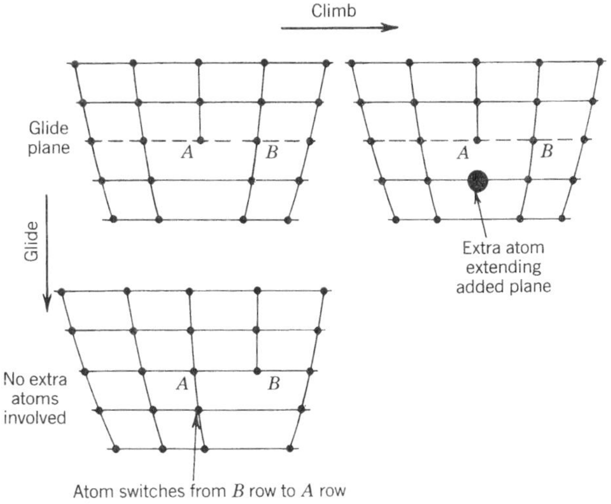

Figure 3.23 Dislocation climb and glide. Glide motion involves no extra atoms; climb needs extra atoms or vacancies. See also Figures 3.24 and 3.27.

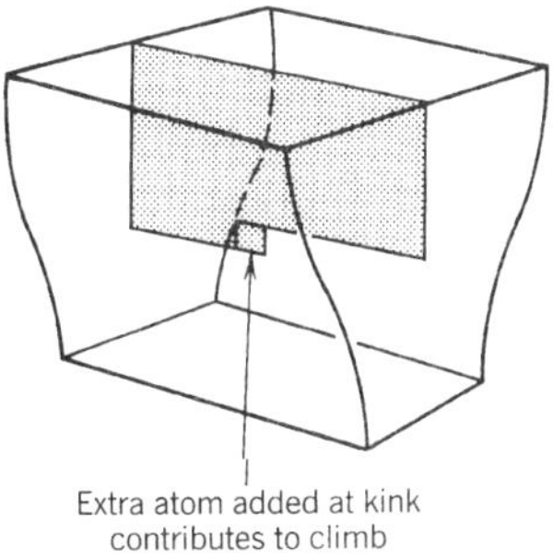

Figure 3.24 Kink on an edge dislocation. Dislocation climb may take place by adding an extra atom at a kink. Just as vacancies and interstitials take part in bulk solid-state reactions, so do kinks take part in dislocation processes.

shown in Figures 3.25 and 3.26. When climb occurs in diatomic crystals, ions of both species are needed, and this can cause complications if only one of the species is available (e.g., in Figure 3.27; see also Section 6.2.2). Dislocation loops are described as perfect if equal numbers of layers of each species are involved (Figure 3.22) and there is then no stacking fault on crossing the plane.

Stacking faults can be generated by the dissociation of a dislocation. The edge dislocation of Figure 3.28, for example, involves the insertion

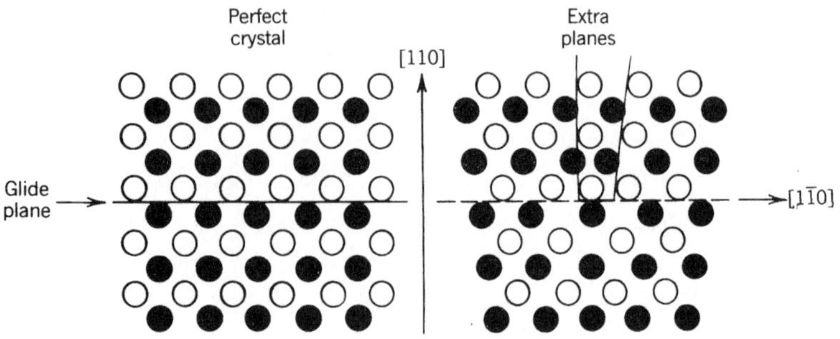

Figure 3.25 A pure edge dislocation in the NaCl structure, with [001] axis and [110] Burgers vector.

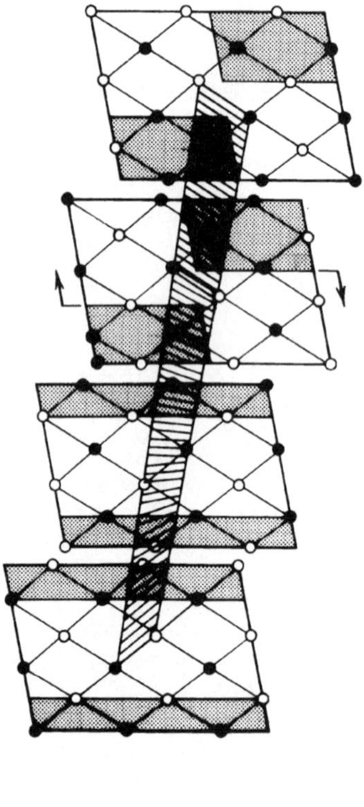

Figure 3.26 Screw dislocation in the NaCl structure. For clarity, the scale along the axis is expanded.

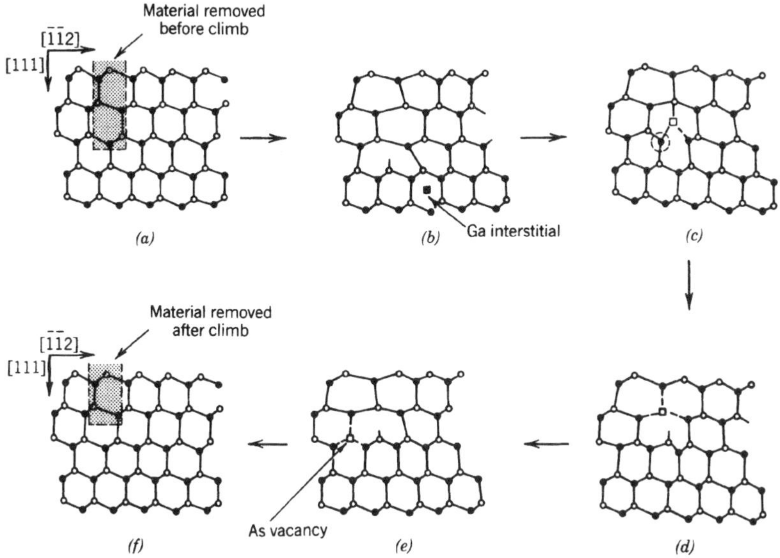

Figure 3.27 Dislocation climb in GaAs [after Petroff and Kimerling (1976)]. Here excess Ga interstitials (solid box) are able to cause climb, involving both Ga (solid circle) and As (open circle) sublattices by causing emission of As vacancies, (*a*) (110) plane, showing material removed in the dislocation. (*b*) Unreconstructed (i.e., with a dangling bond, here on As) 60° dislocation (Burgers vector $\frac{1}{2}a\langle 110\rangle$) with a nearby Ga interstitial. (*c*) Cation interstitial moves to the dangling bond so that, in effect, an As vacancy (open box) is created. In (*d*) and (*e*) two moves of the As vacancy reorganize the anion sublattice close to the dislocation core. (*f*) The new configuration, corresponding to (*a*) but after climb.

of extra cation and anion planes together, so that there are alternate anion and cation planes both above and below the glide plane. If the extra cation plane is separated from the anion plane (Figure 3.28), there is a reduction in elastic energy [this term is of the form Ab^2, where b is the Burgers vector, so two partial dislocations, each with half the Burgers vector, have energy $2A(b/2)^2$, i.e., $\frac{1}{2}Ab^2$] at the cost of stacking fault energy, proportional to the separation of the partials. Since the elastic energy can be calculated, observed separations of the partials give the stacking fault energy. Experimentally, the clearest cases of dissociation are seen in Al_2O_3 and Y_2O_3 (Castaing 1984). Stacking faults are one way of describing the polytype structure of crystals like ZnS and SiC (Figure 3.29).

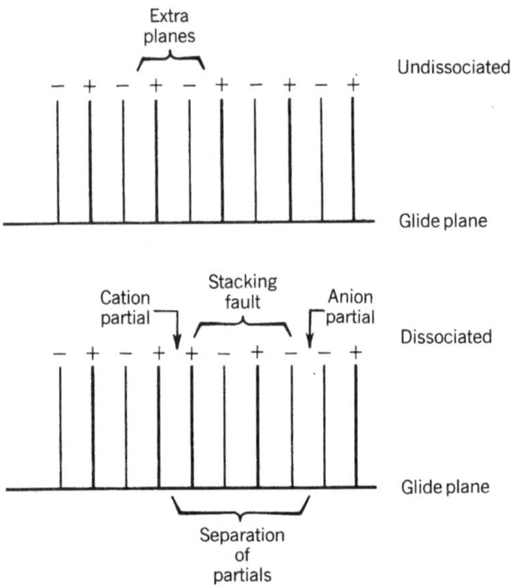

Figure 3.28 Dissociation of edge dislocations into partials. Only the planes above the glide plane are shown. The anion planes and cation planes are shown by lines of different thickness. The partials are separated by a stacking fault.

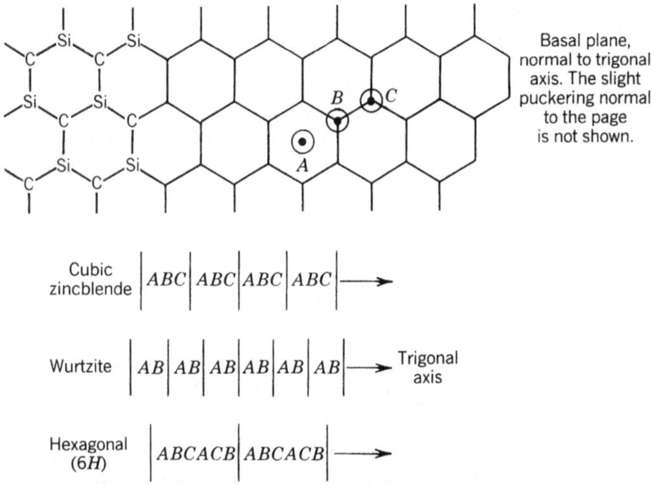

Figure 3.29 Stacking sequences in polytypes, illustrated here for SiC (or ZnS). In SiC all the Si atoms remain tetrahedrally coordinated by four C atoms. Each polytype can be regarded as built of a stack of planes normal to the trigonal axis. However, successive planes along a trigonal axis can be displaced, each centered on one of the three equivalent possibilities *A*, *B*, *C*. The order in which these displacements occur determines which polytype is found. Very many polytypes are known.

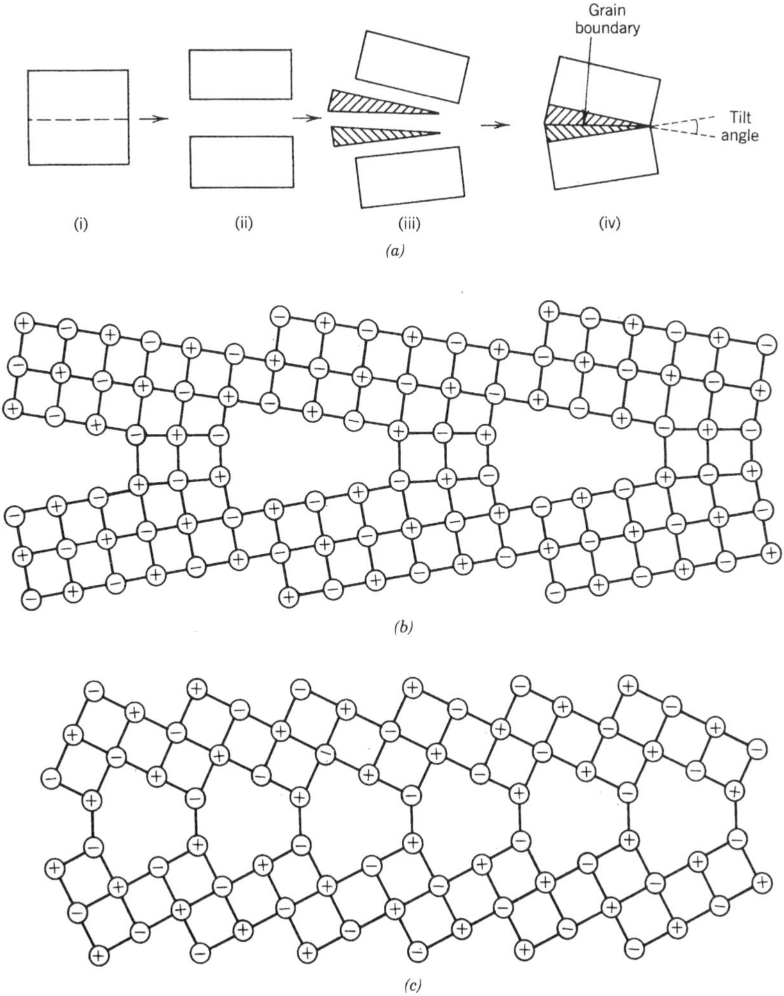

Figure 3.30 (*a*) Schematic representation of formation of a tilt boundary in NiO: (i) Perfect crystal, (ii) cut along a (100) plane, (iii) insert two wedges of perfect crystal each with an angle of $\theta/2$, (iv) rejoin to give a grain boundary with a tilt angle θ. An alternative description may be given in terms of edge dislocations; if a perfect edge dislocation is inserted at every *n* plane along a boundary, an (n10) tilt boundary is generated [see (*b*)]. (*b*) Schematic representation of an array of edge dislocations giving rise to a (610) tilt boundary in NiO [see (*a*)]. (*c*) A (210) tilt boundary in NiO; for the high misorientations shown here the relationship between edge dislocations and tilt boundaries breaks down [after Tasker and Duffy (1983)].

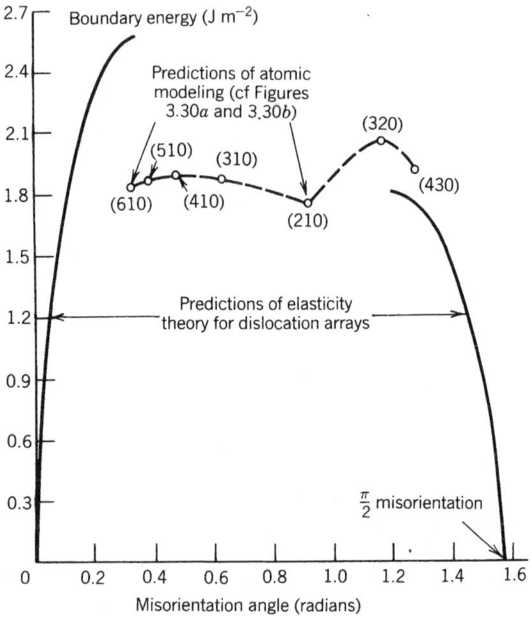

Figure 3.31 Calculated energies for tilt boundaries based on a (100) plane of NiO (see Figure 3.30). Tilts of 0 and $\pi/2$ correspond to perfect crystal and the boundary energy vanishes [after Tasker and Duffy (1983)].

Since surfaces are considered separately in Chapter 7, and since shear planes were described in Section 3.3.2 as a mechanism of nonstoichiometry, we shall finish this section with a brief description of grain boundaries. Suppose one takes a perfect crystal and cuts it on a plane (hkl). If one wants a twist boundary, one half is rotated through an angle θ and the two halves rejoined. If one wants a tilt boundary, one half has material shaved away (or added) to change the faces to angles of $\pm\theta/2$ to the original, and the two halves are rejoined.

For small angles of tilt, θ, tilt boundaries are conveniently considered as a row of edge dislocations, with axes in the boundary plane (Figure 3.30). For larger angles, which would correspond to much closer dislocations, this ceases to be a useful description (e.g., Figure 3.31). There is a very important difference between grain boundaries in metals and in ionic crystals. In metals, short-range forces dominate, and the lowest-energy forms are controlled by packing. In ionic crystals, Coulomb interactions dominate so that the lowest-energy forms place anions close to cations. In semiconductors bonding considerations also count, and one expects configurations to optimize both bond lengths and bond angles.

CHAPTER FOUR

Spectroscopy of Solids

In this chapter we shall be concerned with the absorption and emission of radiation by defects in the visible, near-infrared (Figure 4.1), and microwave regions of the electromagnetic spectrum. The complex line-shapes that are found in the optical spectra of defects are of fundamental interest in themselves, often indicating the way in which a defect couples to its host and even hinting at the way the defect may exhibit nonradiative transitions or participate in optical energy transfer.

For spectroscopy in all regions of the spectrum, absorption and emission are related through the Einstein A and B coefficients. We consider two energy levels, denoted 1 and 2, with populations N_1 and N_2, and with level 2 uppermost. There are three possible radiative processes connecting the two energy levels:

1. Atoms in level 2 decay spontaneously to level 1 with emission of energy $\hbar\omega_{12}$; the number of such transitions is $A_{21}N_2\,\mathrm{s}^{-1}\,\mathrm{cm}^{-3}$.
2. In the presence of radiation of density ρ and frequency ω_{12}, atoms in level 1 may jump to level 2 with absorption of energy $\hbar\omega_{12}$ from the radiation field; the number of such transitions is $B_{12}N_1\rho(\omega_{12})\,\mathrm{s}^{-1}\,\mathrm{cm}^{-3}$.
3. In the presence of radiation $\rho(\omega_{12})$, atoms in level 2 undergo induced transitions to level 1 with emission of energy $\hbar\omega_{12}$; the number of such transitions is $B_{21}N_2\rho(\omega_{12})\,\mathrm{s}^{-1}\,\mathrm{cm}^{-3}$.

In equilibrium the rate at which atoms enter level 1 or level 2 must equal the rate of departure. The principle of detailed balance requires that

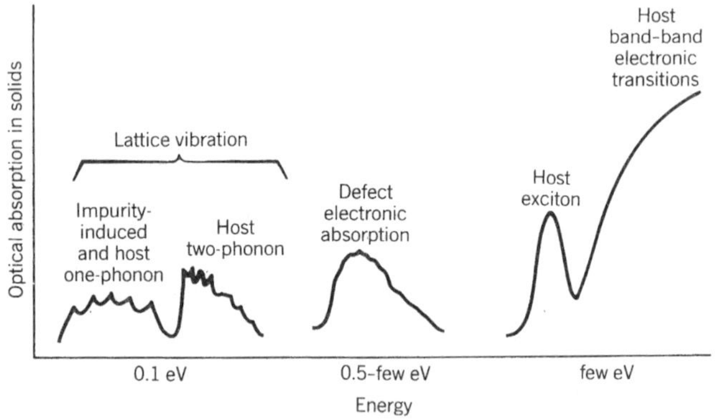

Figure 4.1 The main contributions to optical absorption in solids from lattice vibrations in the region ≲0.1 eV to the intrinsic band-to-band electronic transitions in the region 1–10 eV.

this equality must hold separately for nonradiative and radiative processes and for transitions to and from all other levels of the system. Confining our attention to levels 2 and 1 we therefore arrive at the well-known relationship involving the Einstein coefficients:

$$B_{12}N_1\rho(\omega_{12}) = A_{21}N_2 + B_{21}N_2\rho(\omega_{12}) \tag{4.1}$$

The spontaneous transition rate A_{21} for a particular type of transition (electric dipole, magnetic dipole) can be calculated if the wavefunctions of the states of the transition are known. Spontaneous transitions occur only from higher to lower states, so that A_{12} is zero. This emission has a finite width, related to the lifetime, which may be described by a normalized linehape function $g(\omega)$. To the lowest approximation $g(\omega)$ is a Lorentzian,

$$g(\omega) = (\Gamma/2\pi)[(\omega_0 - \omega)^2 + (\Gamma/2)^2]^{-1} \tag{4.2}$$

where ω_0 is the resonance frequency and Γ is the full width at half the maximum intensity; this is the same lineshape as that of a damped harmonic oscillation. The normalization condition is

$$\int_{-\infty}^{\infty} g(\omega)\, d\omega = 1 \tag{4.3}$$

The lineshape (4.2) may be contrasted with another type of lineshape, referred to as Gaussian. In this case the shape varies as $\exp[-\gamma(\omega_0 - \omega)^2]$, where $\gamma = 4\ln(2/\Gamma^2)$; it may be encountered in solids at high temperatures as a result of interaction with phonons [see Equation (4.17)].

The line broadening described by Equation (4.2) is due to a finite lifetime of the emitting center and is referred to as homogeneous. It is characterized by the fact that all the emitting atoms are indistinguishable and hence all have the same $g(\omega)$. There is another type of line broadening, referred to as inhomogeneous [Stoneham (1969); Figure 4.2], which arises when the emitting atoms are distinguishable. For example, atoms in a particular type of lattice site in a crystal may be distinguishable because crystals are never perfect and imperfections change from one part of a crystal to another. Changes in the lattice environment will affect emitted frequencies and hence contribute to line broadening.

The narrowest optical lines of defects in solids are generally $\gtrsim 1\ \text{cm}^{-1}$ (30 GHz), due to inhomogeneous broadening. We mention here, in passing, that this limit on resolution can be circumvented by a variety of techniques involving the use of narrow band (~1 MHz) tunable dye lasers, to reveal the homogeneous width. The technique of hole burning by lasers (Figure 4.2*c*) in inhomogeneously broadened lines [see, e.g., Macfarlane and Shelby (1979)] can increase the resolution over that obtainable by

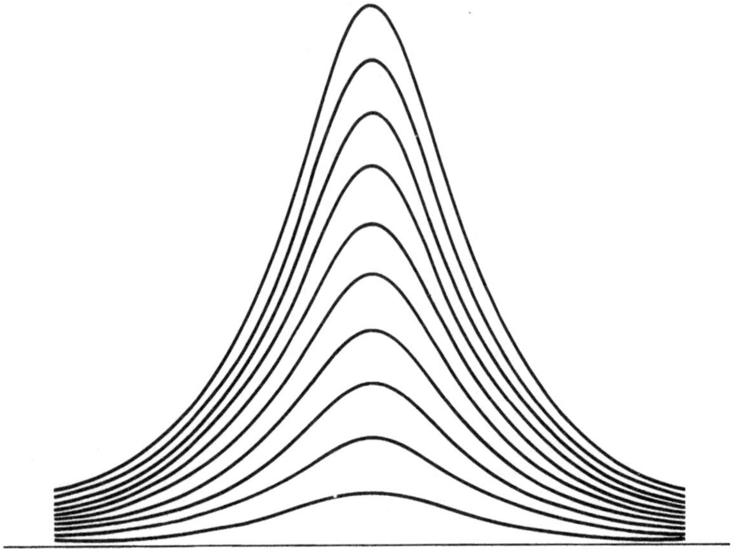
(a)

Figure 4.2(*a*)

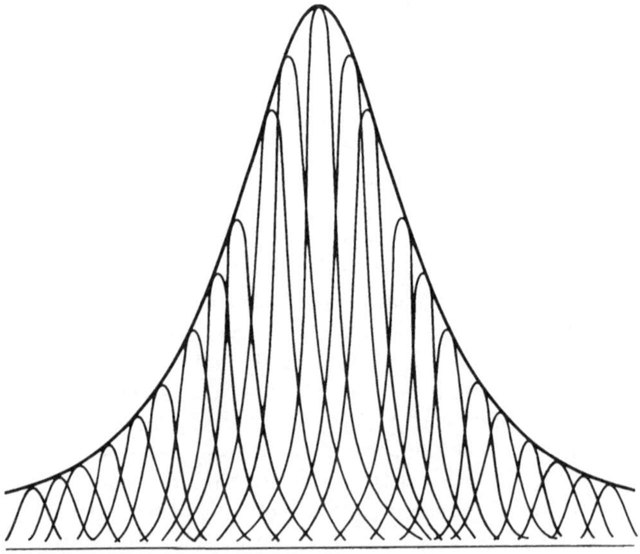
(b)

Figure 4.2(*b*)

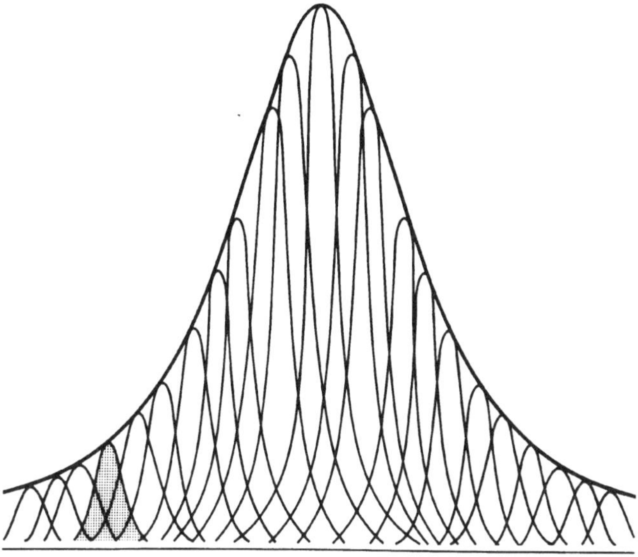

Figure 4.2 (*a*) Homogeneously broadened line shown schematically as a superposition of contributions from many individual centers in the solid, all giving absorption centered on the same frequency. (*b*) An inhomogeneously broadened line represented as a superposition of homogeneously broadened components. (*c*) Hole burning involves removal of absorption of one homogeneous component either by saturation or by photochemical change.

conventional diffraction-grating techniques by more than four orders of magnitude and can ultimately give resolution comparable to that obtainable by magnetic resonance methods (Shelby et al. 1983).

In Section 4.1 we set out some of the basic aspects of optical spectroscopy, and in Section 4.2 we discuss the diversity of optical lineshapes that occur in solids. The transfer of electronic excitation energy to lattice vibrations is dealt with in Section 4.3, and the transfer of electronic excitation energy between defect centers is covered in Section 4.4.

Although the optical properties of defects in solids had been studied extensively since the early 1920s, identification did not emerge with any degree of certainty until the application of electron-paramagnetic-resonance (epr) techniques in the middle 1950s. Because of the importance of these techniques in the evolution of defect studies in insulators and semiconductors, we outline the basic physics in Section 4.5.

4.1 SOME BASIC ASPECTS OF OPTICAL SPECTROSCOPY

It is helpful to start with the basics of spectroscopy. An electromagnetic field couples to an atom or molecule, or solid-state defect, either through the electrical dipole moment ($\mathbf{R} = |e| \sum_i \mathbf{r}_i$) or the magnetic dipole moment ($\mathbf{M} = \sum_i \mathbf{l}_i + 2\mathbf{s}_i$), where $\mathbf{l}_i$ is the orbital angular momentum and $\mathbf{s}_i$ is the spin of the electron i associated with the center. Effects of higher-order multipole terms are generally negligible.

It is the form of these operators that gives selection rules. Transition probabilities are proportional to the square of matrix elements, for example, to $|\langle a|\mathbf{R}|b\rangle|^2$. Symmetry causes some matrix elements to be zero, giving forbidden transitions. The most important rule for allowed transitions comes from inversion symmetry. For a defect with inversion symmetry, $\mathbf{R}$ connects states of opposite parity, for example, $s \leftrightarrow p$ transitions are electric dipole allowed; $\mathbf{M}$ connects states of the same parity, for example, $d \leftrightarrow d$ transitions are magnetic dipole allowed.

Optical absorption can be expressed in several alternative ways, all related. The absorption coefficient k_a measures attenuation of optical intensity with distance $[I \sim I_0 \exp(-k_a z)]$ and can be written in terms of the complex refractive index. The absorption cross section is $\sigma \equiv k_a/N$, where N is the number of absorbing centers per unit volume. The oscillator strength, f, is a dimensionless parameter measuring the strength of a transition and is especially convenient because it satisfies a sum rule. Finally, the lifetime τ in luminescence is related to the oscillator strength through the Einstein coefficients.

Consider an electric dipole transition between states ϕ_a and ϕ_b, where the energy difference is $E_{ab} = \hbar\omega_{ab}$. The transition matrix element is $\mathbf{R}_{ab} \equiv \langle \phi_a|\mathbf{R}|\phi_b\rangle$. The oscillator strength f_{ab} for an atom or molecule in free space is then

$$f_{ab} = \frac{2m\omega_{ab}}{3\hbar} |R_{ab}|^2 \tag{4.4}$$

where m is the mass of the electron, and the absorption cross section σ_{ab} is

$$\sigma_{ab} = \frac{2\pi^2 e^2}{mc} f_{ab}\delta(\omega - \omega_{ab}) \tag{4.5}$$

There are two important rules involving transition strengths, both exact if the wavefunctions are exact. The first states that $|R_{ab}|^2$ is equal to $|\mathbf{P}_{ab}|^2(m\omega_{ab})^2$, where $\mathbf{P}_{ab} \equiv \langle \phi_a|\mathbf{P}|\phi_b\rangle$ is the momentum matrix element of the electron. The second (the Thomas–Kuhn sum rule) tells us that, for a one-electron system, $\sum f_{ab} = 1$, that is, the sum of oscillator strengths over all transitions is fixed at unity. Both rules have wide applications to atomic and solid-state systems.

In a solid, the velocity of light is reduced by the refractive index n and, in addition, the actual electric field coupling to the electric dipole moment of a defect is multiplied by a factor (E_{eff}/E), known as the local-field correction. This factor arises because the effective field, E_{eff}, inducing transitions at the defect is different from the applied field, E, within the crystal, far from the defect. This difference occurs because the polarization of the defect and its immediate environment is different from that of the perfect crystal. Indeed, the interaction between the defect polarization and host polarization complicates matters further. Many workers favor the Lorenz–Lorentz expression, $E_{\text{eff}}/E=(n^2+2)/3$, for the local-field correction, where n is the optical dielectric constant. This value is usually far too high for electronic transitions at defects, especially in semiconductors ($n^2\sim 10$), where experimental values in the range $1<E_{\text{eff}}/E_0\lesssim 2$ are typical. The combined effect of the refractive index and local-field corrections on σ or k_a is to multiply them by $n(E_{\text{eff}}/E)^2$.

The lifetime τ of an excited state is related to the oscillator strength f by the useful expression

$$f\tau=\frac{c}{2e^2}\frac{1}{(2\pi)^2}\{\lambda^2/[n(E_{\text{eff}}/E)^2]\}$$
$$=(4\times 10^{-9}\,\text{sec})(\lambda/5000\,\text{Å})^2/[n(E_{\text{eff}}/E)^2] \tag{4.6}$$

that is, $f\tau$ is proportional to the square of the wavelength λ. A second useful formula is Smakula's expression for an absorption band,

$$N\left(\frac{E_{\text{eff}}}{E}\right)^2 f=An\alpha k_a^{\text{max}}\Gamma \tag{4.7}$$

where N is the number of absorbing centers per cm^3, $A=0.82\times 10^{17}/\text{cm}^3$ is a constant, α is a factor determined by details of the lineshape (usually $1\lesssim\alpha\lesssim\pi/2$), k_a^{max} is the maximum absorption coefficient, and Γ is the full width at half intensity of the band. If the oscillator strength for a transition is known, it is possible to use Equation (4.7) to determine N.

4.2 OPTICAL TRANSITIONS COUPLED TO LATTICE VIBRATIONS

The electronic states and energies of defects in solids are modulated by lattice vibrations. When there is an electronic transition, this defect–lattice coupling changes the delta function $\delta(\omega-\omega_{ab})$ of Equation (4.5) to the lineshape function $G(\omega - \omega_{ab})$. We now show how the coupling alters the observed shape.

The configuration coordinate diagram (Figure 2.14) is a convenient starting point. Within the adiabatic (Born–Oppenheimer) approximation the electrons rapidly reorganize themselves to follow the sluggish atomic motions. In any one-electron state we may write the vibronic wavefunction, that is, the wavefunction for the coupled electronic and vibrational motion, as $\psi_a(\mathbf{r}, \mathbf{Q}) = \phi_a(\mathbf{r}; \mathbf{Q}_{a0})\chi_a(\mathbf{Q} - \mathbf{Q}_{a0})$. Here the electronic part ϕ depends parametrically on the nuclear coordinates $\mathbf{Q}$. The vibrational part χ describes vibrations about a mean position $\mathbf{Q}_{a0}$, which depends on the electronic state. Hence a transition from ψ_a to ψ_b will involve a change in mean displacement, $\mathbf{Q}_{a0} - \mathbf{Q}_{b0}$, a factor determining the lineshape function (Section 4.2.2).

Since we are concerned with both absorption and emission, we mention some general points. First, we shall usually assume that nonradiative thermalizing transitions in excited electronic states take place faster than any luminescence. In some cases so-called hot luminescence is seen from excited vibrational levels of the excited electronic state to the ground electronic state, notably for molecular impurities in crystals. This is a fast, generally weak multistep process, and the system loses memory of the polarization of the incident light. By contrast, resonance Raman scatter retains this memory. Second, we may not see any luminescence at all because of efficient nonradiative transitions from the excited electronic state to the ground electronic state. Third, we shall want to know what relation exists between the shape of an absorption line ($a \rightarrow b$) and the corresponding emission ($b \rightarrow a$).

4.2.1 The Lineshape Function

The matrix element that appears in the oscillator strength for transition between vibronic states a and b is

$$\int d\mathbf{Q}\chi_a^*(\mathbf{Q} - \mathbf{Q}_{a0})M_{ab}^{el}(\mathbf{Q})\chi_b(\mathbf{Q} - \mathbf{Q}_{b0}) \tag{4.8}$$

in which M_{ab}^{el} is the matrix element of the electronic dipole operator:

$$M_{ab}^{el}(\mathbf{Q}) = \int d^3\mathbf{r}\phi_a^*(\mathbf{r}; \mathbf{Q})\mathbf{R}\phi_b(\mathbf{r}; \mathbf{Q}) \tag{4.9}$$

If $M_{ab}^{el}(\mathbf{Q})$ does not depend significantly on $\mathbf{Q}$ (the Condon approximation), several important results follow. First, the transition matrix element factors, giving an electronic factor M_{ab}^{el} and a vibrational factor

$$M_{ab}^{vib} = \int d\mathbf{Q}\chi_a(\mathbf{Q}_a - \mathbf{Q}_{a0})\chi_b(\mathbf{Q} - \mathbf{Q}_{b0}) \tag{4.10}$$

Because the mean displacements $\mathbf{Q}_{a0}$, $\mathbf{Q}_{b0}$ in the initial and final states differ (see Figure 2.14), M_{ab}^{vib} can involve initial and final states that correspond to phonons being emitted or absorbed in the optical transition. This is the origin of the lineshape function $G(\omega - \omega_{ab})$, which replaces $\delta(\omega - \omega_{ab})$ in the expression (4.5) for σ. Like the δ function, the lineshape function also has unit area:

$$\int d\omega G(\omega) = \int d\omega\delta(\omega) = 1 \tag{4.11}$$

Since $\sigma(\omega)$ is proportional to $\omega G(\omega)$ [see Equations (4.4) and (4.5)], we can derive the law of constant area, which tells us that, for a transition centered on ω_{ab},

$$\begin{aligned} \int d\omega\sigma_{ab}(\omega) &\propto \int d\omega\, \omega G(\omega - \omega_{ab}) \\ &\propto \omega_{ab} \int d\omega G(\omega - \omega_{ab}) \\ &\propto \omega_{ab} \end{aligned} \tag{4.12}$$

Hence the integrated absorption is independent of electron–lattice coupling, but is proportional to the mean transition energy; the area will increase under a perturbation, which raises ω_{ab}, a result that can be used as a test of the Condon approximation (Schnatterly 1965).

A second important result is the rule of mirror symmetry. Under conditions that we shall discuss, $G(\omega - \omega_{ab})$ for absorption and $G(\omega - \omega_{ba})$ for emission are identical, even though ω_{ab} and ω_{ba} may differ. Deviations from mirror symmetry usually indicate large changes in interatomic forces between initial and final states or the presence of electronic degeneracy (Section 4.2.3 and Figure 4.4).

4.2.2 Optical Lineshapes: Orbitally Nondegenerate States

We now return to the configuration coordinate diagram of Figure 2.14. The critical quantity determining optical lineshapes is the change $(Q_{a0} - Q_{b0})$ in displacement in the transition. We assume coupling of electrons to a single lattice mode only, of energy $\hbar\omega_0$ and effective mass M. If the interatomic harmonic force constants are unaltered in the transition (a useful and often adequate approximation), then the relaxation energy

$\frac{1}{2}M\omega_0^2(Q_{a0} - Q_{b0})^2$, measured in units of $\hbar\omega_0$, is a convenient estimate of the strength of the electron–lattice interaction:

$$S_{ab} = \frac{\frac{1}{2}M\omega_0^2(Q_{a0} - Q_{b0})^2}{\hbar\omega_0} = \frac{(Q_{a0} - Q_{b0})^2}{2(\hbar/M\omega_0)} \tag{4.13}$$

S_{ab} is known as the Huang–Rhys factor for the transition [see also Equation (2.56)]. Note that $\hbar/M\omega_0$ is the mean square amplitude of the zero-point motion.

The strong- and weak-coupling limits are defined solely by S. When $S \ll 1$, there is only a small change in mean atomic positions. This weak coupling is typical of transitions of rare-earth ions in solids and of aggregate color centers (Section 5.1.1). When $S \gg 1$, there is a massive displacement following a transition. Such strong coupling is shown by F centers in alkali halides (Section 5.1.1) and many other systems.

In the vibrational part of the matrix element, M_{ab}^{vib}, a and b define only the electronic states. Experiment does not monitor the individual vibrational states. Instead, one measures a thermal average over the initial vibrational states and a sum over the final states whose weights are determined by individual matrix elements of the sort

$$M_{ab}^{vib}(n_i, n_f) = \int dQ\chi_a^*(n_i; Q_a - Q_{a0})\chi_b(n_f; Q_b - Q_{b0}) \tag{4.14}$$

The mean number of phonons in the initial state, $\overline{n_i}$, is given by the Bose–Einstein expression $[\exp(\hbar\omega_0/k_BT) - 1]^{-1}$. Exact expressions for M_{ab}^{vib} are available [e.g., Stoneham (1975, p. 291)], which reduce to two important limits for lineshape calculations:

1. *Weak-Coupling Limit.* The lineshape function in this limit is

$$\begin{aligned} G(\omega - \omega_{ab}) = {} & [1 - S]\delta(\omega - \omega_{ab}) \\ & + \tfrac{1}{2}(S + S_0)\delta(\omega - \omega_{ab} - \omega_0) \\ & + \tfrac{1}{2}(S - S_0)\delta(\omega - \omega_{ab} + \omega_0) \\ & + \text{small terms} \end{aligned} \tag{4.15}$$

Here we have distinguished for convenience between S_0, which is just S_{ab} of Equation (4.13), and $S \equiv S_0 \coth(\hbar\omega_0/k_BT)$, which describes the temperature dependence of the electron–lattice interaction. In the weak-coupling limit we see (a) a zero-phonon line, with $n_i = n_f$, and energy corresponding to the minimum–minimum transition AC of Figure 2.14; (b) a one-phonon Stokes band, in which one phonon is emitted, so $n_f = n_i + 1$; and (c) a one-phonon anti-Stokes band, in which a phonon is absorbed. Obviously the anti-Stokes band disappears as the temperature falls, since no phonons are available for absorption. As the coupling S_0 increases, intensity is gradually transferred from the zero-phonon line to the one-phonon transitions. We may anticipate that as the coupling continues to increase, processes involving two and more phonons appear (Figure 4.3).

The line shape described by Equation (4.15) can be pictured without difficulty. In absorption at low temperatures there will be a sharp zero-phonon line, inhomogeneously broadened by random strains and internal electric fields. At higher energies there will be one-phonon transitions. Since many modes of many frequencies will be coupled to some extent, there should be a band of one-phonon transitions whose shape reflects the phonon density of states weighted by the mode Huang–Rhys factors.

2. *Strong-Coupling Limit.* Here $S_0 \gg 1$ and we can conveniently separate the low- and high-temperature limits. At low temperatures ($\hbar\omega_0 \gg k_B T$) a Poisson distribution results for the line shape:

$$G(\omega - \omega_{ab}) = \sum_{\Delta} [\exp(-S_0) S_0^{\Delta}/\Delta!]\delta(\omega - \omega_{ab} - \delta\omega_0\Delta) \qquad (4.16)$$

where $\Delta \equiv n_f - n_i$ represents the net number of phonons emitted. At high temperatures (or, strictly, at high mean values of Δ) a Gaussian lineshape emerges:

$$G(\omega - \omega_{ab}) \propto \exp[-\alpha(S_0)(\omega - \omega_{ab} - S_0\omega_0)^2] \qquad (4.17)$$

This is centered on the energy $\omega_{ab} + S_0\omega_0$, that is, on the energy for the vertical (Franck–Condon) transition (AB in Figure 2.14). The vibrational overlap is greatest in this case because, whereas the most probable ground state Q is near the minimum Q_{a0}, the most probable excited vibronic state is near the classical turning point at B. The linewidth has the root-mean-square value $\hbar\omega_0\sqrt{S} \equiv \hbar\omega_0\sqrt{S_0 \coth(\hbar\omega_0 2 k_B T)}$. From both the temperature dependence and the absolute value of the width one can extract both S_0 and $\hbar\omega_0$. The full width at half-maximum of the Gaussian line is $\sqrt{8 \ln 2}(\hbar^2\omega_0^2 S_0)^{1/2}$ at low temperatures and $\sqrt{8 \ln 2}(\hbar\omega_0 S_0 k_B T)^{1/2}$ at high temperatures.

In the strong-coupling limit, we expect a broad optical absorption band centered on the Franck–Condon energy, E_{abs}. The corresponding emission energy, E_{em}, differs from that for absorption by the Stokes shift, which is the sum of the relaxation energies in the ground and excited states (see Figure 2.14). In thermal transitions, the only matter of consequence is establishment of thermal equilibrium, and hence the

Range of Huang–Rhys Factors	Calculated Lineshape [Equation (4.16) text]
$S_0 \lesssim 1$ weak coupling	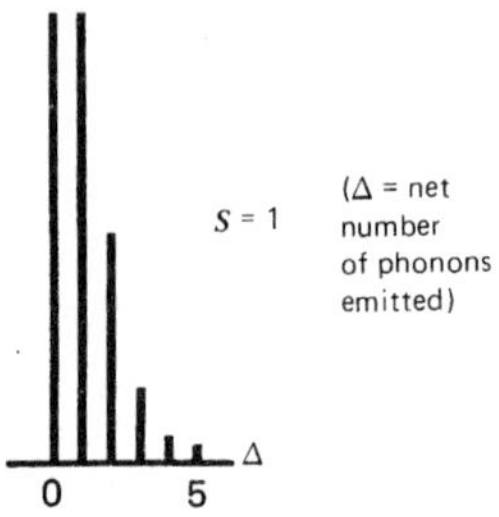
$1 \lesssim S_0 \lesssim 6$ intermediate coupling	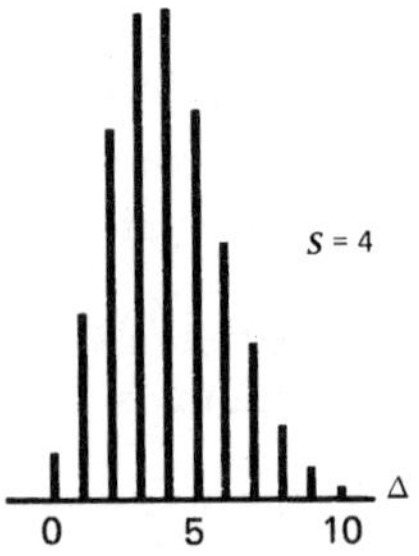
$S_0 > 6$ strong coupling	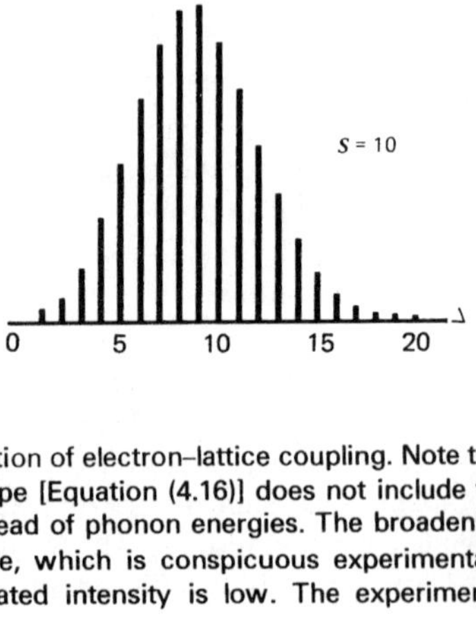

Figure 4.3 Defect optical lineshapes as a function of electron–lattice coupling. Note th the illustration here of the simple Poisson shape [Equation (4.16)] does not include tł broadening resulting from the continuous spread of phonon energies. The broadenir widens all peaks except the zero-phonon line, which is conspicuous experimenta because it is narrow, even when its integrated intensity is low. The experimen

<table>
<tr><th>Observed Spectrum</th><th>Examples of this Class</th></tr>
<tr><td>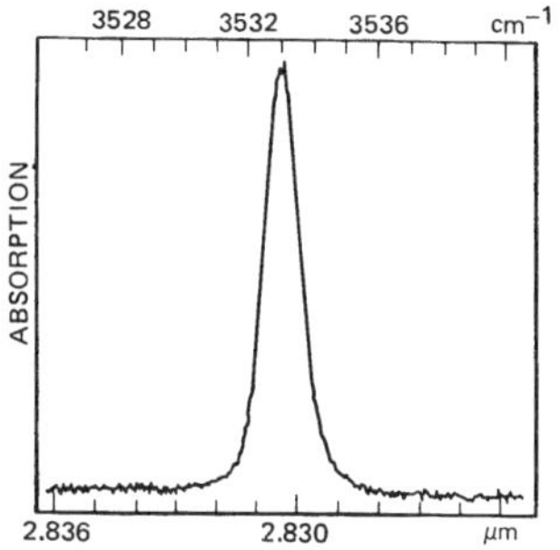

(a)</td><td>Donors and acceptors in Si and Ge; donor–acceptor pairs in semiconductors; rare-earth and some transition-metal ions; molecular impurities like O_2^- in alkali halides</td></tr>
<tr><td>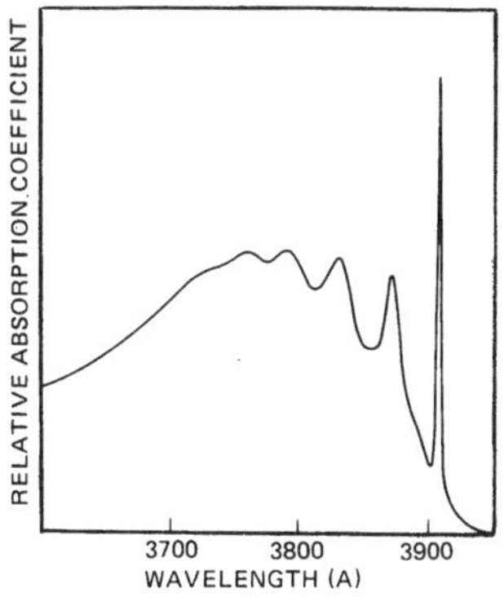

(b)</td><td>Vacancy and impurity centers in diamond; F^+ center in CaO; F center aggregates in alkali halides</td></tr>
<tr><td>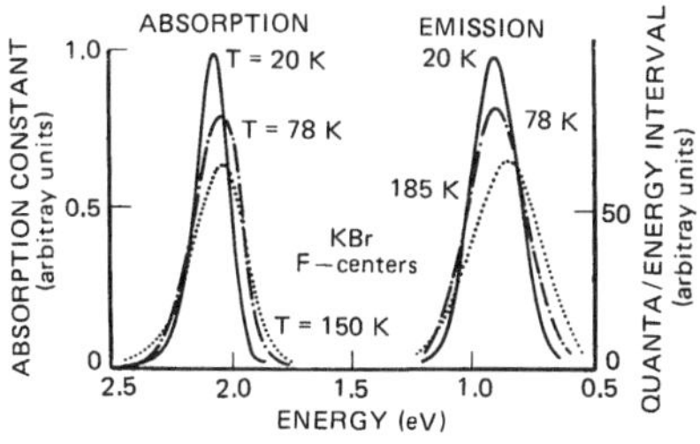
</td><td>F centers in alkali halides; V_k center and self-trapped exciton; V^- center and other charge-transfer bands</td></tr>
</table>

xamples given in order of increasing S are (a) emission of the Ni^+ ion in cubic ZnS at 5 K fter Clerjaud et al. (1984)], (b) absorption of R centers in LiF at helium temperatures [see ection 5.1.1 and Fitchen et al. (1963)], and (c) absorption and emission of F centers in Br as a function of temperature [after Gebhardt and Kühnert (1964); see Section 5.1.1].

overlap of the ground and excited vibrational states is not critical, as in optical transitions, and thermal excitation can readily take place to the lowest vibrational states of the electronic excited state. Thus, we have (see Figure 2.14)

$$E_{abs} = \hbar\omega_{ab} + S_0\hbar\omega_0 = E_B - E_A$$

$$E_{em} = \hbar\omega_{ab} - S_0\hbar\omega_0 = E_C - E_D$$

$$E_{\text{thermal}} = \hbar\omega_{ab} = E_C - E_A$$

The thermal excitation energy at low temperatures is the same as that of the zero-phonon line.

We shall now deal with transitions in nondegenerate electronic states of a defect and begin by considering the mean distortions Q_{a0} and Q_{b0} (Figure 2.14). In many cases Q is best thought of as a breathing mode, that is, the symmetric inward or outward movement of the nearest neighbors. The state-dependent part of the force which drives the distortion mainly comes from the state-dependent charge density of the defect electrons. Formally, we know that the total energy of the crystal is quadratic in Q when there is no defect present. (This is just Hooke's law, giving energy $E_0 + \frac{1}{2}M\omega_0^2Q^2$, where E_0 is the energy of the undistorted crystal.) The defect adds extra terms $E_{a0} - F_aQ$ in state a (ignoring higher-order terms), where E_{a0} is the defect energy when Q is zero and F_a is a defect force [cf. Equation (1.20)]. The total energy is now

$$E_a(Q) = E_0 + E_{a0} - F_aQ + \tfrac{1}{2}M\omega_0^2Q^2 \tag{4.18}$$

which can be rewritten

$$E_a(Q) = E_0 + E_{a0} - (\tfrac{1}{2}F_a^2/M\omega_0^2) + \tfrac{1}{2}M\omega_0^2\left(Q - \frac{F_a}{M\omega_0^2}\right)^2 \tag{4.19}$$

Equation (4.19) shows that there is a defect relaxation energy $E_R = (F_a^2/2M\omega_0^2)$ and a mean distortion of the surrounding lattice, $Q_{a0} \equiv F_a/M\omega_0^2$. The vibrational frequency remains as ω_0, a general feature of linear coupling.

Parallel calculations can be made for the b state. The change in displacement becomes $(Q_{a0} - Q_{b0}) \equiv (F_a - F_b)/M\omega_0^2$, corresponding to [cf. Equation (4.13)]

$$S_0 = (F_a - F_b)^2/2M\hbar\omega_0^3 \tag{4.20}$$

The Huang–Rhys factor depends, therefore, on $F_a - F_b$ and on the force constant $M\omega_0^2$. Large S_0 corresponds to a large change in charge density (e.g., in electronic orbital radius) and to low force constants.

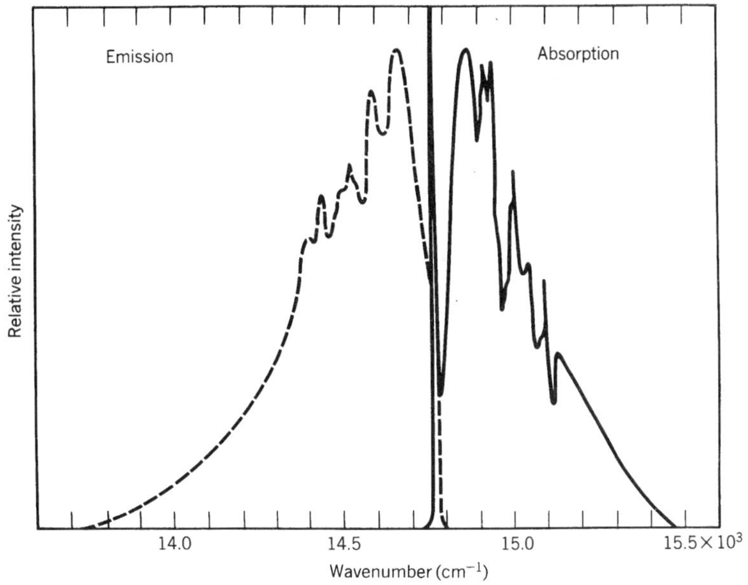

(a)

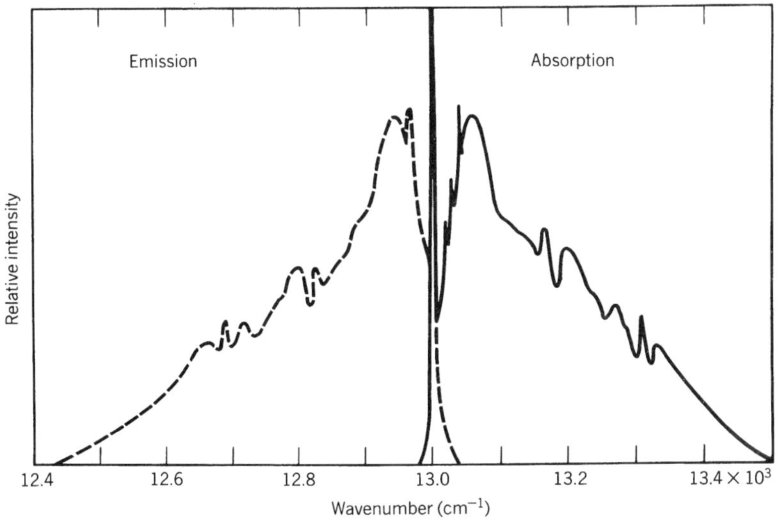

(b)

Figure 4.4 Mirror symmetry between absorption and emission bands for *R* centers in (*a*) CaF_2 and (*b*) SrF_2 at 20 K [after Beaumont et al. (1972)].

We see that S_0 is the same for the $a \rightarrow b$ and $b \rightarrow a$ transitions. This gives the rule of mirror symmetry, mentioned earlier, and is illustrated in Figure 4.4. If the force constants are state dependent (i.e., if ω_0 is different in states a and b), this symmetry is removed. Mirror symmetry is also broken when there is orbital degeneracy (Section 4.2.3).

4.2.3 Optical Lineshapes: Orbitally Degenerate States

Orbital degeneracy, for example, the threefold degeneracy (p_x, p_y, p_z) of p states in cubic symmetry, leads to special behavior when there is linear electron–lattice coupling. This can be traced to the Jahn–Teller theorem, which states that, apart from certain unimportant cases, an orbitally degenerate state in a molecule or a solid always gives rise to a distortion that lowers the local symmetry, thus removing the orbital degeneracy and lowering the energy (see e.g., Sturge 1967). The exceptions include linear molecules and also Kramers' degeneracy, that is, the residual degeneracy of a system with an odd number of electrons, which can only be removed by an applied magnetic field. The origin of the Jahn–Teller theorem is the fact that elastic interactions (quadratic in displacements) cannot restrain a defect in undistorted surroundings when there are linear electron–lattice coupling terms. The proof then rests on symmetry arguments which show that, when there is degeneracy, there will always be linear terms associated with distortions that lower the symmetry. Apart from lowering the local symmetry at a defect the Jahn–Teller effect may result in unexpected differences between matrix elements for the response to external fields (magnetic fields or uniaxial stress) as deduced from zero-phonon lines and from broad Franck–Condon bands (the Ham effect; see below).

We illustrate the Jahn–Teller effect by considering Cu^{2+} ($3d^9, {}^2D$) in sixfold coordination. The undistorted geometry is octahedral and in this symmetric configuration the d orbital splits into a doublet (e) and a triplet (t_2) with the unpaired electron in the doublet (Figure 4.5). The Cu^{2+} $d(e)$ orbitals transform in the same way as $\theta \equiv (2z^2 - x^2 - y^2)/\sqrt{6}$ and $\epsilon \equiv (x^2 - y^2)/\sqrt{2}$ and are said to have e- or E-type symmetry. (Note there are three functions $x^2 - y^2$, $y^2 - z^2$, $z^2 - x^2$ in cubic symmetry, but only two are linearly independent because the sum of three is zero.) These orbitals will couple to tetragonal distortions Q_2 and Q_3 of the neighbors (Figure 4.6); Q_2 transforms like ϵ and Q_3 like θ. Symmetry arguments alone show that the linear electron–lattice coupling matrix must have the form

$$V_{JT} = -A \begin{pmatrix} -Q_3 & Q_2 \\ Q_2 & Q_3 \end{pmatrix} \tag{4.21}$$

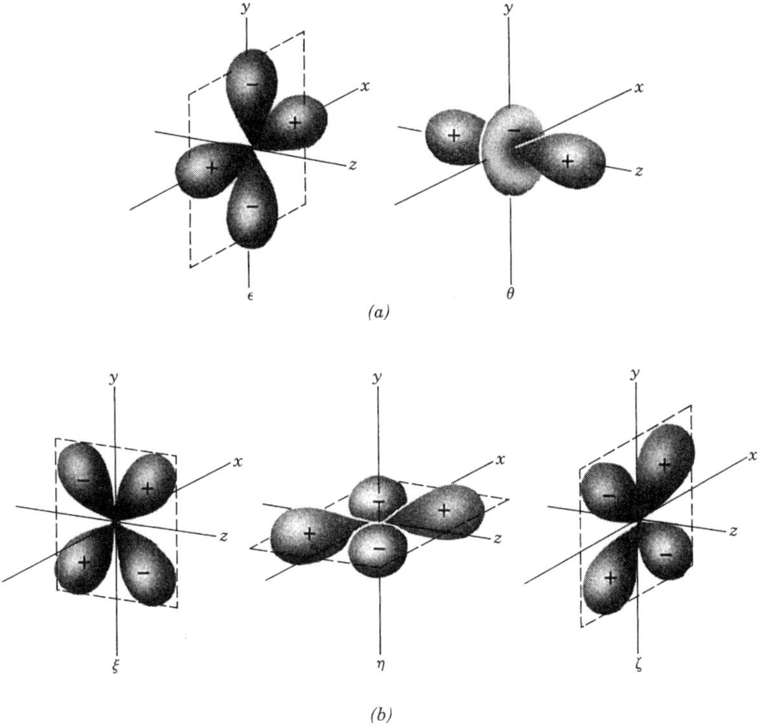

Figure 4.5 Wavefunctions for symmetrized combinations of *d* orbitals. (*a*) Orbitals of *e* symmetry, $\epsilon \sim (x^2 - y^2)/\sqrt{2}$, $\theta \sim (2z^2 - x^2 - y^2)/\sqrt{6}$; (*b*) orbitals of t_2 symmetry, $\xi \sim yz$, $\eta \sim zx$, $\zeta \sim xy$.

that is, $\langle\theta|\mathcal{H}_{JT}|\theta\rangle = -\langle\epsilon|\mathcal{H}_{JT}|\epsilon\rangle = AQ_3$ and $\langle\epsilon|\mathcal{H}_{JT}|\theta\rangle = \langle\theta|\mathcal{H}_{JT}|\epsilon\rangle = AQ_2$. The value of A depends on the details of bonding to the neighbors and is not fixed by symmetry alone.

The adiabatic potential-energy surface which determines the vibronic (vibrational plus electronic) energy is the sum of $\mathcal{H}_{JT}$ and the elastic energy, giving

$$V_{\text{eff}} = -A\begin{pmatrix} -Q_3 & Q_2 \\ Q_2 & Q_3 \end{pmatrix} + \tfrac{1}{2}\mu\omega_E^2(Q_2^2 + Q_3^2), \tag{4.22}$$

in which μ is an atomic effective mass and ω_E is the vibrational frequency (hence $\mu\omega_E^2$ is a force constant). It is convenient to transform

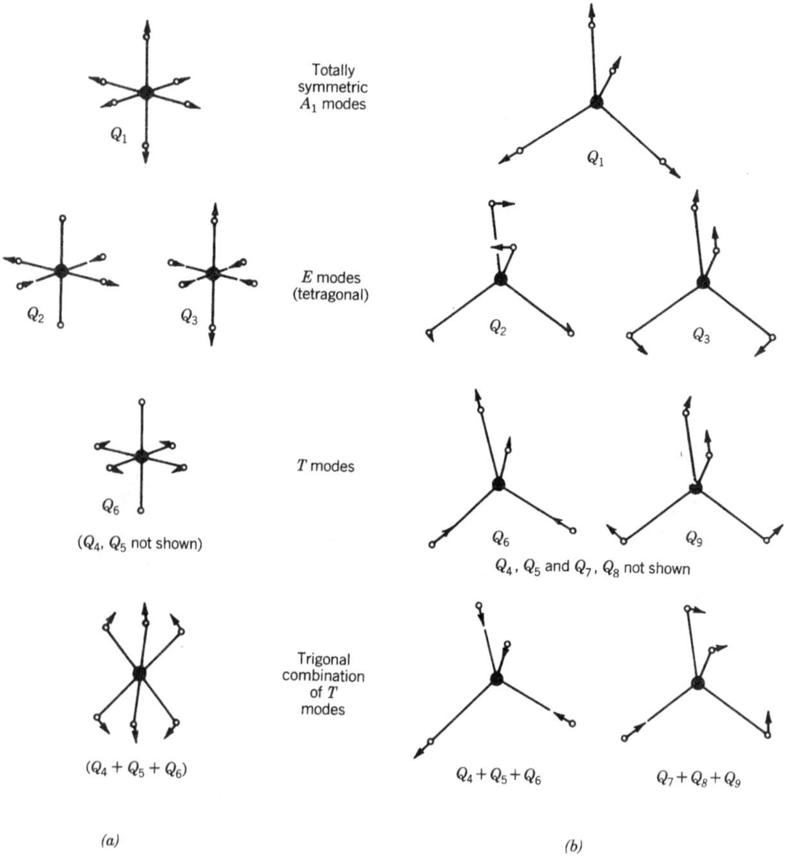

Figure 4.6 Symmetrized displacements of the neighbors of a transition-metal ion in (*a*) an octahedral and (*b*) a tetrahedral environment. These are totally symmetric (Q_1 of A_1 symmetry), tetragonal (Q_2, Q_3 of E symmetry), and combinations of trigonal modes Q_4, Q_5, Q_6, Q_7, Q_8, Q_9 of T symmetry. Also shown is an example of a trigonal distortion formed by superpositions of T modes (for further discussion see e.g. Sturge 1967).

V_{eff} into polar coordinates, writing $Q_3 \equiv \rho \cos\theta$ and $Q_2 \equiv \rho \sin\theta$; ρ is a measure of the absolute magnitude of the distortion from the octahedral and θ is an angle that defines the tetragonal axis of the distortion. We then find V_{eff} takes the form

$$V_{\text{eff}} = -A\rho \begin{pmatrix} -\cos\theta & \sin\theta \\ \sin\theta & \cos\theta \end{pmatrix} + \tfrac{1}{2}\mu\omega_E^2\rho^2 \tag{4.23}$$

with eigenvalues

$$V_{\text{eff}} = \pm A\rho + \tfrac{1}{2}\mu\omega_{\text{E}}^2\rho^2 \tag{4.24}$$

This result has several important features, some general and some specific. First, V_{eff} has two values. For a given distortion there will be two states; one is determined by the upper energy surface and the other (ground) state is determined by the lower energy surface, so that the twofold degeneracy is lifted. Second, the minimum value of V_{eff} is lower in energy than the undistorted configuration by the Jahn–Teller energy $E_{JT} = A^2/2\mu\omega_{\text{E}}^2$, and this minimum energy occurs for displacements $|\rho| = A/\mu\omega_{\text{E}}^2$. Third—and this is a feature special to this case—V_{eff} does not depend on θ; any linear combination of tetragonal distortions is equally effective if the magnitude is right. The potential-energy surface has the so-called Mexican hat structure of Figure 4.7. If our model were complete, there could only be a dynamic Jahn–Teller effect, since rotation in θ would occur without restriction, and any experiment whose timescale was longer than the rotation period would record octahedral symmetry.

In fact, there are important higher-order terms. The values $\theta = 0$, $\pm 120°$ (or $\pm 90°$, $180°$ depending on whether Q_2 or Q_3 gives the lowest energy) correspond to distortions along the cubic axes. The higher-order terms tend to stabilize distortions along these axes, so that three minima occur. For low barriers between the minima, tunneling and rapid reorientation can still happen. In this dynamical regime the higher-order terms act as a perturbation on the freely-rotating "Mexican hat" system. As the barriers rise the static Jahn–Teller limit is reached and the distortion is frozen into one of the minima for times long compared with characteristic experimental timescales. However, at high temperatures thermally activated reorientation can lead to an averaged cubic symmetry once more. As an order of magnitude, the Jahn–Teller energy

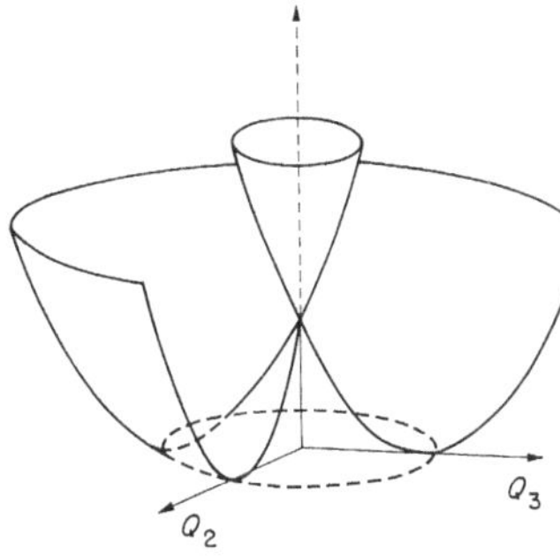

Figure 4.7 Potential-energy (V) surface for a doubly degenerate electronic state of e symmetry coupled to tetragonal (E) vibrational modes (see Figure 4.6). The dotted line (trough) indicates the lowest-energy distortions, degenerate in this ideal case. In real systems there are three equally spaced minima and three maxima around this trough.

$A^2/2\mu\omega_E^2$ may be 1000 cm^{-1}; both the barrier from higher-order terms and the phonon energy $\hbar\omega_E$ might be 100–200 cm^{-1}. For AgCl:Cu^{2+} both the isotropic and the anisotropic epr spectra coexist from 90 to 300 K. Other systems that show this type of behavior include both the neutral vacancy in diamond ($GR1$ center; see Section 5.4) and substitutional nitrogen in diamond.

We could have started equally well from the static Jahn–Teller limit and moved to the dynamic case. This has some conceptual advantages. In the static limit there are three degenerate distorted states, corresponding to x, y, and z axes of distortion. When there is weak tunneling between them, the three states are not exact eigenstates. The true eigenstates in the weak-tunneling limit are a linear combination with A symmetry and energy E_A,

$$|A\rangle = (|X\rangle + |Y\rangle + |Z\rangle)/\sqrt{3} \tag{4.25}$$

and two degenerate combinations of E symmetry and energy E_E,

$$\begin{aligned} |E\alpha\rangle &= (2|X\rangle - |Y\rangle - |Z\rangle)/\sqrt{6} \\ |E\beta\rangle &= (|X\rangle - |Y\rangle)/\sqrt{2} \end{aligned} \tag{4.26}$$

where $|X\rangle$ is the Born–Oppenheimer vibronic state associated with the x minimum, etc. The E states are the lowest in energy, reduced relative to $|X\rangle$ by a tunneling energy Γ, whereas $|A\rangle$ is raised by 2Γ. As can be seen from Figure 4.8 this tunneling picture leads directly to the dynamic limit

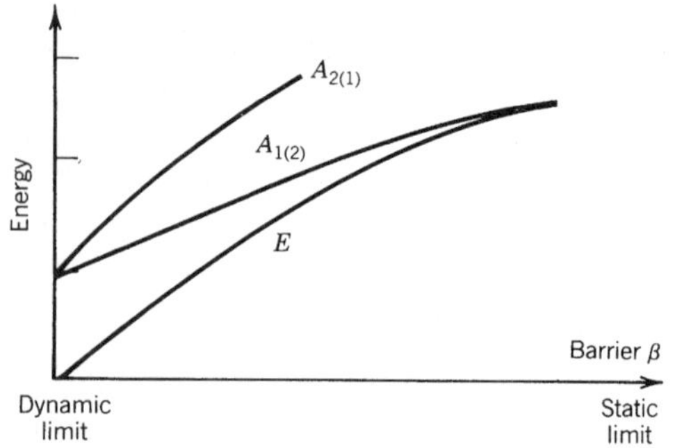

Figure 4.8 Energy levels for an e electronic state coupled to E vibrational modes, showing dynamic and static limits. β is the energy difference between maxima and minima around the energy trough of Figure 4.7. In the static limit, the threefold degeneracy corresponds to the three possible tetragonal distortions (x, y, z). In the dynamic limit, the levels are equally spaced, like those of a free rotor. Compare this figure with Figure 2.13, but note that the Jahn–Teller case has a doubly-valued energy surface (see Figure 4.7 where there is an upper and lower surface for each distortion) [after O'Brien (1964)].

we have discussed, starting from Equations (4.23) and (4.24). In cases like $MgO:Cu^{2+}$ and $CaF_2:Sc^{2+}$ (a $3d^1$, ^2D system) Γ is about $10\,cm^{-1}$, and tunneling between different axes of distortion is observed. When Γ is very small (say $1\,cm^{-1}$), random strains in a crystal have important effects, mixing states and inhibiting tunneling.

We have considered so far only the case of an E electronic state coupled to E vibrations. There are three other important coupling cases in cubic or tetrahedral symmetry:

(a) *T Electronic State Coupled to E Vibrational Modes* (*referred to as* $T \times \varepsilon$). Both T_1 and T_2 electronic states (each with threefold degeneracy) couple to tetragonal distortions. This case is especially simple since the matrix V_{JT} has only diagonal matrix elements.

(b) *T Electronic State Coupled to T Vibrational Modes* (*referred to as* $T \times \tau$). T_1 and T_2 electronic states can exhibit tetragonal or trigonal distortions. The coupling that gives the largest reduction in energy determines the actual symmetry. Mixed trigonal and tetragonal distortions do not occur except in special circumstances of accidental degeneracy.

(c) *Near Degeneracy*. Suppose a T state and an A state are separated by a small energy Δ. There are linear-coupling matrix elements of T distortions that mix the A and T electronic states. If they are large enough, there is a Jahn–Teller distortion even though the degeneracy is not exact. If the Jahn–Teller energy were E_{JT} for $\Delta = 0$, there would still be a distortion so long as $\Delta < 4E_{JT}$. This result is useful also in the case of an isolated T state where spin–orbit coupling splits the state to first order (Öpik and Pryce 1957).

Suppose a T electronic state is subject to a Jahn–Teller interaction with E vibrational modes. Consider the matrix elements of an electronic operator Ω among the lowest vibronic states in each minimum. Each such state can be approximated by a Born–Oppenheimer product (see Section 4.2)

$$|X\rangle = \phi_x(\mathbf{r}; \mathbf{Q}_{x0})\chi_x(\mathbf{Q} - \mathbf{Q}_{x0}) \tag{4.27}$$

so that a matrix element of an operator Ω between vibronic states would be

$$\langle Y|\Omega|X\rangle = \int d^3\mathbf{r}\,\phi_y^*(\mathbf{r}; \mathbf{Q}_{y0})\Omega(\mathbf{r})\phi_x(\mathbf{r}; \mathbf{Q}_{x0}) \times \int d\mathbf{Q}\,\chi_y^*(\mathbf{Q} - \mathbf{Q}_{y0})\chi_x(\mathbf{Q} - \mathbf{Q}_{x0}) \tag{4.28}$$

that is, the product of an electronic matrix element Ω_{yx}^{el} and a vibrational overlap factor

$$\int d\mathbf{Q}\chi_y^*\chi_x \simeq \exp\left(-\alpha\frac{E_{JT}}{\hbar\omega}\right) \tag{4.29}$$

where α depends on the system considered. Clearly the matrix element between off-diagonal vibronic states $\langle Y|\Omega|X\rangle$ is reduced by $\exp(-\alpha E_{JT}/\hbar\omega)$ over the value between purely electronic states Ω_{yx}^{el}. This reduction is referred to as the Ham effect (Ham 1965). Only off-diagonal elements are affected, since the vibrational overlap factor is unity for diagonal elements. (Other types of Jahn–Teller coupling can give similar results but the details are more complicated.)

This result has important consequences. Purely off-diagonal operators, like the orbital Zeeman interaction $\beta\mathbf{H}\cdot\mathbf{L}$, or the spin–orbit coupling $\lambda\mathbf{L}\cdot\mathbf{S}$, or a (111) oriented stress for a tetragonally distorted system, appear to be weakened when one measures vibronic matrix elements. In practice, this means that measurements of perturbations on zero-phonon lines, which give $\langle Y|\Omega|X\rangle$, lead to lower values than Ω_{xy}^{el} obtained from measurements on the corresponding broad-band. This happens because the zero-phonon transition involves vibrational states centered on different displacements Q_{x0} and Q_{y0} (cf. Q_{a0} and Q_{b0} in Figure 2.14), whereas the broad-band transitions are the vertical Franck–Condon transitions in which this displacement is not effective. Of course, in systems with weak Jahn–Teller coupling, for example, rare earths, this distinction is less material [see Equation (4.29)].

To illustrate the effects of Jahn–Teller splitting on optical bands we confine our remarks to ^2S-^2P-type transitions of F centers (Section 5.1.1). The P state can couple to E- and T-type lattice vibrations, resulting in a splitting. We have already pointed out that if the vibronic coupling is strong enough non-Jahn–Teller vibronic bands tend to be Gaussian in shape, and this is true no matter how complex the spectrum of lattice vibrations involved. Bands in which Jahn–Teller coupling is dominant will not, as a rule, be Gaussian in shape. However, symmetric-mode coupling is generally present as well as Jahn–Teller coupling and quite often effects of the latter are smeared out by the former.

We show in Figure 4.9 the emission and absorption bands of the F^+ center in CaO. These bands are due to transitions of the type $^2S \leftrightarrow {}^2P$ of a single electron in an oxygen vacancy. The electron–lattice coupling here is strong enough to produce a characteristic broad-band spectrum but weak enough for a zero-phonon line to be visible. It is apparent that

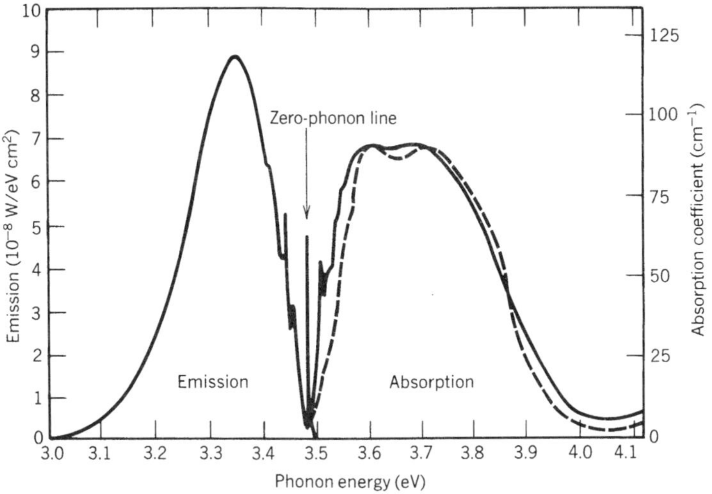

Figure 4.9 Absorption and emission bands of the F^+ center in CaO at 5 K [after Henderson et al. (1972)]. Note the lack of mirror symmetry (contrast Figure 4.4). Theoretical predictions are shown as broken lines [after O'Brien (1976)].

the emission and absorption bands have different shapes. This is characteristic of Jahn–Teller systems, in contrast to non-Jahn–Teller systems, where coupling to symmetric modes produces emission and absorption bands that are closely mirror images of each other (Figure 4.4).

The calculation of optical band shapes of Jahn–Teller systems is generally tackled by numerical methods. Figure 4.10(b) shows approximately the strongest optical transition probabilities for an S-P transition at an octahedral site with the P state equally strongly coupled to two vibrational modes [ϵ_g (or E_g) and τ_g (or T_g); Figure 4.6)] of the octahedron with vibrational energy $\hbar\omega$. It is apparent that (1) the $S \rightarrow P$ transition has two humps since the final degenerate state is split at the most probable configuration and that the spacing of the vibronic levels is unequal and (2) that the $P \rightarrow S$ transition with its nondegenerate final state has a single-peaked shape and that the spacing of the vibronic levels is equal since only the vibrational levels of this terminal S state are involved. A version of Figure 4.10 smoothed by coupling to A_{1g} modes with $S = 2$ is compared with the F^+ absorption in CaO in Figure 4.9. In this particular system the spin–orbit coupling in the ^2P excited state is strongly quenched by the Ham effect; it is not observable in the zero-phonon line.

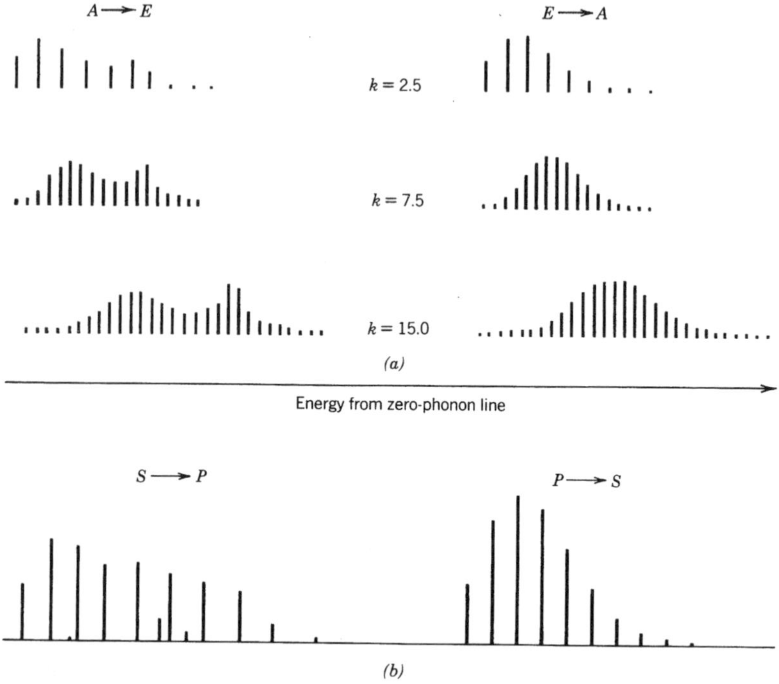

Figure 4.10 (*a*) Optical band shapes for $A \rightarrow E$ and $E \rightarrow A$ transitions. The values of k (defined by $E_{JT} = 2k^2\hbar\omega$) are shown. For $A \rightarrow E$ transitions, the double hump corresponds to final states on upper and lower energy surfaces; for $E \rightarrow A$ transitions, the initial state is always the lower energy surface at low temperatures [cf. Longuet-Higgins et al. (1958)]. (*b*) Optical band shapes for $s \rightarrow p$ and $p \rightarrow s$ transitions for a p state equally coupled to trigonal and tetragonal modes. In the case shown, $E_{JT} = 10\hbar\omega/3$.

4.3 NONRADIATIVE TRANSITIONS

When a solid is excited, by whatever means, its recovery may be radiative, giving rise to luminescence, or nonradiative. Although the nonradiative transitions are of enormous importance, their nature is rarely analyzed in detail since their consequences are usually only seen negatively, through the absence of luminescence.

Nonradiative transitions form a part of many processes; examples include the broadening of sharp optical transitions, spin–lattice relaxation in epr and nmr, a range of degradation mechanisms of devices, and, as an extreme instance, diffusion in its various forms. In most cases nonradiative transitions are undesirable, diverting energy into destructive or irrelevant channels. Occasionally they are eminently desirable, for example, in bolometers, where one wishes to measure incident energy. The most important systems are, nevertheless, those where radiative recovery is desirable, principally phosphors and lasers.

In the case of solid-state lasers, nonradiative transitions lead to a range of misbehavior beyond mere reduction of optical efficiency. Localized regions with high nonradiative recombination rates can give rise to undesirable laser pulsations (Henry 1980). Also, dislocation growth, driven by nonradiative recombination (see Figure 3.27 and Section 6.4.3), can drastically reduce the working lifetime of solid-state lasers.

We may place nonradiative processes in four categories, depending on the way in which the initial electronic excitation energy is dissipated:

1. Transfer of electronic excitation energy to vibrational energy.
2. Transfer to other electronic excitation such as in (a) the Auger process involving only free carriers and electrons at a single defect site (if any) (Bonch-Bruevich and Landsberg 1968), and in (b) energy transfer from one defect site to another (Section 4.4).
3. Transfer to configurational energy. This includes (a) the production of defects during deexcitation (Section 6.2) and (b) diffusion (Section 6.4.6), where the jump process requires a specific configuration to be reached, for example, the saddle point.
4. Transfer of vibrational excitation to other vibrational degrees of freedom, the so-called cooling transitions (see Figure 4.11).

We shall concentrate immediately on processes (1) and (4).

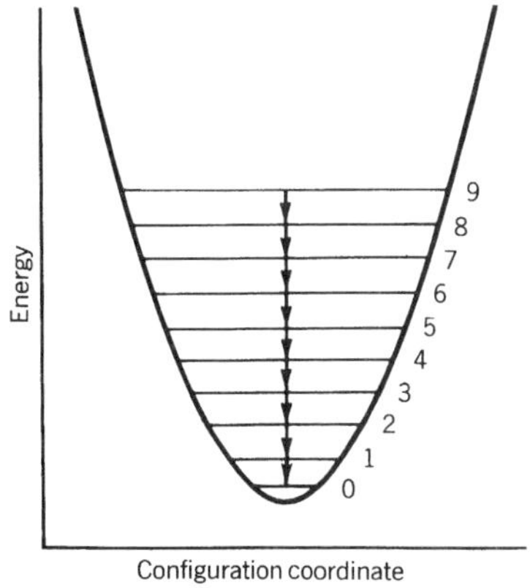

Figure 4.11 Cooling transitions. Only a single configuration coordinate is shown. The degree of excitation is indicated ($n = 0, 1, \ldots$).

4.3.1 Transfer of Electronic Energy to Lattice Vibrations

The various forms of this class of nonradiative transition underly many parts of solid-state science, from the establishment of thermal equilibrium to rather special phenomena like recombination-enhanced defect reactions. The basic processes are ones in which an electronic state changes ($a \rightarrow b$) and lattice vibrational energy is redistributed and may have its total value altered ($\{n\rangle \rightarrow \{n'\}$). In principle, the individual transition probabilities can be written in the form

$$W_{ab} \equiv w_{ab}(\{n\}, \{n'\}) = |M_{ab}|^2 \times (\text{factor for energy conservation}) \qquad (4.30)$$

where M_{ab} is a matrix element (which may depend on the lattice variables) and the energy conservation factor is a generalization of the δ function usually written. Normally one has no way of keeping track of the phonon mode occupancies. In such situations one needs an average over the initial vibrational states $\{n\}$ and a sum over the possible final states $\{n'\}$:

$$W_{ab} = \left\langle \sum_{\{n'\}} w_{ab}(\{n\}\{n'\}) \right\rangle_{\{n\}} \qquad (4.31)$$

Even without detailed analysis, several important points follow. First, the matrix element M_{ab} may depend on the lattice coordinates. Any phonon mode that affects M_{ab} is said to be acting as a *promoting* mode. An important factor is that the symmetry of the two electronic states leads to selection rules and the promoting mode is that with the correct symmetry to give an allowed non-radiative transition. However, almost all modes have some effect, and the extent to which M_{ab} depends on these other modes is described as going beyond the Condon approximation (Section 4.2), that is, the assumption that M_{ab} is independent of lattice geometry. Second, the electronic energy will eventually end up in vibrational modes that, as a rule, are unlikely to be the same as the promoting modes. Those modes that take up electronic energy, and so appear in the factor for energy conservation [Equation (4.30)], are said to be acting as *accepting* modes. Third, this generalized energy conservation factor is also a measure of how readily the lattice takes up vibrational energy in response to a perturbation.

One finds a strong relationship between the nonradiative and radiative transition rates. Suppose, first, that there is no coupling to the lattice. As in Section 4.1, we know that the radiative transition probability is given by an expression like $(2\pi/\hbar)|M_{ab}^{opt}|^2\delta(E-\hbar\omega_{ab})$. Again, as in Section 4.1, when coupling to the lattice is allowed, the δ function is replaced by the lineshape function $G(E-\hbar\omega_0)$, where $\hbar\omega_0$ is the phonon energy. In the corresponding nonradiative transition, the transition rate involves the same lineshape function, although both its argument (which no longer contains a photon energy) and the electronic matrix element are changed, giving $(2\pi/\hbar)|M_{ab}^{NR}|^2 G(E)$ for the nonradiative transition probability.

4.3.2 Cooling Transitions and Absence of Luminescence

In a harmonic lattice it is always possible to consider the vibrations as a superposition of vibrations in dynamically independent normal modes. This is true both for quantum and classical cases, the classical limit corresponding to the well-known situation of small oscillations. Each of the normal modes Q is a linear combination of the displacements of the atoms:

$$\mathbf{Q}_\alpha = \sum_i A_{\alpha j} \mathbf{u}_j \tag{4.32}$$

However, any chosen combination of displacements need not be a normal mode, that is, need not be dynamically independent. We call such a combination a *reaction coordinate* (see Section 2.4.3) since we shall usually be concerned with the most important displacements in a solid-state reaction. The reaction coordinate will usually be a superposition of normal modes over a range of frequencies, and we now show that this has important consequences.

Suppose motion in a normal mode Q is excited. In the absence of anharmonic interactions the excitation will remain within that mode and there will be no trend to equilibrium. Obviously, real crystals are always anharmonic to some extent, and terms in the Hamiltonian like $\mathcal{H}_{anh} = AQQ_BQ'_B$ transfer energy to the heat bath of other modes $Q''_B, Q'''_B, \ldots$. Motion in a reaction coordinate is dissipated even in a purely harmonic lattice, for the motions in the different frequency components move out of phase and the systematic motion is lost. This can be described differently in terms of $\mathcal{H}^1 = BQQ''_B$, where Q is the most important mode in the reaction coordinate. It follows from the fact that both $\mathcal{H}_{anh}$ and $\mathcal{H}^1$ are linear in Q that the rate at which energy is transferred to the heat bath of other modes increases with the level of excitation; roughly, the rate is proportional to the total excitation energy. This result is only true in the usual case of weak damping, in which the vibrational quantum number only changes every few vibrational periods. When much stronger damping is present, that is, a large energy loss per period, a different analysis is needed (Seitz 1940).

One common example (Figure 4.12) illustrates both cooling transitions

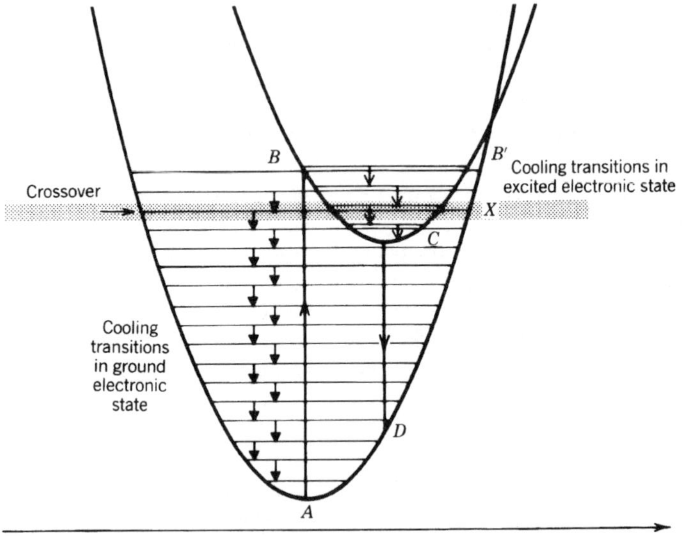

Figure 4.12 Competing radiative and nonradiative transitions following excitation. Cooling transitions in the excited state can lead to C, giving a luminescence transition to D, followed by nonradiative decay from D to A. Alternatively, crossover to the ground state at X can lead to nonradiation decay directly to A, via cooling transitions.

and the competition between luminescence and nonradiative recovery. It is a useful description of many systems, especially the F centers (electrons trapped at anion vacancies) in the many halides and oxides in which they have been observed (Section 5.1.1). It uses the configuration coordinate model (Section 2.4.3), concentrating on two electronic states. In a cycle starting and ending in the ground state one has two optical transitions, absorption (AB) and luminescence (CD), associated with two groups of cooling transitions (BC and DA). But, if the cycle is started by optical absorption (AB), will there be luminescence (CD), or will the cooling transitions take the system directly back to A?

There is a simple and effective rule (Dexter et al. 1956; Bartram and Stoneham 1975) which states that luminescence will occur efficiently if B lies below the crossover X, and that if B is higher in energy, there will be little or no luminescence at zero temperature. The rule can be understood as follows. Near the crossover energy ($E \sim E_x$) there are vibronic states that involve strong admixtures of both the ground and excited electronic states, and at $T = 0$ the dominant nonradiative route from the upper energy surface to the lower is via these crossover states. If the crossover states are not encountered during cooling transitions (as when B lies below X), the system will simply cool to C and luminesce. If, however, B lies above X, then the important factor is the branching ratio; one expects that this ratio of nonradiative to radiative recoveries will be roughly equal to (fraction of cases leaving the crossover states and cooling down the vibrationally excited states of the lower surface)/(corresponding fraction for the upper surface). From our earlier

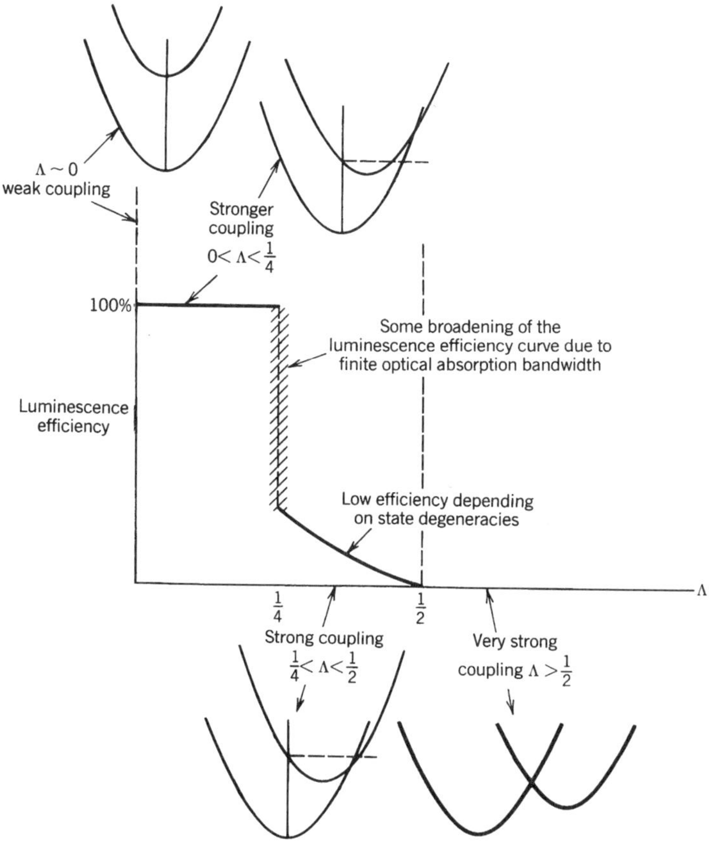

Figure 4.13 Luminescence efficiency for a system like that of Figure 4.12 as the energy of the crossover (X) relative to the initially excited vibronic state (B) is varied. Λ is the ratio of relaxation energy ($E_B - E_C$) to excitation energy ($E_B - E_A$) (see Figure 4.12); for $\Lambda > 1$, X lies below B.

Table 4.1 Luminescence and Absence of Luminescence of *F* Centers[a]

		Luminescence	
Host	Λ	Predicted	Observed
LiI	0.831	No	No
NaI	0.384	No	No
NaBr	0.375	No	No
LiCl	0.371	No	No
LiBr	0.358	No	No
LiF	0.323	No	No
NaCl	0.260	Marginal	Yes
Kl	0.231	Yes	Yes
KBr	0.223	Yes	Yes
RbI	0.211	Yes	Yes
KF	0.189	Yes	Yes
KCl	0.188	Yes	Yes
RbCl	0.182	Yes	Yes
NaF	0.175	Yes	Yes
RbF	0.173	Yes	Yes
RbBr	0.162	Yes	Yes
CsI	0.077	Yes	Yes
CsBr	0.019	Yes	Yes
CsCl	0.018	Yes	Yes
CsF	0.009	Yes	Yes
MgO	0.260	Marginal	Yes
SrO	0.070	Yes	Yes
CaO	0.046	Yes	Yes

[a] After Bartram and Stoneham (1975), who give full references to experiment and results for other systems. The results for NaCl and MgO have Λ close to the critical value of 0.25, so the prediction is less firm.

discussion of cooling transitions we may expect this to be in the ratio $(E_X - E_A)/(E_X - E_C)$, that is, strongly weighted to nonradiative transitions when X lies below B. For more extreme cases, when C lies outside the lower curve, one would always expect nonradiative recovery.

The attractive feature of the Dexter–Klick–Russell rule is that one can predict from absorption data whether luminescence is expected, at least for simple systems. The criterion reduces to the rule that strong luminescence is expected only when $\Lambda=$ (relaxation energy)/(absorption energy) is less than $\frac{1}{4}$. The relaxation energy can be obtained from the linewidth and its temperature dependence; if $\Delta^2(T)$ is the second moment of the lineshape [Equation (4.16)], the relaxation energy is $\Delta^2(T)/8\ln 2k_BT$. This criterion is illustrated in Figure 4.13, with slight generalization. Such results are immediately open to check:

(a) $\Lambda=\frac{1}{4}$ does indeed divide the many F centers into those that luminesce efficiently and those that do not (Table 4.1).
(b) Tuning a laser through the crossover energy shows the expected dependence of luminescence efficiency on laser energy (Hirai and Wakita 1983).
(c) The weak luminescence region (Λ between $\frac{1}{4}$ and $\frac{1}{2}$) is found experimentally, although detailed analysis shows that other mechanisms can contribute to the luminescence observed (Bartram and Stoneham 1983; Luty 1983).

Because of cooling transitions the shapes of energy surfaces determine many features of nonradiative transitions. In many cases one can determine which of several competing processes (luminescence, heat generation, diffusion, or defect creation) dominates without knowing the coupling strength accurately. Thus, for the cooling transitions we find that the condition for luminescence is expressed above in terms of Λ, which is an energy ratio and which does not involve the perturbation that induces the individual transitions from one vibrational state to another.

4.3.3 Relaxation Processes

Among the commonest nonradiative transitions are those induced by lattice vibrations. They can be classified (like the cooling transitions) in terms of values of the Huang–Rhys factor, S_0, and of p, a dimensionless parameter that measures the energy to be degraded into heat in terms of vibrational quanta (see Section 2.4.3). The mean phonon energy $\hbar\omega_0$ enters explicitly as soon as the temperature dependence is needed; indeed, it is often convenient to define a characteristic temperature $T_0\equiv\hbar\omega_0/k_B$. The following cases are among the processes of interest and will be discussed in this section:

(i) S_0 small, $p=1$. Spin–lattice relaxation (direct process; see Figure 4.14b on page 211), giving, for example, homogeneous broadening of zero-phonon lines.

(ii) S_0 small, p large. Weak coupling, affecting efficiency of phosphor luminescence.
(iii) S_0 large, $p=0$. Diffusion, including small-polaron hopping.
(iv) S_0 large, p large. Strong coupling, for example, killer centers in phosphors.

Although in relaxation processes the net effect is conversion of electronic energy into heat, lattice vibrations can, of course, cause electronic excitation. The principle of detailed balance requires a relation between the upward and downward transitions (see p. 180). If the degeneracies of upper and lower electronic levels are d_u and d_l, the relationship can be expressed in the form

$$d_u W_{u\to l} = \exp(-E_{ul}/k_B T) d_l W_{l\to u} \tag{4.33}$$

where W is a transition probability. Generalizations of various sorts are necessary, especially when a magnetic field is involved [see, e.g., Stoneham (1975, p. 289)]. For simplicity, we shall normally consider the downward transition between nondegenerate levels, e.g., the $S_z=+\frac{1}{2}$ and $S_z=-\frac{1}{2}$ Zeeman components of $S=\frac{1}{2}$.

In spin–lattice relaxation of Zeeman-split levels, the electronic energy (of order 1 cm^{-1}) is much less than the largest phonon energies. Hence a single phonon may be emitted in the transition (or absorbed in the reverse transition), in the so-called direct process (Figure 4.14); the recovery to equilibrium varies with temperature as $2n_0+1$, where n_0 is the phonon occupation number for those modes with energy equal to the Zeeman energy (δ in Figure 4.14). Clearly only lattice vibrations with a discrete energy contribute. In special cases n_0 can be far removed from its equilibrium value leading to the so-called phonon bottleneck or the phonon avalanche in suitable circumstances (Brya and Wagner 1967).

Another type of relaxation transition involves the two-phonon process. The absorption of one phonon and emission of another, with the net transfer of energy δ, exploits all the phonon modes and not just a discrete band. The dominant contribution comes either from phonons with energy of order k_BT (the Raman process) or, if there is an excited state with excitation energy in the range of phonon energies, by phonons of energy Δ (the resonant Raman or Orbach process) (Figure 4.14). For the Raman process the rate varies as T^N, with N large, typically $N=7$–9. N depends in part on the phonon density of states and for hemoglobin, the value of N has been used to argue for the low-dimensional (fractal) nature of vibrations (Allen et al. 1982). For the resonant process the temperature dependence is dominated by the factor $\exp(-\Delta/k_BT)$.

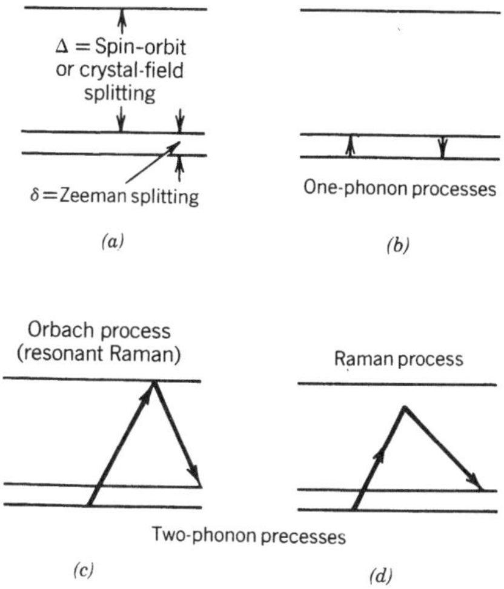

Figure 4.14 Spin–lattice relaxation in a system (*a*) with ground-state Zeeman splitting δ and excited states higher in energy by Δ due to spin–orbit coupling or crystal-field effects. The absorption and emission steps of the direct process are shown in (*b*), each involving a single phonon. Two-phonon processes are shown in (*c*) and (*d*). These are sometimes referred to as Raman processes, being a phonon analogy of the optical Raman effect. Only nonresonant Raman processes are possible if Δ exceeds the maximum phonon energy.

The weak-coupling systems, typically phosphors of rare earths in oxides, also show a T^N dependence for the rate of nonradiative transitions at optical energies but for quite different reasons. Here the phonons with largest energy dominate, not those with energy around k_BT. Suppose the rates of N and $N-1$ phonon processes are related by $W_N \sim \alpha W_{N-1}$, where α is small and roughly independent of N. Suppose, too, that the relaxation process of lowest order always dominates. In this case N will be equal to (energy released, E_0)/(largest phonon energy) or $N = p$ with phonon energy $\hbar\omega_{max}$. Consequently, noting that $\alpha = \exp \ln \alpha$ we find

$$
\begin{aligned}
W_N \sim A\alpha^N &= A \exp[-N \ln(1/\alpha)] \\
&= A \exp(-\gamma E_0/\hbar\omega_{max})
\end{aligned}
\tag{4.34}
$$

where we have used $N = E_0/\hbar\omega_{max}$ and where $\gamma \equiv \ln(1/\alpha)$ can be shown to lie in the range 1–2, as a rule. Equation (4.34) is a statement of the energy-gap rule, that W depends exponentially on the energy released, E_0. This is illustrated in Figure 4.15 for experimental systems. The temperature dependence can be understood if α is proportional to the

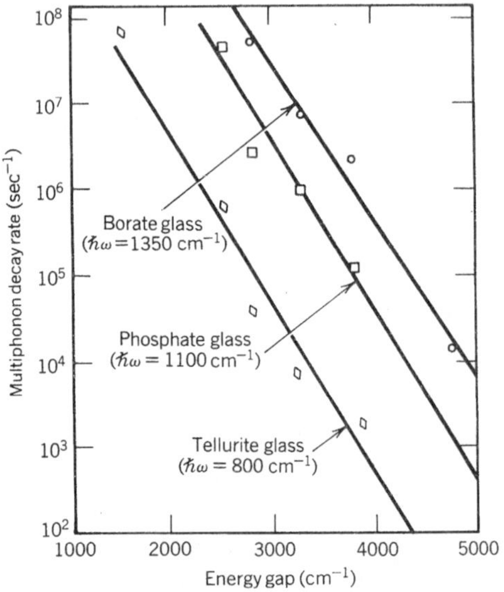

Figure 4.15 Energy gap rule. The multiphonon decay rate (sec^{-1}) is shown as a function of the energy gap to the next lower level for rare-earth ions in glasses [after Layne et al. (1977)].

phonon occupation number n (or strictly to $n+1$ since only emission is involved) (Stoneham 1975, p. 477) and hence to T, at high temperatures. In most practical situations Equation (4.34) leads to $W_N \sim T^N$.

At $T=0$ there is a useful generalization of Equation (4.34) due to Mott (1977) which is valid for both strong and weak coupling:

$$W(T=0) = \tilde{W} \exp[-S_0 + p \ln(S_0 p)] \tag{4.35}$$

Comparison of Equations (4.34) and (4.35) gives $\gamma \equiv \ln(E_0/E_R)$ since the energy released is $E_0 = N\hbar\omega$ and the relaxation energy is $E_R = S_0\hbar\omega$. The preexponential factor in Equation (4.35) has the dimensions of a frequency. Typically it is of the order of lattice phonon frequencies or rather less. As a rough estimate, taking $\tilde{W} \sim 10^{12}$ sec^{-1} and noting that allowed optical transitions have rates of $\sim 10^8$ sec^{-1}, one expects at $T=0$ that luminescence will dominate when $\gamma E_0 \gtrsim 10\hbar\omega$, that is, $p \gtrsim 5$–10. In general, the nonradiative transitions become stronger and luminescence weaker as the transition energy, and hence p, decreases.

In the strong-coupling limit the downward transition rate from the lowest vibrational level of the excited state takes the form $W \sim \exp(-E_A/k_BT)$ at high temperatures, where E_A is the classical activation

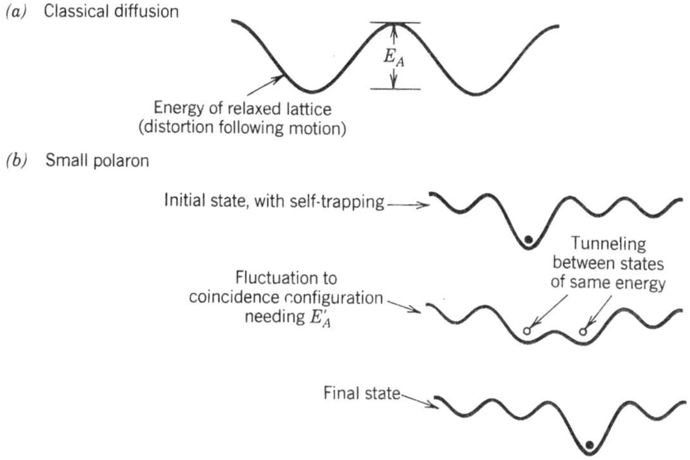

Figure 4.16 (*a*) Classical diffusion and (*b*) small-polaron motion. In (*a*) the energy surface shown is that in which the lattice relaxes around the moving particle at each state, and E_A is the activation energy for motion. In (*b*) there is a thermal fluctuation (requiring energy E'_A) from the initial, self-trapped, state to the coincidence configuration, in which tunneling can occur between states of the same energy on two adjacent sites.

energy to a cross-over point (see Figure 2.14). Both classical arguments and quantum treatments agree that this transition rate is expected from thermal excitation to the crossover of the two energy surfaces. In diffusion (Figure 4.16) the crossover corresponds to the saddle point as the diffusing particle moves from one site to another. In small-polaron hopping the crossover corresponds to the coincidence configuration in which the initial and final sites appear to be equivalent, to the diffusing small polaron.

Nonradiative transitions at optical energies are important technologically because they limit device behavior. They do this in two main ways, (a) by competition with desirable processes (e.g., luminescence, or current control within a semiconductor device) and, (b) by production or growth of defects that have harmful effects (e.g., dislocation climb in semiconductors). However, the rate-determining process may vary widely. In some cases, such as the so-called large-dark-spot phenomena (Henry and Logan 1977), carrier diffusion to a nonradiative region is rate determining. In other cases a macroscopic nonradiative region may be critical, of the type that causes pulsations in injection lasers (see Section 4.3).

So-called killer centers, with fast nonradiative transitions, reduce the efficiency of optical systems such as phosphors and optical semiconductor devices. It is useful to identify some of the defect types likely to act as killer centers and we list four examples:

1. Defects with many electronic states close in energy, so that rapid one-phonon processes can occur.
2. Defects with favorable vibrational properties, that is, with large-amplitude modes promoting the transitions, and large-energy modes to take up the electronic energy.
3. Defects with very strong coupling to lattice distortions such as certain dislocations and some vacancy centers.
4. Defects where rapid Auger processes [see, e.g., Stoneham (1975, Chapter 14)] are possible. Here the energy goes to other electronic terms, for example, electronic kinetic energy. For rapid transitions one needs localized, correlated carriers. Colloid centers (Section 6.3.4) may be an example (Stoneham 1981).

4.3.4 Deep Level Transient Spectroscopy (DLTS)

This form of spectroscopy (Lang 1974) exploits nonradiative transitions between localized electronic states and band states in solids, and it therefore measures thermal rather than optical transition energies (see Section 4.2.2). This is, in fact, an advantage because most of the systems for which it is used are semiconductor device materials, where thermal transitions may determine operational behavior.

The technique of DLTS uses the capacitance of a *p-n* junction with good transient response for studying deep traps in semiconductors. A capacitance change is caused by using a short bias pulse to introduce carriers and thus change the electron occupation of a trap from its steady-state value. As this population returns to equilibrium the capacitance returns to its equilibrium value. The presence of each trap is indicated by a positive or negative peak on a flat baseline, plotted as a function of temperature. The height of each peak is proportional to the associated trap concentration and the sign of each peak indicates whether it is due to a majority or minority carrier trap. By suitably choosing experimental parameters it is possible to measure the capture rate of carriers of each trap, the thermal emission rate of carriers, and the position of the trap relative to the valence or conduction band.

Suppose the emission rate per defect $e(T)$ and the capture rate $c(T)$ per defect are measured separately. The capture rate per defect is usually $\langle v\sigma(v)\rangle$, with a thermal average over the carrier velocity v and the capture cross section $\sigma(v)$. The principle of detailed balance (see p. 180) tells us that

$$e(T) = \gamma c(T)\exp(-\Delta G/k_B T) \tag{4.36}$$

where γ is the ratio of electronic degeneracies $d_{\text{upper}}/d_{\text{lower}}$ of the states and ΔG is the free energy difference for the thermal transition. We can always write $\Delta G \equiv \Delta H - T\Delta S$ in terms of the changes in enthalpy ΔH and entropy ΔS. An important case has an upper state that is ionized. In that case γ becomes N_c/g, where N_c is the conduction band density of states and g is the bound level degeneracy; in simple instances $N_c \propto T^{3/2}$. If we write $\langle v\sigma(v)\rangle$ as the product of a temperature independent cross section σ and the root-mean-square velocity $\sqrt{\overline{v^2}}$ then $c(T) = \sigma\sqrt{\overline{v^2}}$, and if we also note that $\frac{1}{2}m\overline{v^2}$ is $\frac{3}{2}k_B T$ in thermal equilibrium (so $\sqrt{\overline{v^2}} \propto T^{1/2}$), we see that we have a T^2 factor overall in Equation (4.36) which we should check before extracting ΔG (and hence ΔH or ΔS). One finds finally, with T in degrees K, σ in Å^2, and the emission rate in sec^{-1};

$$[e(T)/T^2\sigma(T)] = 3.25 \times 10^5 (m^*/m_0)\exp(-\Delta G/k_B T) \tag{4.37}$$

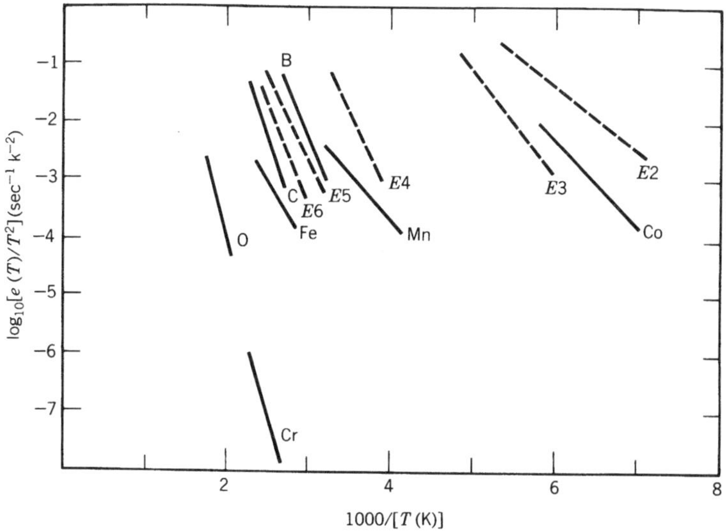

Figure 4.17 DLTS results for electron emission rates $e(T)$ from deep levels in GaP. Since GaP has a complex conduction band, Equation (4.37) can only be used qualitatively. If the capture cross section is assumed to be independent of temperature, steeper lines (e.g., O, Cr, $E5$, $E6$) correspond to deeper traps and flatter lines (e.g., Co, Mg, $E2$) to shallower traps. Large intercepts at $T = \infty$ correspond to high capture cross sections (O, C, Cr) and low intercepts to small cross sections (Fe, Mn) (for a discussion of defect notation see the Appendix).

Equation (4.37) is used to extract defect parameters from DLTS data (Figure 4.17).

4.4 ENERGY TRANSFER

When an atom or molecule is excited, it may transfer energy to another atom or molecule before luminescence or nonradiative relaxation takes place. This behavior is very common, ranging from critical steps in photosynthesis to the behavior of phosphors of many kinds. A general review is given by Levine and Jortner (1976). Often it is described as exciton migration, a description that is technically correct (at least in the sense of Chapter 1, where we related the exciton to the excited states of a solid), although the picture of a free exciton moving does not always help.

The earlier studies of energy transfer were of the gas phase, where the study of the so-called resonance radiation of mercury vapor lamps (Mitchell and Zemansky 1934) introduced many important ideas. One was the significance of spectral diffusion as well as of spatial diffusion. The spatial form exactly parallels particle diffusion, with a random sequence of jump directions and, to some extent, of distances. Spectral diffusion is best understood from the optical transfer case and recognizes that, if emission with an energy E_e is very strong, absorption at the same energy $E_a = E_e$ may also be very strong. The major spatial transport can come from the wings of the line, away from the peak E_e. In such cases, any change in energy from absorption to emission is critical, giving spectral diffusion of energy within the lineshape as the rate-determining factor in spatial diffusion. This idea has had many subsequent applications, e.g., to phonon systems [see, e.g., Anderson 1959].

The main processes of energy transfer can be described by considering just two impurities, a sensitisor and an activator, in an otherwise perfect solid. The sensitisor S is excited by incident light, or by some indirect means (e.g., involving free-carrier capture). Energy is passed to the activator A which may luminesce subsequently. The energy transfer from sensitisor to activator will be concentration dependent, unlike the decay of an isolated imperfection. We assume the sensitisor has a radiative lifetime τ_s, a normalized absorption lineshape $g_s(E)$, and a quantum efficiency η_s in the absence of the activator. For the activator we assume a lineshape $g_A(E)$ and an integrated optical cross section σ_A. Energy transfer takes place by three main mechanisms:

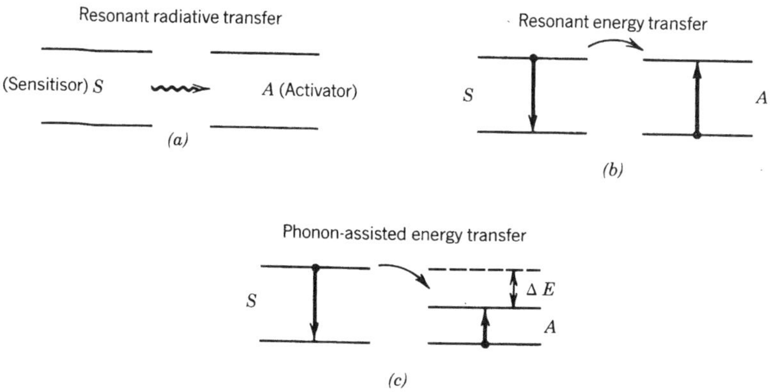

Figure 4.18 (*a*) Resonant radiative energy transfer involving emission of a photon from sensitisor *S* and absorption by an activator *A*. (*b*) Resonant energy transfer (without explicit transfer) of a photon and (*c*) phonon-assisted energy transfer.

(a) Resonant radiative transfer, which proceeds via photons (Figure 4.18*a*) with a rate W_{RRT} given by

$$W_{RRT} = \frac{\sigma_A}{4\pi R_{SA}^2} \cdot \frac{1}{\tau_s} \int dE g_S(E) g_A(E); \tag{4.38}$$

here R_{SA} is the separation of the sensitisor and activator. A mechanism of this type is observed for singlet excitons in aromatic crystals. The dependence of Equation (4.38) on the overlap of the sensitisor and activator lines is typical of energy transfer generally. The long range of Equation (4.38) leads to special features, for example, dependence on sample shape when the photon travel distance is long enough. There is also the possibility of photon trapping of resonance radiation, mentioned earlier.

(b) Resonant energy transfer, which does not involve the explicit emission and absorption of photons. In the simplest case (Förster 1948) one can regard the interaction inducing the transition as between the instantaneous dipole d_S on the sensitisor and the dipole it induces on the activator. Since the induced dipole d_A will be proportional to $d_S/\epsilon R_{SA}^3$, where ϵ is the dielectric constant, and the interaction will be proportional to $d_A d_S/\epsilon R_{SA}^3$, one expects overall a rate W_{RET} for resonant energy transfer of

$$W_{RET} = \frac{1}{\tau_s} \cdot \frac{A}{\epsilon^2 R_{SA}^6} \tag{4.39}$$

This has been verified as a function of R_{SA} by Stryer and Haugland (1967) for the transfer of energy from α-naphthyl to dansyl. It is usual to write $A\epsilon^{-2} \equiv R_0^6$, where R_0 is typically tens of angstroms. R_0^6, and hence W_{RET}, is proportional to the overlap of the lineshape functions, $\int dEg_s(E)g_A(E)$.

The dipole–dipole contribution may need to be supplemented by higher-order terms, so that the rate becomes

$$\tilde{W}_{RET} = \sum_m \frac{1}{\tau_m}\left(\frac{R_{0m}}{R_{SA}}\right)^m \tag{4.40}$$

with $m = 6$ (dipole–dipole), 8 (dipole–quadrupole), etc. The leading term at large distances depends on the symmetry of the activator and sensitisor states between which the transitions occur. At short distances, several terms in this multipole expansion may contribute significantly. These higher-order terms can be identified in exciton diffusion in molecular crystals, a point discussed by Flynn (1972, pps. 492, 526).

(c) Phonon-assisted energy transfer. The discussion so far has ignored coupling to lattice vibrations. If there is an energy mismatch ΔE between the level spacings of sensitisor and activator, it may be taken up by phonons when energy is transferred (Figure 4.18*c*). The main effect (Mizakawa and Dexter 1970) is to introduce an additional factor $\exp(-\tilde{\beta}\Delta E/\hbar\omega)$ into the rate for the process, where

$$\tilde{\beta} = \beta' - \beta''$$

with

$$\beta' = \ln[(\Delta E/\hbar\omega)/S_0(n+1)] - 1$$

and

$$\beta'' = \ln[1 + (S_{0A}/S_{0S})];$$

here S_{0A} and S_{0S} are the Huang–Rhys factors for the activator and sensitisor, respectively.

We have so far neglected energy transfer among the sensitisors. If there is rapid diffusion of energy between sensitisors, the rate of decay of luminescence is exponential in time after a pulse of excitation:

$$I(t) = I_0 \exp(-t/\tau_D) \tag{4.41}$$

with a rate proportional to the activator concentration, that is, $\tau_D^{-1} \propto N_A$ When diffusion of energy limits the decay, one can still find exponential decay, although the rate is now also proportional to the sensitisor concentration, that is, $\tau_D^{-1} \propto N_A N_S$. If there is no diffusion, however, nonexponential decay is expected, because of the range of sensitisor–activator spacings.

4.5 ELECTRON PARAMAGNETIC RESONANCE (epr)

We shall assume at the outset that we are dealing with point defects in which the orbital angular momentum is totally quenched (but see later) so that we have only spin magnetism to consider. The magnetic moment of a single defect can then be written

$$\boldsymbol{\mu} = -g\beta \mathbf{S} \tag{4.42}$$

where g is closely equal to 2. If the z axis of a coordinate system is specified by the direction of an applied magnetic field $\mathbf{B}$, the interaction between $\boldsymbol{\mu}$ and $\mathbf{B}$ is given by

$$\mathcal{H}_{el} = -\boldsymbol{\mu} \cdot \mathbf{B} = g\beta BM \tag{4.43}$$

where $M = S_z$ has $2S+1$ values. {We ignore for the moment the effect of the crystal environment on $\boldsymbol{\mu}$ [Equation (4.42)]}. The selection rule for allowed magnetic dipole transitions between the Zeeman levels (4.43) is $\Delta M = \pm 1$. Transitions are caused by an rf field perpendicular to $\mathbf{B}$ when the frequency of the rf field obeys the resonance condition

$$h\nu_0 = g\beta B \tag{4.44}$$

For point defects in thermal equilibrium the lower Zeeman levels have a greater population than the upper levels, and the result of both absorption and induced emission will be a net absorption of rf power at ν_0. To observe resonance it is customary to keep the frequency fixed and to vary B until Equation (4.44) is satisfied. The sample is placed in a tuned microwave cavity and resonance is generally detected by a fall in cavity Q. With $g = 2$ and $\nu_0 = 10$ GHz (X band), the value of B at resonance is 0.35 T; with $\nu_0 = 24$ GHz (K band), $B = 0.85$ T. Many defects are unstable at room temperature and provision is normally made to cool the cavity to temperatures as low as 4 K. Crystals can be readily irradiated in a cavity at 4 K with X-rays, fast electrons, or light.

When a paramagnetic impurity has a nuclear moment

$$\boldsymbol{\mu}_n = g_n \beta_n \mathbf{I} \tag{4.45}$$

where $\mathbf{I}$ is the nuclear spin, interactions between $\boldsymbol{\mu}$ and $\boldsymbol{\mu}_n$ may give rise to resolved hyperfine structure. If this interaction is isotropic, it may be written in the scalar form

$$\mathcal{H}_{hf} = A\mathbf{I} \cdot \mathbf{S} \tag{4.46}$$

where A contains the magnitude of the nuclear dipole moment. Generally, however, A depends on the orientation of $\mathbf{B}$ relative to crystal axes and Equation (4.46) is written

$$\mathcal{H}_{hf} = \mathbf{I} \cdot \mathbf{A} \cdot \mathbf{S} \tag{4.47}$$

The total Hamiltonian is now

$$\mathcal{H} = \mathcal{H}_{el} + \mathcal{H}_{hf} \tag{4.48}$$

and to first order in perturbation theory the energy levels are

$$E \simeq g\beta BM + AMm \tag{4.49}$$

where m has the $2I+1$ values of $I_z, I_{z-1}, \ldots, I_{-z}$. A schematic energy level diagram for a paramagnetic impurity with $S=\frac{1}{2}$ and $I=\frac{3}{2}$ is shown in Figure 4.19; the selection rules for microwave transitions are $\Delta M = \pm 1$, $\Delta m = 0$, giving a four-line hyperfine structure.

In the case of paramagnetic point defects in solids, such as F centers (Section 5.1.1) or neutral donors in semiconductors (Section 5.3), the unpaired electron is delocalized and will interact with the nuclei of

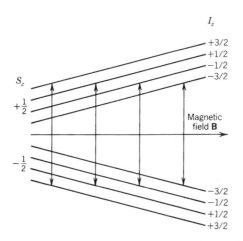

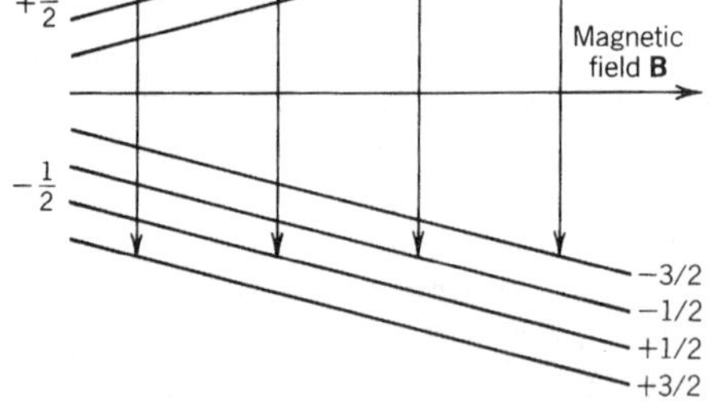

Figure 4.19 Schematic representations of the Zeeman splitting of an electron spin ($S=\frac{1}{2}$) interacting with a nucleus with $I=\frac{3}{2}$, showing four allowed hyperfine transitions associated with a fixed microwave frequency.

neighboring ions in the crystal. This is taken into account by writing Equation (4.47) in the form

$$\mathcal{H}_{hf} = \sum_n \mathbf{I}_n \cdot \mathbf{A}_n \cdot \mathbf{S} \tag{4.50}$$

where the summation is over the interacting nuclei. We show in Figure 4.20*a* the epr spectrum of the *F* center in CaF_2 where the unpaired

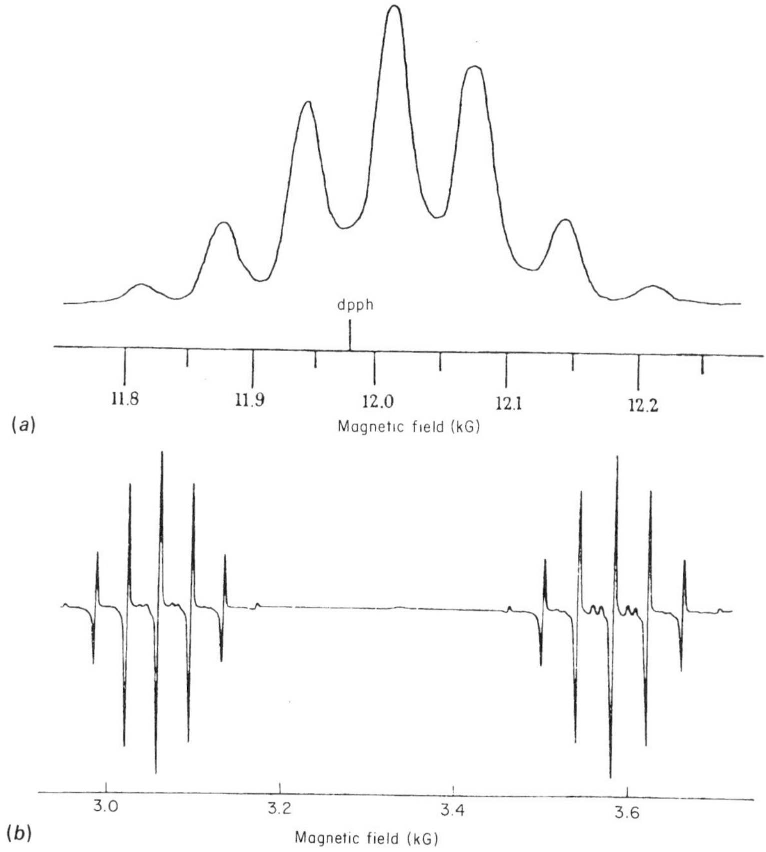

Figure 4.20 (*a*) Epr spectrum of *F* centers in CaF_2 with $B\|[111]$ showing a seven-line hfs [after Hayes and Stott (1967)]. (*b*) First derivative of the epr absorption spectrum of H_s^0 centers in CaF_2 with $B\|[111]$ showing a large two-line hfs due to interaction with the proton and a seven-line hfs due to interaction with the nearest fluorines [after Bessent et al. (1967)].

electron in a fluorine site is surrounded by six nearest-neighbor fluorines, each with $I=\frac{1}{2}$. When B is in a $\langle 111\rangle$ crystallographic orientation, the strength of the hyperfine interaction of the F electron with each of the fluorine nuclei is the same, and we observe seven hyperfine structure lines corresponding to $\sum_n M_n = 3, 2, 1, 0, -1, -2, -3$. A little consideration will show that whereas there is only one way of producing $\sum_n M_n = 3$, there are 20 different ways of producing $\sum_n M_n = 0$. This accounts for the observation of a binomial intensity distribution in the hyperfine structure of 1:6:15:20:15:6:1.

We have already discussed the vibrations of H^- ions dissolved in fluorine sites in CaF_2 (Section 2.4.2). X-irradiation of CaF_2:H at low temperatures in a microwave cavity results in the production of neutral hydrogen atoms, H_S^0, in substitutional fluorine sites and we show in Figure 4.20b the epr spectrum of H_S^0 with B parallel to a $\langle 111\rangle$ crystallographic orientation. Because the proton has nuclear spin, $I=\frac{1}{2}$ we find a two-line hyperfine structure whose magnitude is closely equal to that of the free hydrogen atom. However, the hydrogen electron spreads on to the six nearest fluorines so that each of the hydrogen hyperfine lines is split further into a seven-line hyperfine structure, as with the F center (Figure 4.20a). For both the F center and H_S^0 in CaF_2 the unpaired spin does not interact equally with each of the six nearest-neighbor nuclei when B is not in $\langle 111\rangle$ directions, and the fluorine hyperfine structure is then more complex in appearance.

The wavefunctions of unpaired spins in point defects extend, to varying degree, beyond the nearest-neighbor shell. However, the hyperfine interactions with nuclei of next-nearest-neighbor and further shells is not generally resolved in epr experiments; they contribute only to the inhomogeneous linewidth. Such weak interactions can be resolved, however, by the technique of electron nuclear double resonance (endor). In an endor experiment nuclear resonance [$\Delta m = \pm 1$; see Equation (4.49)] is performed on distant nuclei by applying an additional rf magnetic field in the megahertz region while simultaneously observing epr. Nmr is detected as a change in signal intensity of an epr line. Since the resolution of endor is about four orders of magnitude greater than that of epr, the approximations used in deriving Equation (4.49) are no longer sufficiently accurate. It is now necessary to include the nuclear Zeeman energy $g_n\beta_n\mathbf{B}\cdot\mathbf{I}$ and also nuclear quadrupole interactions for nuclei with $I>\frac{1}{2}$. It is possible with this technique to explore the environment of paramagnetic point defects to distances of $\gtrsim 10$ Å from the defect.

We have so far considered only defects with $S=\frac{1}{2}$. Occasionally, as with excitons (Section 5.1.3), we come across paramagnetic defect centers with $S > \frac{1}{2}$ and we now find additional structure in the epr spectrum, referred to as fine structure. This structure is associated with higher-order terms in S, the most important being the quadratic term

$$\mathcal{H}_{fs} = \mathbf{S} \cdot \mathbf{D} \cdot \mathbf{S} \tag{4.51}$$

where **D** is the fine-structure tensor. Let us take the simple Hamiltonian

$$\mathcal{H} = g\beta \mathbf{S} \cdot \mathbf{B} + \mathbf{S} \cdot \mathbf{D} \cdot \mathbf{S} \tag{4.52}$$

as an example. When $D \ll g\beta B$, we find that the energy levels are given, to first order in perturbation theory, by

$$E(M) \simeq g\beta BM + \tfrac{1}{2}\left[\sum_{n=1}^{3} D_n(3\cos^2\theta_n - 1)\right] M^2 \tag{4.53}$$

where the θ_n are the angles between B and the principal values of the **D** tensor. The effect of the fine-structure contribution to Equation (4.52) is to remove the equal spacing of the Zeeman levels provided by the first term alone, giving $2S$ lines spaced by $(m - \frac{1}{2})\sum_n D_n(3\cos^2\theta_n - 1)$, associated with transitions $M \leftrightarrow (M-1)$ (Figure 4.21).

We have assumed so far absence of orbital angular momentum in the magnetic state of defects and this is usually a good approximation. We have already pointed out in Section 4.2.3 that an orbital D state in a cubic-crystal environment is split into an orbital doublet and an orbital triplet by the lattice potential. We have also discussed the Jahn–Teller effect, a spontaneous distortion of the local lattice environment to remove residual orbital degeneracy. These effects are referred to as

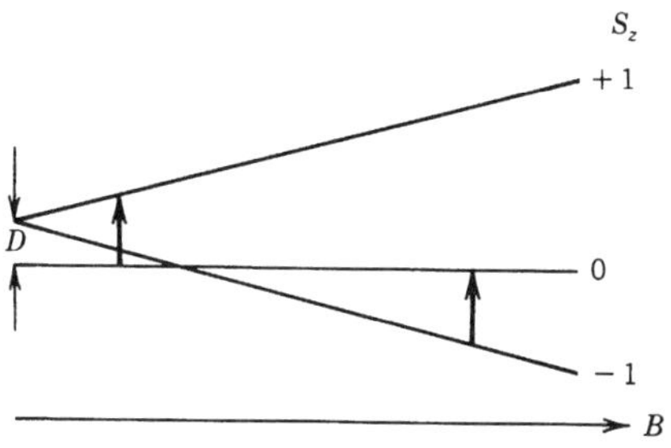

Figure 4.21 Zeeman splitting of an $S = 1$ state in an environment with axial (z) symmetry and with $B \| z$. The two allowed epr fine structure transitions are shown. The fine structure splitting in zero magnetic field is D.

quenching of orbital angular momentum. However, despite the quenching tendency in solids, orbital angular momentum is preserved in many ground states, notably with transition-metal and rare-earth ions, leading to g values substantially different from 2. Even with an orbital singlet ground state, the spin–orbit interaction $\lambda \mathbf{L} \cdot \mathbf{S}$ will admix excited states into the ground state giving a second-order contribution to the g value of magnitude $\sim \lambda/\Delta \sim 0.1$, where Δ is the energy of the admixed excited state. This means that, in general, the g value must be treated as a tensor quantity, so that Equation (4.43) is properly written as

$$\mathcal{H} = \beta \mathbf{S} \cdot \mathbf{g} \cdot \mathbf{B} \tag{4.54}$$

The principal values of the tensors **g**, **A**, and **D** are determined by measuring the variation in the resonance value of **B** as the angle between **B** and the crystal axes is varied. Such measurements can quite often provide unequivocal identifications of point defects, especially when hyperfine structure is resolved.

We have confined our discussions so far to paramagnetic ground states. It is possible to study the structure of excited electronic states, using conventional Zeeman spectroscopy, if sharp zero-phonon lines occur in optical transitions involving these states. However, most defects in ionic solids are strongly coupled to the lattice, optical bands are broad, and zero-phonon lines are not observed (Section 4.2). Even in such situations it is sometimes possible to carry out epr studies in excited electronic states using the technique of optical detection of magnetic resonance (odmr) [for a more detailed discussion of this technique see, e.g., Davies (1976)]. We take as an example a spin triplet state decaying radiatively to a singlet ground state, a situation sometimes observed in exciton decay (Section 5.1.3) (Figure 4.21)). In the presence of a magnetic field B, decay from the $S_z = +1$ and $S_z = -1$ components of the spin triplet to the singlet ground state gives rise to circularly polarized emission in the Faraday configuration and emission from $S_z = 0$ to the ground state is forbidden. There is therefore a tendency for population to accumulate under excitation in the $S_z = 0$ state and, hence, microwave-induced transitions between $S_z = 0$ and $S_z = \pm 1$ cause increase in emission of right and left circularly polarized light. Detection of these changes in emission make observation of epr in the triplet state possible (Section 5.1.3). Transitions $S_z = 0 \leftrightarrow S_z = \pm 1$ also cause an increase in the total intensity of emitted light, so that use of polarizers is not essential for odmr.

In concluding this section we mention briefly the technique of muon spin rotation (μsr) in which short pulses ($\lesssim 10^{-8}$ sec) of positive muons are implanted in solids (Schenk 1976). The muons are created by pion decay such that, initially, their spins are aligned along the beam direction. Once in the sample, the muon spins precess in any internal or applied magnetic fields. They then decay, after a time that can be measured, emitting positrons, principally along the muon spin direction at the time of decay. The direction of emission of the detected positron and the time lapse together allow one to measure the way the muon spin direction changes with time. The behavior of the muon can be studied in this way on time scales in the range from 10^{-9} sec (limited instrumentally) to 10^{-5} sec (limited by the 2.2-μsec lifetime of the muon). Analysis of the data gives the sign and strength of local magnetic fields at the muon site and, in suitable cases, the local distortion. In addition, information about the muon motion from site to site and the dynamics of trapping may be obtained. Important advantages of μsr are that muons can be implanted into any solid, that they carry their own spin, and that (apart from small and irrelevant losses) every muon is detected. An example of these advantages is shown by the observation that, in diamond, silicon, and germanium, two forms of muonium (i.e., the $[\mu^+ e]$ complex) are seen: normal muonium with cubic symmetry, and anomalous muonium, with trigonal symmetry. Atomic hydrogen, however, is not seen in these crystalline semiconductors, probably because spin-zero molecular hydrogen is formed (Section 5.5). Thus, in this case, μsr can give unique information about an important class of defect.

CHAPTER FIVE

Electronic Properties of Point Defects

In the previous chapter we discussed the spectroscopic techniques that have been most usefully applied to the study of the electronic properties of defects. In this chapter we outline results obtained from spectroscopic studies, using illustrative examples.

We shall confine our attention here to defects with optical transitions within the bandgap of the crystal. Such transitions are observed, for example, as intraconfigurational excitations within the valence shell of transition-metal ions ($3d \leftrightarrow 3d$) and rare-earth ions ($4f \leftrightarrow 4f$). Whether or not an internal transition of an impurity lies within the bandgap depends on the size of the bandgap (Figure 1.4) and the oxidation state of the impurity. For example, Cu^+ ($3d^{10}$) substitutes for host cations in alkali halides and has a characteristic absorption spectrum at energies less than that of the adsorption edge of the crystal, due to transitions $3d^{10} \rightarrow 3d^9 4s$. In the compensated phosphor material ZnS:(Cu, Al) the transitions of Cu^+ lie above the band edge and the green luminescence is due to donor–acceptor recombination. However, if ZnS:Cu is not codoped with aluminum donors, the Fermi level drops and copper is present as $Cu^{2+}(3d^9)$. Both absorption and luminescence now occur below the band edge and are again associated with transitions within the copper energy levels.

In Section 5.1 we deal with intrinsic point defects in simple ionic solids, for example, the F center (an electron trapped in an anion vacancy) and the self-trapped hole (generally called the V_K center). The F center has been very extensively studied, both experimentally and theoretically, and is well understood.

In recent years there has been a renewal of interest in the properties of transition-metal impurities in semiconductors, particularly in III–V compounds. These impurities give rise to defect levels deep in the bandgap of the crystals and in this respect their behavior is analogous to that of the F center. We discuss the electronic structure of these deep levels in Section 5.2, with some emphasis on application of epr techniques.

Section 5.3 deals with the energy levels of shallow donors and acceptors in semiconductors, with a discussion of effective mass theory and its limitations. This section also covers donor–acceptor recombination and the behavior of excitons bound to impurities. It is generally assumed that excitation of impurities using energies greater than the bandgap is due to formation of bound excitons. This may result from carrier trapping at the impurity site followed by trapping of a carrier of opposite sign.

Section 5.4 contains a description of vacancy centers in semiconductors, emphasizing silicon and diamond. Finally, in Section 5.5 we catalog some interesting crystal impurity systems of a general kind, with particular attention to GaP:O.

Table 5.1 contains a résumé that includes all the defect systems discussed in this chapter.

5.1 DEFECTS IN IONIC CRYSTALS

5.1.1 The *F* Center

The *F* center, an electron trapped at an anion vacancy (Figure 5.1), is the best-understood point defect in ionic crystals. Indeed, because of the range of systems and processes for which it is characterized in detail, it is

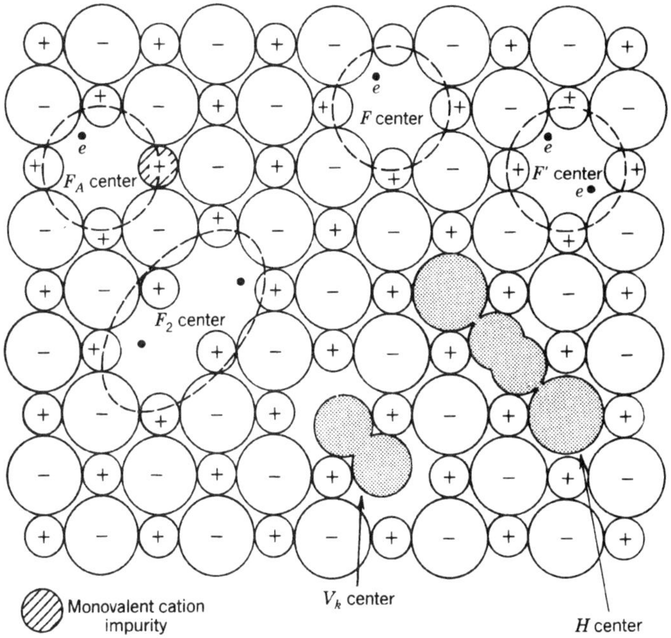

Figure 5.1 Common color centers in alkali halides *MX*. These include electron-excess centers, notably the *F* center and related defects (the *F′* center has an extra electron; the F_A center is an *F* center with an adjacent cation impurity; the F_2 center is two adjacent *F* centers), and two centers involving X_2^- molecular ions, namely, the self-trapped hole (V_K center) and neutral anion interstitial (*H* center).

Table 5.1 Systematics of Defects in Solids[a]

	Ionic			Covalent	
Type of Defect	Alkali Halides	Oxides	Other II–VI	III–V	Group IV
Anion vacancy	$V^- = F'$ $V^0 = F$ V^+ = simple vacancy	V^- reported $V^0 = F'$ $V^+ = F^+$ V^{++} = simple vacancy	F', F^+ and "simple vacancy" reported sometimes associated with impurities	No confirmed reports	Diamond (V^+), $V^0 = GR1$, $V^- = ND1$; Silicon (V^{2+}), V^+, V^0, V^-, V^{2-}; germanium V^0, V^-, not V^+(?)
Cation vacancy	$V^0 = V_F$ V^- = simple vacancy	V^0 V^- V^{2-} = simple vacancy	V^0 V^- V^{2-} = simple vacancy	V^0 in GaP	
Polyvacancies	Divacancy; M, R, N centers; colloids	Anion vacancy aggregates; voids		Possibly colloids	Divacancies in various charge states; chains of vacancies; voids

Interstitials	$I^0 = H$ center I^+ = simple cation interstitial I^- = simple anion interstitial	O^{2-} in fluorite structure oxides	Evidence very indirect, although interstitial dislocation loops are easily found		Inferred from reactions under irradiation and trom defect aggregates at high temperatures
Other intrinsic defects	Halogen molecule in anion–cation divacancy; self-trapped hole; self-trapped exciton	Vacancy-interstitial complexes; shear planes, small polarons in some systems		Antisite defects	Atoms with nonstandard coordination in amorphous materials
Important impurities	Molecular ions: OH^- O_2^-; substitutional Tl, halogen, and alkali ions; hydrogen (many forms)	Alkalis (acceptors) and halogens (donors); transition-metal ions (many charge states)		Shallow donors, acceptors, and isovalent (e.g., N in GaP) impurities; deep centers (transition metals, Au, etc.); "benign" impurities like H or O	

[a] This table compares defects by type in the best-known crystalline solids. It oversimplifies in several respects, notably by ignoring other crystal structures. Charge states of vacancies are defined by V^{N+} meaning an ion of charge ($N-$) has been removed; for interstitials, I^{M+} means an ion of charge ($M+$) is inserted. Traditional labels are given when appropriate (see Appendix).

arguably even better understood than the donors in semiconductors. The F center, with a single electron localized by the net positive charge of the missing anion, has obvious analogies with the hydrogen atom. These can be exploited in several ways. First, one can label states in the same way, referring to the $1s$, $2s$, $2p, \ldots$ states. This needs slight generalization, because of lattice relaxation. Thus it is usual to use the convention $2p$ for the state reached by a Franck–Condon transition (constant lattice geometry) from the $1s$ ground state; the corresponding state following lattice relaxation in the vicinity of the excited F center is referred to as $2p^*$. A second use of the analogy is to give rough estimates of transition energy and orbital radius. A simple change of dielectric constant gives

$$\Delta E = 3/(4\epsilon_\infty^2)\text{Ry} \tag{5.1}$$

for the energy of the $1s$–$2p$ transition of the F center (F band), where $\frac{3}{4}$ Ry is the energy of the corresponding transition in the hydrogen atom. Ignoring the change in effective mass of the F electron leads to an effective radius of ϵ_∞ Bohr radii [see also Equation (5.18)]. Equation (5.1) works very well, especially with a correction due to the fact that ϵ_∞ is dominated by the anions, whereas the cation neighbors dominate the behavior of the F center electron. Third, most calculations on F centers use a variational method, guessing a physically reasonable wavefunction $\psi(\mathbf{r}, \lambda)$, and minimizing the energy with respect to a variational parameter λ. Hydrogen atom wavefunctions scaled in the manner mentioned above are indeed quite a good starting point for a trial function $\psi(\mathbf{r}, \lambda)$.

It is found experimentally for F-band absorption (but not emission) that the relation

$$\Delta E = (16.75\ \text{eV})/(a\ \text{Å})^{1.772} \tag{5.2}$$

holds quite well for alkali halides (Figure 5.2), where a is the cation–anion spacing measured in Angstrom units. Equation (5.2) is referred to as the Mollwo–Ivey rule. It suggests strongly that the dominant energies depend only on lattice geometry. This can only occur (if it is not fortuitous) if the point-ion potential dominates, that is, if [cf. Gourary and Adrian (1957, 1960)] one can replace anions and cations by point charges $\pm|e|$. Indeed, the point-ion model proves a very respectable and successful approximate description. It does have its limitations, and these are conveniently estimated as ion-size corrections in the pseudopotential formulation of Bartram et al. (1968). Figure 5.3 gives some idea of the accuracy of these approaches, which we now discuss.

We know that the full wavefunction $\psi(\mathbf{r})$ of the F-center electron must be orthogonal to all the occupied core states $|c\rangle$ of the crystal ions. This

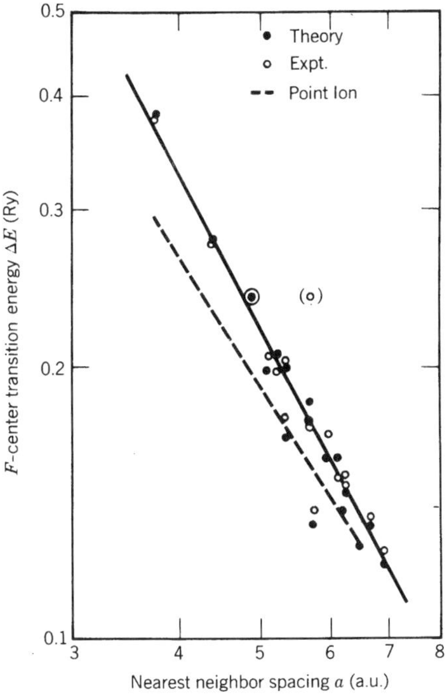

Figure 5.2 *F*-center transition energies (in rydbergs) as a function of nearest-neighbor distance in atomic units. Experiment (open circles, Dawson and Pooley (1969)] and theory are compared in the point-ion limit (dashed line, Gourary and Adrian (1957)] and with ion-size corrections (solid circles, Bartram et al. (1968)]. The experimental value for LiI is not certain and is shown bracketed. The solid line represents the Mollwo–Ivey rule (see text), that is, the best straight-line fit to experiment.

requirement can be included in variational calculations in two main ways. One is to make the variational function $|\psi(r)\rangle$ explicitly orthogonal to the core states, for example, by writing

$$|\psi\rangle = N\left(|\tilde{\psi}\rangle - \sum_c |c\rangle\langle c|\tilde{\psi}\rangle\right) \tag{5.3}$$

The expectation value $\langle\psi|\mathcal{H}|\psi\rangle/\langle\psi|\psi\rangle$ is to be minimized and the relatively complex form of $|\psi\rangle$ [Equation (5.3)] does not help. In the pseudopotential method, (Section 1.1.1), instead of attempting to solve

$$\mathcal{H}|\psi\rangle = E|\psi\rangle, \tag{5.4}$$

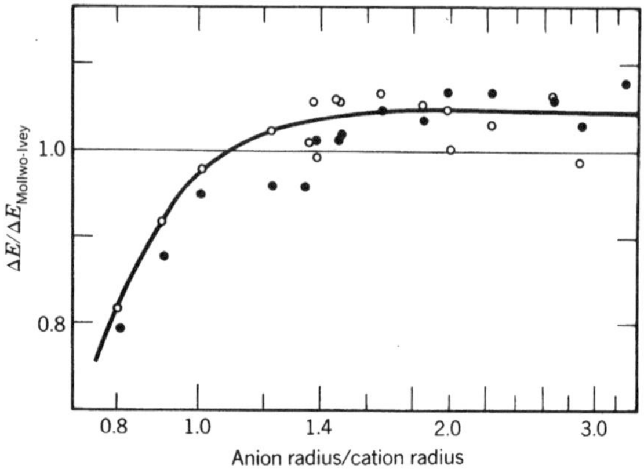

Figure 5.3 Ion-size corrections. This replots the data of Figure 5.2 so as to show explicitly the dependence of *F*-center transition energies ΔE on the ratio (anion radius/cation radius). The energies are given in terms of the corresponding Mollwo–Ivey rule prediction, so that direct dependence on the lattice parameter is eliminated: open circles, experiment (Dawson and Pooley 1969); light line, Mollwo–Ivey rule; bold line, simplified analytic form of theory (Bartram et al. 1968); solid circles, detailed theory (Bartram et al. 1968).

subject to the orthogonality constraint, one incorporates the constraint in an extra term giving

$$(\mathcal{H} + h)|\phi\rangle = E|\phi\rangle \tag{5.5}$$

This is simply rewriting and no new approximation is involved; h does not need to be small in any sense. However, because of h, $|\phi\rangle$ is no longer subject to the orthogonality constraint. Note that the energy eigenvalues of Equations (5.4) and (5.5) are the same. In fact, one has some freedom in choosing h. One good choice makes $|\phi\rangle$ as smoothly varying as possible, since this helps in choosing a suitable variational function.

Bartram et al. (1968) adopted this smoothly-varying criterion for h, so that their pseudopotential has the energy-dependent form $h = \sum_c |c\rangle\langle c|(\bar{V} - V)$ with V the potential energy and $\bar{V}$ the expectation value of $V + h$. To make useful application further approximations are needed. These assume that the core orbitals on different ions do not overlap and that the pseudo wavefunction $|\phi\rangle$ varies only slowly over the core of each ion. One can simplify the resulting expressions to show that the effective potential $V_{\text{eff}} = (V + h)$ has the form

$$V_{\text{eff}}(\mathbf{r}) = V_{PI}(\mathbf{r}) + \sum_i C_i \delta(\mathbf{r} - \mathbf{R}_i) \tag{5.6}$$

with $C_i = A_i + B_i(\bar{V} - U_i)$. Here A_i and B_i only depend on the ionic species; U_i is the point-ion potential at site i, and $V_{PI}(\mathbf{r})$ is the point-ion potential at a general point $\mathbf{r}$.

Several points follow immediately. First, the point-ion approximation will be a good approximation whenever the C_i are small. Second, both A_i and B_i increase with ionic radius and are genuine ion-size effects. Third, the ion-size term is not just a superposition of one-ion terms, since $\bar{V}$ depends on all ions; the ion-size term is electron-energy dependent. In fact, the main effect of the ion-size term is to replace the point-ion value by $\bar{V}$ within a sphere centered on each ion. Thus, near anion sites the ion-size term reduces the repulsion of the F-center electron and near cation sites the attraction is reduced.

Our scaled hydrogen-atom model suggests a number of experiments in which applied electric or magnetic field perturb the F center. The F band is very broad, typically with full width of 0.2 eV at half-maximum. Hence measurements of field effects concentrate on changes induced in the moments $M_N \equiv \int dE\, E^N \sigma(E)$ of the absorption band (see Henry et al.

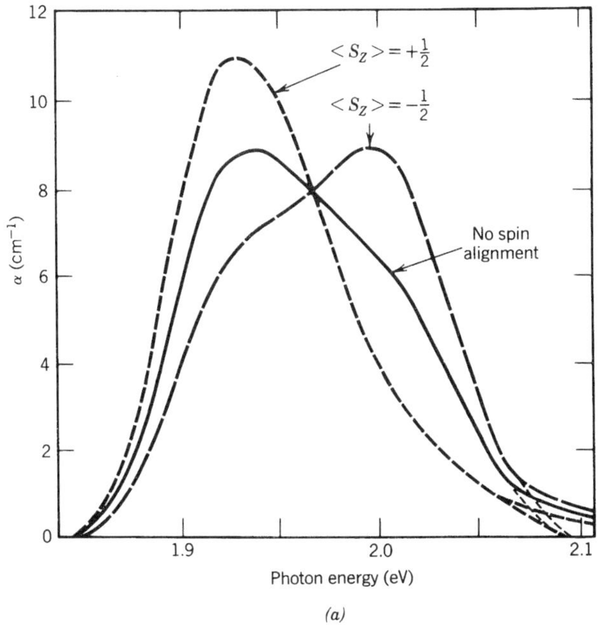

Figure 5.4(a).

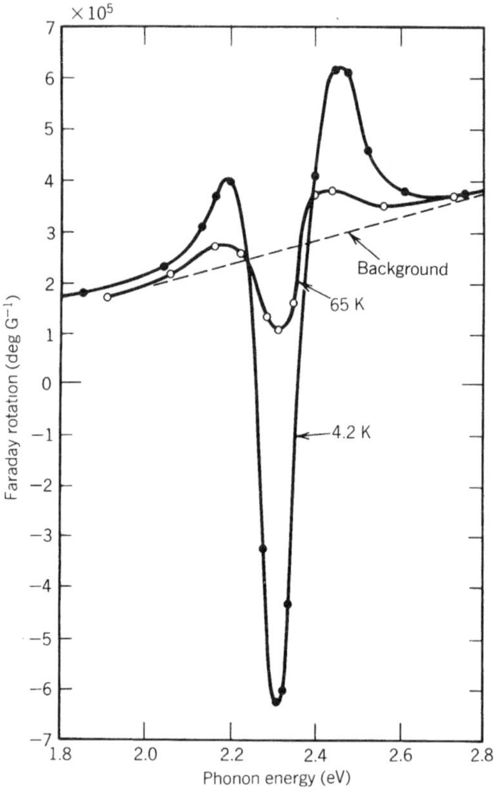

(b)

Figure 5.4 (*a*) Magnetic circular dichroism of the *F* band in CsBr at 1.86 K with $B = 8.6$ T (dashed line, absorption of right circularly polarized light; dotted line, absorption of left circularly polarized light; solid line, absorption of unpolarized light) [after Margerie and Romestain (1964)]. (*b*) Faraday rotation of the *F* band in KCl at 65 and 4.2 K with $B = 5$ T [after Luty and Mort (1964)].

1964). The zeroth ($N = 0$) moment of the band is the area; the first ($N = 1$) moment measures the mean energy; and the square root of the second ($N = 2$) moment gives a measure of the width. Detailed information about the nature of the ^2P state, including phonon coupling, may be obtained by studying magnetic circular dichroism (i.e., the difference in absorption of right and left circularly polarized light traveling through the crystal parallel to a magnetic field B) or the equivalent

Faraday rotation. There is no change in the area of the F band induced by B for right (+) and left (−) circularly polarized light:

$$\langle \Delta A \rangle_{\pm} = 0 \tag{5.7}$$

assuming no mixing of other states by B. The change in the mean energy of the two circularly polarized components is

$$\langle \Delta E \rangle_{\pm} = \pm \langle x | L_z | y \rangle (\beta B + \lambda \langle S_z \rangle) \tag{5.8}$$

where $\langle x | L_z | y \rangle$ is the matrix element of orbital angular momentum within the ^2P state, λ is the spin–orbital coupling constant of the ^2P state, and $\langle S_z \rangle$ is the thermal average value of the spin in the ^2S ground state. The first term on the right-hand side of Equation (5.8) is small and not easy to measure accurately. The second term is temperature dependent, through $\langle S_z \rangle$, and is large at low T for large λ. Measurement of this term (Figure 5.4) gives $\langle x | L_z | y \rangle \lambda = \lambda'$, providing a check on theoretical wavefunctions (see below). Measurements of changes of higher moments induced by B

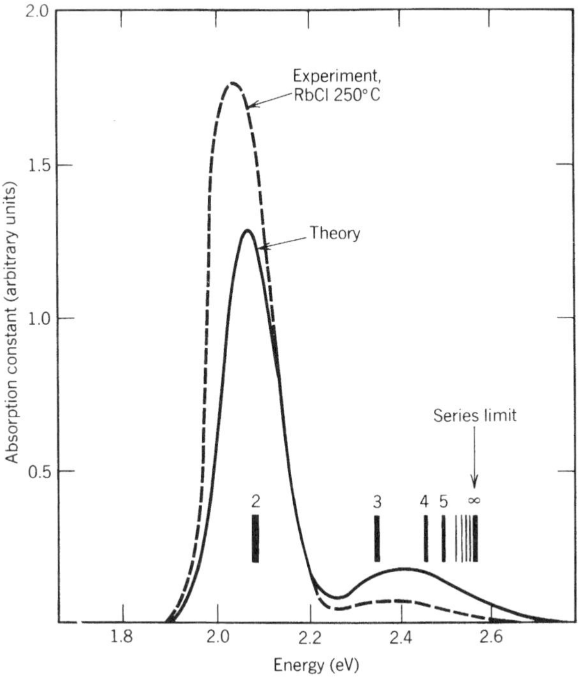

Figure 5.5(a).

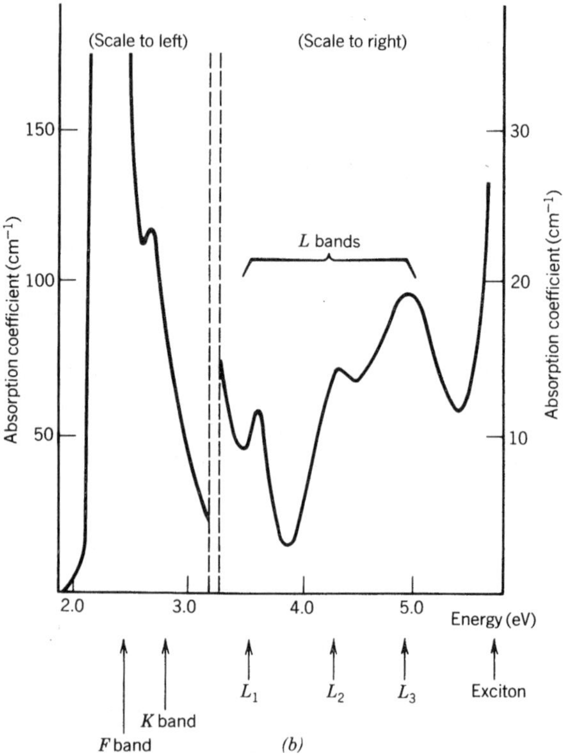

Figure 5.5 (*a*) Comparisons of measured and calculated *F* and *K* bands in RbCl. The vertical lines refer to transitions from the 1*s* ground state to 2*p*, 3*p*, . . . excited states. The width of each line is an indication of the calculated relative absorption strength [after Smith and Spinolo (1965)]. (*b*) Optical absorption of the *F*, *K*, and *L* bands in KCl at 93 K [after Luty (1969)].

give information about coupling to phonons of different symmetry but, again, are difficult to measure with precision. Direct measurement of the phonon coupling can be obtained using Raman scattering.

Measurements of magnetic circular dichroism show that λ' for the ^2P state is negative (e.g., for KCl, $\lambda' = -58\ \text{cm}^{-1}$) and increases in magnitude with increasing atomic weight of the halogen. The spin–orbit coupling is dominated by regions of high electric field gradient, for example, near nuclei, and the negative sign of λ' can be explained as a result of orthogonalization of the F-center wavefunctions to nearest-neighbor ions.

Continuing the analogy between the F center and the hydrogen atom we expect an infinite series of excited levels of the F center converging on a continuum, in this case the ionized states with the electron delocalized in the conduction band. A relatively weak asymmetric absorption band tailing to high energy occurs on the high-energy side of the F band (Figure 5.5a). This band, which has about 10% of the strength of the F band, is referred to as the K band; it consists of a superposition of transitions $1s \rightarrow 3p, 4p \ldots np$. There is a high quantum efficiency for photoconductivity associated with excitation in the high-energy tail of the K band.

Still weaker absorption bands of the F center, called the L bands, occur a few eV higher in energy than the K band and have an intensity of ~1% of the F-band intensity (Figure 5.5b). Excitation in these bands gives photoconductivity with a higher quantum efficiency than in the K band, and there seems little doubt that the L bands are conduction-band states perturbed by the F center.

The luminescence of F centers raises two important points. First, does one expect to see luminescence at all? It appears that the arguments of Section 4.4 work rather well and one can predict from absorption data alone which F centers luminesce (see Table 4.1). The second point concerns the F-center lifetime. Measurement of the oscillator strength of the F band in absorption and application of the principle of detailed balance [Section 3.3; Equation (4.33)] suggests that the luminescence of the F center in KCl, assuming a $^2P \rightarrow {}^2S$ transition, should have a lifetime of $\sim 10^{-8}$ sec and a quantum efficiency near unity. The measured quantum efficiency of luminescence is near unity but the measured lifetime is much longer than expected, being $\sim 10^{-6}$ sec. This result may be explained by assuming a more extended wavefunction in the relaxed $2p^*$ state of the F center, resulting in smaller overlap with the $1s$ ground state. However, an additional contribution to the lifetime is expected to come from admixture of the relaxed $2s^*$ and $2p^*$ states by local electric fields. Stark-effect measurements of luminescence indicate that these relaxed states are only a few tens of meV apart and appreciable admixture will increase the lifetime of the radiating state from that expected for a pure $2p^*$ state.

Aggregates of F centers are stable in ionic solids in units of two (called M centers), three (called R centers), and four (called N centers). In more extreme situations aggregation leads to formation of colloidal particles of the metallic constituent of the crystal. The optical properties of these aggregates have been thoroughly described (Hughes and Jain

1979); we shall discuss the similar colloid system seen in silver halides in Section 6.3.4.

The F center in halides is electrically neutral. However, it may lose its electron forming an empty anion vacancy (the α center) or gain another electron forming a charged center (the F' center). F' centers are stable only at low temperatures (<100 K) and have a broad absorption band extending through the visible region, usually at lower energies than the F band itself. The α center gives rise to an absorption band in the ultraviolet; in KI, for example, the α band occurs at 238 nm, close to the first exciton peak, at 230 nm. It is generally assumed that this band is due to creation of a perturbed exciton near the ionized F center. A perturbed exciton band associated with the (neutral) F center, the so-called β band, occurs between the α band and the first exciton peak of the perfect crystal.

In oxides an oxygen vacancy may trap a single electron forming the F^+ center, which closely resembles the F center in halides. In addition, the oxygen vacancy readily traps two electrons forming the F^0 center (occasionally misleadingly written as F center) and this corresponds to the two-electron F' center in halides.

Color-center systems have been operated as infrared lasers, broadly tunable in the range 0.4–1 eV (depending on host and center) and capable of continuous operation. The color centers involved are all related to the F center. They include the F_A center, for example, the F center in KCl with an Li or Na neighbor, the F_B center, for example, the F center in KCl with two Na neighbors, and the F_2^+ center, that is, two adjacent anion vacancies with a trapped electron.

Various laser systems with enhanced performance have been developed, based on the F^+ center in CaO (Henderson 1981), on the $F_A(Tl^+)$ and $F_A(Ag^+)$ centers, and the F_2^+ center stabilized by Mn^{2+} or OH^-. The effectiveness of these systems (Mollenauer et al. 1982, Gellermann et al. 1981, Gellermann and Pollock 1981) can be understood from features of the F center itself. Excitation takes place at a convenient energy (0.8–1.5 eV) for optical pumping, and the wide, inhomogeneously broadened line gives useful tunability. The states initially excited relax rapidly to states that luminesce efficiently, but at an energy changed enough to avoid stimulated emission from the pumping beam. The power output varies with power input P_{in} and threshold power P_{th} as $A(P_{in} - P_{th})$; the low threshold P_{th} and high slope efficiency A make color-center lasers extremely effective, especially F_2^+ center lasers. However, the F_2^+ laser tends to degrade in operation, as solid-state reactions destroy the lasing centers. The development of efficient, non-degrading color-center lasers remains a technical challenge.

One might expect a close parallel in ionic solids between trapped electron centers and trapped hole centers, for example, one might expect to find the antimorph of the F center where a cation vacancy traps a hole, giving the V_F center in halides or the V^- center in oxides. However, the difference between electron and hole centers are more striking than the parallels. In halides the V_F center is relatively hard to produce, and has not been intensively studied. In oxides, however, the V^- center is one of the most studied intrinsic defects (Section 5.1.2).

5.1.2 The V^- Center

The cation vacancy in MgO with a single trapped hole (i.e., an MgO crystal from which Mg^+ has been removed) has an effective single negative charge in the crystal and is an example of a V^- center. Similar centers occur in CaO, SrO, and BaO, in the fourfold coordinated oxides like BeO and ZnO, and in related sulfides, selenides, and tellurides (Watkins 1977, Norgett et al. 1977), not to mention some complex oxides. However, when the cations can easily change valence state (FeO, NiO, CoO, MnO), quite different defects result. The V^- center readily associates with other defects, for example, substitutional Al^{3+}, and easily captures hydrogen or alkali ions; it may also capture another hole to form the neutral V^0 center.

The V^- center has been studied extensively by epr and optical methods. The results show dramatic differences from the F^+ center in the same host. First, epr spectra at low temperatures demonstrate that the hole is localized on a single neighboring oxygen ion, so that in MgO there are one O^- and five O^{2-} ions next to the magnesium vacancy. As the temperature rises the hole moves more and more rapidly among the equivalent oxygens until it appears delocalized at room temperature. Second, there are two types of optical transition (Table 5.2). One consists of transitions internal to the O^- ion on which the hole is localized (the so-called crystal-field transitions, since the transition energy is determined almost entirely by the electrical field at the O^- ion caused by the absence of Mg^{2+}). The other, more important, transitions are the charge transfer type in which the hole moves to other neighbors of the vacancy. Similar charge transfers are important in determining the color of gemstones and in donor–acceptor pair recombination (Section 5.3.2).

Why is the hole initially localized on one oxygen? If it were delocalized, forming bandlike states over all six neighbors, without asymmetric distortion, there would be an energy gain 5Γ, where Γ is the matrix element for exchange between oxygens. This gain competes with possible gains in polarization energy associated with charge localization.

Table 5.2 Optical Transitions of the V^- Center and Related Centers in MgO[a]

	Experiment	Theory
V^- center (Mg^{2+} vacancy with one hole trapped)	2.15	2.3 (1.5 for crystal-field transition)
V^0 center (Mg^{2+} vacancy with two holes trapped)	2.16	2.3
$[Na]^0$ (Na^+ substituting for Mg^{2+}, with one hole trapped)	1.51	1.58
$V^-(Al)$ (V^- center with Al^{3+} at the (200) Mg site)	2.54	2.32

[a]The predicted energies, in eV, refer to the charge-transfer calculations of Norgett et al. (1977) unless otherwise stated.

We can illustrate this latter effect by supposing that a charge q is divided into N portions, q/N, placed far apart in a polarizable crystal. For each charge the polarization energy is proportional to the square of the charge. Suppose it has value βq^2 for the whole charge q. The N fractional charges have polarization energy gain $N \times \beta(q/N)^2$, that is, $\beta q^2/N$, so that polarization always favors localization. Both detailed qualitative analysis and accurate quantitative estimates (Norgett et al. 1977) agree that for a hole trapped by a cation vacancy the polarization terms win, giving localization. For a free hole in an otherwise perfect crystal the balance is marginal, but delocalized holes in oxides tend to be generally more stable by a few tenths of an eV (Colbourn and Mackrodt, 1981).

Exactly the same polarization terms and similar (but not identical) tunneling terms enter in the charge-transfer energy. The initial and final oxygens involved in the charge-transfer transitions are exactly equivalent apart from distortion and polarization; the charge-transfer energy between such equivalent ions would be zero in an unrelaxed, unpolarized host. Charge transfer can occur at more than one transfer range. It is a general rule that the longer the range the weaker the transition and the higher its energy. We see too that one type of transition (Figure 5.6) has a final state

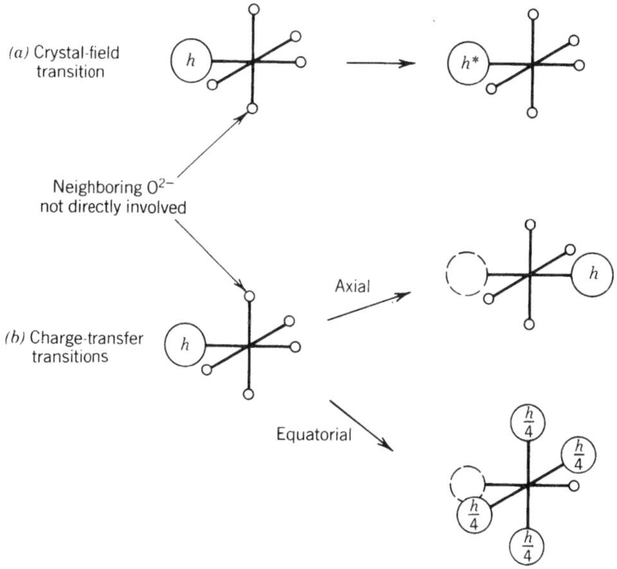

Figure 5.6 Transitions of the V^- center in MgO. In (*a*), h^* indicates an excited state of O^- reached in a crystal-field transition. In (*b*), the hole fractions on each neighbor in the excited state are shown before lattice relaxation. These charge-transfer transitions dominate the optical absorption. The equatorial transition will be split by electron exchange terms between the four oxygens. Note that h always indicates that the lowest crystal-field state of oxygen is occupied or partly occupied; h^* indicates that an excited crystal-field state is occupied.

which is split by exchange. The splitting (where the exchange matrix element Γ' is altered from Γ because of lattice distortion) is easily discerned in the fourfold coordinated systems like ZnSe. The effect of nearby impurities leads to shifts of transition energies consistent with the model (Table 5.2). We note too that the charge-transfer picture does not depend on the host being ionic. Exactly parallel qualitative descriptions can be given from a covalent model (as for vacancies in Si) and from an ionic picture (as for vacancies in BeO), [see, e.g., Stoneham (1975, page 618)].

5.1.3 The Self-Trapped Hole (V_K Center), The Self-Trapped Exciton, and the Self-Interstitial (H Center)

Irradiation of simple halide crystals at low temperatures with ionizing radiation produces F centers and trapped hole centers. Two intrinsic hole centers, the so-called *H* centers and V_K centers, have somewhat similar structures (Figure 5.1) and are stable at low temperatures. In addition a transient center, the self-trapped exciton, also forms at low temperatures. We shall discuss each of these centers in turn.

(a) The V_K Center

The V_K center can be thought of as a hole localized between two adjacent halide ions. To a good approximation it is simply an X_2^- molecule ion, where X is a halogen, created from two X^- ions and the hole. Usually it is oriented along the closest-packed row of halogens, that is, the ⟨110⟩ orientation in the case of alkali halides (Figure 5.1).

The V_K center is a good example of a molecule-in-a-crystal defect, that is, the host weakly perturbs the X_2^- molecular ion. We may construct a molecular-orbital picture based on the valence orbitals of the halogens, for example, $3p$ for chlorine. If we ignore the departure from axial symmetry of the V_K center in the crystal, we obtain the schematic energy-level diagram of Figure 5.7. Before the hole is trapped, the twelve $3p$ electrons of two Cl^- ions completely fill the molecular orbitals, giving no net bonding. However, when a hole is trapped in the σ_u orbital, its lowest-energy state, an antibonding electron is lost. This gives a stable X_2^- ion with ground-state configuration $\sigma_g^2\pi_u^4\pi_g^4\sigma_u(^2\Sigma_u^+)$. The hole may be excited to the π_g, π_u, or σ_g orbitals giving excited states $^2\Pi_g$, $^2\Pi_u$, and $^2\Sigma_g^+$, respectively. The transition $^2\Sigma_u^+ \to {}^2\Sigma_g^+$ is electric dipole allowed, giving rise to an intense, broad, absorption band in the ultraviolet (Figure 5.8) polarized parallel to the molecular axis. The infrared transition, $^2\Sigma_u^+ \to {}^2\Pi_g$, has dipole polarization normal to the molecular axis. The transition $^2\Sigma_u^+ \to {}^2\Pi_u$ is parity forbidden. No luminescence due to the V_K center has been reported, although the exact mechanism of nonradiative recovery is unclear. V_K centers can be preferentially oriented at low temperatures by pumping in the optical absorption bands with linearly polarized light. This occurs because of preferential excitation of centers with favored orientations, followed by recovery randomly to the various equivalent crystal orientations that are possible. The centers reorient from the directions most easily excited to those least readily excited.

One may obtain some insight into the energetics of V_K formation by looking at the energy required to localize a charge at a point in the lattice and the consequence of relaxation of neighboring ions into an energetically favorable configuration. The relative stability of self-trapped holes (as in alkali halides) and delocalized holes (as in silicon) can be seen from arguments set out by Norgett and Stoneham (1973). As for the V^- center (Section 5.1.2), there is a polarization term favoring localization and

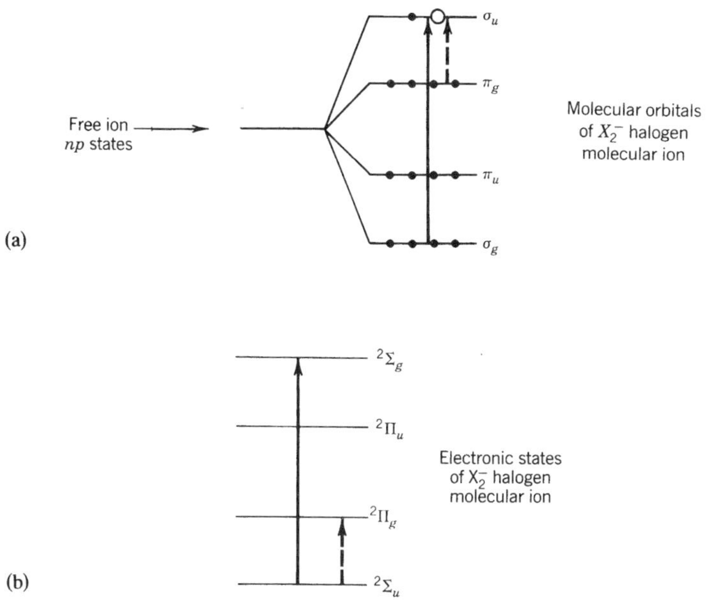

Figure 5.7 (*a*) Distribution of electrons in the $p\sigma$ and $p\pi$ orbitals of the ground electronic configuration of X_2^-; note the hole, ○, in the σ_u orbital. (*b*) Ground and excited electronic states derived from (*a*), showing optical transitions; the transition $^2\Sigma_u \rightarrow {}^2\Sigma_g$ is electric-dipole allowed (see Figure 5.8).

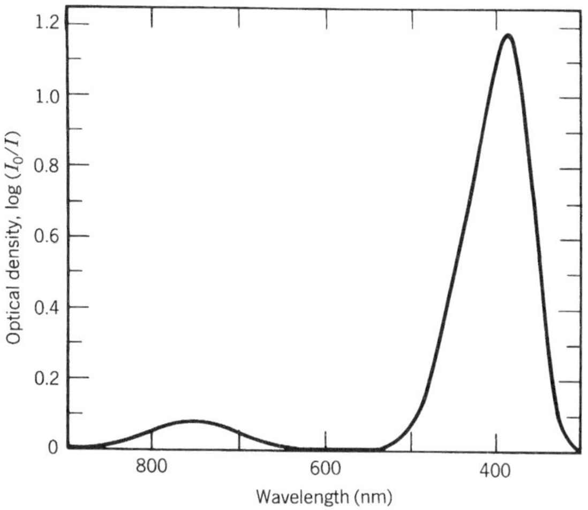

Figure 5.8 Optical absorption of the Br_2^- ion in KBr showing the (strong) $^2\Sigma_u \rightarrow {}^2\Sigma_g$ and the (weak) $^2\Sigma_u \rightarrow {}^2\Pi_g$ transitions (see Figure 5.7) [after Delbecq et al. (1961)].

a band term favoring delocalization. The latter is about half of the (unrelaxed crystal) valence bandwidth. The polarization term is enhanced here by the strong chemical bonding of X^0 and X^- to form X_2^-. The calculations show that the localized, self-trapped form is stable by several eV.

Similar calculations show that the V_K center is very well represented by the molecule-in-a-crystal model. This can be verified by comparing predicted optical transition energies and predicted epr parameters for an

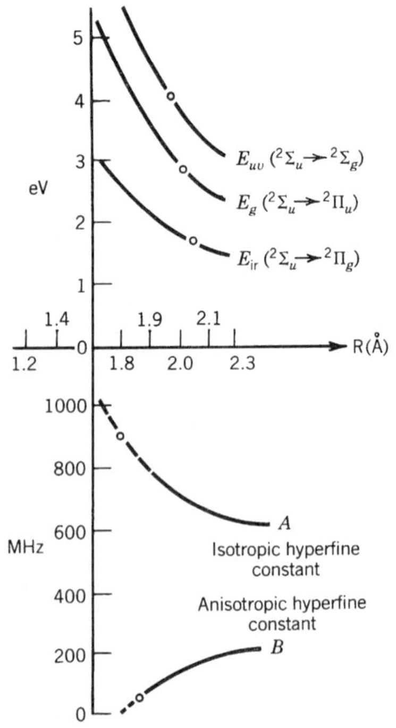

Figure 5.9 Energy levels of the V_K center in CaF_2 related to predicted energies of the F_2^- molecular ion at different spacings R. For each transition I the observed energy in the solid E^I_{obs} and predicted energy of the free ion $E^I_{theory}(R)$ indicate a molecular ion spacing R_I in the ground state. If the V_K center were accurately an F_2^- ion in the solid with spacing fixed by the environment, all predicted values of R_I should be the same, irrespective of transition. Since the points do lie reasonably close to a single vertical line, it is clearly a good first approximation to regard the V_K center here as an F_2^- molecular ion. The same conclusion is arrived at by comparing observed and predicted epr hyperfine constants (see the lower part of the figure).

X_2^- ion as a function of X–X spacing, together with observed V_K center data. This comparison leads to a predicted internuclear separation for the V_K center in its ground state, which is close to that for the free molecular X_2^- ion. Accurate calculations (see Section 3.4) also predict essentially the same spacings. It appears that the crystal environment increases the spacing of the X ions in X_2^- by less than 5% of that of the free X_2^- ion (Figure 5.9).

(b) The Self-Trapped Exciton

The self-trapped exciton in alkali halides is essentially a self-trapped hole to which an electron is bound, that is, $[V_K e]$ or $[he]$. It is clear that there are at least three types of excited state of the self-trapped exciton: electron-excited states $[he^*]$, hole-excited states $[h^* e]$, and states with both components excited $[h^* e^*]$. These excitations are distinct in energy and readily characterized. Nonradiative transitions can take the center from one type of state into another. The hole excitations are only weakly perturbed from those of the V_K center by the presence of the electron. The electron excitations, with the electron moving in the field of a localized positive charge, resemble those of a hydrogen atom, apart from the axial perturbation that splits p states into axial σ and equatorial π states. Also, singlet and triplet excitons can occur, depending on whether the electron and hole have antiparallel or parallel spins. The triplet form can only recombine by a spin-forbidden transition. The lowest triplet state has a long lifetime, of order 10^{-3} sec, and a whole range of spectroscopic measurements can be made on it, including optical absorption, luminescence, and optical detection of magnetic resonance (odmr).

The various transitions observed or inferred for self-trapped excitons show enormous variety. The two principal luminescence transitions give the so-called π and σ bands (Figure 5.10), with dipoles, respectively, normal and parallel to the axis of the self-trapped hole. The long-lifetime π band is seen in all alkali halides, and is produced by decay of the lowest triplet state to the crystal ground state (electron–hole recombination); in the triplet, the hole is in essentially a V_K center ground state and the electron in an A_{1g} ($1s$-like) excited state. The rapid σ transition is seen in some, but not all, alkali halides. In this case theory showed that the luminescence came from an excited singlet state $[he^*]$, with the electron in an A_{1g} ($2s$-like) state, and that there was negligible luminescence from the lowest excited singlet.

There are several types of nonradiative transitions involving self-trapped excitons:

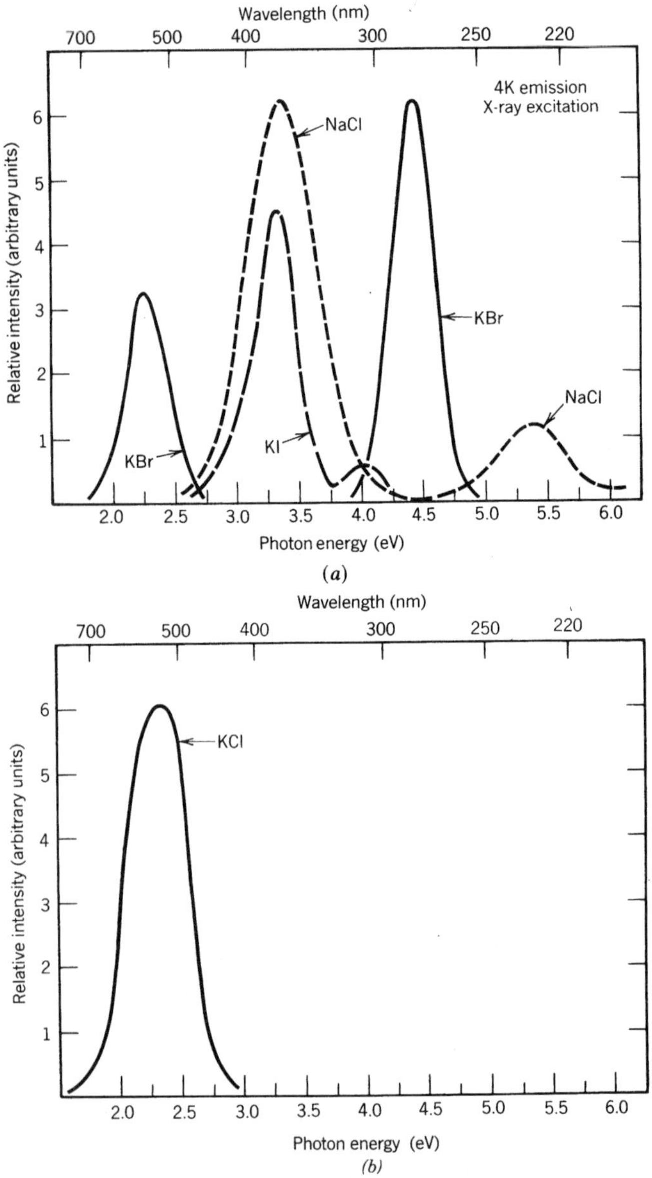

Figure 5.10 X-ray-induced emission spectra of self-trapped excitons in alkali halides at 4 K. Note that (*a*) NaCl, KBr, and KI show two peaks, and (*b*) KCl shows only one peak, lacking the σ luminescence (see text) (after Kabler 1964).

(a) The formation of the self-trapped exciton from the unrelaxed exciton, involving thermal excitation over a barrier ($E_B \sim k_B\theta$ with $\theta \sim 10$–100 K), so that self-trapping can be inhibited and band edge luminescence enhanced at low temperatures.

(b) The normal multiphonon processes between levels of the self-trapped exciton, leading to electron–hole recombination. It is believed that the σ band is missing in some hosts because of rapid nonradiative recombination of the excited singlet states. The triplet exciton has spin-forbidden recombination; for practical purposes all triplet excitons reach the lowest triplet state before recombination. A special case, which combines the multiphonon transitions with the Auger transitions induced by the electron–electron interaction, transfers energy from electron-excited ($[he^*]$) to hole-excited ($[h^*e]$) states. In states like $[h^*e]$, other nonradiative transitions may cause reorientation of the self-trapped exciton.

(c) Formation of F centers and H centers, that is, of neutral halogen vacancies and interstitials from nonradiative decay of self-trapped excitons. This formation, which does not take place from the lowest triplet state, is central to the phenomenon of radiolysis in ionic solids (Chapter 6 and below).

(c) Self-Interstitials: The H Center and Other Interstitials

Both the self-trapped hole (V_K center) and the neutral anion interstitial (H center) in alkali halides can be regarded as X_2^- halogen molecular ions (Figure 5.1). Both have (110) symmetry, since the H center is also aligned along the close-packed row of anions. Both V_K centers and H centers have rather similar epr and optical spectra. The H center, like the V_K center, is created by exposure to ionizing radiation at low temperatures.

The H center in alkali halides is an example of a split interstitial (Figure 3.1), with two ions replacing one at the same site. This form is also found in metals and probably occurs for most important self-interstitials in semiconductors. Two (110) oriented H centers may unite to form a neutral di-interstitial in alkali halides, and this appears to adopt a split form also. This di-interstitial can be regarded as an X_3^- ion, probably oriented along (111).

X_2^- and similar molecular ions are common basic hole units in halide crystals. However, they can take various forms. In the CaF_2 structure (Figure 3.4) the F_2^- ion (V_K center) has a (100) molecular axis, involving two nearest fluorines on a cube edge. However, the H center has a (111) molecular axis with two inequivalent fluorine ions, one almost on site and the other nearer the center of the normally empty cube in the fluorite structure (see Figure 6.12). The X_2^- species are also seen in solution and in glasses.

The negative interstitial in alkali halides, X_i^-, is formed by ionizing radiation at low temperatures (Chapter 6) and is believed to occupy the natural interstitial site, that is, at the center of an anion and a cation tetrahedron. In the fluorite structure the corresponding interstitial occurs at the center of the empty cube of fluorines. Thus both the site occupancy and the structure of interstitials can depend on charge state, and in semiconductors this may lead to the Bourgoin–Corbett mechanism of recombination-enhanced diffusion (Section 6.4.6).

It happens that many interstitials are elusive, with low activation energies of motion, so that they aggregate or else annihilate with vacancies, even at low temperatures. Such behavior is evident in some semiconductors (see Section 6.4).

5.2 TRANSITION-METAL IMPURITIES

The transition metals are those from the three series in the periodic table with incomplete inner shells: the iron group ($3d^n$), the rare earths ($4f^n$), and the actinides ($5f^n$). Since many of their properties of interest are governed by the partly filled shell of inner electrons, they are often conveniently characterized by the behavior of that shell alone. A second simplification is that the $3d$ shell of an iron-group ion might be considered as a component of a perturbed free ion. It is important to realize that both simplifications have more to do with symmetry (i.e., there is a simple correspondence in symmetry between the actual wavefunctions and these simple approximations) than with precise quantitative description.

We shall discuss the broad features of transition-metal ions in octahedral coordination (e.g., in MgO) and tetrahedral coordination (e.g., in Si, or in a III–V compound like GaAs), with special reference to the iron group ($3d$ series). In cubic or tetrahedral symmetry the one-electron d orbitals fall into two groups: the triply degenerate t_2 orbitals transforming as xy, yz, zx, and the doubly degenerate e orbitals transforming as $3z^2 - r^2$, $x^2 - y^2$ (Figure 4.5). The crystal cubic-field splitting Δ, sometimes written $10Dq$, gives the one-electron energy difference

$$\Delta = |E(e) - E(t_2)| \tag{5.9}$$

The key factors determining the distribution of electrons over the e and t_2 states are the electron–electron interaction, the spin–orbit splitting (a secondary factor in the $3d$ series, but dominant for rare earths), and the size of the cubic-field splitting. A simple electrostatic model, with negative electrical point charges $-Z|e|$ at nearest neighbors (the ligands), correctly predicts that the e states have higher energy in octahedral coordination, and that the t states have lower energy; in tetrahedral or eightfold coordination the situation is reversed. The same point-charge model (which is deceptive, since weak covalency is actually important) predicts:

$$E(e) - E(t_2) = \frac{3}{5}\frac{Ze^2}{R}\frac{\langle r^4 \rangle}{R^4} \qquad \text{(octahedral)}$$
$$= -\frac{4}{15}\frac{Ze^2}{R}\frac{\langle r^4 \rangle}{R^4} \qquad \text{(tetrahedral)} \tag{5.10}$$

where R is the distance to the ligands and $\langle r^4 \rangle$ is the expectation value of r^4 for the $3d$ wavefunction. As is customary, we assume the radial parts of the e and t_2 functions are the same. Typical values of Δ are about 1 eV for iron-group ions. Only nearest neighbors affect values of Δ appreciably and there are systematic observed trends (Jorgensen 1958): (1) increasing covalency usually enhances Δ, (2) Δ is bigger for transition-metal ions of higher charge (e.g., 3+ versus 2+), and (3) Δ is bigger for higher transition series (e.g., Δ increases as one goes from $3d^n$ to $4d^n$ to $5d^n$).

The dependence of Δ on charge state is apparent both from optical spectroscopy ($d^n \rightarrow d^{n^*}$ transitions) and from comparisons with other techniques in which the number of d electrons changes. Thus, in photoelectron spectroscopy, in which a d electron is removed ($d^n \rightarrow d^{n-1^*}$) the higher charge of the final state leads to a larger value of Δ than in the optical case.

Suppose we look at $3d^n$ states in octahedral coordination, with the t_2 states lowest. If Δ is small, the electron–electron interaction dominates and this favors parallel spin orientations, giving the occupancies shown under the so-called weak-field limit in Table 5.3. When Δ is large, the t_2 states fill up first, irrespective of spin. For some cases ($3d^n$ with $n = 1, 2, 3, 8, 9$ in octahedral symmetry, or $n = 1, 2, 7, 8, 9$ in tetrahedral symmetry) both limits give the same occupancies. For the other cases that is, for d-shells roughly half-full, there is a balance resulting in high-spin (weak-field) or low-spin (strong-field) configurations.

Experimentally, transition-metal ions have been studied systematically in many hosts using epr (see e.g., Abragam and Bleaney 1970) and optical (see e.g., McClure 1959) spectroscopy. Although $d \leftrightarrow d$ electric dipole transitions are forbidden in octahedral symmetry, one sees forced electric dipole transitions involving symmetry-breaking phonons, and also the weaker magnetic dipole transitions, and charge-transfer transitions between

Table 5.3 Occupancies of *d* Orbitals for Octahedral Coordination

Number of *d* Electrons		Weak-Field Limit: Strong Electron–Electron Interaction t_2	e	Strong-Field Limit: Weak Electron–Electron Interaction t_2	e
1		↑		↑	
2		↑↑		↑↑	
3		↑↑↑		↑↑↑	
4	Critical region	↑↑↑	↑	↑↓↑↑	
5	Critical region	↑↑↑	↑↑	↑↓↑↓↑	
6	Critical region	↑↓↑↑	↑↑	↑↓↑↓↑↓	
7		↑↓↑↓↑	↑↑	↑↓↑↓↑↓	↑
8		↑↓↑↓↑↓	↑↑	↑↓↑↓↑↓	↑↑
9		↑↓↑↓↑↓	↑↓↑	↑↓↑↓↑↓	↑↓↑

the transition-metal ion and the host. Nonradiative transitions are important, both as spin-lattice relaxation (when Zeeman energy is converted to heat) and because the complex energy-level schemes often allow efficient degradation of energy to phonons after optical excitation.

Transition metals in insulators are one of the main causes of crystal color (Nassau 1980). The precise hue depends on many factors, including the color of the illuminating light, on whether one.looks in transmission or reflection, and on the degree of scatter or of subsequent luminescence. In ruby, for example, the crystal-field levels of Cr^{3+} in the Al_2O_3 matrix give absorption in the violet and the green/yellow regions of the spectrum, with the characteristic red seen in transmission. Sapphire, also basically Al_2O_3, appears blue because of charge transfer of an electron from the impurity Fe^{2+} to a Ti^{4+} impurity. Metal colloids (Section 6.3.4) are often the origin of color in stained glasses. While most semiprecious gems are colored by transition metals, many other mechanisms can be involved. Non-transition-metal impurities color diamond; boron gives a bluish tinge and nitrogen gives a yellow color at very low concentrations, changing to green at the tens of ppm level. Color centers, aptly named, are simple intrinsic defects in many lattices. Band-to-band excitation leads to the yellow color of CdS ($E_g = 2.6$ eV) and to the vermillion color of the pigment HgS ($E_g = 2.1$ eV). Quite different mechanisms are found in living creatures, as in the colors of birds or butterflies, when interference effects can be important.

As in the case of intrinsic point defects, epr has provided the key to detailed understanding of transition-metal behavior. We have already used the concept of the spin Hamiltonian (Section 4.5), which provides the meeting point between theory and experiment. Experiment fits a spin Hamiltonian phenomenologically to describe observed spectra, whereas theory attempts to predict the parameters derived. Spin Hamiltonians involve two independent ideas. At low temperatures only a small group of N states are occupied; excited states $|x\rangle$, split off by crystal-field energies of order Δ, or by the spin–orbit coupling, are unoccupied. The first idea is that we may concentrate only on the N states, including the effect of all higher states by modifying the matrix elements. Thus, if the N states are labeled $|I\rangle$, with $I = 1, \ldots, N$, then the matrix element $\langle I|\Omega|J\rangle$ of some operator Ω becomes, to second-order perturbation:

$$\langle I|\Omega|J\rangle_{\text{eff}} = \langle I|\Omega|J\rangle - \sum_x \frac{\langle I|\Omega|x\rangle\langle x|\Omega|J\rangle}{E_x - E_J} \tag{5.11}$$

If Ω is the Hamiltonian $\mathcal{H}$, the $N \times N$ effective matrix will be Hermitian; there are thus N^2 independent components of the matrix, that is, N real diagonal ones and $\frac{1}{2}N(N-1)$ complex off-diagonal ones such that $\mathcal{H}_{ij} = \mathcal{H}_{ji}^*$.

We can now pretend that the states $|I\rangle$ are the eigenstates of the z component of an effective spin $S' = \frac{1}{2}(N-1)$. We can then write all the matrix elements in terms of (effective) spin operators, which is convenient for many practical reasons, for example, because the manipulation of the spin matrices is a standard textbook problem. We now have a spin Hamiltonian, expressed in terms of functions of a fictitious spin $\mathbf{S}'$, but whose matrix elements exactly reproduce those of the full Hamiltonian, including admixture of excited states. Three qualifications should be mentioned. First, real spin and effective spin may differ. Thus Co^{2+} has real spin $S = \frac{3}{2}$. However, in MgO the lowest state of Co^{2+} is a doublet ($N = 2$), so $S' = \frac{1}{2}$. Second, S' could change with temperature if N increases, as higher states become populated. Third, in rare cases it is helpful to choose a larger effective spin than that determined by N, simply to make the symmetry of the spin Hamiltonian more transparent. This is only rarely necessary for states in cubic symmetry.

The system $Cu^{2+}(3d^9)$ with octahedral coordination shows many interesting features. We have already seen (Section 4.2.3) that the hole (and correspondingly the one unpaired electron) is in an e state. The Hamiltonian contains two important terms, namely, a Zeeman term $\beta\mathbf{H} \cdot (\mathbf{L} + 2\mathbf{S})$ (where β is the Bohr magneton, $\mathbf{L}$ is the orbital angular momentum, and $\mathbf{S}$ is the spin) and the spin–orbit coupling $\lambda\mathbf{L} \cdot \mathbf{S}$. In the present case both real spin S and effective spin S' are $\frac{1}{2}$, since the Jahn–Teller distortion splits the doubly degenerate e state (Section 4.2.3). We shall assume the state $|\epsilon\rangle \equiv |x^2 - y^2\rangle$ is lower in energy and anticipate that there will be anisotropy as the magnetic field direction changes. If we consider only the two spin states for the $|\epsilon\rangle$ orbital, we find a spin Hamiltonian of $2\beta\mathbf{H}\cdot\mathbf{S}$, since $\mathbf{L}$ has no diagonal matrix elements. However, the $\beta\mathbf{H}\cdot\mathbf{L}$ and $\lambda\mathbf{L}\cdot\mathbf{S}$ terms together admix a contribution from the t_2 states at $E_x \simeq \Delta$ higher in energy, and the spin Hamiltonian now becomes

$$\mathcal{H} = g_{\parallel}\beta S_z + g_{\perp}\beta(S_x H_x + S_y H_y) \tag{5.12}$$

with $g_{\parallel} = g_0 - 8\lambda/E_x$ and $g_{\perp} = g_0 - 2\lambda/E_x$, where $g_0 = 2.0023$ is the free electron g value. Clearly, the Zeeman splitting will show the ground-state anisotropy. The subtle but important point should be noted that λ is positive for electrons ($g < 2$) and negative for holes ($g > 2$) and hence the g values for Cu^{2+} given above are greater than the free-electron g value (2.0023).

When the effective spin is greater than $\frac{1}{2}$, several new terms appear of which the most important is the zero-field splitting, $D[S_z^2 - \frac{1}{3}S(S+1)] + E(S_x^2 - S_y^2)$. This splitting vanishes for cubic symmetry, but for large enough spin, for example, for Mn^{2+} ($3d^5$, ${}^6S_{5/2}$), there is an extra term,

$$\tfrac{1}{6}a[S_x^4 + S_y^4 + S_z^4 - \tfrac{1}{5}S(S+1)(3S^2 + 3S - 1)],$$

where a is called the cubic-field splitting parameter. These terms come from dipole–dipole interactions and from spin–orbit coupling to second order. In suitable cases experimental values of the parameters can be used to get spin–spin distances, using $R^3 \simeq 3(g\beta)^2/2|D|$. This approximate form, valid at large distances, tends to overestimate the spacing R.

Many defects, especially transition-metal ions, have multiple charge states (Table 5.4) giving a range of levels in the gap separated by a few tenths of an eV, and this deep-level spectrum may be complex (see also Section 5.5). Although deep levels can be characterized by activation energies, using electrical methods such as deep-level transient spectroscopy (DLTS) (Section 4.3.4), conventional spectroscopic methods are required to establish its chemical nature, symmetry, and charge state. The deliberate use of deep-level impurities such as Cr to produce insulating GaAs and InP substrates for microwave field-effect transistors [known as FET's; Sze (1982)] has stimulated interest in deep levels, and we shall concentrate here on III–V compounds. In these materials $3d$ ions occupy metal sites and the electrically neutral charge state of the defect corresponds in the ionic limit to a trivalent ion, represented as A^0 ($3d^n$) (see

Table 5.4 Transition-Metal Ions: Charge State Stability[a]

Host	$n = 1$	2	3	4	5	6	7	8	9
MgO and CaO substitute for Mg^{2+} or Ca^{2+}		V^{3+}	Mn^{4+} Cr^{3+} V^{2+} Ti^{+}	Mn^{3+} Cr^{2+}	Fe^{3+} Mn^{2+} Cr^{+}	Fe^{2+}	Ni^{3+} Co^{2+} Fe^{+}	Ni^{2+} Co^{+}	
$SrTiO_3$ substitute for Ti^{4+}	Cr^{5+} V^{4+}		Fe^{5+} Mn^{4+} Cr^{3+}	Fe^{4+}	Co^{4+} Fe^{3+} Mn^{2+}	Ni^{4+} Co^{3+} Fe^{2+}	Ni^{3+}	Ni^{2+}	
GaP, GaAs, InP substitute for group-III element		Cr^{4+} V^{3+}	Cr^{3+} V^{2+}	Mn^{3+} Cr^{2+}	Fe^{3+} Mn^{2+} Cr^{+}	Co^{3+} Fe^{2+}	Ni^{3+} Co^{2+} Fe^{+}	Cu^{3+} Ni^{2+}	Cu^{2+} Ni^{+}

[a]The charge states are listed in columns according to n, the number of $3d$ electrons. Highly positive states are limited by capture of electrons from the valence band; low positive charges are limited by loss of electrons to the conduction band. Cr^{+} has not been reported in CaO; in GaAs it may be unstable. Some identifications are by Mössbauer methods; the others are from epr.

Appendix). Most $3d^n$ ions act as deep acceptors, so that the first ionization state is A^- $(3d^{n+1})$ and the second is A^{2-} $(3d^{n+2})$.

We shall take as an example nickel dissolved in Ga sites in GaP. This is a persistent inadvertent impurity in device-grade GaP where the neutral acceptor Ni^{3+} is reputed to be present. However, there is no optical

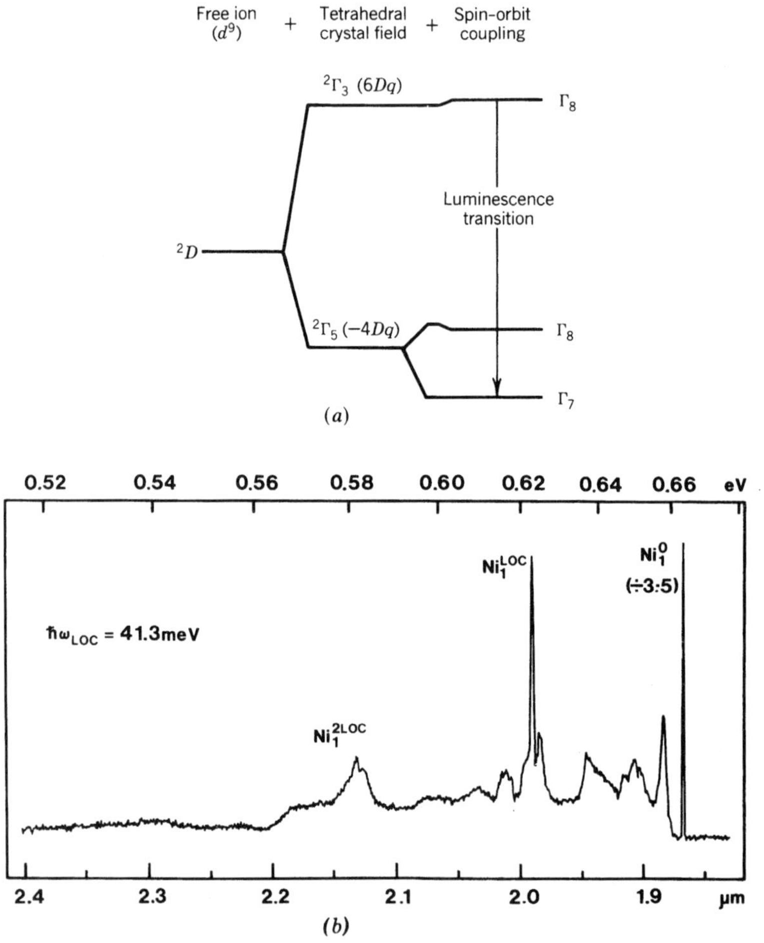

Figure 5.11 (*a*) The energy-level scheme for Ni^+ $(3d^9)$ in a weak tetrahedral crystal field, showing spin–orbit effects. (*b*) Photoluminescence of Ni^+ in GaP at 2 K showing the zero-phonon line at 5354 cm^{-1} (0.6636 eV) [corresponding to the transition $\Gamma_8 \rightarrow \Gamma_7$; see (*a*)] and phonon replicas at lower energy [after West (1980)].

evidence for the presence of Ni^{3+} ($3d^7$, ^4F), although an epr spectrum with a partly resolved hyperfine interaction with four phosporus nuclei has been tentatively assigned to this charge state. There is, however, convincing spectroscopic evidence for the presence of Ni^{2+} ($3d^8$, ^3F) and Ni^+ ($3d^9$, ^2D) in n-type GaP. The Ni^+ ion gives rise to a readily detectable luminescence in the infrared, which arises from transitions between the crystal-field-split components of the ^2D ground state (Figure 5.11); the spectrum consists of a zero-phonon line at 5354 cm^{-1} and phonon replicas on the low-energy side. This type of spectrum is characteristic of electron–phonon coupling of intermediate strength (Section 4.2). Zeeman and uniaxial stress studies on this zero-phonon line have shown conclusively that it is due to the $\Gamma_8 - \Gamma_7$ transition of Ni^+ in a tetrahedral environment (Figure 5.11a).

The photoconductivity response of GaP:Ni has an onset at 0.62 eV that is attributed to the charge-transfer reaction

$$Ni^{3+} \xrightarrow{h\nu} Ni^{2+}\ (^3\Gamma_4(F)) + h \qquad (5.13)$$

resulting in a release of a hole into the valence band. The absorption spectrum of Baranowski et al. (1968) and photoconductivity studies can be used to position the excited levels of Ni^{2+} in the bandgap. The additional reaction

$$Ni^{2+} \xrightarrow{h\nu} Ni^{+}(^2\Gamma_5(D)) + h \qquad (5.14)$$

places the ground state of Ni^+ 1.57 eV from the valence band.

Table 5.4 shows that most transition metals can exist in three or four stable charge states in crystals. Since free-ion ionization potentials, and Madelung potentials where relevant, can be tens of eV, whereas bandgaps are only a few eV, a striking cancellation is involved. The precise nature of the cancellation is still controversial, but important components are undoubtedly covalent hybridization (Section 8.4) and especially the polarization and distortion of the host lattice.

5.3 SHALLOW IMPURITY CENTERS IN SEMICONDUCTORS

5.3.1 Shallow Donors and Acceptors

Modern electronics technology is based on ability to produce high-purity semiconductors and to control their electrical properties by deliberately doping with impurities. The case of phosphorus impurity in silicon is a good example. Phosphorus dissolves substitutionally in Si sites or can be created there by neutron transmutation (Si(n, β)P). For an impurity A dissolved in a site S we use the notation A_S, for example, P_{Si}. The substitutional phosphorus has the valence configuration $3s^2 3p^3$. Four of the valence electrons form saturated bonds with the four nearest silicons, in sp^3-type hybrids (see Section 8.4), and the fifth electron is readily ionized into the conduction band. At low temperatures this donor electron is bound by the attractive interaction with the positively charged phosphorus core. Again, as in the case of the F center (Section 5.1.1) there are obvious analogies with the hydrogen atom.

As another example, consider boron impurity in Si. Boron has the valence configuration $2s^2 2p^1$, one electron short of the four required for tetrahedral bonding of the sp^3 type. This extra electron for full bonding may be obtained from silicon neighbors, creating a hole in the valence band. Again there is an attractive interaction, localizing the hole near the negatively charged boron core at low temperatures.

If we confine our attention to shallow donors and acceptors, for example, P_{Si} in Si and B_{Ge} in Ge, the so-called effective-mass approximation (Kohn 1957; Stoneham 1975) may be used. This makes formal the analogy with the hydrogen atom. Two distinct approximations are involved, (I) that the defect potential varies slowly over the unit cell and (II) that the envelope function, $F(\mathbf{r})$ (see below), which modulates the Bloch functions to give the full wavefunction, varies very slowly over the unit cell. There is no requirement that the defect potential $V(\mathbf{r})$ should be weak. In effective-mass theory we are able to simplify our equations to concentrate on just the single donor electron, or acceptor hole, moving with the correct effective mass (Section 1.2) in a potential $V(\mathbf{r})$ which is the correction to the perfect-lattice potential. The host lattice enters in three ways. First, in the effective-mass equation there is a mass m^* instead of the free-electron value m_0. Second, there is screening of $V(\mathbf{r})$. Hence, if there is only a Coulombic defect potential, we have

$$\left(-\frac{\hbar^2}{2m^*}\nabla^2 - \frac{e^2}{\epsilon_0 r}\right) F(\mathbf{r}) = E(\mathbf{r}) F(\mathbf{r}) \tag{5.15}$$

Third, the envelope function $F(\mathbf{r})$ modulates a host-crystal band function to give a total wavefunction $\psi(\mathbf{r})$. The wavefunction of a donor in a crystal with a nondegenerate conduction band whose minimum is at the zone center is

$$\psi(\mathbf{r}) = F(\mathbf{r}) u(\mathbf{k} = 0, \mathbf{r}) \tag{5.16}$$

where $\mathbf{u}(\mathbf{k}, \mathbf{r})$ is the Bloch function. In any predictions of epr parameters, or transition matrix elements, ψ must be used in full.

These equations have some obvious consequences. The solutions are simply scaled solutions for the hydrogen atom, with an altered effective Rydberg and Bohr radius,

$$\begin{aligned} &\text{Ry } (=\text{rydberg}) &&\rightarrow \text{Ry}^* = \frac{m^*/m_0}{\epsilon_0^2} \cdot \frac{m_0 e^4}{2\hbar^2} = \text{Ry}[(m^*/m_0)/\epsilon_0^2] \\ &a_0 \ (=\text{Bohr radius}) &&\rightarrow a^* = \frac{\epsilon_0}{m^*/m_0} \frac{\hbar^2}{m_0 e^2} = a_0[\epsilon_0/(m^*/m_0)] \end{aligned} \tag{5.17}$$

There will be an infinite number of bound states, with a series of optical transitions corresponding to both the bound–bound and bound–free transitions of the H atom. For typical values of ϵ_0 ($\sim$10) and of m^* ($\sim 0.1\, m_0$), the transition energies will be tens of meV and a^* will be $\sim$50 Å, much larger than interatomic distances.

One other important result follows. When the defect electron changes state, the defect electron charge density changes from one very diffuse, spread-out, form to another. There is no significant driving force for local lattice distortion; in most cases distortions will be nearly identical in ground and excited states, giving a very small Huang–Rhys factor, and optical spectra will be dominated by zero-phonon lines (see p. 262 and Figure 5.15).

Matrix elements of $\psi(\mathbf{r}) = F(\mathbf{r})u(\mathbf{r})$ appear in two main forms. If we want an optical transition intensity, we need $\langle \psi_a | \mathbf{r} | \psi_b \rangle$, where the operator (here $\mathbf{r}$) varies slowly in space. It is readily shown that $\langle \psi_a | \mathbf{r} | \psi_b \rangle \simeq \langle F_a | \mathbf{r} | F_b \rangle$, that is, we may forget that $F(\mathbf{r})$ is merely an envelope function. However, in epr where the isotropic hyperfine interaction is proportional to $\langle \psi_a | \delta(\mathbf{r}) | \psi_a \rangle$, the value $|u(0)|^2 |F_a(0)|^2$ explicitly includes the band function. This appears in just the same way that pseudopotential expressions for epr parameters contain core orbitals explicitly.

So far we have assumed a conduction band, with no degeneracy, with its one minimum at the zone center (the Γ point) and with at least tetrahedral symmetry. GaAs and InSb satisfy these criteria. Most other important systems, notably Si, Ge, and GaP, are more complex and some of the consequences which emerge are the following:

1. Anisotropy of a single, nondegenerate minimum in a crystal with axial symmetry (e.g., CdS). The appropriate measure of anisotropy with axial symmetry is $\alpha \equiv 1 - \epsilon_\perp m_\perp^* / \epsilon_\| m_\|^*$. The main effects (Henry and Nassau 1970) are to split the hydrogenic $2p$ states clearly, giving the singlet $2p_0$ and the doubly degenerate $2p_{\pm 1}$ (see Figure 5.15), and to separate them from the $2s$ state.

2. Nondegenerate band with several equivalent minima at $\mathbf{k} \neq 0$ in a cubic crystal, for example, Si (100 minima), Ge (111 minima), and GaP (100 minima very close to the zone edge). Two main changes appear. First, since all these minima are anisotropic, the features mentioned in item 1 are present. Second, the wavefunction is now a sum of contributions from each extremum j,

$$\psi(\mathbf{r}) = \sum_{j=1}^{n} \alpha_j F_j(\mathbf{r}) u(\mathbf{k}_{j0}, \mathbf{r}) \tag{5.19}$$

where j in Si, for example, runs from 1 to 6. Qualitatively, the main changes come from the intervalley terms that connect components of Equation (5.19) with different values of j. If we are not to violate the assumption (I) of effective-mass theory (see p. 258), these intervalley terms must be small. Symmetry gives most of the important results. For the sixfold spatial degeneracy in Si, the original $1s$ state is split into a singlet (Γ_1), a triplet (Γ_5), and a doublet (Γ_3). This is shown in Figure 5.12, together with the splitting of the fourfold degeneracy in Ge into a singlet (Γ_1) and a triplet (Γ_5). There are similar, but smaller, splittings in excited states, partly from anisotropic minima away from $\mathbf{k} = 0$.

3. Degenerate-band extrema, for example, valence bands in almost all hosts (Figure 5.13). Here the effective-mass equation becomes a matrix equation. Further complexity comes from the spin–orbit coupling, which raises valence-band degeneracy even in the perfect crystal. In Si this splitting separates the valence-band extremum at the zone center into an upper $\Gamma_8(p_{3/2})$ quartet separated by 4.8 meV from a lower $\Gamma_7(p_{1/2})$ doublet, and this complicates calculation of acceptor energies. In Ge the larger splitting, 290 meV, makes matters slightly simpler.

Lipari and Baldereschi (1970) have devised a convenient scheme for characterizing and calculating the dependence of acceptor levels on the valence-band structure. They exploit symmetry arguments in handling

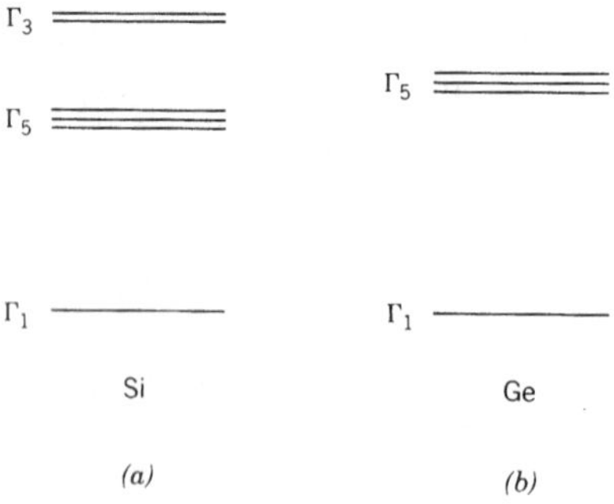

Figure 5.12 Schematic representation of intervalley splitting of ground-state donor levels in semiconductors with multi-valleyed conduction bands. The total number of levels [six in (*a*) for Si, four in (*b*) for Ge] corresponds to the total number of conduction-band minima.

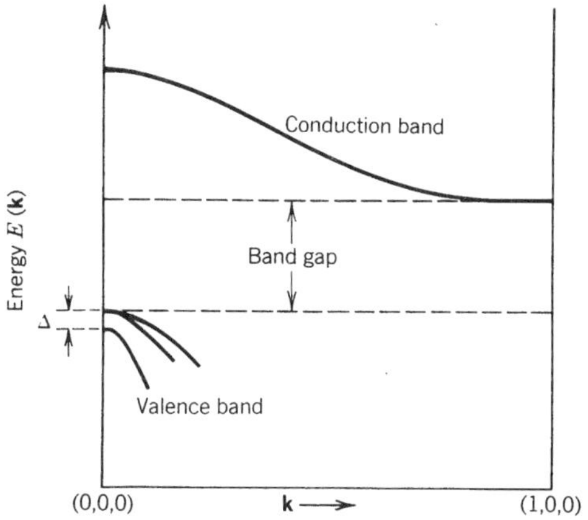

Figure 5.13 Schematic representation of the electronic energy-band structure of Si along a ⟨100⟩ direction, showing the spin–orbit splitting, Δ, of the valence band.

the complicated equations appropriate for a degenerate valence band. Two parameters (μ, δ) define the valence band of the perfect crystal in the usual limit of strong spin–orbit coupling. One parameter, μ, measures the spherical term in the spin–orbit coupling, while the other, δ, measures the cubic deviations from spherical symmetry. Usually δ is small, so the solutions with $\delta = 0$ provide a good zero-order result, both to classify the states and to estimate eigenvalues.

In strong spin–orbit coupling the usual sixfold degeneracy arising from twofold spin degeneracy and threefold orbital degeneracy is lifted, giving twofold and fourfold degenerate levels (see above). Suppose we characterize the important group of four levels by an effective angular momentum $J = \frac{3}{2}$. If $\mathbf{L}$ characterizes the angular momentum of the envelope function, $\mathbf{F} = \mathbf{J} + \mathbf{L}$ is an approximate constant of the motion. Hence the nth state with an S-like envelope function ($L = 0$) leads to states $nS_{3/2}$, with the subscript $F_z = \frac{3}{2}$. Similarly, the state with a p-like envelope function ($L = 1$) and principal quantum number n gives states $nP_{1/2}$, $nP_{3/2}$, and $nP_{5/2}$. The cubic spin–orbit term σ splits the $2P_{5/2}$ states into Γ_8 and Γ_7 states. Figure 5.14 shows results for both $\delta = 0$ and various μ, and also the splittings induced by small values of δ.

The ionization energies, E_D, of some donors in Si and Ge are given in Table 5.5 and of some acceptors, E_A, in Table 5.6. The ionization energies of shallow donors and acceptors in Ge are rather similar, being about 11 meV, whereas the corresponding ionization energies in Si show a spread, covering the range 40–160 meV. Because ionization energies

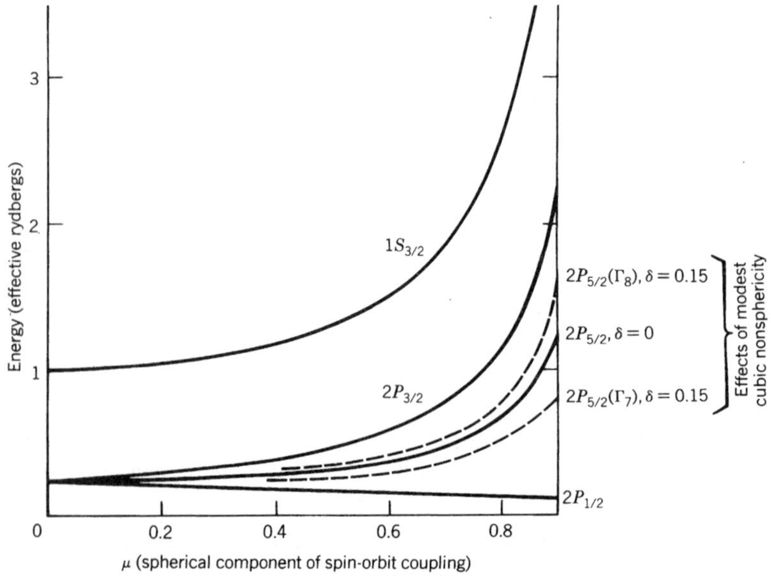

Figure 5.14 Acceptor levels for a general model, as calculated by Lipari and Balderes-chi (1970). The binding energy is given in effective rydbergs, Ry*, (i.e., scaled by the dielectric constant and by the effective mass in the limit $\mu = 0$). Here μ is a measure of the spherical component of the spin–orbit coupling and δ is a measure of the nonspherical components. Values of the parameters for common semiconductors are these:

	Si	GaP	ZnTe	Ge	GaAs	InSb
Ry* (meV)	24.8	28.0	35.7	4.3	11.3	1.2
μ	0.483	0.661	0.755	0.766	0.767	0.935
δ	0.249	0.162	0.152	0.108	0.114	0.036

are low [Equation (5.17)], samples have to be measured at low temperatures (generally $T < 70$ K) to prevent thermally induced ionization. Some of the transitions of the donor electron P_{Si} at 4 K measured by infrared absorption spectroscopy are shown in Figure 5.15. As mentioned earlier, such spectra illustrate the limit of negligible electron–phonon coupling.

In the III–V semiconductors a group-VI impurity, for example, S, substituted in a group-V site will act as an electron donor whereas a group-II impurity, for example, Zn, in a group-III site acts as an electron

Table 5.5 Ionization Energies E_D (meV) of Neutral Donors in Si and Ge

Donors	Si	Ge
P	45.31	12.76
As	53.51	14.04
Sb	42.51	10.19
Bi	70.47	12.68
Ideal (effective-mass theory)	31.27	9.81

Table 5.6 Ionization Energies E_A (meV) of Some Neutral Acceptors in Si and Ge

Acceptor	Si	Ge
B	45	10.47
Al	57	11.80
Ga	65	10.97
In	160	11.61
Ideal (effective-mass theory)	37.1	9.28

acceptor. A group-IV impurity, for example, Si, may dissolve in a group-III site acting as a donor and also in a group-V site acting as an acceptor. Impurities that behave as both donors and acceptors are referred to as amphoteric.

The wide variations of orbital radii have important consequences. In Si, where the conduction band has anisotropic minima away from $k = 0$, one finds a longitudinal effective mass $m_l^* = 0.98$ and a transverse effective mass $m_t^* = 0.19 m_0$, where m_0 is the free-electron mass. Inserting an appropriate average into Equation (5.18) gives $a^*(\text{Si}) = 20$ Å for the ground state of a neutral donor. The corresponding values for GaAs, InAs, and InSb are ~100 Å, ~350 Å, and ~650 Å respectively. Hence overlap of wavefunctions occurs even at low concentrations, leading to hopping between donor levels and eventually, with increasing concentration, to formation of impurity bands. We shall see later (Section 8.2) that the onset of a metal–insulator transition (generally known as a Mott transition) occurs for Si:P at a critical phosphorus concentration $n_c \sim$

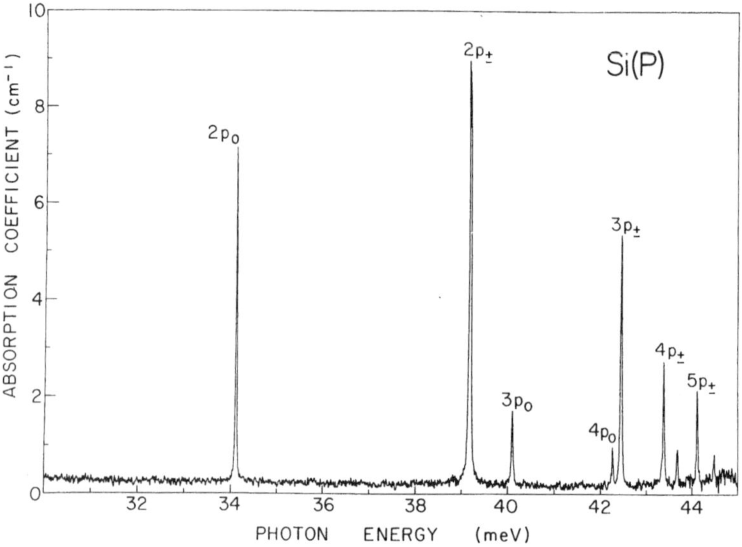

Figure 5.15 Absorption spectrum of P donors in Si at 4 K [after Jagannath and Ramdas (1981)]. Transitions from the valley-orbit-split ground state (Figure 5.12) to excited *p* states are split because the conduction-band minimum is along ⟨100⟩ (Figure 5.13), giving tetragonal symmetry.

3×10^{18} cm^{-3}. The corresponding critical concentration for Ge:As is $n_c \sim 2.8 \times 10^{17}$ cm^{-3}.

Very different behavior is found for diamond, an interesting example of a group-IV semiconductor with a wide bandgap (Figure 1.4), although there has been little successful work on controlled doping. Some natural diamonds, referred to as type II, have resistivities of ~300 Ω cm, as compared to ~10^{14} Ω cm for the majority of diamonds. Acceptor centers in the concentration range ~5×10^{16} cm^{-3} are responsible for this behavior. Synthetic semiconducting diamonds have been produced with B and Al as acceptors. The ionization energy E_A is ~380 meV. The rather deep acceptor centers in diamond might be expected to be outside the range of effective-mass theory, although effective-mass theory does lead to successful predictions for excited states.

The need for corrections to effective-mass theory is most clearly evident in the properties of ground states of donors and acceptors. In these states the bound electron or hole has appreciable amplitude at the core, leading to ground-state ionization energies different from values extrapolated from positions of measured excited states. The overlap with the core also results in valley-orbit splitting (see Figure 5.12) and in different measured ionization energies for different impurities (Tables 5.5 and 5.6).

This brings us naturally to the so-called central-cell correction. Even for P_{Si} and B_{Si}, evaluation of the potential energy for the dopant–carrier interaction is a complex problem. It is normally written

$$V(\mathbf{r}) = U_c(\mathbf{r}) - \frac{e^2}{\epsilon_0 r} \tag{5.20}$$

where $U_c(\mathbf{r})$ is referred to as the central-cell contribution. Since it is negligible outside the central atomic volume, it will usually vary rapidly in space near the defect and violate the effective-mass theorem. For isovalent substitutional impurities $U_c(\mathbf{r})$ is the only term in Equation (5.20). Some impurities of this type, for example, O_{Te} in ZnTe and N_P in GaP, tend to form bound exciton states, and are discussed in Section 5.3.3. In silicon the divalent interstitial Mg and substitutional S form double donors. An electron will be bound by a net charge $2|e|$ in the singly ionized state $[\oplus\!\oplus e]$, and will have a more compact wavefunction, enhancing the importance of the central-cell corrections $U_c(\mathbf{r})$. Central-cell corrections are especially small when the Coulomb term dominates, as for interstitial Li, which has a small, compact core. Lithium diffuses rapidly, by an interstitial mechanism, often forming complexes with acceptors in *p*-type Si.

All careful analyses of central-cell terms show that many factors contribute [Stoneham (1975, page 762) gives a comprehensive list] and, despite the enthusiasm and optimism of many workers, full understanding has not been reached. For practical purposes, the most important points are these. First, if one regards $U_c(\mathbf{r})$ as an unknown short-range potential, one can obtain a consistent phenomenological $U_c(\mathbf{r})$ from observed epr and optical data. This deduced value does not explain, however, why Sb has an anomalously low central-cell correction in Si and Ge compared with values for the adjacent elements, As and Bi, in the same column of the periodic table. Second, one can attempt to predict $U_c(\mathbf{r})$ from basic electronic structure. The difficulties here stem partly from the nonapplicability of effective-mass theory to this term and partly because cancellations between large terms have to be made accurately. There is modest success. Yet neither the problem of Sb in Si and Ge nor whether or not a donor will be deep or shallow can be calculated reliably. Why is ZnS:F_S deep and CdS:Cl_S shallow? Why are ZnS:Ga_{Zn} and ZnS:In_{Zn} deep, CdF_2:In_{Cd} both deep and shallow, and CdS:Ga_{Cd} shallow? We cannot yet provide convincing answers.

Both shallow and deep levels have application in semiconductor devices, and it is useful to make some general comments that contrast their behavior and applications. For shallow donors and acceptors, the main role is to provide carriers. Donors give electrons and n-type conduction; acceptors give holes and p-type conduction. If the concentration of donors alone (or of acceptors alone) is increased, there is a semiconductor-to-metal transition (see Section 8.2). If one starts with n-type material and adds acceptors, the carrier densities are altered as compensation proceeds, with a switch to p-type occurring in due course. The deep impurities, like Ni in GaP (Section 5.2), however, act as traps and as convenient intermediate states for electron–hole recombination. They allow one to increase resistivity; for example, we have already seen in Section 5.2 that Cr in GaAs helps in the compensation of residual shallow donors in the semi-insulating GaAs used as substrate for microwave-field-effect transistors. As more deep-level impurities are added, there need be no change of type of conduction.

The easier electron–hole recombination via deep levels can be either an advantage or a disadvantage. The reduction in minority-carrier lifetime can be useful in power devices, like switching transistors, rectifiers, or thyristors, where Au, Pt, or radiation-damage defects are incorporated to obtain the best compromise between operational parameters. However, deep states can reduce the efficiency of light-emitting diodes and cause noise and degradation in other cases. Optically, the shallow donors and acceptors are desirable because they can bind excitons to give near-bandgap luminescence with high oscillator strength (Section 5.3.3), that is, essentially the highest optical energies available in a given crystal. Deep levels, by introducing states in the middle of the gap, may be used to provide optimum sensitivity to light in a chosen part of the spectrum, for example, for extrinsic photoconductive devices like CdSe:In, Si:In, or Ge:Hg.

5.3.2 Donor–Acceptor Recombination

Donors and acceptors present in the same crystal tend to compensate each other. Indeed, semiconductors will generally show some degree of compensation, even if inadvertent. In an n-type semiconductor with N_D donors and N_A acceptors there will be N_A ionized donors and acceptors at equilibrium and $N_D - N_A$ neutral donors at temperatures too low to cause ionization. Suppose excess electron–hole pairs are injected into the crystal, so that some electrons are trapped by ionized donors and some holes by ionized acceptors, to create neutral impurities. The trapped electron–hole pairs will recombine to restore equilibrium by the charge-transfer process:

$$\underset{\text{Neutral donor}}{[\oplus e]} + \underset{\text{Neutral acceptor}}{[\ominus h]} \rightarrow \underset{\text{Ionized donor}}{\oplus} + \underset{\text{Ionized acceptor}}{\ominus} + E_r \tag{5.21}$$

with emission of recombination energy E_r. Nonradiative multiphonon processes are unlikely if the recombination energy greatly exceeds the largest phonon energy (Section 4.3.2), and, indeed, one generally finds that radiative recombination is the rate-limiting process for dilutely doped semiconductors at high excitation intensities. To a first approximation the recombination energy is (Figure 5.16)

$$E_r = E_g - (E_A + E_D) \tag{5.22}$$

Equation (5.22) is exact at infinite separation, where donor and acceptor do not interact.

The earliest evidence for the existence of luminescence from donor–acceptor (D–A) recombination came from work on ZnS phosphors. ZnS is a relatively ionic semiconductor with a bandgap of 3.8 eV (Figure 1.4). Ionization energies of donors and acceptors are ~70 meV, and the ground states of the neutral centers are fairly localized. This means that

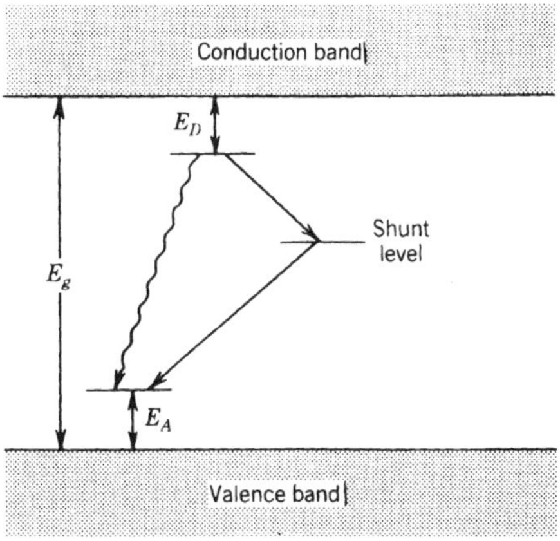

Figure 5.16 Schematic representation of the energy levels of donors (ionization energy E_D) and acceptors (ionization energy E_A) in a crystal with bandgap E_g, showing that the donor–acceptor recombination energy is $E_R = E_g - (E_A + E_D)$. Gap states associated with other defects (referred to as shunt levels) can provide an alternative path (possibly nonradiative) for recombination.

D–A luminescence is dominated by transitions of a few relatively closely spaced pairs (see below), and since the electron–phonon coupling is strong, the emission consists of broad featureless bands in the blue, green, and red (Figure 5.17). These are very similar to charge-transfer bands in minerals, and certainly involve phonon cooperation.

The presence of both copper and chlorine in ZnS gives rise to green luminescence. Copper dissolves as Cu_{Zn}^{+} and is charge-compensated by Cl_S^{-}. Copper is referred to as the activator (cf. Section 4.4) and is an acceptor; chlorine is referred to as the coactivator, and is here a donor. It seems that close D–A pairs form during firing at ~1000°C and that luminescence originates in an electron transition form an excited state of the neutral donor to the ground state of the neutral acceptor. The peak wavelength of the green emission is not changed even if Cl is replaced by the donors Br_S or Al_{Zn}. Further evidence for the D–A nature of the green copper emission has been obtained by odmr measurements by Patel et al. (1981), who also conclude that the efficiency of the green emission is reduced by a shunt path provided by another defect (Figure 5.18). Similar studies have been carried out on other II–VI compounds such as ZnSe and CdS.

More striking results are found in less polar materials, such as GaP. High-resolution studies of suitably doped crystals show the existence of a wealth of sharp lines that can be confidently assigned to D–A pairs with a range of separations. Energy levels of donors (e.g., S, Se, Te on P sites)

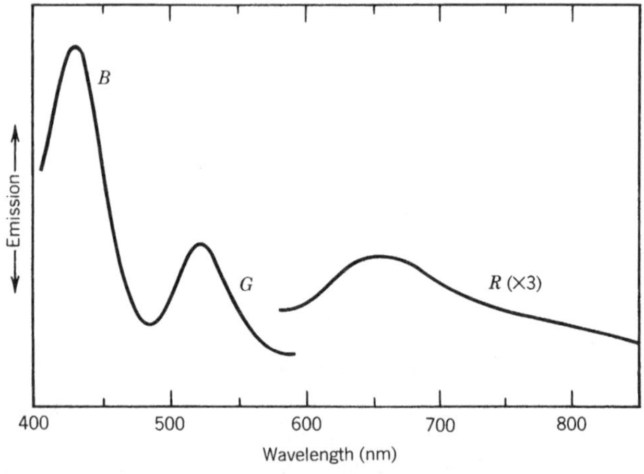

Figure 5.17 Emission from a ZnS crystal at 2 K excited with 351-nm laser radiation, showing emission in the blue (*B*), green (*G*), and red (*R*) [after Patel et al. (1981)].

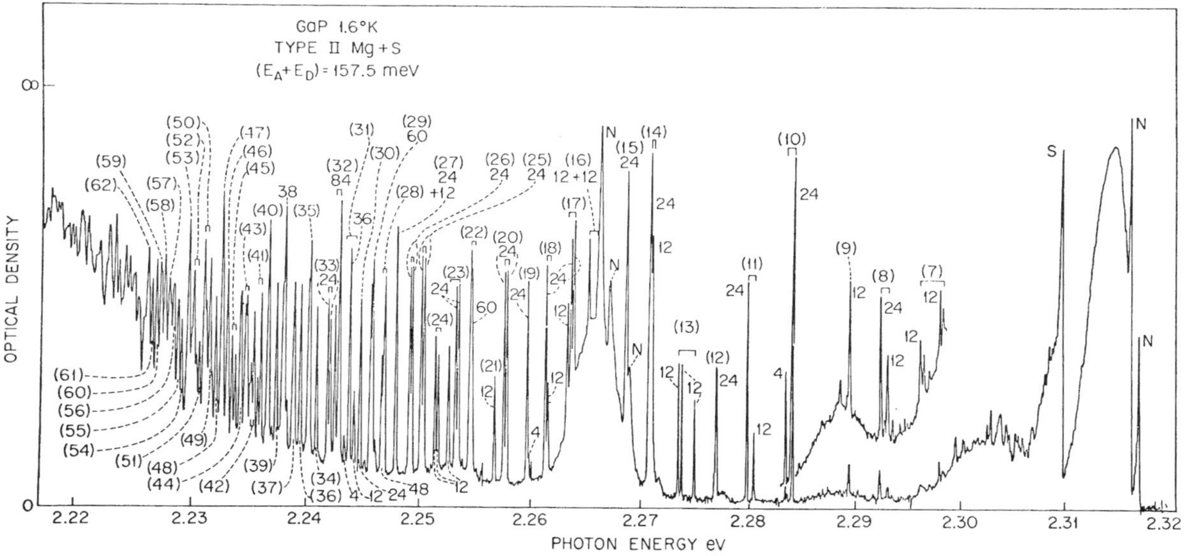

Figure 5.18 Part of the photoluminescence spectrum in the green of GaP at 1.6 K showing recombination levels due to S_P donors and Mg_{Ga} acceptors. The D-A pair shell numbers are in brackets. Some luminescence lines due to excitons bound to isoelectronic N traps are also present and are labeled N [after Dean et al. (1970)].

and acceptors (e.g., Zn, Mg on Ga sites) in GaP have been studied by light scattering (Hayes and Loudon 1978) and by infrared absorption (Ramdas and Rodriguez 1981). However, much of the interest in donors and acceptors in this material has centered on D–A pair spectra. Ionization energies of isolated donors and acceptors in GaP are given in Table 5.7, and it is apparent from Equation (5.22) that emitted light is expected to be in the green–yellow (2.25 eV) region of the spectrum.

The rich harvest of lines shown in Figure 5.18 for Mg_{Ga} acceptors and S_P donors in GaP can be interpreted accurately using a modified form of Equation (5.22). If the separation of neutral donors and acceptors is great enough to give only a small overlap of wavefunctions, the transition energy can be written more accurately in the form

$$E_r(R) = E_g - (E_A + E_D) + \frac{e^2}{\epsilon_0 R} - f(R) \tag{5.23}$$

The last two terms give corrections arising from the interaction between the charged (ionized) donor and acceptor, regarded as point charges, and a smaller overlap term $f(R)$. We expect the transition energy to decrease as R increases, so that ultimately the lines associated with individual D–A pairs will merge into a band at lower energy (Figure 5.18). The recombination transition probability decreases with increasing separation of donors and acceptors as the overlap of donor and acceptor wavefunctions decreases, so emission from closer pairs will occur more rapidly after excitation than from more distant pairs. This is verified by time-resolved spectroscopy. Figure 5.19 shows how accurately recombination energies can be predicted. Even at the largest spacings the overlap of the donor electron and acceptor hole have an appreciable effect.

Table 5.7 Ionization Energies (meV) of Donors and Acceptors in GaP

Donors		Acceptors	
O_P	896	C_P	46
S_P	104	Si_P	202
Se_P	102	Ge_P	257
Te_P	89		
Si_{Ga}	82	Mg_{Ga}	52
Ge_{Ga}	201	Zn_{Ga}	62
Sn_{Ga}	69	Cd_{Ga}	94

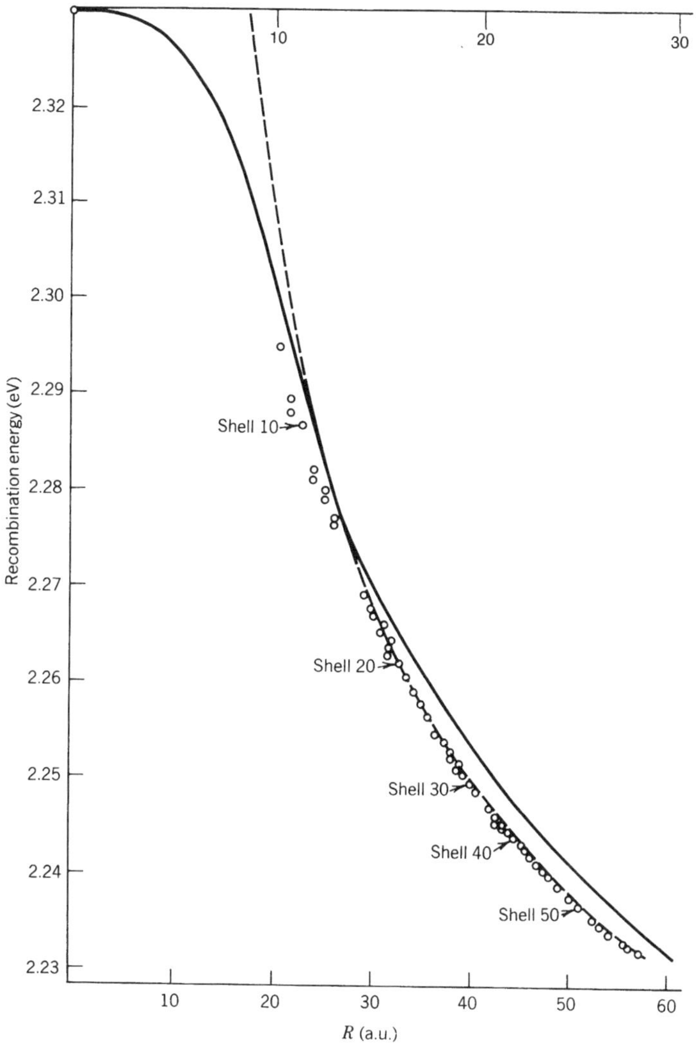

Figure 5.19 Experiment and theory for donor–acceptor pair recombination in GaP, shown here for the C acceptor and S donor on the P sublattice. The broken line shows Equation (5.23) fitted to the data of Vink et al. (1973) but with $f(R) = 0$. This rather good fit is slightly misleading because it ignores donor–acceptor overlap, and this overlap is still significant even beyond the 50th shell. The solid line (Stoneham and Harker 1975) shows the detailed theory, including central-cell corrections, for the same assumed value of $E_g - E_D - E_A$. This greatly improves agreement at small values of R, although there is now an apparent discrepancy at large R. This discrepancy implies a small error in the assumed value of $E_g - E_D - E_A$.

The GaP lattice may be regarded as two interpenetrating fcc sublattices, with Ga ions on one sublattice and P ions on the other, and there are two distinct types of donor–acceptor pairs. We may have a pair spectrum arising from donors and acceptors on the same sublattice (referred to as type-I spectrum, as in Figure 5.19). If we label shells of acceptor ions centered on a specific donor ion by the index m, the radius R_m is given by

$$R_m = (m/2)^{1/2} a \qquad \text{type I} \tag{5.24}$$

where m is a simple integer and a is the lattice constant (5.45 Å for GaP). However, care has to be used, for Equation (5.24) correctly describes the lattice structure only when one omits certain values of m, namely, $m = 14, 30, 46, \ldots, \frac{1}{2}(8k+7)4^l$ with integers $k \leq 0$, $l > 0$. Gaps are indeed observed at these values.

If the donors and acceptors are on separate sublattices (type-II spectrum), no gaps occur and the relation between R_m and a is

$$R_m = \left(\frac{m}{2} - \frac{5}{16}\right)^{1/2} a \qquad \text{type II} \tag{5.25}$$

Excellent fits can be obtained to lines of type-I and type-II spectra using Equations (5.23)–(5.25).

5.3.3 Impurity-Bound Excitons

Semiconductors have potential application as luminescent diodes (LEDs), in which electrical energy is converted to light at low voltages and low powers. If the emission is to be at wavelengths at which the eye is sensitive, relatively large bandgaps are needed, and emission energies E_r should be close to the gap energy E_g. Free excitons have recombination energies given by

$$E_r \text{ (free exciton)} \simeq E_g - \frac{\mu}{m_0} \cdot \frac{\text{Ry}}{\epsilon_0^2} \equiv E_g - E_X \tag{5.26}$$

where Ry is the Rydberg (13.6 eV) and μ is the reduced mass of the exciton [see Equation (1.28)]. The exciton binding energy, E_X, is often only a few meV. However, the efficiency of radiative recombination of the intrinsic material is generally poor, especially for indirect-gap materials. This situation can usually be improved by weak binding to impurities, with binding energy E_{BX}, when the impurity potential rather than a momentum-conserving phonon renders the transition allowed.

Such bound excitons* were first noted by Haynes (1960). The spectral lines of his observations for Si suggest a small binding, $E_{BX} \lesssim 10$ meV. Clearly, thermal release of excitons from impurities is expected for $k_B T \gtrsim E_{BX}$.

To understand bound exciton systems most readily we follow Hopfield's (1964) notions of quantum chemistry, combined with effective-mass theory. Bound exciton systems have molecular analogs just as a simple donor is analogous to a hydrogen atom. We shall consider transitions involving carriers (e, h) combined in various ways with ionized donors ($\oplus$), ionized acceptors ($\ominus$), or isoelectronic defects ($\circledcirc$). Thus [see Equation (5.21)],

$$[\oplus e] + E_D \rightarrow \oplus + e \tag{5.27}$$

represents a typical reaction. Other reactions are expressed in the more general form

$$[\alpha\beta\gamma] + E_{\alpha(\beta\gamma)} \rightarrow [\alpha] + [\beta\gamma] \tag{5.28}$$

with $E_{\alpha(\beta\gamma)}/E_D$ given as some function of $\sigma = m_e/m_h$ (see also Section 1.3.2). Thus, the acceptor binding energy is $E_A = E_D/\sigma$; similarly, the donor may bind an extra electron with energy $E_{(\oplus e)e} = 0.055 E_D$ from the known ionization energy (0.055 Ry) of the H^- ion. The exciton binding energy, E_X [Equation (5.26)], is $E_D/(1+\sigma)$.

Excitons bound to ionized donors, giving $[\oplus eh]$, or ionized acceptors, giving $[\ominus eh]$, have two modes of decay. It is readily shown that the lowest-energy decay is the loss of one carrier, not two, that is, $[\oplus eh] \rightarrow [\oplus e] + h$. Indeed, there is only a stable bound state provided σ is less than a critical value, that is, $m_h > \sigma_{\text{crit}} m_e$, since the kinetic energy of the hole must not overwhelm the weak binding to the neutral donor component $[\oplus e]$.

Excitons bound to neutral donors, giving $[\oplus eeh]$, have binding energies readily estimated in two limits. For a massive hole (σ small) the analogy is with the H_2 molecule. The free H_2 molecule has a dissociation energy of 0.33 Ry, so we have

$$E_{(\oplus e)(eh)} \simeq 0.33 E_D, \qquad \sigma \rightarrow 0 \tag{5.29}$$

for the dissociation energy of the neutral donor and the exciton. For a light hole, the electron component of the exciton is bound as in $[\oplus ee]$, with the hole contributing little; in this case the dissociation energy is

$$E_{(\oplus e)(eh)} \simeq 0.055 E_D, \qquad \sigma \rightarrow \text{large} \tag{5.30}$$

Various calculations that interpolate are available, although they are not

*When there is strong binding, there is no point in using the term bound exciton, since one has simply an excited state of the impurity.

all consistent in assumptions or accuracy [see Stoneham (1975, p. 831), for a summary of results].

In the simple form given above the quantum chemistry approach ignores central-cell corrections and some other important details. Experimentally, it is found that excitons bound to neutral donors and acceptors have a linear variation of exciton binding energy, E_{BX}, with increasing impurity binding energy, E_D:

$$E_{BX} = E_{(\oplus e)(eh)} = a + bE_D \tag{5.31}$$

with a corresponding result for acceptors. Equation (5.31) is referred to as Haynes' rule (Figure 5.20). In the case of neutral donors in Si, $a \sim 0$ and $b \sim 0.1$; in GaP, a is nonzero and appears to have opposite signs for donors and acceptors. It is remarkable that the exciton binding energy and donor or acceptor binding energy can remain proportional, even when there are substantial central-cell corrections. The simple relationship (5.31) does not hold for some III–V compounds, such as GaAs and InP.

In the case of isoelectronic defects, where there is a central-cell term,

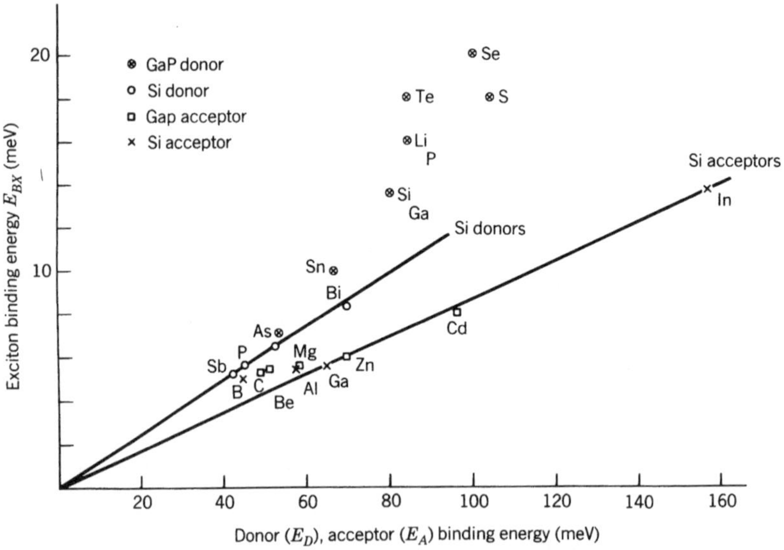

Figure 5.20 Haynes' rule, showing the relation between the binding energy E_{BX} of an exciton to a neutral donor (acceptor) as a function of donor (acceptor) binding energy E_D (E_A). Energies are in meV.

but no core coulombic contribution, it is often useful to regard exciton binding as consisting of (i) localization of one carrier because the electronegativity difference of host and impurity is large enough and (ii) the Coulomb binding of the other carrier to the one first localized.

The optical spectra of excitons bound to impurities in semiconductors have been reviewed by Dean and Herbert (1979). As an illustration we

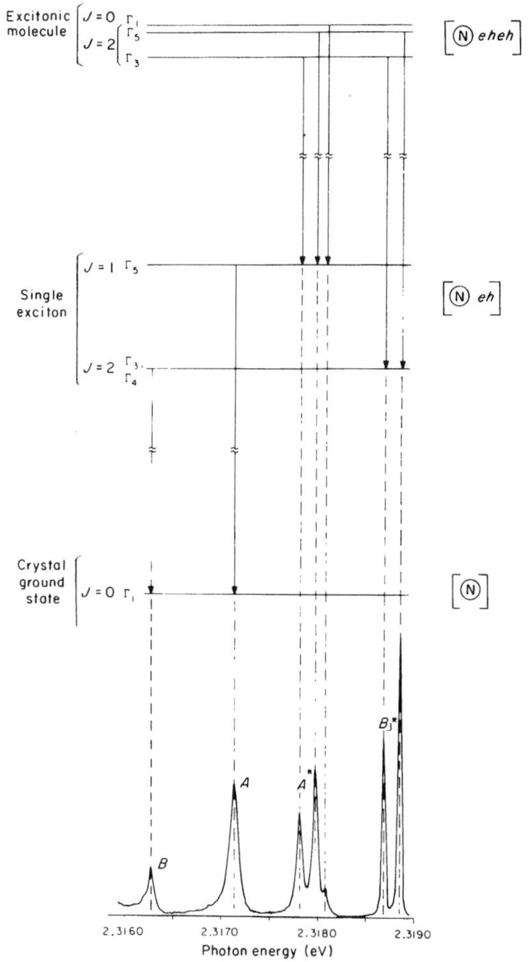

Figure 5.21 Zero-phonon lines in the recombination luminescence of GaP:N_P showing the spectrum of single excitons (lines *A*, *B*) and of molecular excitons (lines *A**, *B**) bound to N_P [after Merz et al. (1969)].

consider here the case of the isoelectronic impurity N_P in GaP. The bound excitons contain $J=1$ and $J=2$ states arising from j–j coupling of a $j=\frac{3}{2}$ hole and a $j=\frac{1}{2}$ electron. The luminescence spectrum therefore shows two lines, A and B, terminating on the $J=0$ ground state of the exciton (Figure 5.21). The $J=2$ level is, in fact, split into a doublet (Γ_3) and a triplet (Γ_4) by the tetrahedral lattice environment, but the splitting is too small to be resolved in this particular case. This splitting is, however, resolved in the case of the isoelectronic impurity Bi_P in GaP, for example. In GaP:N_P, trapping of an electron is the first stage of exciton binding, whereas in GaP:Bi_P the first stage is trapping of a hole.

With higher excitation intensities additional luminescence lines appear represented as A^* and B^* in Figure 5.21. The intensity of the A^* and B^* emissions increases at low temperatures with optical excitation roughly as the square of the intensity of A and B, and the properties of the A^*, B^* emission, in general, can be understood in terms of decay of a second exciton bound to the N_P trap with an energy ~15% less than that of the first (see Section 1.3.2 for a discussion of excitonic molecules).

There is also evidence for the presence of multiple bound excitons. The emission spectrum of recombination lines associated with neutral P donors in Si shows lines apparently characteristic of six bound excitons. However, the interpretation of these weak spectra is not uncontroversial [see Dean and Herbert (1979) for a discussion].

5.4 VACANCY CENTERS IN SEMICONDUCTORS

We have already discussed anion and cation vacancies in ionic crystals (Chapter 3). Vacancies in nonpolar crystals have features in common with ionic solids, but also show their own distinct behavior. Here we shall concentrate on vacancies in silicon and diamond. However, vacancies in SiC, Ge, and Si:Ge have also been studied. Many centers involving vacancy aggregates and impurity–vacancy complexes are also known.

Watkins (1964, 1976) has given a very useful framework in which to describe vacancy behavior in semiconductors. This is not the only possible description, but it is both clear and consistent. As in Coulson and Kearsley's (1957) pioneering calculations on diamond, the first assumption is that one may concentrate on the four dangling bonds on the four neighbors to the vacancy. The vacancy can exist in the charge states V^{2+}, V^+, V^0, V^-, and V^{2-}, depending on the position of the Fermi level. In the neutral vacancy, V^0, there will be four electrons associated with the four dangling bonds. One can construct symmetrized combinations of the dangling bonds, giving a totally symmetric one-electron orbital $|a_1\rangle$, and a threefold degenerate orbital of t_2 symmetry. Just as in the F center, the $|t_2\rangle$ state lies above the $|a_1\rangle$ state. We may now ask how the vacancy electrons, four for V^0, are distributed over the one-electron orbitals. The important factors are:

(a) The a_1 orbital, with lower energy, is always full. Indeed, the a_1 orbital probably lies within the valence band, giving a resonance-type peak in the density of states instead of a localized state in the bandgap.
(b) Electrons in the t_2 orbitals prefer to have parallel spins and a charge distribution such that the electrons are concentrated on different neighbors as far as possible.
(c) The Jahn–Teller energies are large and, in some cases (notably V^0), decisive in determining the ground state.

Both theory and experiment are broadly in agreement, qualitatively at least. In silicon, regrettably, many detailed theoretical studies have not attempted to calculate the quantities observed experimentally, such as epr parameters. The picture that emerges for the various charge states of the vacancy in silicon is the following:

$V^{2+}(a_1^2)$. The a_1 orbital is full and there are no t_2 electrons. This center, whose existence is only inferred indirectly, presumably has $S = 0$ and the full symmetry of the vacancy site.

$V^{+}(a_1^2 t_2^1)$. This center has $S = \frac{1}{2}$ due to one t_2 electron, which drives a tetragonal Jahn–Teller distortion. Epr studies show that reorientation of the tetragonal axis requires only the small activation energy of 0.013 eV.

$V^{0}(a_1^2 t_2^2)$. There are two t_2 electrons causing a tetragonal Jahn–Teller distortion, which stabilizes an ^1E ground state rather than triplet states like $^3T_1$. The neutral vacancy has an activation energy of 0.33 eV for low-temperature migration and an activation energy of 0.23 eV for reorientation of the tetragonal Jahn–Teller axis. If we make the working approximations (i) that each t_2 electron contributes an amount to the force driving the Jahn–Teller distortion which is independent of the vacancy charge state and (ii) that the elastic restoring forces are likewise independent of charge state, we conclude from theory (Baraff and Schluter 1978) that $2\,V^+$ should disproportionate into $V^0 + V^{2+}$. This prediction of a negative U reaction (Section 8.1.4) was subsequently identified by experiment. It should be added that application of assumptions (a) and (b) above to

V^+ would predict a trigonal, rather than the observed tetragonal, Jahn–Teller distortion from the measured behavior of V^-.

$V^-(a_1^2t_2^3)$. The three t_2 electrons give a resultant spin of $S = \frac{1}{2}$. Epr studies show that the symmetry is reduced by a mixed trigonal and tetragonal distortion. It should be noted that the simple Jahn–Teller theorem (Section 4.2.3) does not permit mixed symmetry distortions. This restriction can be evaded either by a qualitative argument for successive distortions (Watkins 1964) or by noting that Jahn–Teller coupling to other electronic configurations can lead to mixed distortions (Stoneham and Lannoo 1969). The V^- center reorients easily. Since this center shows mixed trigonal and tetragonal distortions, there are two distinct reorientations with activation energies of 0.008 and 0.072 eV, the higher corresponding to a change of tetragonal axis.

$V^{2-}(a_1^2t_2^4)$. The four t_2 electrons have $S = 0$ so that epr techniques cannot be used; the (inferred) activation energy for migration of V^{2-} is 0.18 eV.

In the case of diamond, detailed optical data are available for the neutral vacancy V^0 (the so-called *GR*1 center) and the negatively charged vacancy V^- (the so-called *ND*1 center). The *GR*1 transition gives rise to a broad band at about 2 eV with an associated zero-phonon line at 1.673 eV; there are also some satellites, notably one about 70 cm^{-1} from the main zero-phonon line, arising from a state just above the ground state (Figure 5.22). This satellite is a consequence of a dynamic Jahn–Teller effect in the ^1E ground state. In terms of Figure 4.8, the main zero-phonon line comes from a transition from the lowest E vibronic state and the satellite line from the lowest thermally excited A vibronic state.

Experimentally and theoretically the situation for V^0 is probably better understood for diamond than for silicon. This is because the structure in the *GR*1 optical absorption band can be analyzed to give many details of the ground-state energy surface. This band corresponds to a redistribution of the two t_2 electrons in V^0 over the t_2 orbitals, that is, it is a transition that cannot be represented in a simple one-electron scheme.

The optical spectrum of diamond containing vacancies is rich, showing a number of sharp lines labeled *GR*2 to *GR*8 (Figure 5.22). It is known that the *GR*2 to *GR*8 lines arise from the same ground state as the *GR*1 line. Yet these many sharp, close, lines have no simple explanation. They give photoconduction, yet are not one-electron levels at a band edge. Their sharpness and symmetries likewise preclude other

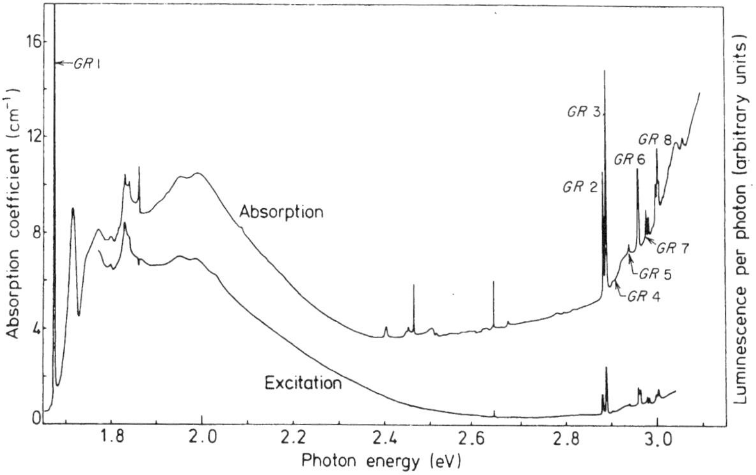

Figure 5.22 The *GR*1 line and the *GR*2–*GR*8 series of lines associated with the neutral vacancy in diamond [after Collins (1981)]. Also shown is the excitation spectrum of the *GR*1 line, that is, the intensity of *GR*1 luminescence as a function of excitation energy. The *GR*2–*GR*8 lines contribute to the excitation spectrum of the *GR*1 line but show no direct luminescence themselves. The sharp lines between 2.4 and 2.7 eV are not related to the vacancy.

simple suggestions. The absence of a convincing explanation of these lines in terms of the electronic structure of the vacancy is still a major challenge.

We mention in passing that the ideas of dangling-bond states of vacancies have been used with profit in understanding deep impurity levels in semiconductors (Section 5.2; Vogl 1981). The basic idea is that there are two main types of states, namely, atomic-like impurity states and vacancy-like dangling-bond states. When these are close in energy, strong hybridization (see Section 8.4) can occur.

5.5 OTHER IMPURITY CENTERS IN SEMICONDUCTORS AND INSULATORS

5.5.1 Isoelectronic Defects

The subject of isoelectronic defects has already been touched on in Section 5.4. Here a lattice ion is substituted by an impurity ion of the same valence; the latter is sometimes referred to as an isovalent impurity. Examples are wide ranging, including such systems as rare gases in dilute solid solution (e.g., Ar:Xe), ZnTe:O_{Te}, GaP:N_P, Si:Ge, KCl:Tl, KCl:Ag, and AgCl:Br. Almost all such cases show exciton binding at the impurity site.

In KCl:Tl one finds four absorption bands below the crystal band edge (8.7 eV), labeled *A* (4.38 eV), *B* (5.06 eV), *C* (5.30 eV), and *D* (5.50 eV). The *A*, *C*, and *D* bands are strong and rather temperature dependent, whereas the strength of the weaker *B* bands is temperature independent. The ground state of the free Tl^+ ion is $6s^2(^1S_0)$ and the next-highest levels come from the configuration $6s6p(^3P_{0,1,2}, ^1P)$. A qualitative explanation of the observed absorption spectrum was put forward by Seitz (1938), who suggested that the energy levels of the free

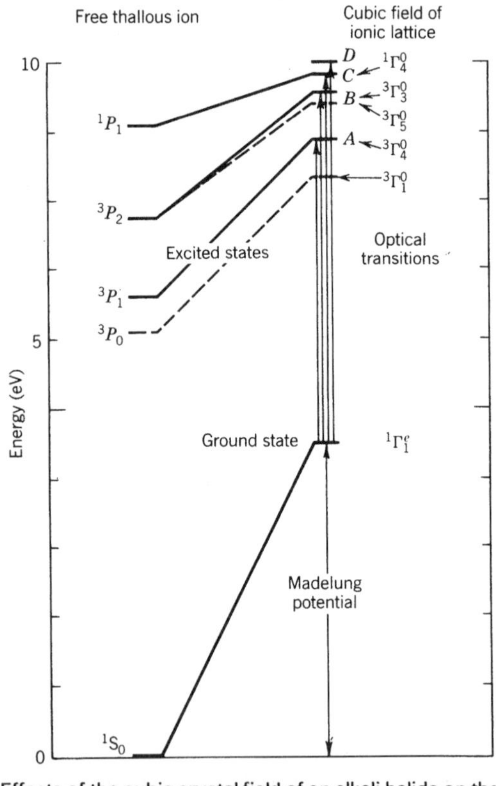

Figure 5.23 Effects of the cubic crystal field of an alkali halide on the low-lying energy levels of the Tl^+ ion. All states are raised in energy by the Madelung potential, the ground state probably more than the excited states [after Seitz (1938)].

ion would be raised in the crystal by the Madelung potential, the ground state probably more than the excited states (Figure 5.23). The transitions ${}^1S_0 \rightarrow {}^3P_1$ (*A* band) and ${}^1S_0 \rightarrow {}^1P_1$ (*C* band) are allowed, their relative intensities depending on the size of the spin–orbit coupling. Transitions ${}^1S_0 \rightarrow {}^3P_2$ (*B* band) are forbidden but will be allowed by mixing of other states by phonons (the ${}^3P_2$ state, which is fivefold degenerate, will be split into a triplet and a doublet by the cubic lattice environment) (Figure 5.23). The most plausible explanation of the *D* band is a charge-transfer transition involving excitation of an electron from a ligand Cl^- to Tl^+, as with the *F*-center-induced β band (Section 5.5.1). The *A* and *C* bands show partly resolved structure due, very likely, to a dynamical Jahn–Teller effect.

The problems of computing the energy levels of Tl^+ in KCl are so formidable that no quantitative treatment has been made. The situation is much more complex than for the *F* center because Tl^+ has itself many electrons and exchange and spin–orbit effects are important. Indeed calculations of the energy levels of even the free Tl^+ ion have not given good agreement with experiment.

Thallium-doped alkali halides fluoresce in the visible and ultraviolet regions, and there is an extensive literature dealing with this emission (Fowler 1968). In heavily doped alkali halides additional absorption and emission bands are found. In NaI:Tl scintillators the production of scintillation by high-energy particles may be due to exciton creation as the particles slow down, followed, perhaps, by preferential trapping and deexcitation at thallium dimers.

Tl^+ ions in alkali halides may be converted photochemically into Tl^0 and Tl^{2+} by trapping of electrons and holes at low temperatures. The Tl^{2+} center in KCl has absorption at 77 K at 3.4, 4.2, 4.7, and 5.6 eV, but the nature of the transitions is not well understood. Tl^0 in KCl gives rise to structured absorption in the infrared, close to the separation of the ${}^2P_{3/2}$ and ${}^2P_{1/2}$ components (0.97 eV) of the ground $6s^26p$ configuration of the free Tl^0 atom.

The Ag^+ ion also gives rise to absorption bands in the bandgap of alkali halides. The bands shown in Figure 5.24 are relatively weak ($f \sim 10^{-3}$–10^{-2}) and the temperature dependence of their absorption strength suggests that they arise from the parity-forbidden transitions. $4d^{10}({}^1S_0) \rightarrow 4d^95s({}^1D_1, {}^3D)$, being permitted by odd-parity vibrations. The theoretical situation here is at least as complicated as that for Tl^+ since crystal-field splitting of the *D* levels must be considered.

As in the case of Tl^+, Ag^+ in alkali halides may be photochemically converted at low temperatures to Ag^0 and Ag^{2+}. Ag^0 gives rise to an absorption at 2.9 eV in KCl due to a transition of the type $4d^{10}5s({}^2S) \rightarrow$

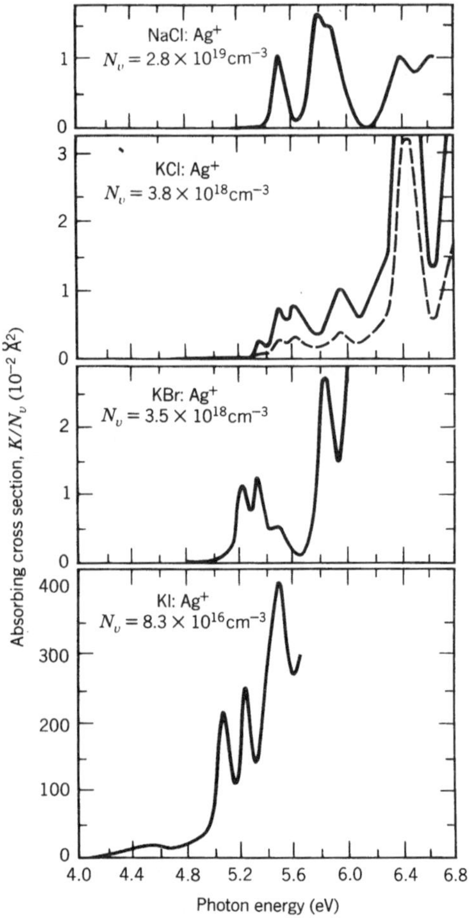

Figure 5.24 Absorption in the bandgap of Ag^+-doped alkali halides at 8.5 K. For KCl:Ag the dashed curve is reduced by a factor of 2.5 [after Fussgänger et al. (1965)].

$4d^{10}5p(^2P)$ and Ag^0 may be regarded as a heavy-atom analog of the F center. In free Ag^0 this transition occurs at 3.7 eV, and the spin–orbit splitting of the ^2P state is 0.11 eV. This splitting is not directly resolved in the absorption of Ag^0 in KCl, but it may be determined by magnetic-circular-dichroism measurements (Brown et al. 1967). In KCl the splitting is 0.068 eV, in KBr 0.019 eV, and in KI it is too small to be detected. This change is due to the presence of two opposing contributions: a positive contribution from the silver nucleus and a negative contribution resulting from orthogonalization to states of nearest-neighbor halogens. This latter contribution is analogous to the term leading to the negative spin–orbit splitting in the ^2P state of the F center and its magnitude is larger for heavier halogens (Section 5.1.1).

The Ag^{2+} center in KCl has the configuration $4d^9$, ^2D and, like Cu^{2+} ($3d^9$, ^2D), forms a Jahn–Teller distorted complex with an $|x^2-y^2\rangle$ ground state (Section 5.2), strongly bonded to the four chlorine ligands in the xy plane. The structure of this complex is remarkably similar to that of the self-trapped hole in AgCl (Section 6.3.1).

Self-trapping of excitons occurs in AgCl as well as self-trapping of holes (Section 6.3.1). This behavior contrasts with that of the more covalent AgBr, which does not show the self-trapping phenomenon. In AgCl:Br trapping of holes occurs preferentially at the bromine impurity sites because Br is more electronegative than chlorine. There is also a preferential trapping of excitons at the Br site at concentrations of Br of up to 45 mole%; at higher concentrations of Br, excitons cease to be localized (Hayes 1982).

The isoelectronic impurity system CsI:Na is an important X-ray phosphor, since the high atomic numbers of Cs and I lead to effective absorption of X-rays and the sodium impurity generates efficient luminescent emission in the blue. As in the case of KCl:Tl and KCl:Ag the sodium ion can trap an electron (because Na^+ is more electronegative than Cs^+) or a hole (because of elastic interaction between the impurity and the self-trapped hole). The trapping of either carrier may eventually lead to bound-exciton formation followed by emission of luminescence associated with radiative recombination [see Ong et al. (1979)].

5.5.2 Hydrogen Impurity

Hydrogen is a pervasive, if often harmless, impurity. It is present in crystals in a variety of forms:

1. In alkali halides three forms are prevalent:

   U center (H_s^-, substituting for an anion).
   U_1 center (H_i^-, the interstitial ion).
   U_2 center (H_i^0, the interstitial atom produced by ionizing radiation).

The U center has optical transitions that obey a Mollwo–Ivey type of rule [like Equation (5.2)], that is, that are relatively localized but depend strongly on the lattice spacing. The U_2 centers, however, appear to show charge-transfer transitions. These various forms of hydrogen center are especially interesting as probes of lattice dynamics, since isotope effects are both dramatic and easily observed and anharmonic effects are pronounced (Section 2.4.2).

2. In oxides the $(OH)^-$ ion is the commonest manifestation of hydrogen. This shows readily in near-infrared absorption. More recently, μsR (Section 4.5) has produced clear evidence of $(O\mu)^-$, the analogue of $(OH)^-$. For MgO, $(OH)^-$ appears unstable in isolation, and hydrogen is usually associated with cation defects like the V^- center (Section 5.1.2).

3. In the III–V compounds hydrogen is implanted as a way of electrically isolating different regions of semiconductor devices. In GaAs it is found, instead of a monotonic trend, that deuterium is far more effective in producing electrical isolation than either proton or triton implantation. The explanation seems to be that the local-mode frequency of deuterium is almost exactly an integral fraction of the bandgap of GaAs. The deuteron can therefore cause efficient nonradiative recombination, thus minimizing the recombination-enhanced annealing of the damage centers (see Section 6.4.6), which are responsible for effective electrical isolation. The large phonon energies of hydrogen impurity make it especially effective in promoting nonradiative transitions.

4. In crystalline Si, unlike amorphous Si (see Figure 2.18 and also Section 8.1.3), direct observation of H seems to involve other defects (e.g., vacancies). Many experiments suggest that hydrogen produces neither electrical nor optical activity. This may happen because hydrogen is present in molecular form, and not because it is a deep donor (Mainwood and Stoneham 1983). The properties of normal and anomalous muonium in Si (see Section 4.5) have been reviewed by Patterson (1984).

5.5.3 Oxygen in GaP and Light Emitting Diodes (LEDs)

Oxygen in GaP has proved to be one of those impurity centers that test and extend our knowledge of semiconductor defects. It has been observed as a component of donor–acceptor pairs [e.g., GaP:(Zn, O); see Section 5.3.2], but the isolated impurity itself is of special interest because of the many complex ways in which the electron–lattice interaction occurs.

The isolated impurity exists in two charge states, the neutral one-electron center, which (formally at least) is obtained by removing a P atom and replacing it by an O atom, and the negative two-electron center, which may be obtained by replacing the P atom by an O^- ion. The one-electron center corresponds to a simple donor $[\oplus e]$, with modest lattice relaxation; the two-electron center $[\oplus ee]$ shows strong relaxation, although it also appears to exist in a metastable, centered (virtually unrelaxed) form $[\oplus ee]_M$, which we discuss below.

Even the one-electron center is deep, that is, both the process of capture of a valence electron by the ionized center and of photoionization of the neutral center have energies of about half of E_g (here $E_g \sim 2.27$ eV):

$$e_v + [\oplus] \to [\oplus e], \qquad 1.453 \text{ eV threshold}$$

$$[\oplus e] \to [\oplus] + e_c \qquad 0.90 \text{ eV threshold}$$

with the bandgap excess $(1.45 + 0.90 - 2.27)$ eV corresponding to a relaxation energy of about 0.1 eV. The internal transitions are like those in other donors, for example, the luminescence from transitions between valley-orbit states may be represented as

$$[\oplus e^*]_{1s(E)} \to [\oplus e]_{1s(A_1)} \quad \text{at} \quad 0.841 \text{ eV}$$

and the hydrogenic absorption as

$$[\oplus e]_{1s(A_1)} \to [\oplus e^*]_{2p_0} \quad \text{at} \quad 0.863 \text{ eV}$$

In addition, a full range of other transitions are identified or inferred, including Auger nonradiative transitions and optical recombination transitions involving both near and distant acceptors.

The two-electron center has a photocreation threshold

$$[\oplus e] + e_v \to [\oplus ee]_G \quad \text{at} \quad 1.65 \text{ eV}$$

in its stable ground state G and a photoionization threshold

$$[\oplus ee]_G \to [\oplus e] + e_c \quad \text{at} \quad 1.4 \text{ eV}$$

indicating a large bandgap excess of $(1.65 + 1.4 - 2.27) \sim 0.8$ eV and hence a large lattice relaxation. However, most of the other transitions of the two-electron center are most easily understood in terms of a metastable state with minimal relaxation. The reasons underlying the different relaxations in the one-electron center and the two types of two-electron center and the experimental consequences are still actively debated. These consequences include a luminescence transition at 0.528 eV between the $1s(E)$ and $1s(A)$ valley-orbit states,

$$[\oplus ee^*]_{M,1s(E)} \to [\oplus ee]_{M,1s(A)}$$

and transitions involving an exciton bound to the one-electron center; for the latter a good description appears to be a hole bound to the two-electron center, that is, $[(\oplus e)(eh)] \equiv [(\oplus ee)_M h]$. Thus, absorption with photocreation of the exciton,

$$[\oplus e] \rightarrow [(\oplus ee)_M h]$$

occurs at 1.738 eV. A full description of work on the many interesting features of this complex system is given by Jaros and Dean (1983), who show how isotope effects ($^{16}O \rightarrow {}^{18}O$), Zeeman spectroscopy, and other techniques have been exploited.

Since oxygen is a deep donor in GaP, with $E_D \simeq 0.9$ eV (Table 5.6), the D–A pair spectra involving oxygen are in the red to near-infrared regions. Usually close pairs do not emit radiation. However, the nearest-neighbor pairs O–Zn and O–Cd are exceptions, producing the red luminescence used in LEDs (Section 5.3.2); the former emits a broad band at 1.86 eV and the latter at 1.83 eV. These close D–A pairs may be regarded as a single isoelectronic defect and the emission as the decay of an impurity-bound exciton.

We now turn briefly to device applications and to comment on LEDs (Sze 1981). We have already seen (Section 5.3.2) that if the density of carriers in an intrinsic semiconductor is altered in such a way that Δn is the change in the number of electrons in the conduction band and Δp the change in the number of holes in the valence band, equilibrium is restored by recombination. It is not difficult to show that the radiative lifetime for recombination is controlled by the change in the density of minority carriers. Nonequilibrium carriers may be generated conveniently using forward-biased *p–n* junctions. The bias voltage raises the potential energy of electrons in the *n*-type regions, reducing the height of the potential barrier in the depletion region, which limits drift of electrons into the *p*-type region. The barrier preventing holes drifting into the *n*-type region is reduced similarly, and we obtain the usual type of diode current flow. Since the barrier controls carrier densities, we see that the optical output can be controlled by applied voltages. If *p*-type GaP containing Zn_{Ga}–O_P complexes is used, for example, current flow in the *p–n* junction results in red emission. The color of the emission is controlled by the choice of dopant. Doping with N_P gives rise to green emission at the junction, characteristic of exciton decay at the defect. However, efficiencies are often low, for the radiative decay of minority carriers injected through a *p–n* junction competes poorly with the nonradiative decay processes often caused by deep traps of unknown origin.

Visible luminescence requires crystals with $E_g \gtrsim 1.8$ eV. Wide gap materials like diamond and SiC are, unhappily, not easy to produce commercially as suitably doped crystals. In general, the wider the bandgap the higher is the melting temperature and the more difficult is the control of electrical doping. The LEDs made from $GaAs_{1-x}P_x$:N are very useful over the red–yellow range. The most generally useful crystals for LEDs are GaP ($E_g = 2.26$ eV) and the alloy $GaAs_{1-x}P_x$. GaAs is a direct-gap material with $E_g = 1.4$ eV. With $x = 0.4$ the bandgap of the alloy is direct, giving relatively efficient radiative recombination, and E_g is sufficiently large to give red luminescence.

CHAPTER SIX

Radiation-Induced Defect Processes

In Chapter 3 we considered point defects present in thermal equilibrium, where formation enthalpies and entropies determined defect concentrations. Under irradiation with X-ray or electrons, where there is an externally controlled production of defects, the final point defect populations may be far from thermodynamic equilibrium. This has several important consequences. First, one may find defect species not normally present. In a crystal that normally shows cation Frenkel disorder, for example, one might be able to bombard it so as to produce displacements only on the anion sublattice. Second, transient defects can be important. Some defects only last while the irradiation proceeds, for some characteristic decay times are only of the order of picoseconds. Others survive until the crystal is heated, but are always eventually annealed out at sufficiently high temperatures (whereas equilibrium defect concentrations usually rise with temperature). Third, there can be effects in the production of defects by irradiation that are highly anisotropic.

In Section 6.1 we outline the classical theory of collisions between particles, with emphasis on energy transfer. We also discuss the phenomenon of channeling of a particle by a lattice and focused collision sequences; these are special cases of anisotropic particle–lattice interactions. In addition we emphasize the concept of threshold displacement energy for removal of an atom from its lattice site and the special consequences of irradiation by electrons, protons, and neutrons.

Some of the consequences of irradiation are of practical importance. The photographic process, like photosynthesis, is based on optically induced chemical changes. Ion implantation, possibly followed by treatments like laser annealing, is a part of the new technology of microelectronics. Fission tracks in minerals offer a record of geological history. At the other extreme, the degradation of materials exposed to particle or optical radiation can be minimized when the underlying damage processes are understood. This is the basis of much work on plastics, on components for nuclear reactors, and on solar cells for satellite applications.

The relaxation of electronic energy following bandgap excitation of ionic crystals is a complex process. In some materials, for example, alkali halides, an important intermediate stage involves the self-trapped exciton (Section 5.1.3). This may form directly from an optically created free exciton; it may also be formed by the initial self-trapping of a hole, followed by capture of an electron. Whichever mode occurs, self-trapping usually takes place in a time period of $\lesssim 1$ psec. The self-trapped exciton can decay either nonradiatively, within a few picoseconds of formation, possibly producing vacancies and interstitials, or it can decay radiatively with a lifetime as long as a few milliseconds. The production of point defects in ionic solids by bandgap excitation is referred to as photolysis if conventional light sources are used for excitation. If X-rays are used the term radiolysis is general, even though the same self-trapping mechanism may be involved. The phenomenon of radiolysis has been intensively studied in alkali halides and, to a lesser extent, in other materials such as alkaline-earth fluorides (Section 6.2).

Photochemical decomposition of a different type is found in silver halides (Section 6.3), leading to the formation of the photographic latent image and, in more extreme situations, to silver colloids. These colloids are responsible for the color change in silver-containing photochromic glasses. Neither bandgap energy nor X-rays are very effective in displacing atoms in semiconductors such as Si and GaAs. In these materials radiation damage is associated with particle irradiation. In Section 6.4 we discuss consequences of radiation damage on the performance of solar cells and other devices and also laser annealing of damage caused by ion implantation. The enhancement of defect processes by the energy released by recombination radiation is also discussed.

In the studies embraced by this chapter the optical and epr techniques described in Chapter 4 have proved very useful. In addition, many of the processes we shall discuss happen on a picosecond time scale and the special techniques of picosecond spectroscopy have proved very informative. Quantitative theory has also proved important, since it may handle behavior hard to isolate experimentally, and led directly to the correct identification of the route by which defect production occurred in the radiolysis of alkali halides. And, at the other extreme, when diffusion-controlled processes and defect aggregation are under study, electron microscopy plays a major role.

6.1 COLLISIONS BETWEEN PARTICLES

Many of the important ideas and phenomena in radiation damage stem from simple basic results of classical dynamics. Relativistic effects will not be important in most situations we shall consider. Conservation of momentum and energy control the behavior. Since the results are readily derived, we state most without proof.

6.1.1 Classical Dynamics

In a particle–particle interaction most energy is transferred in a head-on collision. Suppose the incident particle, of mass M and energy E, strikes a stationary particle of mass M_T. The maximum energy transferred is

E_{max}, where

$$E_{max} = E[4MM_T/(M+M_T)^2] \tag{6.1}$$

The largest fractional transfer, E_{max}/E, occurs when $M = M_T$. As the difference in mass increases, the fractional transfer falls rapidly. If one particle is an electron (mass $\frac{1}{1836}$ amu) and the other an atom or ion of atomic weight A, then E_{max}/E is $4/(1836A)$. Table 6.1 shows how pronounced this mass effect is.

In glancing collisions the energy transfer $E(\theta)$ is reduced from the maximum value E_{max} according to the relation

$$E(\theta) = E_{max} \sin^2 \frac{\theta}{2} \tag{6.2}$$

where θ is the scattering angle in the center-of-mass reference frame (for light incident particles, this is the laboratory frame of reference). In a head-on collision $\theta = \pi$ since the incident particle is reflected back on its track.

In collisions which are almost head-on, the incident particle is returned at a small angle α to its original path and the target particle moves forward at a small angle β to the initial particle motion. Momentum conservation for forward motion and transverse motion shows that $\beta \simeq \alpha f/(1+f)$, where f is the ratio of the final velocity of the incident particle to its initial velocity. If $\beta < \alpha$, we have a so-called focusing collision (see Section 6.1.2).

A third type of collision occurs in which the incident particle changes

Table 6.1 Calculated Energy Transferred in the Collisions of 1-MeV Projectiles with a Hydrogen Atom and with a Copper Atom

	H Atom Target	Cu Atom Target		
Incident Particle	Maximum Percentage Energy Transfer	Maximum Percentage Energy Transfer	Maximum Energy Transfer	Average Energy Transfer
1-MeV electron	0.043	0.007	70 eV	
1-MeV neutron	100	6.1	61 keV	30 keV (hard sphere)
1-MeV proton	100	6.1	61 keV	230 eV (Rutherford)
1-MeV Cu atom	6.1	100	1 MeV	

direction by a small angle ϕ in the laboratory frame. The energy transfer becomes a fraction $(M/M_T)\phi^2$ of the incident energy and the target particle moves away approximately perpendicular to the incident particle's direction.

The results just described do not depend on the precise scattering mechanism. However, if we want to know what fraction of incident particles is scattered through a given angle, we need a cross section $\sigma(\theta)$ in the center-of-mass frame of reference. Since $\sigma(\theta)$ is defined per unit solid angle, a weighting factor $2\pi \sin\theta\, d\theta$ regularly appears. The two limiting cases are hard-sphere scattering, where $\sigma(\theta)$ is independent of θ, and Rutherford scattering. Hard-sphere scattering occurs when fast neutrons and slow-moving heavy ions collide with crystal atoms or ions. Rutherford scattering is especially important for protons or α-particles colliding with nuclei, when the repulsive Coulomb interaction gives

$$\sigma(\theta) = R^2/(2\sin\theta/2)^4, \tag{6.3}$$

where R is the distance of closest approach in a head-on collision, that is, the point at which the Coulomb repulsion energy (ZZ_Te^2/R) is equal to the initial (center of mass) kinetic energy $E_0[M_T/(M+M_T)]$; here E_0 is the energy in the laboratory frame.

One important result for any case is the probability that a given energy E is transferred in a particular collision. We give here the still more useful value of $\sigma_>$, the integrated cross section for energy transfers exceeding a minimum transferred energy $E_{\min}$. For the hard-sphere case, this value is

$$\sigma_> = \sigma_T(1 - E_{\min}/E_{\max}) \tag{6.4a}$$

and

$$\left|\frac{d\sigma_>}{dE_{\min}}\right| = \sigma_T/E_{\max} \tag{6.4b}$$

where σ_T is the total cross section $\int_0^\pi d\theta \sin\theta \int_0^{2\pi} d\phi\sigma(\theta)$. For Rutherford scatter one finds

$$\sigma_>(E_{\min}, E_{\max}) = \tfrac{1}{4}\pi R^2(E_{\max}/E_{\min} - 1) \tag{6.5a}$$

and

$$\left|\frac{d\sigma_>}{dE_{\min}}\right| = \frac{\pi}{4} R^2 \frac{E_{\max}}{E_{\min}^2} \tag{6.5b}$$

The total cross section is much larger for Rutherford scatter, although collisions with low-energy transfer dominate, simply because the long-range potential allows some energy transfer in relatively distant encounters.

6.1.2 Focusing, Channeling, and Straggling

Suppose one member of a row of identical atoms is struck so that it moves at a small angle α_1 to the axis of the row. It strikes another atom that, as we have seen, moves at a smaller angle $\alpha_2 = \lambda\alpha_1$ [with $\lambda \simeq f/(1+f)$ as earlier defined] to the axis. As more and more collisions occur, the momentum transfer becomes more and more axial and the sequence of collisions is focused. For hard spheres of radius R separated by a distance D, one can show that focusing requires $4R > D$, so that close packing (small D) encourages the effect. This phenomenon, first noted by Silsbee (1957), is important in radiolysis of ionic crystals, for it allows an interstitial to be produced at some distance from the vacancy simultaneously created (Section 6.2) (Figure 6.1).

Suppose a fast particle is moving down one of the empty channels that occur in some crystal structures. The collisions are principally those in which only small deflections of direction occur, guiding the particle along the open channels. The trajectory is determined by the crystal structure, a phenomenon referred to as channeling. We shall consider the applications of channeling in Section 6.4.

Channeling and focusing have their similarities, for both concentrate particle motions into well-defined directions and both reduce radiation damage. Channeling occurs mainly with high-energy particles. Focusing is especially effective for low-energy collisions.

Channeling has important applications (Swanson 1982) that can be understood from two examples. First, suppose one measures the backscattered intensity when a particle beam is incident near to a channeling direction of a perfect crystal (Figure 6.2a). What is seen (Figure 6.2b) is a pronounced, characteristic dip, because particles that channel are not backscattered effectively. When a defect or impurity blocks a channel, however, backscattering will be seen with a characteristic structure (Figure 6.2c). Moreover, if the experiment is repeated for several channels that are inequivalent by symmetry, one can often tell which interstitial site is occupied (Figure 6.3a). A similar application tells one about

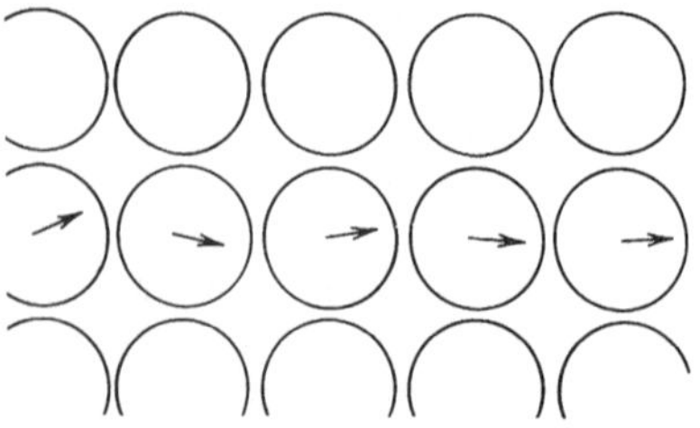

Figure 6.1 Focused collision transferring energy along a row of atoms. With each successive collision, the directions of motion become increasingly aligned.

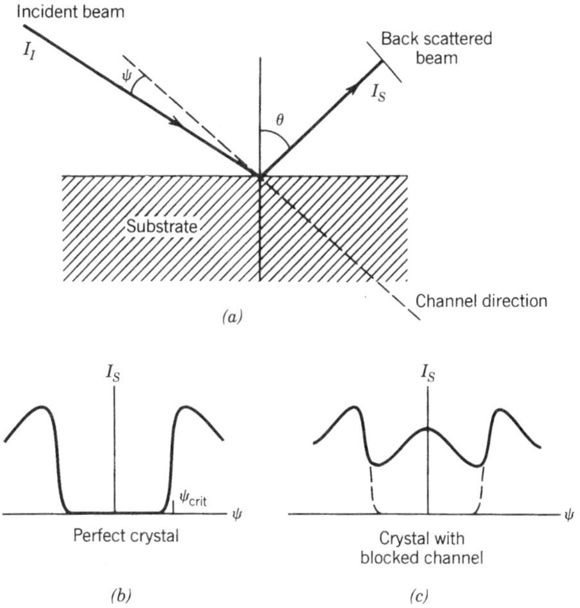

Figure 6.2 (*a*) Backscattering of a particle beam close to a channeling direction. ψ is the deviation of the incident beam, I_I, from the direction for ideal channeling, and ψ_{crit} is the effective angular width of the channel [see (*b*)]. (*b*) For a perfect crystal, the backscattered intensity I_s shows a dip for $|\psi| < |\psi_{crit}|$. (*c*) Blocking of the channel increases the backscattered yield for $\psi \sim 0$.

the matching of interfaces, for example, oxide/silicon or in various layered structures, for the channels in one phase may not all match those on the other side of the interface (Figure 6.3*b*). The characteristic angle ψ_{crit} within which backscattering is insignificant in the perfect crystal is given approximately by $(2ZZ_Te^2/dE_{inc})^{1/2}$, where E_{inc} is the energy of the incident particle of nuclear charge Ze and d is the atomic spacing along the channel of scattering centers of charge Z_Te. The channeling particle can often be treated as moving in a sum of so-called string potentials $V_T(\rho)$ (Lindhard 1965), longitudinally averaged and depending on the perpendicular distance ρ from the string:

$$V_T(\rho) = \frac{2ZZ_Te^2}{d} \ln \sqrt{1 + 3b^2/\rho^2} \tag{6.6}$$

with b a characteristic screening length. If one wishes to predict angular distributions of channeling particles and their transverse oscillations within the channel, it is very convenient to use this string potential, rather than the full complexity of the exact potential.

The second important application concerns the energy of backscattered particles. Suppose heavy impurity atoms of mass M_T block channels. The particle of mass M, incident normally with energy E_0, is backscattered at angle θ to the incident beam with energy E_0K^2, where

$$K = \left\{ \frac{M\cos\theta}{M+M_T} + \left[\left(\frac{M\cos\theta}{M+M_T} \right)^2 + \left(\frac{M_T - M}{M_T + M} \right) \right]^{1/2} \right\}, \qquad (6.7)$$

ignoring energy losses to the lattice. The backscattered yield [Equation (6.7)] has a distinctive high-energy peak, which can help to identify the species blocking the channel. This peak is broadened by losses because of the spread of penetration depths before scattering by the heavy impurity occurs (Figure 6.4).

Dechanneling of particles within the critical angle occurs principally by two mechanisms: the thermal vibrations of atoms near the channel may scatter the particle, or static perturbations due to nearby impurity atoms or defects may suffice. As the temperature rises, thermal dechanneling increases and, at the same time, reduces the effectiveness of defect dechanneling.

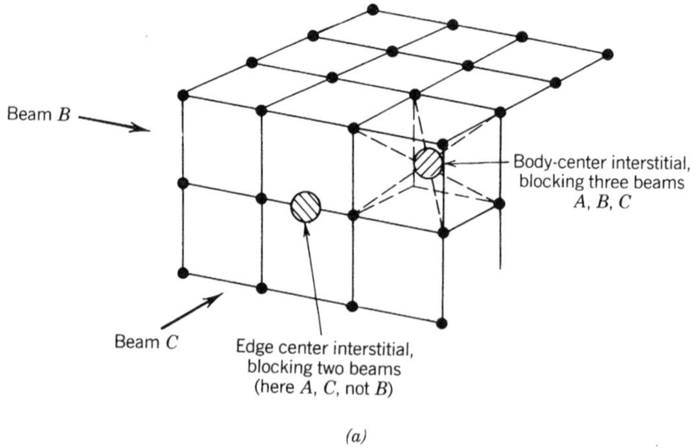

Figure 6.3(a).

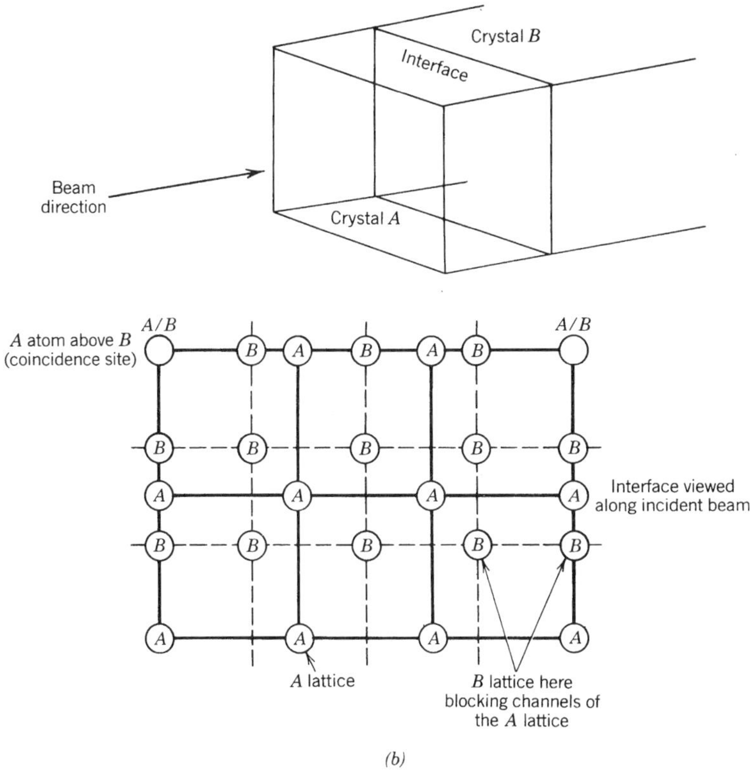

Figure 6.3 (*a*) Interstitial sites can be identified by the extent to which they block channels. The edge-centered interstitial in this simple cubic lattice obstructs beams along two of the three ⟨100⟩ directions, namely, beam *A* and beam *C*; the body-centered interstitial blocks all ⟨100⟩ directions, but not ⟨111⟩. (*b*) Interfaces between crystal structures often give blocked channels because of mismatch between the two crystals.

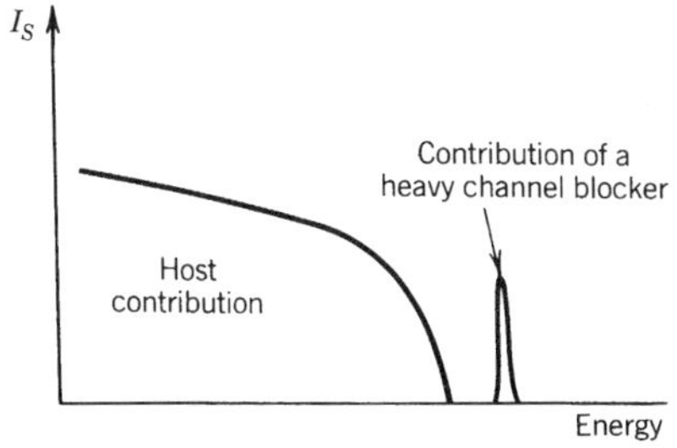

Figure 6.4 Schematic representation of the energy distribution of back-scattered particles (intensity I_s). Heavy interstitials blocking channels lead to backscattered particles with a high-energy peak.

We now turn to straggling, which describes the variation in ranges of particles. Each time a moving particle with velocity v traverses a plane of atoms, it will be deflected. The precise amount depends on many details. However, for fast heavy particles, which are only slightly deflected, the average transverse momentum received will be directly proportional to the Coulomb force (i.e., to ZZ_T) and the time for which it is appreciable (i.e., proportional to $1/v$). The angle of deflection is determined by the ratio of this transverse momentum to the initial forward momentum Mv, and so the angle will vary as (ZZ_T/Mv^2): fast, heavy, low-charged particles incident on low-atomic-number targets are deflected least.

The cumulative effect of uncorrelated deflections by successive planes is a random-walk effect, with a mean angle of scatter of θ per step, and with N steps with random directions of scatter. The effect of this multiple scattering is a change in direction proportional to $\theta\sqrt{N}$. The observed effect is slight for heavy particles but marked for light ones. Indeed, electrons are strongly deflected, so that some electrons in a beam incident on a foil will fail to pass through, even if the foil is thinner than the maximum range of electrons in that material. By contrast, nearly all particles in a beam of protons penetrate the foil if it is thinner than the particle range. The changes of direction lead to the phenomenon of straggling, so that particles have a distribution of depths of penetration. The spread depends both on θ and on the rate of energy loss, the latter determining the number of steps before coming to rest.

6.1.3 Displacement of Host Atoms and Electron Excitation

One simplifying assumption in the discussion of radiation damage is that of a threshold for displacement of atoms in a crystal, that is, that a well-defined minimum energy transfer is needed. This is called the displacement energy, E_D. Energy transfers exceeding E_D will cause atom displacement, either primary displacement, when a host ion is struck by one of the incident particles, or secondary displacement, when energy transfer is from a host atom previously struck. Values of E_D for a range of materials are listed in Table 6.2.

Irradiation of a solid with different types of particles results in distinct types of damage (Figure 6.5).

1. *Electrons*. Since the electron mass is so small, only very-high-energy electrons displace any atoms, and those displaced have such small kinetic energy that they produce no secondaries. Here one expects to find well-separated point defects (Figure 6.5*a*).

Table 6.2 Measured Displacement Energies (eV) of Ions in Crystals[a]

Oxides	II–VI (fourfold coordination)	III–V	Group IV
MgO 52/54	ZnO 30–60/60–120	GaAs 9/9.4	C 25 graphite
CaO 50/50	ZnS 7–9/15–20	InP 6.7/8.7	35–80 diamond
Al_2O_3 18/75	ZnSe 7–10/6–8	InAs 6.7/8.3	Si 13
	CdS 2–7/8–25	InSb 5.7/6.6	Ge 13–16
	CdSe 6–8/8–12		Sn 12
	CdTe 5.6–9/5–8		

[a]When experiments do not give a definitive answer, the range is shown (e.g., 35–80 eV for diamond). Where there are two host species, the value for the cation is shown first (e.g., 30/120 for ZnO means 30 eV for Zn, 120 eV for O). Some cases exhibit anisotropy, although this is usually slight and not shown explicitly here.

2. *Protons.* Rutherford scatter occurs and is dominated by glancing collisions, transferring relatively little energy. The cross section for transfer of energy is given by $\sigma_T(E)\,dE = [4\pi\sigma_{\max}E_{\max}]dE/E^2$, with $E_{\max}$ being the maximum energy transfer and $\sigma_{\max}$ being the maximum cross section (the apparent problem at $E=0$ is prevented by screening of the potential at large distances). In the case of a 1 MeV proton incident on copper the average energy transfer per primary scatter is 230 eV (Table 6.1). Primary damage events occur ~10 times over the proton range of ~5 μm and, on average, there are only a few secondaries for each primary collision (Figure 6.5*b*).
3. *Neutrons.* The hard-sphere scattering of fast neutrons leads to quite large energy transfers, for $\sigma_T(E)\,dE \simeq [4\pi\sigma_0/E_{\max}]dE$ with $\sigma_0 \sim 1$ barn (10^{-24} cm^2). In the case of a 1 MeV neutron incident on copper the average energy transfer per primary event is 30 keV (Table 6.1). A 30 keV copper atom has a range of about 0.1 μm in copper, giving a compact, highly damaged region containing a few hundred vacancies and interstitials.

Other distinctive forms of damage include tracks left by fission fragments. Heavy ions, with atomic weights in the range 80–160, and with perhaps 100 MeV of energy, are produced by fission. These particles leave heavily damaged tracks, perhaps 10 μm long, as they deposit their energy in the lattice (Figure 6.5*c*).

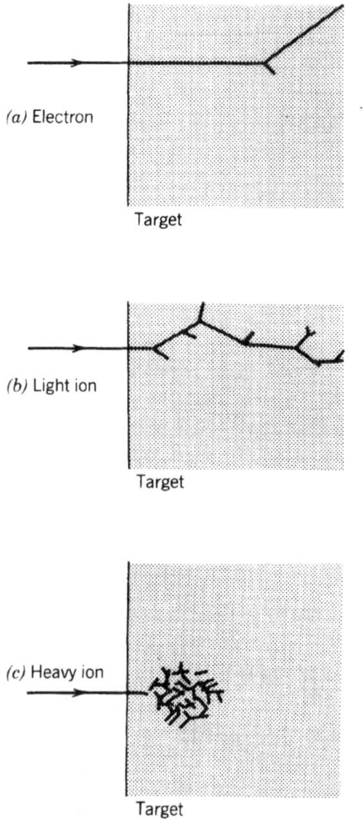

Figure 6.5 Damage distributions arising from irradiation with (*a*) electrons, (*b*) light ions, and (*c*) heavy ions.

When host atoms are knocked on with a large energy, the number of atoms displaced subsequently can be estimated by the Kinchin–Pease formula. This is usually written (in a form slightly different from the original) giving N_c, the number of defects in a cascade, as $\frac{1}{2}\eta$(energy lost in nuclear collisions)/(displacement energy). Here η is an efficiency factor, allowing for recombination. For Si, $\eta \sim 0.8$, although in other systems much smaller efficiencies are appropriate.

In systems like Si, a dose of radiation exceeding a critical value ϕ_c causes amorphization, that is, a continuous disordered layer results whose thickness is determined by the penetration depth. The dose, expressed in terms of the energy transferred in nuclear collisions, seems to depend little on the particle involved (Vook 1972). One simple picture regards the amorphous form as appearing when cascades from different primaries start to overlap. For Si this requires about 13 eV/atom to be lost in atomic collisions (Table 6.2). Parallel behavior is found in SiO_2, where the various crystal forms (with the possible exception of coesite) transform to a common glassy structure. This amorphous structure (known as a metamict state in mineralogy) has a unique density, different from vitreous silica. The metamict form retains the crystal habit of the unirradiated crystal, but is less anisotropic in its properties and does not have long-range order.

For the electrons in a solid the closest parallel to atomic displacement corresponds to excitation to the conduction band. Atom-induced excitation is inefficient, for we have noted (Section 6.1.1) that the maximum energy transfer is $4/(1836A)$ of the initial energy of the particle of atomic mass A. On the other hand, electrons can respond rapidly to a sudden perturbation, so that high-energy charged particles tend to excite electrons preferentially. Typically, bandgap excitation needs charged particles with energies of a few keV. This may be contrasted with electromagnetic radiation; bandgap optical excitation ($\hbar\omega = E_g$) suffices to excite an electron, but for higher-energy sources like X-rays about $3E_g$ of energy is dissipated for each electron–hole produced (Alig and Bloom 1975).

In contrast to metals [e.g., Thompson (1969)], where radiation damage is a relatively simple process, nonmetals exhibit a variety of behavior:

1. Different species occur depending on which sublattice is involved. In polar crystals, whether NaCl, UO_2, or GaP, the two different sublattices will have different properties. While this is partly a question of crystal structure, it is mostly a matter of mass. Thus, although Mg and O ions in MgO both have a displacement energy of about 53 eV (Table 6.2), electrons of 480 keV are needed to displace Mg and of only 330 keV to displace O. Hence knock-on damage can be selective.

2. Charged defect species can occur. The anion vacancy in KCl, for example, can have net charge 0, $\pm|e|$. Charged molecular species (e.g., Cl_2^- or Cl_2^0 in KCl) also occur. Charge conservation and charge compensation are very important in determining radiation damage in ionic solids, since long-range Coulomb forces can overwhelm elastic interactions.

3. Damage initiated on one sublattice tends to stay on that sublattice. If anions are displaced initially, the final damage (with one important exception, Section 6.2) will be on the anion sublattice. This follows from several factors. Focused collision sequences tend to occur only on rows containing a single species, and diffusion may be by processes involving

only a single type of ion. A major factor, however, is simply that Madelung energies in ionic crystals inhibit many of the possible reactions. In III–V compounds, where antisite defects have relatively low formation energies, the situation may differ.

4. Excited states of defects may be important. This is closely related to the fact that ionization damage is found, in addition to knock-on (displacement) damage. Ionization damage takes several forms (Sections 6.2, 6.3, and 6.4) and also tends to be species selective. For most hosts affected (e.g., many alkali halides), the anion sublattice is damaged. Typical time scales are shown in Figure 6.6. Ionization damage has an important experimental consequence, for electron-microscope beams themselves generate some of the damage they are used to study. This is a problem, especially in ionic crystals and molecular crystals.

Annealing of damage is important as a tool for removing damage, as an ultimate limit on numbers of displacements, and additionally as a probe from which defect properties can be determined. From the diffusion equation (Section 3.2.1) it follows by a simple scaling that concentration variations of defects over a distance L are largely eliminated in a time $t \sim L^2/D$, where D is the diffusion constant for the defects. If annealing involves atoms separated by distance L, they will disappear effectively in an anneal time t provided that $D(T) \gtrsim L^2/t$. For 10 ppm of defects (with 5×10^{22} atoms cm^{-3} one has 5×10^{17} defects cm^{-3} and volume $\frac{4}{3}\pi L^3$ per defect) $L \sim 10^{-6}$ cm and $D(T)$ should exceed 10^{-16} cm^2/sec to complete an anneal within 1 hr. There are several annealing strategies which allow one to obtain activation energies and diffusion constants for radiation-induced defects (Thompson 1969).

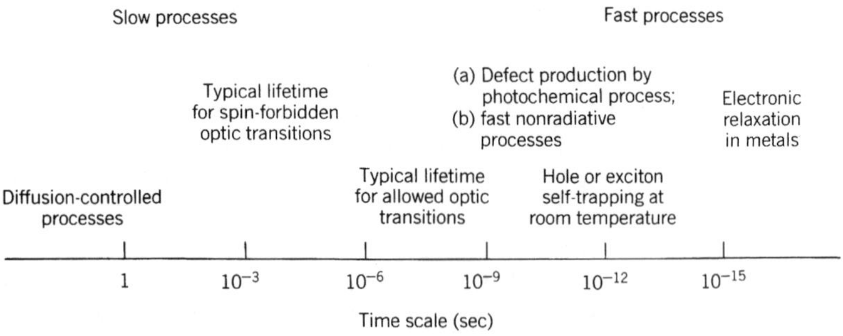

Figure 6.6 Time scales for the various processes of radiation damage.

6.2 RADIATION DAMAGE IN ALKALI HALIDES

6.2.1 Stages of Radiation Damage

In this section we shall consider both the initial damage events and the subsequent developments. Suppose radiation damage is produced photochemically, or perhaps by electron irradiation, at very low temperatures. As the temperature is raised, a sequence of secondary reactions occurs. Each stage involves a new set of defect reactions or solid-state processes.

(a) Primary Processes at Low Temperatures (Approximately 4–200 K)

At the lowest temperatures there are two distinct aspects. One is the production of close anion vacancy–interstitial pairs, at least transiently. The other is the separation of the vacancy from the interstitial so that they do not recombine rapidly. This is achieved partly by a focused collision sequence (see Section 6.1.2). The main defects at this stage are neutral anion vacancies (*F* centers) and anion interstitials (*H* centers; Section 5.1.3). What we have said so far refers to radiolysis at helium temperatures, where *F* centers and *H* centers are immobile. At somewhat higher temperatures secondary reactions occur and the radiolysis process becomes complex (see below). In a crystal such as KCl the production rate of *F* centers by ionizing radiation increases by about an order of magnitude between 100 and 200 K, with an activation energy of 0.07 eV. This is no doubt due in part to increased separation of *F*-*H* pairs in the primary event, thus inhibiting recombination.

(b) Initial Defect Reactions

The interstitials (*H* centers) become mobile below liquid–nitrogen temperatures and undergo several types of reaction to form the so-called *V* centers (V_1, V_2, V_3, V_4) and to recombine with vacancies. Some centers are trapped by impurities, for example, by alkali impurities to form V_1 centers. These complexes dissociate at about 50 K, forming a variety of simple complexes that are not paramagnetic. In KI, for example, Raman spectroscopy suggests that the V_2 and V_3 centers are associated with I_3^- species, that is, interstitial pairs (see Figure 6.7). There is also evidence that V_4 centers consist of pairs of *H* centers. Very little is known about the detailed structure of other aggregates, although apparently they dissociate above about 100 K.

With increasing temperature the anion vacancies also form small clusters. These processes occur both under irradiation and after it has ceased. The vacancies (*F* centers) aggregate to form complexes of two (F_2 or *M* center), three (F_3 or *R* center) or four (F_4 or *N* center) neutral anion vacancies, that is, cation-rich regions. The aggregation takes place by diffusion in the vicinity of room temperature and above. *F* center diffusion rates depend on both the charge state (i.e., the vacancy formed by removing Cl^-, the *F* center, and the F' center all have different rates) and on the electronic state [e.g., excited *F* centers (F^*) diffuse faster than those in the ground state].

(c) Aggregation Leading to Dislocations

From about 200 K to about room temperature, where the interstitials are mobile and the vacancies (e.g., *F* centers) immobile, the interstitials can aggregate to form dislocation loops. As part of this process some *H* centers make a series of reactions (Figure 6.7) leading to the formation of a neutral halogen molecule (e.g., Cl_2^0) in a cation–anion vacancy pair. This defect is most important and, since it is almost impossible to detect (it shows no epr, no distinctive optical absorption, and has no infrared activity), most evidence for it comes from detailed theory (Catlow et al.

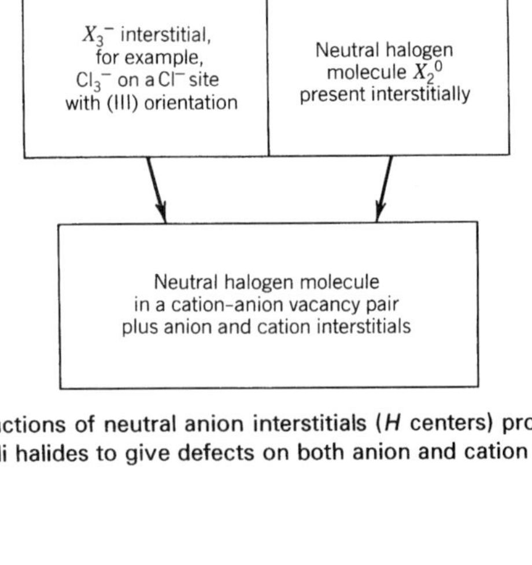

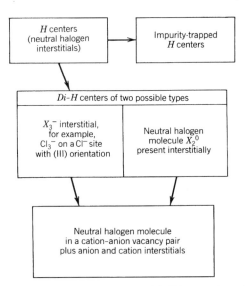

Figure 6.7 Reactions of neutral anion interstitials (*H* centers) produced in radiation damage in alkali halides to give defects on both anion and cation sublattices.

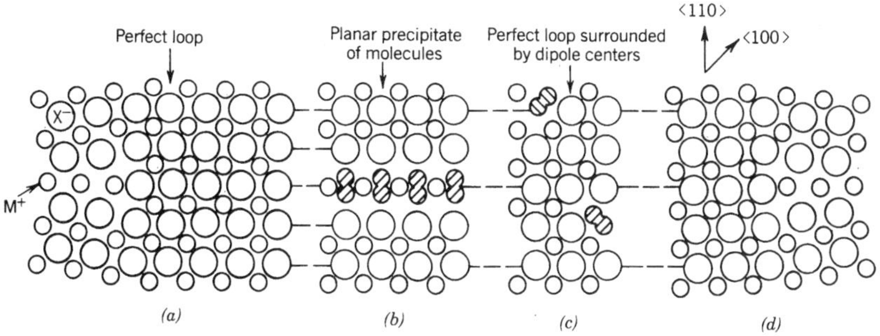

Figure 6.8 Perfect interstitial dislocation loop in alkali halides (see also Figure 3.22). (*a*) and (*d*) represent the edges of the perfect loop. (*b*) and (*c*) indicate two possible ways of incorporating halogen molecules into the dislocation loop. In (*b*), halogen molecules are precipitated at the loop; in (*c*) the dipolar halogen molecular centers surround the perfect loop. Alkali ions M^+ are small circles, halogen ions X^- large circles [after Hobbs et al. (1973)].

1980). The importance of this invisible center is that now there are displacements on both anion and cation sublattices, even though all the significant damage initially was on the anion sublattice. This feature is verified in the next stage where anion and cation interstitials punched out by the halogen molecules can form perfect interstitial loops and the corresponding divacancies can form perfect vacancy loops (Figure 6.8).

(d) Aggregation Leading to Formation of Colloids

If irradiation is carried out at a few hundred degrees centigrade, then the F centers are also mobile. Growth of metal colloids from the anion vacancy aggregates may occur, along with more complex dislocation loops from the clustering of interstitials, as already discussed (see Figure 6.8). The halogen molecule defects formed, along with the dislocation loops, may aggregate to form gas bubbles.

(e) Complete Annealing of Radiation Damage

Finally, at the highest temperatures, full recovery from radiation damage can occur. The colloids anneal first, just below 200°C in NaCl. This probably happens at least partly by the molecule-in-a-divacancy defect diffusing to the colloid at a rate limited by cation jumps. The mechanism does not seem to be solely evaporation of F centers from the colloid. Dislocation loops anneal only at high temperatures, just above 300°C in

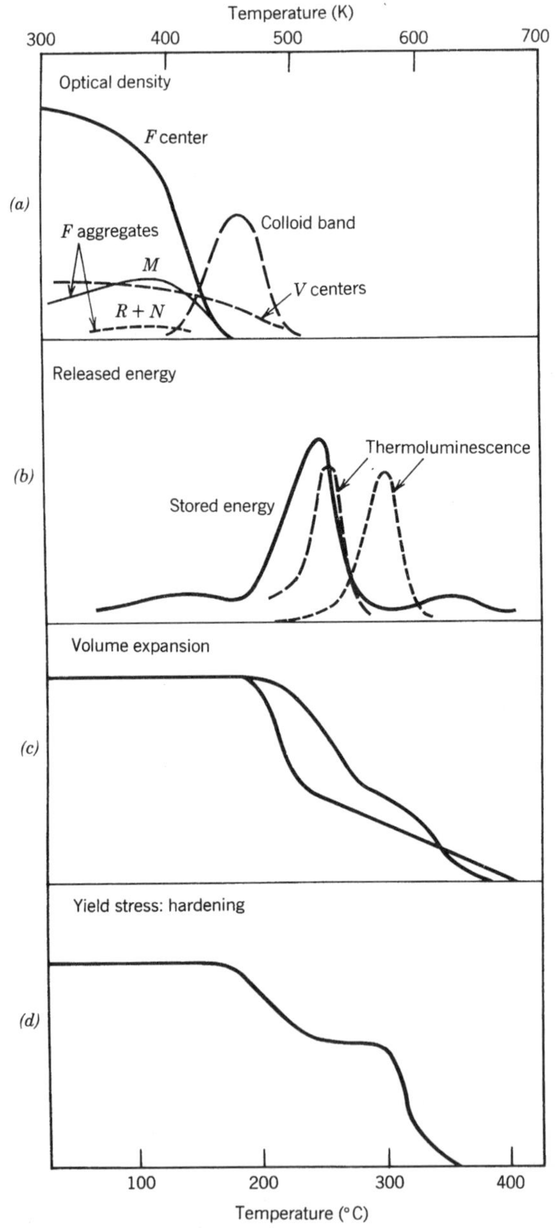

Temperature (K)
300
400
500
600
700
Optical density
F center
(a)
Colloid band
F aggregates
M
V centers
R + N
Released energy
Thermoluminescence
(b)
Stored energy
Volume expansion
(c)
Yield stress: hardening
(d)
100
200
300
400
Temperature (°C)

NaCl. Thermally induced defects are still unimportant at this temperature, so that the rate is possibly limited by the diffusion of divacancies.

Figure 6.9 shows the consequences of some of these defect processes for NaCl as annealing occurs after room-temperature irradiation. The optical intensities shown in Figure 6.9*a* indicate defect reactions as aggregates and then colloids form; at still higher temperatures, the colloids disappear. These defect reactions have several consequences. One is stored-energy release as defects recombine nonradiatively (Figure 6.9*b*). Another is thermoluminescence, where either defect reactions or thermally induced redistribution of electronic carriers leads to luminescence. Since thermoluminescence is a relatively complex phenomenon, whose details depend both on the precise defect and impurity centers present and on experimental details like heating rates, it is not clear whether the two separate peaks (Figure 6.9*b*) do indeed correspond; nevertheless, they indicate important reactions between 200 and 300°C in NaCl. Both the volume expansion (Figure 6.9*c*) and the yield stress (Figure 6.9*d*), which reflects the influence of defects on dislocation motion (Section 3.7), show partial recovery as the colloids anneal, and final recovery at the higher temperatures at which dislocation loops anneal. This final recovery is associated with stored-energy release; the maximum values encountered are of order 0.05–0.1 eV per molecular unit in the most heavily damaged crystals.

6.2.2 Photochemical Damage and Mechanisms of Radiolysis

Photochemical processes, in which light causes a solid-state reaction, are not uncommon. The production of vacancies and interstitials by mere bandgap optical excitation is much rarer and there are good reasons for this. First, the energy supplied, E_g, must exceed the sum of the formation energies of the defects; indeed, it must exceed the energy of any transient intermediate stage of the photochemical process. Narrow-gap materials, therefore, will not be expected to show photochemical damage. Second, there must be some significant reaction after excitation, preventing total recombination. In halides the fact that the reaction $X^- + X^0 \rightarrow X_2^-$ is

---

Figure 6.9 Recovery of physical properties of NaCl on annealing: (*a*) optical density, (*b*) energy release and thermoluminescence, (*c*) volume expansion from two independent measurements and (*d*) yield stress. The *H* centers are mobile below 100 K; *F* centers aggregate between 370 and 420 K, releasing about 0.3 eV per *F* center. All *F* and *V* centers anneal between 470 and 520 K, including the annihilation of Cl_2^0 in divacancies with *F* centers, to produce anion–cation divacancies [after Hughes (1978) who gives full references].

exothermic is an important factor. In oxides, $O^{2-} + O^- \rightarrow O_2^{3-}$ is endothermic, and only when the Coulomb repulsion can be reduced or bypassed does one expect oxides to show photochemical processes. Indeed, it is the relative ease with which anions bond to form molecular units and the unwillingness of cations to do so which causes preferential damage on the anion sublattice. Third, even if a vacancy and an interstitial are formed, they are of restricted interest if they survive only for a short time. Recombination can be inhibited in two ways. There may be a local barrier but, more effectively, a focused collision sequence can take the interstitial far from the vacancy (Section 6.1.2). Focusing depends strongly on crystal structure. There is no problem for the (110) rows of the NaCl structure but, for instance, the rutile structure of MgF_2 lacks suitable focusing rows. Focusing also depends on the interstitial geometry. If the interstitial has a symmetry axis different from that of the close-packed row, as in CaF_2 (Section 5.1.3), the interstitial's own defect forces lead to disruption of focusing before the vacancy and interstitial are properly separated.

Although bare anion vacancies (called F^+ or α centers in alkali halides; see Section 5.1.1) and negatively charged anion interstitials (called I centers; Section 5.1.3) can be observed after X-irradiation at low temperatures, there is now general agreement that the primary products of radiolysis are F centers and the complementary H centers. The production efficiency of primary F-H pairs in alkali halides at helium temperatures is high, having a value of $>10\%$ per ionizing event. However, the majority of F-H pairs recombine within 20 μsec in KCl at low temperatures and the production efficiency of stable F-H pairs is in the region of 1.0 to 0.1%. It seems that most of the F^+ and I centers are formed by secondary reactions (see below), which may be complex, since H centers, in contrast to V_K centers, do not appear to trap electrons.

Initially ionizing radiation produces free electrons and holes that thermalize very rapidly through phonon emission. It is assumed that holes in alkali halides self-trap within a time interval of 0.1–1.0 psec although this trapping time has not been directly measured. Fast-pulse optical studies show that the F centers form in their ground state within 10 psec of bandgap excitation, a time short compared with the radiative lifetime of excited F centers ($\sim$600 μsec; Section 5.1.1). However, this fast formation time is characteristic of relatively high carrier concentrations ($\sim 5 \times 10^{17}$ cm^{-3}) and could become longer as excitation intensity, and hence carrier concentration, decreases. Experiment indicates that V_K centers have a capture cross sections for electrons of $\gtrsim 2 \times 10^{-14}$ cm^2.

The facts that F centers can be produced within 10 psec of excitation and that the lowest triplet state of the self-trapped exciton has a lifetime of ~0.1 msec (Section 5.1.3) suggest that this triplet state is not an important precursor state for radiolysis. Indeed, excitation from the triplet state, populated by a pulsed electron beam, with a synchronized pulsed ruby laser ($\lambda = 694$ nm), results in increased $F + H$ production, indicating that higher excited states of the self-trapped exciton are involved.

Suppose a free exciton has been created by optical absorption. The next stage is self-trapping of the hole. This takes place rapidly in many cases. However, in others [notably iodides; see Rashba (1982) for a review], there is a barrier to self-trapping, so that it may be almost completely inhibited at low temperatures. Once self-trapping has occurred, the self-trapped exciton will normally be in an excited state, which will be one of three types, that is, $[e^* h^*]$ (electron and hole excited), $[e^* h]$ (electron alone excited), or $[eh^*]$ (hole alone excited).

At this stage a complex series of processes follows. The simplest (see Figure 6.10) is radiative recombination, giving the π band (a transition from the lowest triplet state $[eh]$) and the σ band (if seen, a transition from an electron-excited state $[e^* h]$) (Figure 5.10). There are also

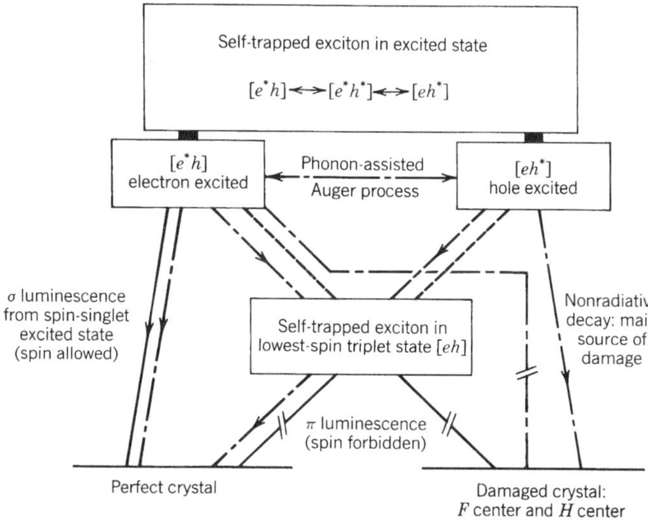

Figure 6.10 Principal recombination processes of self-trapped excitons. Only the main transitions are shown, and the many excited states are indicated schematically (e.g., $[e^* h]$ for all electron-excited states). Strong optical transitions are indicated by solid lines, weak transitions by broken lines, and transitions forbidden either by selection rules or energy conservation by -//-. Note the various types of nonradiative transitions indicated by ——

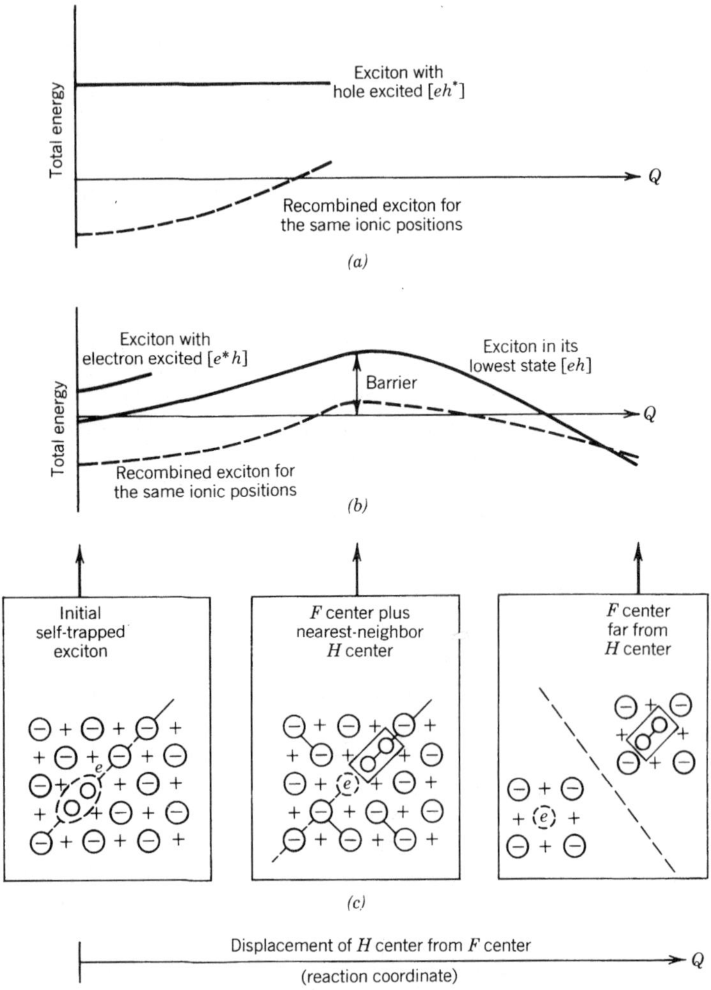

Figure 6.11 Photochemical damage associated with exciton decay, in alkali halides like KCl. The diagram compares energies for hole excited {case (*a*), [*eh**]} and electron excited {case (*b*), [*e***h*] and also [*eh*]} as the reaction (*c*) proceeds [after Itoh et al. (1977)]. The important point is that there is no barrier to *F* and *H* center production (*Q* increasing) when the hole is excited [as in (*a*)], whereas there is a large barrier [as in (*b*)] when the hole is in its lowest state. For reference, the broken line shows the energy for the same geometry when the exciton has recombined.

nonradiative transitions of three main types that do not produce defects. The first type is simple multiphonon processes $[e^{**}h] \rightarrow [e^*h]$, degrading electronic energy into heat. The second type gives rise to nonradiative recombination of electrons and holes. The third type includes Auger processes, notably $[e^*h] \rightarrow [eh^*]$, in which excitation energy of the electron is transferred to the hole (some energy is also degraded into heat in this case).

The final class of nonradiative process gives defect production. In KCl, for which experimental data are more complete, this step is $[eh^*] \rightarrow F + H$, involving excited hole states. No defect formation is possible from the lowest self-trapped exciton state, and electron-excited states $[e^*h]$ are apparently less effective than hole excited states. A major factor favoring hole states is that no significant energy barrier exists for defect formation from the hole-excited state (Figure 6.11). For the electron-excited states the diffuse electron orbital does little to lower the barrier to defect formation [for a full review see Itoh (1982)].

6.2.3 Radiolysis in Other Ionic Solids

Many types of ionic solids other than alkali halides show color center production by radiolysis. In alkaline-earth fluorides, for example, a first-stage radiolysis is observed at helium temperatures, involving production of F centers and self-trapped holes. This is an efficient process, but it saturates, suggesting that an impurity is involved. This extrinsic stage is followed by a slower process involving production of F and H centers, although there are some differences in detail from the alkali halide case. The self-trapped hole in CaF_2, for example, is located primarily on two equivalent nearest-neighbor fluorine ions, giving an F_2^- ion with its molecular axis aligned along the close-packed ⟨100⟩ directions (Figure 6.12*a*). Some of the changes result from different defect symmetries. However, the neutral interstitial (H center, Figure 6.12*b*) is oriented along ⟨111⟩, with a pronounced asymmetry between the two components of the F_2^- ion (the ion near the cube center is more nearly $F^{-1/3}$ and that nearly on-site roughly $F^{-2/3}$). The self-trapped exciton ($V_K + e$) decays into a close F-H pair (Figure 6.12*c*; see also Section 5.1.3). Radiative recombination then occurs in a time of 5–10 msec, restoring the perfect lattice. Although nearly every electron–hole recombination results in formation of the F-H complex, well-separated F-H pairs, and hence observable radiolysis, are less likely at low temperatures than in alkali halides (Section 6.1.2). There is, in fact, no evidence of collision sequences of any length in fluorites, although observations of colloid

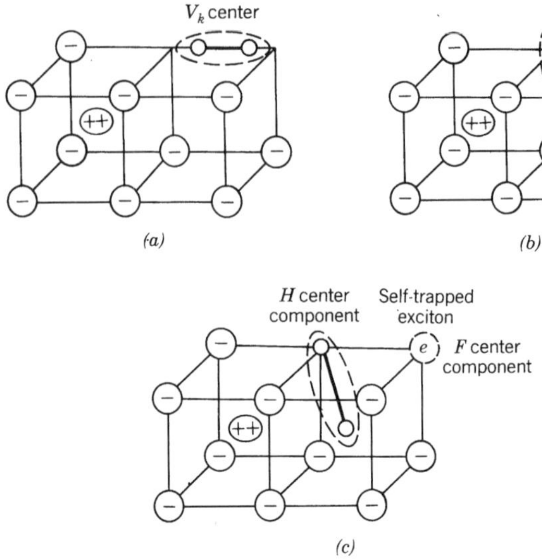

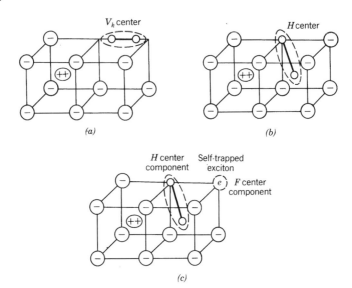

Figure 6.12 Schematic representation of (*a*) the self-trapped hole (V_K center), (*b*) the *H* center, and (*c*) the self-trapped exciton in the fluorite structure. The self-trapped exciton is represented as a close *F*-*H* center pair.

superlattices have been mistakenly interpreted as evidence for such sequences.

Although *F* centers and self-trapped holes can be produced by X-irradiation of $SrCl_2$ at 4 K, radiolytic decomposition is completely absent in zone-refined $SrCl_2$ at 4 K. This observation emphasizes the importance of impurity effects when considering mechanisms of radiolysis generally.

Radiolysis of oxides such as CaO, MgO, and Al_2O_3 does not result in production of vacancies and interstitials. In these materials X-irradiation appears to give rise only to transfer of charge between existing lattice defects, for example, the Fe^{2+} ion present as an impurity in MgO will trap both holes and electrons giving rise to Fe^{3+} and Fe^{+} charge states. Existing positive ion vacancies in materials such as CaO trap holes (Section 5.1.2), and existing negative ion vacancies may trap one electron to form paramagnetic F^+ centers or two electrons to form diamagnetic *F* centers. In silica-based glasses ionization encourages motion of alkali atoms (notably Na) with a variety of indirect consequences.

Transient radiolysis of silica (SiO_2) (which we shall regard as ionic for present purposes) appears to be possible by an excitonic mechanism at low temperatures (Tanimura et al. 1983, Hayes et al. 1984). However, the nature of hole-lattice and exciton lattice interactions in SiO_2 is not yet fully understood (see also Mott 1977 and Hughes 1977).

6.3 PHOTOLYSIS OF SILVER HALIDES AND THE PHOTOGRAPHIC PROCESS

6.3.1 Self-Trapping of Holes and Excitons in AgCl

The phenomenon of hole self-trapping can be seen in AgCl at low temperatures. Hole self-trapping in AgCl is readily observed by ultraviolet excitation of Cu^{2+}-doped AgCl crystals in a microwave cavity. Epr shows that at temperatures below about 77 K the electrons produced by bandgap excitation are captured by the copper impurities ($Cu^{2+} + e \rightarrow Cu^{+}$) and that the holes self-trap on silver ions ($Ag^{+} + h \rightarrow Ag^{2+}$), giving distorted complexes. Just as $Cu^{2+}(3d^9)$ shows a strong Jahn–Teller effect (Section

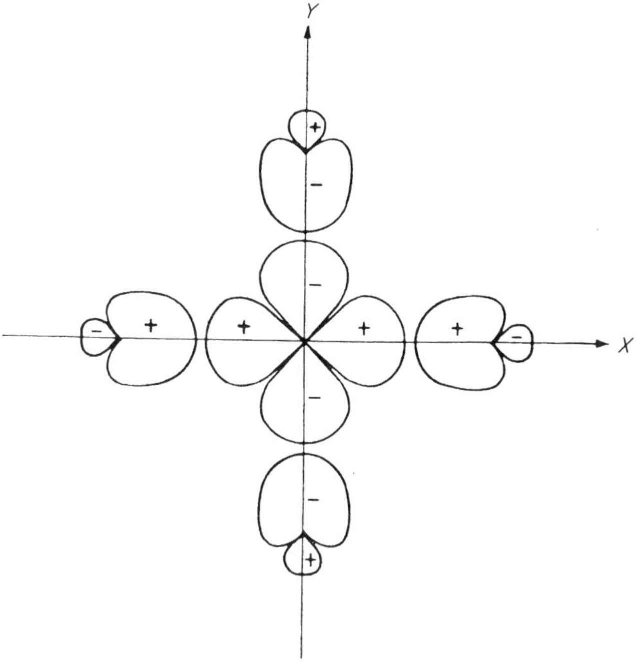

Figure 6.13(a).

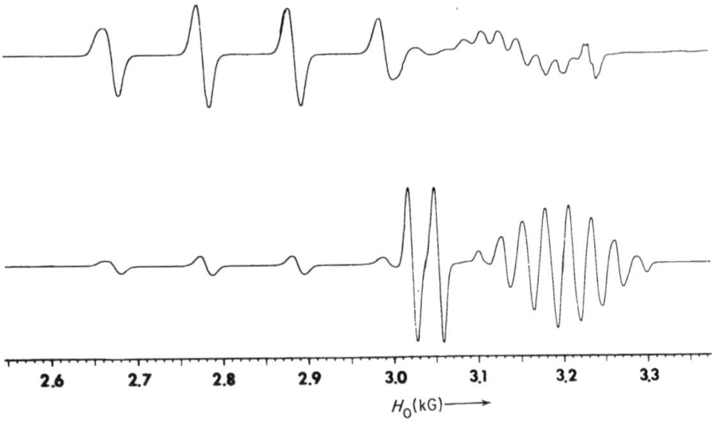

(b)

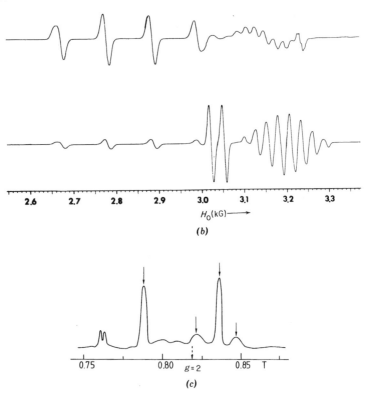

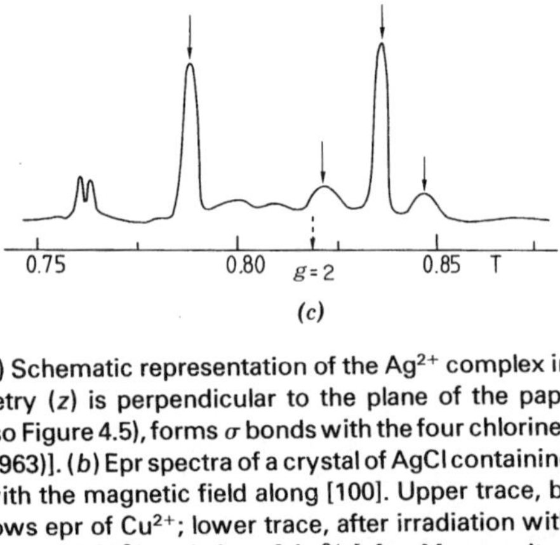

(c)

Figure 6.13 (*a*) Schematic representation of the Ag^{2+} complex in KCl whose principal axes of symmetry (z) is perpendicular to the plane of the paper. The ground state $|x^2 - y^2\rangle$ (see also Figure 4.5), forms σ bonds with the four chlorines in the xy plane [after Delbecq et al. (1963)]. (*b*) Epr spectra of a crystal of AgCl containing about 3 ppm of Cu^{2+} taken at 15 K with the magnetic field along [100]. Upper trace, before irradiation with green light, shows epr of Cu^{2+}; lower trace, after irradiation with green light shows a weak epr spectrum of Cu^{2+} and also of Ag^{2+} [after Marquardt et al. (1971)]. (*c*) Odmr spectrum with the magnetic field along [100] or AgCl at 4 K under X-irradiation. The four lines indicated by solid arrows are fine-structure transitions of the self-trapped exciton [after Hayes et al. (1977)].

5.2), so does $Ag^{2+}(4d^9)$, and this is the main source of the self-trapping distortion. As for Cu^{2+}, the crystal-field splitting of the d level, combined with a Jahn–Teller distortion of the surrounding octahedron of chlorines along a cube-edge (z) direction, leaves an orbital singlet lowest with an angular dependence of the form $|x^2 - y^2\rangle$. This forms weakly covalent bonds with the four chlorines in the x, y plane (Figure 6.13a), and the epr spectrum of the self-trapped hole in AgCl shows resolved hyperfine interaction with the silver nucleus and with the four bonded chlorines (Figure 6.13*b*). The structure of the Ag^{2+} impurity complex in KCl (see Section 5.5) is remarkably similar to that of the self-trapped hole in AgCl. Self-trapping of holes does not occur in AgBr.

As in the case of alkali halides, the self-trapped exciton in AgCl may be regarded as an electron trapped in the Coulomb field of the self-trapped hole. This self-trapped exciton emits in a band at 2.47 eV at low temperatures, about 0.25 eV wide at half height, by radiative decay from its lowest triplet state to its singlet ground state. Odmr studies have used this emission to give detailed information about the structure of the self-trapped exciton (Hayes 1983). Epr studies show also that there is an energy barrier to self-trapping of holes in AgCl.

6.3.2 Mobility of Carriers in AgBr

Optical absorption by AgBr at energies above the exciton region results in the very efficient production of free carriers, with quantum efficiency of order unity at room temperature. Recombination is slow, partly because AgBr has an indirect gap, so that the wavevector selection rule for optical recombination limits the rate (see Section 1.3.1). The indirect gap is itself a consequence of a valence band comprising both Ag^+ 4*d* orbitals as well as the halogen *p* orbitals which alone dominate the valence band in alkali halides.

Shallow traps can reduce the drift mobility μ_D so that it is much less than the microscopic mobility μ. An electron drifting in an electric field spends part of its time immobilized in traps and part in band motion with mobility μ. If there is a single set of shallow traps of density N_t and depth E_t, the two mobilities are related by

$$\mu_D = \mu/[1 + A_t \exp(E_t/k_B T)] \tag{6.8}$$

Here A_t is the product of two factors. One, g_i/g_0, is the ratio of statistical weights for filled and empty traps; it is 2 for AgBr. The other gives the ratio of the trap density N_t to the density N_0 of conduction-band states; in AgBr, $N_0 \sim 5 \times 10^{18}\ \text{cm}^{-3}$. Clearly μ_D is always less than μ, which is effectively the Hall mobility.

Results for AgBr are summarized in Table 6.3 for both bulk crystals and for the crystallites typical of photographic emulsions. Hole mobilities are smaller than electron mobilities, both because the intrinsic mobilities μ are rather smaller and mainly because hole traps are generally deeper ($E_{th} \sim 0.35$ eV, $E_{te} \sim 0.05$ eV). The precise nature of the traps is not

Table 6.3 Drift Mobilities and Lifetimes of Electrons and Holes in AgBr at Room Temperature

	Drift Mobility μ_D (cm^2/volt sec)		Carrier Lifetime τ (msec)	
	e	*h*	*e*	*h*
Bulk crystals	60	1.1	10–1000	10–1000
Emulsion grains	0.2	0.001	3000	15,000

clear; for electrons, interstitial Ag^+ is likely, whereas, for holes, impurities like I^- or transition-metal ions, which can change charge state, may be involved. The results for crystallites show strong effects of their surfaces, since (see Section 7.5) one expects both higher defect densities and a significant electric field (which will separate electrons and holes spatially, hence reducing recombination) close to the surface.

6.3.3 Some Properties of Ionic Defects in Silver Halides

The silver halides AgCl and AgBr have the sodium chloride structure. Unlike the alkali halides, however, where Schottky disorder dominates, the dominant thermally induced defects in silver halides are cation Frenkel pairs. In ideally pure crystals (and, in practice, in reasonably pure crystals above room temperature) the numbers N_V of Ag^+ vacancies and N_I of Ag^+ interstitials are given by (Section 3.1.1)

$$N_I = N_V \propto \exp(-G_F/2k_BT) \tag{6.9}$$

where G_F is the free energy of formation of the pair. Values for the enthalpy of formation [see Equation (3.1)] are given in Table 6.4. In bulk AgBr, the interstitial concentration rises from about 10^4 cm^{-3} near room temperature to around 10^{20} cm^{-3} just below the melting point.

Diffusion and ionic conductivity are dominated by the motion of the very mobile silver interstitial. This can be seen from the activation energies for motion (Table 6.4). In AgCl the diffusion constant is $\sim 10^{-6}$ cm^2/sec at the melting temperature. The interstitial moves by an interstitialcy motion (see, e.g., Figure 3.4) in which one ion moves from its interstitial site into an occupied substitutional site, displacing the ion in residence to an interstitial position. The ionic conductivity in crystallites may be up to two orders of magnitude higher than in a bulk crystal

Table 6.4 Enthalpy (h_f) and Entropy (s_f) of Formation of Frenkel Pairs and Enthalpies of Migration for Vacancies (h_v) and Interstitials (h_I) in AgCl and AgBr, obtained from Experiment[a]

	h_f	s_f	h_v	h_I
AgCl	1.7	~9	0.4	0.15
AgBr	1.2	~7	0.35	0.15

[a]Enthalpies are in eV. Entropies of formation s_f are in units k_B. No allowance has been made for interaction of defects in analysis of experimental data.

This, a characteristic near-surface effect (Section 7.5), is a result of easier generation of interstitials at special surface sites.

6.3.4 Photolysis and Formation of Colloids

Optical absorption in the fundamental band of AgCl (416 nm) at 300 K produces colloidal silver particles. This photochemical reaction involves several steps, presumably (a) electron and hole production, (b) capture of an electron to generate silver atoms ($Ag^+ + e \rightarrow Ag^0$ or $Ag^+ \rightarrow Ag^0 + h$), and (c) aggregation of silver atoms. The colloids form more readily in AgCl than in KCl (see Section 5.1.1) because of the ready mobility of silver ions (see below). As in the case of alkali halides (see Section 6.2.1) the precise structure of the hole centers produced at room temperature is uncertain. In the silver halides the colloidal particles tend to separate out along dislocations, providing one of the earliest examples of microscopically visible dislocation decoration. The optical absorption band of the colloids is broad and is independent of temperature, and its peak varies between 450 and 500 nm. The position of the band can be understood qualitatively from simple dielectric theory. We regard the colloid as a sphere of radius R and dielectric constant ϵ_c in a host continuum of dielectric constant ϵ_H. In a long-wavelength electric field E_0, the sphere has an electric dipole moment $\bar{\mu}$ given by (Coulson 1956)

$$\frac{\bar{\mu}}{E_0} = \frac{\epsilon_c - \epsilon_H}{\epsilon_c + 2\epsilon_H} R^3 \tag{6.10}$$

Clearly $\bar{\mu}/E_0 \equiv \alpha$ is the polarizability. If there are N colloidal particles per unit volume, the overall dielectric constant (for $NR^3 \ll 1$) is given by

$$\epsilon_{\text{eff}} = \epsilon_H + 4\pi N\alpha$$
$$= \epsilon_H + 4\pi NR^3 \frac{\epsilon_c - \epsilon_H}{\epsilon_c + 2\epsilon_H} \tag{6.11}$$

Optical absorption is proportional to $\text{Im}[\epsilon_{\text{eff}}(\omega)]$. For the host insulator, ϵ_H is real at frequencies less than the bandgap. The metal colloid has, however, a complex dielectric constant $\epsilon_c = \epsilon'_c - i\epsilon''_c$, with

$$\epsilon'_c \simeq 1 - \omega_p^2/\omega^2 \tag{6.12}$$

where ω_p is the plasma frequency. The imaginary part of ϵ_{eff} contains a denominator $(\epsilon'_c + 2\epsilon_H)^2 + (\epsilon''_c)^2$, which has a resonance when $\epsilon'_c + 2\epsilon_H = 0$. Simple manipulation shows that the resonance frequency, for which optical absorption is a maximum, is given by

$$\omega = \omega_p/(1 + 2\epsilon_H)^{1/2} \tag{6.13}$$

that is, at the plasma frequency of the colloid metal, as modified (in a shape-dependent way) by the host insulator. For a full discussion of this simple picture and more general models, see Hughes and Jain (1979).

6.3.5 Photolysis and Formation of the Latent Image

The mechanisms of latent-image formation in silver halide photographic emulsions are still a matter of scientific and commercial interest after more than 40 years of intensive research. A modern emulsion consists of a suspension in gelatine of small crystallites of AgBr or AgBr:I with dimensions varying from a few hundredths of a micron to several microns. Growth of these microcrystals is carefully controlled, both to avoid structural defects like dislocations, which may trap photoelectrons in the bulk of the crystal, and to encourage growth of silver specks only on the surface of the grains.

All models of latent-image formation invoke two distinct types of process. The electronic process includes the creation of carriers by light and possible carrier movement in the grain. The ionic process involves the motion of Ag^+ ions or Ag atoms to form surface specks. When the surface specks contain three to six silver atoms they form latent-image centers, and these can catalyze the chemical reduction of the crystallite by reducing agents known as developers. Although the mechanism of catalysis is not fully understood, it seems that the reducing agent transfers electrons to the latent-image catalyst, and these electrons react with silver ions at the silver-to-silver-halide interface. With normal development time those grains containing the latent image are reduced completely to metallic silver, whereas those grains not containing the latent image are unaffected. Since only about ten photons need to be absorbed in a grain to produce a latent image, and since a typical crystallite contains $\sim 10^9$ atoms, a gain (i.e., the number of silver atoms produced per incident photon) of as much as 10^8 results. This enhancement accounts for the high sensitivity of photographic emulsions. Although there is intriguing science associated with the preparation, spectral sensitization, and development of modern emulsions, we shall concentrate here on the basic properties of silver halide crystals relevant to latent-image formation.

A solid-state reaction, discussed by Mott and Gurney (1948), is still the basis of any discussion of latent-image formation. In the electronic process the optically-produced electron migrates via shallow traps. It forms an Ag atom either by direct trapping at an interstitial silver ion near the surface of a grain or by an ionic process in which the Ag^+ interstitial moves to the trapped electron. This is followed by arrival and trapping at this sublatent image of electrons and silver interstitials in alternating sequence, leading to the build up of the latent-image speck. Baetzold's (1973, 1975) calculations suggest that possible reactions might be

$$Ag^+ \underset{(e)}{\rightarrow} Ag^0 \underset{(e)}{\rightarrow} Ag^- \underset{(i)}{\rightarrow} Ag_2^0 \underset{(i)}{\rightarrow} Ag_3^+ \underset{(e)}{\rightarrow} Ag_3^0 \underset{(e)}{\rightarrow} Ag_3^- \underset{(i)}{\rightarrow} Ag_4^0 \cdots$$

at perfect AgBr surfaces as interstitials Ag_i^+, referred to as (i), and electrons (e) are added in sequence. He argues that the sequence is different, although similar, when surface defects are involved (see also Section 7.5):

$$Ag^+ \underset{(e)}{\rightarrow} Ag^0 \underset{(i)}{\rightarrow} Ag_2^+ \underset{(e)}{\rightarrow} Ag_2^0 \underset{(e)}{\rightarrow} Ag_2^- \underset{(i)}{\rightarrow} Ag_3^0 \underset{(e)}{\rightarrow} Ag_3^- \underset{(i)}{\rightarrow} Ag_4^0 \cdots$$

Clearly, the important factors are the mobile ionic defects, Ag_i^+, and electrons. The relatively high stability of Ag_2^0 is also significant. The photoproduced holes are not involved to any real degree, although, clearly, any electron–hole recombination will reduce efficiency. Many holes are trapped at iodine sites, which act as nonradiative recombination centers at room temperature (but luminescing centers at low temperatures).

Hamilton (1984) has recently noted two further factors. First, kink sites at surfaces always have a long-range Coulomb field which cannot be saturated by capture of ions or electrons. This can be seen from an argument due to Seitz (1940). Suppose that a kink site at a ledge ends with a Br^- ion and has a charge $-q|e|$. If an Ag^+ ion is captured the kink charge, by symmetry, will be $+q|e|$. Since the extra charge introduced is $+|e|$ we have $-q|e|+|e|=+q|e|$, giving $q=\frac{1}{2}$. The one-half integral kink charge encourages alternate capture of e^- and Ag^+. The second point is that, even though only shallow intrinsic electron traps have been observed in AgBr, deep traps are necessary since the photographic process occurs at room temperature. One possibility is that the shallow traps are merely metastable, separated by a potential barrier from a deep stable state (but see also Section 7.5).

The silver halides give a negative picture, that is, dark where illuminated. This is characteristic of crystals where an electron–cation reaction produces metal atoms; other examples include mercury halides and the oxalates of iron and palladium.

The lead halides exhibit a photographic process too, but based on an electronic process with mobile holes and an ionic process with rapid anion vacancy motion. In the lead halides, therefore, one gets a positive image, with the illuminated regions light. Similar systems include thallium halides and metal layers on As_2S_3 (see Jaenicke 1977).

6.3.6 Photochromic Glasses

Silver halide photochromic glasses are an example of a wide range of materials that undergo a reversible color change upon optical irradiation. If we assume irradiation with light of wavelength λ_1 at t_1, the absorption coefficient at λ_2 may vary with time as shown in Figure 6.14. At t_2 an equilibrium is reached in which forward and reverse reactions are equal. On removal of the exciting light at t_2, the original situation is gradually restored.

By suitable heat treatment it is possible to prepare an oxide glass containing a suspension of silver-halide-enriched particles containing

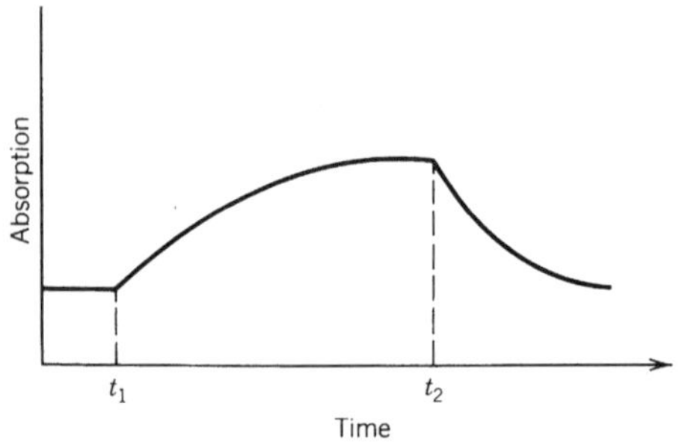

Figure 6.14 Absorption of a photochromic material, as a function of time, at wavelength λ_2 following switch-on of light of wavelength λ_1 at t_1 and switch-off at t_2.

traces of copper in the form Cu^+. These glasses are transparent because the suspended nonmetallic particles are too small (~100 Å) to scatter light appreciably. However, on exposure to uv light the glass becomes dark due to formation of silver colloids and Cu^{2+}, both of which absorb in the visible over a wide region. In the absence of uv light the transparency of the glass is restored by visible light that is absorbed by the colloids and Cu^{2+}, releasing free carriers. Such glasses have commercial applications, for example, as sunglasses.

We mention in passing that the effects described above are a specific example of a more general photochromic phenomenon in solids. Wide-bandgap solids may be colored by absorption bands associated with defects or impurities, and the absorption spectrum (and hence the color) may be changed reversibly by optical bleaching, because of reversible changes in oxidation states. Information storage and retrieval therefore becomes possible (Kiss 1970).

6.4 RADIATION DAMAGE IN SEMICONDUCTORS

6.4.1 Radiation Damage in Silicon

Most of the important experiments to understand the initial steps of radiation damage in Si have used epr, following irradiation with electrons at low temperatures. Electron damage ensures relatively widely spaced primary events (Section 6.1.3) and epr gives very detailed information about specific defects with unpaired spins. With ingenuity, for example, by changing the Fermi level by doping, some features of the defects with zero electron spin (see e.g., Section 5.4) can also be inferred.

No optical bands have been identified firmly with either vacancy or interstitial centers in Si or Ge; in diamond, both the neutral vacancy ($V^0 \equiv GR1$ center) and the negative vacancy ($V^- \equiv ND1$ center) have characteristic absorption (see Figure 5.23). Despite the lack of optical spectra for Si and Ge a consensus has emerged for the mechanism of electron damage at liquid-helium temperatures; it involves rapid production of vacancies and interstitials followed by rapid motion of interstitials. In *p*-type Si, interstitials displace substitutional acceptors A ($A =$ Al, B, also probably Ga; C is also displaced but not Ge) (Watkins 1964, 1967):

$$\mathrm{Si}_i + A_{\mathrm{Si}} \rightarrow \mathrm{Si}_{\mathrm{Si}} + A_i \tag{6.14}$$

Thus, in Al-doped Si, the numbers of Al interstitials and Si vacancies observed after irradiation at 4 K are about equal. The interesting features are partly the energetics of the displacement reactions and partly the mobility of the interstitial. Almost certainly the interstitial moves by an ionization-enhanced mechanism, driven by the ionization induced by the radiation that causes damage. Ways in which this can occur are described in Section 6.4.6. Since charge states may change many times in the presence of ionization, one cannot say which vacancy and which interstitial charge states are first produced.

In n-type Si and Ge, impurity trapping or the formation of small aggregates seems the likely fate of those interstitials that do not recombine with vacancies.

As the temperature rises vacancies in Si become mobile, with a low activation energy that depends on the charge state (Figure 6.15). In p-type Si the neutral vacancy, V^0, is the stable form; it becomes mobile at 150–180 K, with a characteristic migration activation energy of

T (K)	Species appearing	Species disappearing		
700				
		V-O		
600				
		C_S V-V		
		V-Al		
500				
	Al_i-Al_S, Ga_i-Ga_S	Al_i V-Sn	V-Sb	
			V-As	
400				
			V-P	
300				
	C_i	B_i; V-B		
		V-Ge		
200				
	(Pairs from V^0)	V^0		
100				
	(Pairs from V^{2-})	V^{2-}		

Figure 6.15 Appearance and disappearance of various single and paired defect species in the annealing of silicon after electron irradiation at liquid-helium temperatures. For clarity, acceptor species disappearing are shown to the left of the second column and donor species disappearing are shown to the right.

0.33 eV. Impurity-trapped vacancy centers emerge as isolated vacancies disappear. In n-type Si, V^{2-} is the stable form, moving at 70–80 K with a migration energy of 0.18 eV (Section 5.4).

Vacancy clusters have been observed too, although it is hard to know how representative are those that can be observed in epr. One interesting cluster comprises (110) chains of vacancies with a spin of 1 (Corbett et al. 1976). These clusters apparently need oxygen for their formation. The most striking feature is that the splitting of the $S = 1$ state in zero magnetic field is determined by the magnetic dipole–dipole interaction between the two unpaired spins at the ends of the chain. The length of the chain (hence the number of component vacancies) can be deduced from the epr data. Similar (110) oriented centers are seen in diamond.

6.4.2 Radiation Damage in III–V Compounds

Radiation damage in III–V semiconductors is not as well understood as in Si or in more ionic crystals. This stems partly from a lack of unambiguous epr and optical data. Many experiments use deep-level transient spectroscopy (Section 4.3.4) to determine ionization energies and to decide whether electrons or holes are bound. Such data enable one to catalog defects rather than to identify structure (see e.g., Bourgoin 1984 and Newman 1984).

The III–V compounds show several differences from Si. First, the defects are more stable. This can be understood if ions tend to remain on their own sublattice, since jump distances in III–V materials correspond to second neighbor distances in Si. Second, the III–V compounds exhibit more lattice strain per defect than Si or Ge, although the mechanisms of strain are unclear; about one atomic volume per defect is observed, comparable to those in alkali halides. Third, there are strong orientational effects; there is a marked asymmetry in damage produced between [111] and $[\bar{1}\bar{1}\bar{1}]$ incident electron beams, reflecting a difference in structure (Figure 6.16) and in cation and anion masses.

Defects induced by electron or neutron damage cause carrier removal in n-GaAs. This appears to be an intrinsic effect, since the rate of removal depends on the fluence of electrons or neutrons, but not on the initial carrier concentration.

Both GaAs and GaP contain a range of electron traps after irradiation (Lang 1976). In GaAs at least one important hole trap is formed. These traps are usually labeled $E1, E2, \ldots$ for the electron traps, with ionization energies increasing along the sequence. Corresponding labels $H1, \ldots$ are given to the hole traps (see Appendix). These defects can act as very efficient minority-carrier traps, with capture cross sections of

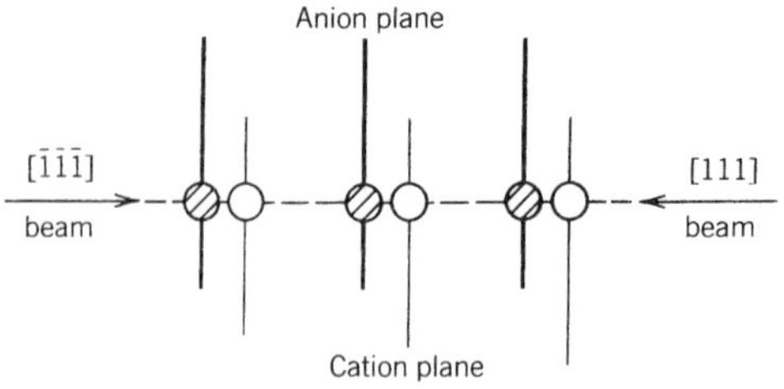

Figure 6.16 Asymmetry of the zincblende structure, showing how anions and cations are differently exposed to [111] and [$\bar{1}\bar{1}\bar{1}$] particle beams.

about 100 Å^2. They also show recombination-enhanced annealing in many cases. This phenomenon (Section 6.4.6) is interpreted as a local heating effect, in which recombination energy is efficiently converted into vibrational motion, aiding passage over a diffusion-jump barrier. This is a different mechanism from the local excitation process in F-center formation (Section 6.2.3), where there is a much lower barrier to defect formation in an excited electronic state.

6.4.3 Radiation Damage of Solar Cells and Devices

Solar cells have been most successful in special applications where low power and low voltage have been compensated by a special advantage of lifetime or light weight (Sze 1981). Solar cells for earth satellites have been especially successful, and are mainly based on silicon p-n junctions. The main source of degradation here is radiation damage, which reduces the minority-carrier lifetime and diffusion length. In the inner van Allen belt, a satellite encounters perhaps 10^4 protons/cm^2/sec at energies from 20 to 200 MeV; in the outer belt the protons have lower energies (1–5 MeV) but higher fluxes (10^7–10^8 protons/cm^2/sec) with, additionally, an electron flux (10^4–10^5 electrons/cm^2/sec with energies of 1–2 MeV). The effects of these particles can be minimized by a combination of techniques. A simple coverslip reduces the number of particles reaching the silicon, although this coverslip must be chosen to resist darkening under ultraviolet radiation. Low dislocation densities in silicon are an advantage. Surprisingly, Li doping of the base of p-n Si cells improves their radiation tolerance, especially their recovery. This appears to be associated more with indirect effects on oxygen impurity than with the direct effect of the Li itself.

Solar cells are not the only silicon devices to deteriorate in the radiation environment experienced by spacecraft. Anomalous behavior of satellite components is seen, for example, switching on or off unexpectedly. These anomalies, the so-called single-event upsets, are of two main types. Soft errors allow a device to continue to function, but it processes false information because a memory store has been changed. Hard errors prevent the proper functioning of a device, for example, a logic circuit may be stuck in one mode. The malfunction can be due to charges distorting electric fields in the devices or from effects on minority-carrier lifetimes. It might result from a heavy cosmic-ray nucleus or perhaps from a proton that causes a nuclear reaction and a recoil fragment. Clearly, the difficulties in finding a strategy to avoid the problems increase with the complexity of the integrated circuit involved, and also become worse as the size of the components of the circuit is reduced. Indeed, microminiaturization has led to similar hard- and soft-error problems for entirely terrestrial use. Traces of uranium, ubiquitous on earth, and tenacious radioactive Xe, used in tests of encapsulation, have both caused problems.

6.4.4 Radiation Techniques in Semiconductors

The semiconductor industry is dominated by devices based on crystalline silicon. Throughout the extensive developments and new approaches, the industry has maintained a consistent demand for accurate doping. Rapid, cheap doping, fully controlled as regards species, and precisely placed in a chosen complex pattern on a very fine scale, is a continuing demand. Success has made large-scale integrated circuits practical, with the advantages of the low power consumption, high performance, and cheapness.

Two methods are widespread in impurity doping. One is high-temperature diffusion. Donors and acceptors may be diffused in at 1000–1300°C within tens of minutes. However, there are problems associated with this technique. First, the edge of the diffused zone is not very sharp. This is a disadvantage for some specific devices, for example, for so-called hyperabrupt Schottky diodes (Figure 6.17), which exploit the variation of capacitance with voltage for frequency multiplication and/or automatic frequency control circuits. The blurred edge to the diffused zone also sets limits to the complexity of the microcircuitry that can be achieved. The second disadvantage of diffusion is that each time the device is processed (e.g., metallization to put on contacts or conducting strips, or passivation by oxidation) there can be further diffusive motion.

Ion implantation, in which the dopants are injected using an accelerator, has important advantages [see, e.g., Dearnaley et al. (1973)].

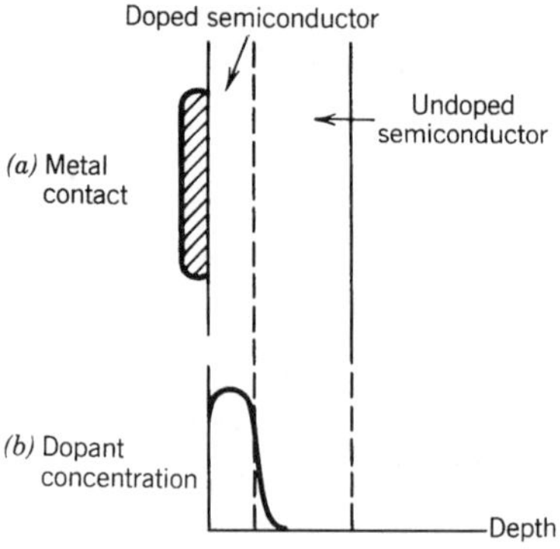

Figure 6.17 (*a*) Typical hyperabrupt Schottky barrier. In (*b*) the dopant concentration is shown.

Even at modest energies, useful penetration of silicon can be achieved; for example, the concentration of 30-keV boron ions peaks at a depth of between 1000 and 2000 Å (see Figure 6.18). The advantages are the relatively sharp edge to the dopant distribution, control of depth distribution by control of the incident energy, and control of lateral distribution by masking. A further advantage is dopant purity (isotopic purity if need be). Obviously, ion implantation has problems associated with the radiation damage caused by the implantation. However, this can be annealed out thermally at 600–700°C, when only limited dopant motion occurs. In special cases laser annealing (Section 6.4.5) may be used as an alternative.

A completely different, if somewhat restricted, method is important if uniform doping is needed. Transmutation doping exploits the reaction $^{30}Si + n \rightarrow {}^{31}Si + \gamma$, with thermal neutrons, which is followed by β decay

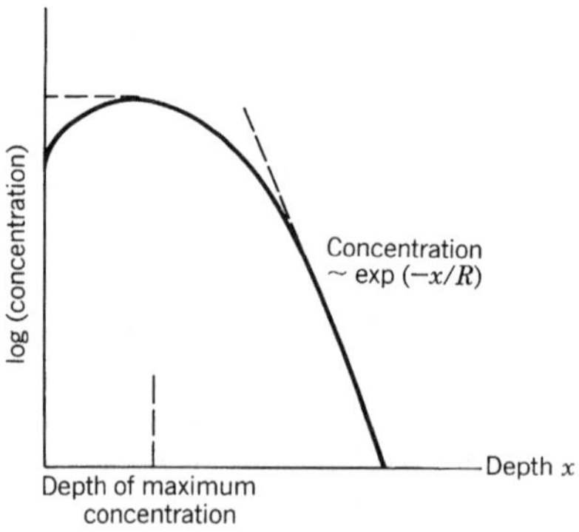

Figure 6.18 Distribution of implanted ions relative to the surface. The depletion near the surface occurs because the ions are fired in at finite energy. At large depths x the concentration falls off as $\exp(-x/R)$, where R is a measure of the range.

of ^{31}Si to ^{31}P. Some radioactivity is induced at the same time, but this decays rapidly to well below acceptable levels.

6.4.5 Laser Annealing

Semiconductor devices are prepared by a complex series of dopings and heat-treatment sequences. However, the two principal aims of compactness (and hence very well-defined boundaries between regions) and of freedom from defects introduced during implantation, are fundamentally incompatible. Diffusion doping leads to fuzzy boundaries; implantation gives sharp boundaries that are smeared by thermal anneals to remove damage (Section 6.4.4).

Laser and electron-beam anneals offer a real possibility of avoiding the diffusion smearing of implanted material. There are three main forms of anneal:

1. *Pulsed Laser Anneal.* Short laser pulses (perhaps 10 nsec long) locally melt the implanted device; in the subsequent rapid epitaxial resolidification the damage is eliminated. Repeated pulses over adjacent regions cover the area of device to be treated. The energy per pulse, typically 1.5–2 J/cm^2 of above bandgap light, is carefully adjusted to be above the minimum for effective annealing and below the minimum for damage. Since the melt is achieved by massive transient electronic excitation, it may differ in some ways from the normal thermally melted material. The distinctions are a subject of scientific controversy but not of major practical importance.

Since molten Si is significantly more dense than the solid, one can imagine the pulse causing melting to a depth of 1 μm (at which stage the liquid surface has receded about 1000 Å below the surrounding solid). The resolidification occurs in about 1 μsec, with the solid/fluid boundary moving out at some 500 cm/sec.

2. *Continuous Laser Anneal.* Here a scanned laser beam raises the surface to just below the melting temperature, but (unlike the pulsed beam) does not supply the latent heat of melting. Annealing takes place locally, in the transiently hot solid. Energy inputs of around 270 kW/cm^2 are typical.

3. *Electron-Beam Heating.* Here a current of around 1000 A/cm^2 of 10-keV electrons is used to give the correct energy input. Both continuous (scanned) and pulsed forms are possible. Since electrons penetrate much farther than above-bandgap light, a low angle of incidence is needed.

Laser and electron-beam anneals are still mainly novel science, rather than technological practice. Both cost and technical difficulties restrict current applications to special cases. The advantages are:

1. One can treat local regions of devices that are unstable at high temperatures, such as Li-drifted silicon or germanium. Likewise, anneals can be carried out in an atmosphere that prevents loss of volatile constituents (like As in GaAs, or like Hg or Cd in $Cd_{1-x}Hg_xTe$) without the need for coating.

2. One can produce situations that cannot be obtained conventionally. Thus pulsed anneals allow one to incorporate dopants substitutionally to well above normal equilibrium solubilities (see Table 6.5), a possibility that is useful in solar cells, ultraviolet photodetectors, and junction field-effect transistors. Also special profiles can be achieved. For example, if one implanted Si with both P and Al, the P moves more readily from the liquid into the solid as the melt boundary moves to the surface, resulting in *p*-type Si (Al rich) on the surface side of *n*-type Si (P rich).

Against these advantages, there are problems associated with annealing of implanted solids. Resolution of laser annealing in space is limited to one or two optical wavelengths laterally. Pulse annealing produces structure in the doping profile owing to a complex mixture of factors involving the melt-front dynamics, concentration fluctuations, thermal inhomogeneity because of laser beam structure, sound-wave generation, and defects in the underlying solid. This type of undesirable structure may also be associated with local mechanical damage or with separation of silicon from a covering oxide layer and can have undesirable electrical effects.

6.4.6 Recombination-Enhanced Diffusion and Related Phenomena

High densities of carriers affect the rates of many processes in solids, including diffusion rates and defect production rates. This is especially important in systems that normally operate with high carrier densities, including some semiconductor lasers and solar cells used with light intensifiers. Systems like these may degrade very rapidly in operation, although there are no comparable effects when not in use, even at elevated temperatures.

The source of enhanced diffusion or defect production is the recombination energy, E_r, released by electron–hole annihilation, or by capture of a carrier at a defect. There are three main mechanisms [see, e.g., Stoneham (1981)]:

1. *Local Heating*. The recombination energy released is transformed into vibrational energy so that normal thermal vibrations have to supply only a part of the usual activation energy E_M for motion. The observed activation energy will be $E_M^* = E_M - (E_r - L)$, where L represents the losses because the recombination energy is only partly converted into atomic motions of the right sort. Indeed, unless the reaction coordinate (the atomic motions giving the diffusion jump; Section 2.4.3) is a localized mode or a local resonance, the losses L may be considerable. The local-heating mechanism has been verified for Al interstitials in silicon and for hydrogen trapped at silicon vacancies in SiC.

2. *Local Excitation*. The recombination energy excites the defect into a different electronic state in which thermal activation energy is reduced. This is most easily observed when there is an actual change in charge state of the defect, since the resulting state may be stable for some time; vacancies in Si and anion vacancies in alkali halides show this effect. However, a transient excitation to another electronic state with the same charge can suffice. This is possibly the case for interstitials in diamond and is observed also in photochemical damage in KCl (Section 6.2).

3. *Bourgoin–Corbett Mechanism*. This mechanism requires the defect to occupy different sites in its different charge states. For example, when the + state $\oplus$ occupies site 1, and the neutral state $\circledcirc$ site 2, then alternate capture of electrons and holes moves the defect through the lattice via alternate sites 1 and 2:

$$\oplus_1 + e \rightarrow \circledcirc_2; \qquad \circledcirc_2 + h \rightarrow \oplus_1, \qquad \text{etc.}$$

This process appears to underly rapid self-interstitial motion in silicon and has been observed for the boron interstitials in silicon, where the motion involves the +, 0, and − charge states and is thus slightly more complicated.

While these three types cover most of the phenomena observed, there are situations where a different description is useful. One such case is the degradation of the (Zn, O) pair center in GaP, where prolonged excitation causes some of the original pairs to separate. These donor–acceptor pairs can be monitored uniquely by their luminescence energies, which depend on separation in a well-defined way (Section 5.3.2). The easiest way to understand the observed behavior is to assume that one of the substitutional components (Zn or O) becomes highly excited, moving into a mainly interstitial position, that the host atom (Ga for Zn, P for O) moves into the site vacated, and finally that the displaced impurity finds another substitutional site, most probably the one from which the host atom has moved. In a situation like this it is not clear whether the route by which the impurity is first moved from its site involves local heating or excitation.

Finally, we mention that dislocation climb (Section 3.7) can be caused by recombination and affects the performance of devices such as light-emitting diodes (LEDs). In binary compounds like III–Vs two defects are needed for jog motion, just as defects on both sublattices were needed to form perfect dislocation loops in alkali halides (see Section 6.2.1(c)). In thermal equilibrium, one is unlikely to find both present to the same supersaturation. Petroff and Kimerling (1976) proposed that if, for instance, there was an interstitial Ga excess in GaAs (as is usual in growth), this would lead to the creation of As vacancies (and hence As interstitials) and to dislocation climb involving the Ga_i and the As_i produced (see Figure 3.27). This process can be enhanced by recombination, since both interstitials and vacancies, as inferred from dislocation and loop growth, respectively, are mobile under carrier injection. In alkali halides there is a parallel process, with anion interstitials generating cation interstitials in the damage process (Section 6.2.1(c)).

CHAPTER SEVEN

Properties of Surfaces

The surface of a crystal, which is where it makes contact with its environment, can be regarded as one of the most important solid-state defects. Mechanical properties, whether deformation, fracture, or friction, depend on the surface topography, that is, the character of roughness and the existence of any small cracks. Electrical properties are easily influenced by surfaces, especially in semiconductors, where miniaturization leads to many interfaces in relatively small regions. Both the benefits of chemical catalysis, which underlies the chemical industry, and the problems of general corrosion, which is enormously wasteful, involve processes at surfaces, although, as we shall see, the surface phenomena may not be rate determining. Given the many practical consequences of surfaces, it is not surprising that there is an immense literature on the physics and chemistry of surfaces. The literature, it should be said, ranges from the scholarly to the empirical, and is occasionally controversial.

It is useful to begin by imagining the physical appearance of a solid as it is examined under successively higher magnifications. We imagine this in five stages (Figure 7.1):

(a) $\times 1$. Even to the naked eye, many single crystals have very well-defined shapes. The characteristics are known as the crystal habit, which defines the types of facet commonly seen (Figure 7.1*a*).

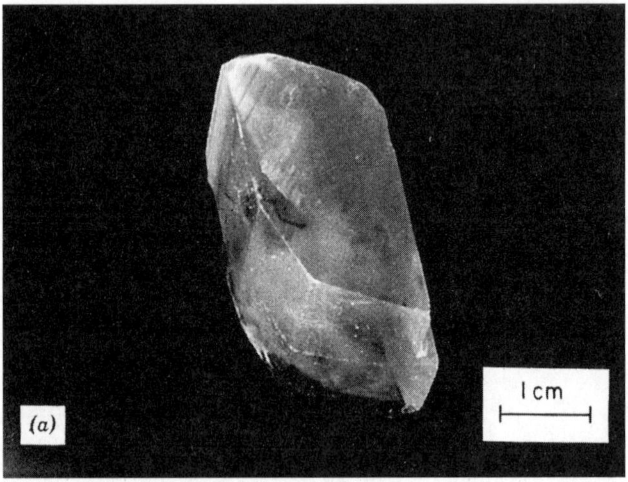

Figure 7.1(a) Surface features at various magnifications: (*a*) Facets of a crystal of copper sulfate (grown by Robert Hayes).

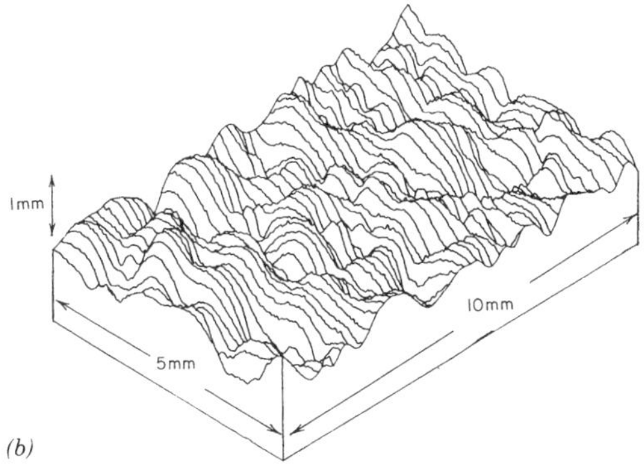

(b)

Figure 7.1(b) Surface features at various magnifications: (*b*) Surface topography of a steel plate, after sandblasting (Thomas 1982).

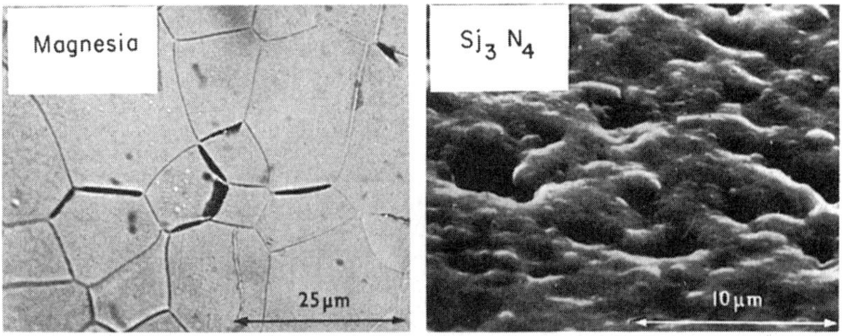

(c)

Figure 7.1(c) Surface features at various magnifications: (*c*) Grain boundaries in polycrystalline magnesia (R. W. Davidge, private communication). Note the cracks produced by plastic deformation. This may be contrasted with reaction-sintered Si_3N_4, which shows surface pores.

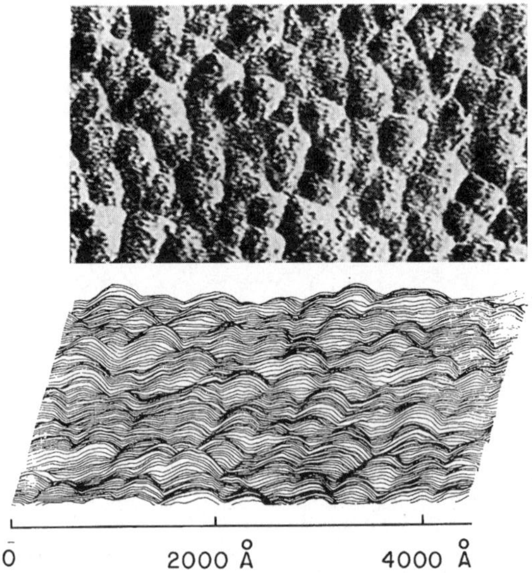

(d)

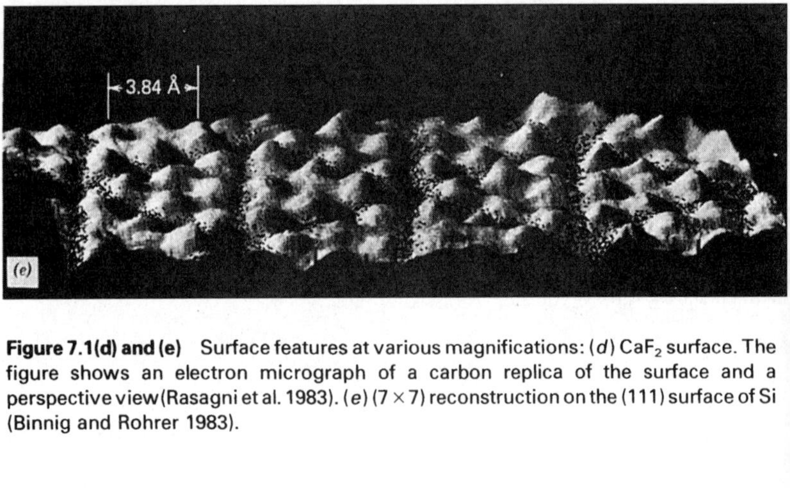

Figure 7.1(d) and (e) Surface features at various magnifications: (*d*) CaF_2 surface. The figure shows an electron micrograph of a carbon replica of the surface and a perspective view (Rasagni et al. 1983). (*e*) (7 × 7) reconstruction on the (111) surface of Si (Binnig and Rohrer 1983).

We shall talk of facets and faceting whenever the surface consists of crystallographic planes of high symmetry.

(b) $\times 10^2$. On this scale (Figure 7.1*b*) one can see surface roughnesses (which may depend on how the surface was cut or polished, or may be a result of the way it was grown) and whether it is a single crystal or a conglomerate of small crystals. If it is an aggregate of some sort, there will be areas where two microcrystals are firmly bonded, and there will also be pores of various sorts through which gases and liquids can penetrate.

(c) $\times 10^4$. On this scale grain boundaries (Figure 7.1*c*) and dislocations become visible. At a grain boundary one has a misorientation between the two crystalline regions on either side. This misorientation is usually described as twist, or as tilt or as some combination of these (see Section 3.7). We can see a qualitative distinction between *polycrystalline* materials, where the components are joined over macroscopically intermittent regions and *polygrannular* materials, where there is essentially continuous contact of misoriented regions. Typical densities of dislocations in nonmetals are $10^6\ \text{cm}^{-3}$ (and much higher if mechanical deformation has occurred), corresponding to dislocation spacings of about 10 μm. Specially prepared semiconductors may be dislocation free.

(d) $\times 10^6$. On this scale (Figure 7.1*d*) some of the atomic structure may become evident, but is often hard to discern. One possible feature is *patchiness*. For example, on a nearly (111) surface of NaCl, there may be patches where only anions are visible in the outer layer and patches where there are only cations visible. There may be similar areas, tens or hundreds of angstroms across, where atoms or molecules (intrinsic or extrinsic) are adsorbed on the surfaces. Likewise, there may be precipitates of other phases.

(e) $\times 10^8$. The atomic structure is clearly visible (Figure 7.1*e*). Almost all the bulk defects have analogs that can be seen. Individual impurity atoms will be present, often in concentrations enhanced over the bulk levels. This is known as impurity segregation. Point defects will be discerned, especially vacancies and adsorbed atoms (adatoms) or molecules, and may occur in any of several charge states. Thus, on MgO (100) surfaces, the F_s^+ center has been very clearly identified, where (like the bulk F^+ center: Section 5.1.1) an electron is trapped at an anion vacancy. Adsorbed species may be impurities, or may be intrinsic, like O_2^- on MgO. Line defects, for example, ledges or steps, will be clear, and may themselves contain point defects just as a bulk dislocation may show kinks (see Figure 3.24). Finally, the ions of the surface itself will have moved slightly from the sites expected from the perfect bulk crystal. They will show *relaxation* (i.e., a change in mean separation of the outer ionic planes) and *rumpling* or *puckering* (i.e., variations in the relative normal displacement of ions in the outer plane). Some surfaces may even show extensive reconstruction, changing their overall periodicity.

These observations have referred mainly to spatial features observed as one moves across a surface. We are also concerned with the effective depth of a surface. For some properties the surface is merely an atomic layer deep. For neutral or isovalent defects and for adatoms, either static or moving over the surface, the surface can be thought of as a single ionic layer. This, perhaps unexpectedly, is even true for the Madelung potential; the bulk value is reached quite accurately by the second atomic layer.

For a second group of properties one may need to consider perhaps tens of layers of atoms. Exciton behavior in semiconductors is modified over such ranges, and also the rumpling falls to zero within a few layers. Electron tunneling is possible through layers tens of angstroms thick. In radiation damage, focused collision sequences started within the outer few layers may reach the surface, giving particle emission from the surface (sputtering) along preferred directions (Itoh 1976). Coatings to protect a surface against its environment may only need to be tens of angstroms thick.

For a third group of properties, even deeper regions may be involved. The most important effects concern enhanced defect concentrations near the surface, space charge (Section 7.5) being a good example. In the bulk, electrical neutrality must be maintained when defects are generated thermally. This condition is relaxed at the surface, and hence a region occurs, usually tens or hundreds of atomic layers deep, with higher concentrations of defects or impurities. This gives rise to what we shall call *near-surface* effects (Section 7.5; see also Section 3.1.1). There is an electric field normal to the surface from the different spatial distributions of positively charged and negatively charged defects, and this affects electrical contact phenomena. In addition, the vacancy concentration is generally larger in the space-charge region at the surface, so that diffusion parallel to the surface by a vacancy mechanism will be enhanced. Finally, we note that dislocations and grain boundaries that intersect the surface may produce fast-diffusion routes into the bulk (see also Section 3.7).

One major concern in surface physics is cleanliness. The older standard methods of polishing and electrochemical treatment leave highly imperfect and contaminated surfaces. Really clean surfaces are rarely achieved, although two techniques for cleaning surfaces often give acceptable approximations. One is cleavage in an ultra-high vacuum ($\lesssim 10^{-10}$ torr). Even here, a monolayer of residual gas may be adsorbed in a few hours. It is customary to distinguish between *physisorption* and *chemisorption* of impurities, depending on whether purely physical interactions (e.g., van der Waals forces, or dipolar polarization by surface charges) or chemical reactions are involved (Section 7.7).

The second technique used to clean surfaces is erosion by bombardment with argon atoms, again in ultra-high vacuum. Here impurities in the bulk may diffuse and recontaminate the surface during the thermal anneal required to remove damage caused by the argon beam. Several bombardment/anneal cycles may be needed until an acceptably clean surface is achieved. The manifest problems of possibly dirty experimental conditions and the complexities of theory for rumpled or reconstructed surfaces have led to many misunderstandings between theory and experiment. In making comparisons, it is well to remember Table 7.1. Unless otherwise stated, we shall only discuss experiments on properly cleaned surfaces and theories which include serious estimates of surface relaxation, rumpling, or reconstruction.

In this chapter we shall discuss some of the important properties of surfaces and the processes that occur at or near them. Some of the consequences of the discontinuity at a surface are obvious. Many of the attractive interactions (or bonds) in a perfect crystal are eliminated when the crystal is cut in two and the halves separated to form surfaces. One expects the energy cost per surface atom to be comparable with the cohesive energy, that is, up to a few eV per atom (a few J/m^2), although much smaller values are found when only weak van der Waals forces are involved. We discuss surface energies in Section 7.1. The fact that the interatomic forces are now imbalanced (they were, of course, in equilibrium in the perfect solid) means that there will be surface relaxation. Indeed, just as for liquids, there will be a surface tension tending to reduce the surface area.* As we have mentioned already, the precise surface geometry may involve rumpling and inward or outward relaxation, as well as a simple change of scale. This surface reorganization is described in Section 7.2. Not only are there altered forces at the surface, but there are also altered force constants that restore surface atoms to their equilibrium

*We shall not use the phrase *surface tension* again, since different usages conflict. In Section 7.1 we shall define surface energy (σ) and surface stress (γ). Early work called σ surface tension; recent work labels γ surface tension. For a liquid there is no confusion, since the two cannot be distinguished, so $\sigma = \gamma$.

Table 7.1 Real and Unreal Surfaces[a]

Uncleaned surface	Very dirty surface, that is, surface layers of oxides, grease, etc.		
Surface cleaned by polishing or electrochemically	Dirty surfaces, that is, significant numbers of adsorbed molecules and perhaps oxide layers		Experiment
Ar-bombarded or vacuum cleaved surface in ultra-high vacuum	Possibly clean surface	Realistic comparison may be made here between theory and experiment	
Relaxed, polarized, and reconstructed surface	Possible model of real surface		Theory
Unrelaxed perfect surface	"Science fiction"–ideally terminated surfaces do not exist		

[a]This table emphasizes the distinction between real surfaces and the experimental and theoretical idealizations often met in practice. So-called cleaned surfaces are not necessarily clean; ideally terminated models are not realistic. The closest contact of theory and experiment corresponds to experiments on surfaces vacuum cleaved or cleaned by Ar bombardment, and to the theory of relaxed, reconstructed surfaces.

positions after mechanical perturbation. The surface waves, which show this most clearly (Section 7.3), vary from the long-known Rayleigh waves of continuum elasticity to the complex waves in piezoelectric crystals, which, because they can be controlled by applied electric fields, have led to the important field of acoustoelectronics.

In Section 7.4 we shall consider the electronic states associated with surfaces. Some are properties of the defect-free surface; there may be resonances, altering the density of band states locally, or bound electronic states at energies in the forbidden bulk gap. Other surface states, usually more important, arise from surface defects. These are widely believed to play an important part in catalysis, although many other processes are vital too (Section 7.9). Surface defects may also be the cause of the rapid recombination of electrons and holes at semiconductor surfaces.

Both impurity atoms and defects exhibit surface diffusion (Section 7.6). This is especially important in changing the topography of a surface, for example, recovery to eliminate grooves, or in the development of microstructure. In particular, diffusion usually controls the nature of ceramics, for example, the oxides for high-temperature applications. In making oxides, their high melting points can be a major problem for practical purposes. It is common to make a powdered material (if necessary involving several oxides) and then *sinter* it to form the appropriate ceramic. In sintering, heating to a temperature below the melting temperature of the major component allows surface diffusion to occur so as to eliminate pores and voids and to consolidate the powder into a coherent solid. Surface diffusion can involve several quite different processes. The motion on or in the outer atomic layer is one form. Another involves the enhanced vacancy concentrations near to surfaces. These, and related, near-surface effects (see above) are surveyed in Section 7.5.

In Section 7.8 we discuss the various processes involved in oxidation and, to a lesser extent, corrosion. These processes involve adsorption and diffusion, and electric fields generated as oxidation proceeds, the way the grain structure emerges, and the way in which the growing oxide responds to the stresses generated in its development.

7.1 SURFACE ENERGY AND SURFACE STRESS

The surface energy σ is a measure of the energy per unit area needed to create new surface. If a large crystal of N atoms or molecules has a surface area A [we presume for the moment that only one type of surface is involved, e.g., (100) surfaces in alkali halides], then there will be two main terms in the total crystal energy. One is NE_{coh}, where E_{coh} is the bulk cohesive energy per atom or molecule, that is, exactly the value one would expect in an infinite solid. The other term, σA, gives the surface correction to the energy. Since both E_{coh} and σ involve the forces binding atoms or ions in the solid, we may anticipate that they will show similar trends from crystal to crystal.

The surface energy σ quantifies the cost in creating new surface, that is increasing the surface area by arranging that more atoms become part of the surface. In contrast, the surface stress, γ, measures the cost of creating a new surface by stretching the existing surface, that is, changing the area at constant number of surface atoms. They are related by the equation

$$\gamma = \frac{d}{dA}(\sigma A) = \sigma + A\frac{d\sigma}{dA} \tag{7.1}$$

For liquids, γ and σ are identical, since the atoms in the liquid state are free to adjust both numbers and positions to minimize the excess energy of the surface. In a solid, γ is strictly a tensor. However, this property has not proved important in experiments on surfaces.

Surface energies are measured in two main ways. The first involves controlled cracking to introduce a new surface. There is a problem of possible plasticity, since deformation beyond the yield point can generate dislocations as well as a fresh surface. One can therefore easily obtain values that overestimate the true surface energy. Clearly brittle fracture, rather than a dislocation-producing ductile fracture, is desired in such experiments. There are, in fact, several extra complications. One is that cleavage may create a metastable surface, that is, a surface of higher energy (Section 7.2). A related question arises from the suggestion that cleavage may proceed so fast that somehow an unrelaxed, unpolarized surface is produced. Careful enquiry shows that this is not so. Only in the fluorites are good estimates of σ from controlled cleavage available (Table 7.2). These agree very well with the predictions of theory using a shell model (see Section 2.2.3).

The second approach to surface energies uses calorimetric methods. Such values are not notably accurate. Available data for alkali halides (Figure 7.2) suggest that theory is probably more reliable than experiment and also verify systematic trends of σ with the cohesive energy E_{coh} from one material to another.

The idea that controlled brittle fracture, that is, without plastic deformation, can be used to measure surface energies may also be applied to put a limit on the stress a brittle solid can withstand. The well-known Griffith

Table 7.2 Comparison of Calculated and Measured Surface Energies (J/m²) of Fluorites (Tasker 1980)

Crystal	Theory	Experiment
CaF_2	0.472	0.51, 0.45
SrF_2	0.407	0.42, 0.356, 0.26, 0.465, 0.43
BaF_2	0.349	0.350, 0.280

criterion uses the idea that a crack will propagate so long as the increase in surface energy from new surface generated is less than the decrease in stored elastic strain energy. Suppose there is a preexisting crack of elliptical cross section, having a width (i.e., a major axis) 2δ and a much smaller separation of the surfaces. The strain energy has the form $\alpha\delta^2\tilde{\sigma}^2/c$, where $\tilde{\sigma}$ is the stress, α is a geometric constant, and c is an elastic constant. The surface energy is $4\sigma\delta$ (note σ, here the surface energy,

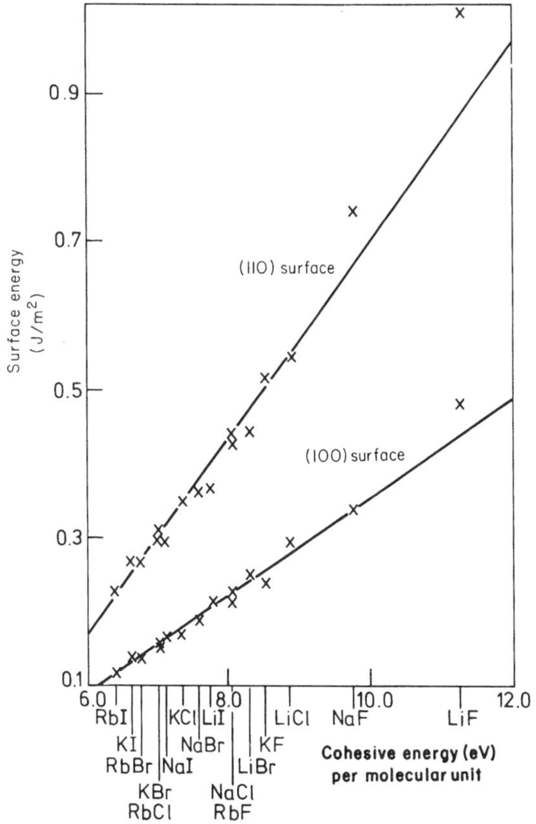

Figure 7.2 Correlation of experimental cohesive energy and predicted surface energy of various alkali halides [after Tasker (1980)]. Both energies are dominated by Coulomb interactions. Note that, since the surface energy is per unit *area* (not per unit cell on the surface) and the cohesive energy is per molecular unit, the lines do not pass through the origin.

differs from $\tilde{\sigma}$, here the stress). The failure stress, $\hat{\sigma}_f$, which is the maximum stress that fails to extend the crack, is given by

$$\frac{d}{d\delta}[\alpha\delta^2\tilde{\sigma}^2/c - 4\sigma\delta] = 0, \quad (7.2)$$

so that the failure stress has the value $\sqrt{(2/\alpha)\sigma c/\delta}$. From the same equation we see that the longest crack stable against extension in a stress $\tilde{\sigma}$ has length $\delta_f = (2/\alpha)\sigma c/\tilde{\sigma}^2$. These results and their generalizations have very wide applications in engineering design.

Surface energies determine many important crystal properties, partly because both σ and γ [Equation (7.1)] refer to specific crystal planes, and so vary from face to face. The first property affected is crystal habit, that is, the crystal faces that are found. Since it costs energy to create a surface, one expects the lowest-energy surfaces to dominate. This is seen whenever conditions are such that thermal equilibria can be established. In MgO smokes, formed by burning of Mg, one does indeed find mainly (100) faces, whereas, in diamond, the (111) faces are common. In polycrystalline samples one expects the pores and voids to favor low-energy surfaces increasingly, with more heat treatment. However, many crystals grow in a way that precludes equilibrium. Often (and especially in solids crystallized from the solution at low temperatures) crystal growth takes place far more rapidly at the kink sites of screw dislocations. In such cases kinetic considerations dominate and the relative energies of different surfaces are secondary.

The corollary to the point that surfaces of low energy should dominate is that surfaces of very high energy should not be seen. One expects high energies for surfaces that have high concentrations of the same charge, like the (111) faces of NaCl. This is a slight oversimplification, although the correct picture is easily described [Tasker's rule (Tasker 1983c)]. In a crystal, moving in from the surface, one encounters a series of atomic planes, possibly *AB AB AB* ... or *ABC ABC ABC* .... If the repeating unit (*AB* or *ABC* here) has a dipole moment normal to the surface, the electrostatic part of the surface energy is infinite. The point is that, in such cases, there is an electric field throughout the whole crystal, and the sum of its interactions with layer after layer of dipoles is infinite (or strictly infinite for an infinitely deep crystal). The result is summarized in Figure 7.3. Despite the general and clearcut nature of the argument against such surfaces [e.g., the (111) face of MgO], experiments often claim to find them. The explanation in such cases is almost certainly electronic screening due to carriers localized near the surface, or impurity adsorption, or

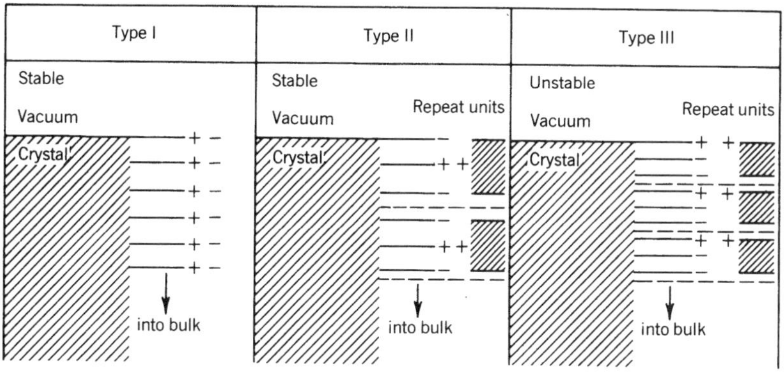

Figure 7.3 Classification of surfaces by surface energy. Unstable means that surface energy is infinite; stable indicates finite values. For type I, there is no dipole moment normal to the surface in any atomic layer parallel to the surface of the unrelaxed crystal [e.g., the (100) surface for NaCl]. For type II, there is no dipole moment normal to the surface in any repeat unit [e.g., the (111) surface of CaF_2 that terminates in F]. For the unstable type III, the repeat unit has a net dipole moment normal to the surface [e.g., the (111) surface of CaF_2 terminating in Ca, or the (111) surface of NaCl]. Repeat units are separated by broken lines for types II, III.

surface defects, or the fact that the surface is constructed of small facets of stable faces, or of patches of opposite charge [e.g., O or Mg patches outermost on a (111) face of MgO] that are present to avoid divergence in the energy.

A second point is that one expects cracks to propagate along a path of low energy. Cleavage should take place on low-energy planes. This is clearly so for mica and for graphite, where there is a spectacular anisotropy, with very low-energy interplanar cleavage. For diamonds, the (111) planes cleave easily, a feature exploited by the diamond industry. In contrast, diamond (100) faces are difficult to cleave, and the machining necessary for diamond polishing typically wastes some 5% of the diamond. The III–V semiconductors cleave most easily on (110) faces.

Cleavage in polygrannular materials can take place along grain boundaries (intergrannular) or through the grains themselves (transgrannular). The mode depends on the relative surface (E_s) and grain-boundary (E_{gb}, Section 3.7) energies. Fracture should be increasingly transgrannular along the sequence LiF, NaCl, Al_2O_3, ZrO_2 since the grain-boundary binding energies $2E_s - E_{gb}$ increase.

One consequence of a finite surface energy is that a solid cannot be ground up into an infinitely fine powder. This can be seen most easily by realizing that, for fracture to occur, there must be a strain energy E_{strain} in one of the existing particles that exceeds the energy E_{fract} needed for fracture. However, if the particle can deform plastically, with a lower energy E_{plast} than that needed for fracture, then the particle will merely deform and not fragment. We now argue that for sufficiently small particles, deformation will always occur first. If the particle has radius R, the energy for fracture will be of order $2\pi R^2\sigma$, with the factor 2 because two surfaces are created. Plastic deformation, however, occurs when the strain ϵ reaches a critical value ϵ_{crit}. There will be a corresponding stress $c\epsilon$, where c is an elastic constant, and the strain energy per unit volume will be $\frac{1}{2}c\epsilon_{\text{crit}}^2$. Thus plastic deformation needs an energy of order $\frac{4}{3}\pi R^3 \cdot \frac{1}{2}c\epsilon_{\text{crit}}^2$, which will always be less than the cleavage energy $2\pi R^2\sigma$ for $R < 3\sigma/c\epsilon_{\text{crit}}^2$.

Our final points on surface energies concern solid surfaces in contact with liquids. One example is the mode of growth from the melt. It is found that the way a crystal grows from the melt depends on the product of two factors, namely, a material-dependent entropy of melting and a crystallographic (structural) term. When the product is small, rounded surfaces are formed, nucleating without problems. For most nonmetals, however, the product of these factors is large, specific facets are favored, and surface steps will usually determine growth rates. A second example involves liquid drops (and their equivalent solid forms) on surfaces, for example, particles of Pt on Al_2O_3 used as a catalyst (Section 7.9). The droplet shape in such cases is determined by a balance between the surface energies (not surface stresses) of the three types of interface [insulator–metal (*im*), insulator-vapor (*iv*), metal–vapor (*mv*)]. The critical features (Figure 7.4) are the *wetting angle* θ and the *work of adhesion* w_a:

$$\cos\theta = (\sigma_{iv} - \sigma_{im})/\sigma_{mv} \tag{7.3a}$$

$$w_a = \sigma_{mv} + \sigma_{iv} - \sigma_{im} \tag{7.3b}$$

Measurements of θ and w_a have been surveyed by Naidich (1981).

So far we have concentrated on surface energies σ. The surface stress determines rather different properties, and these are far less conspicuous. First, it determines the lattice parameter in microcrystals. Just as the surface tension in liquids produces a pressure inside liquid droplets, so does the surface stress in a solid. The pressure is $2\gamma/R$ for a sphere of radius R (faceting causes only minor changes). If the compressibility is β, there is a lattice parameter change $\Delta a/a$ given by

$$\frac{\Delta a}{a} = -\frac{2}{3}\frac{\gamma\beta}{R} \tag{7.4}$$

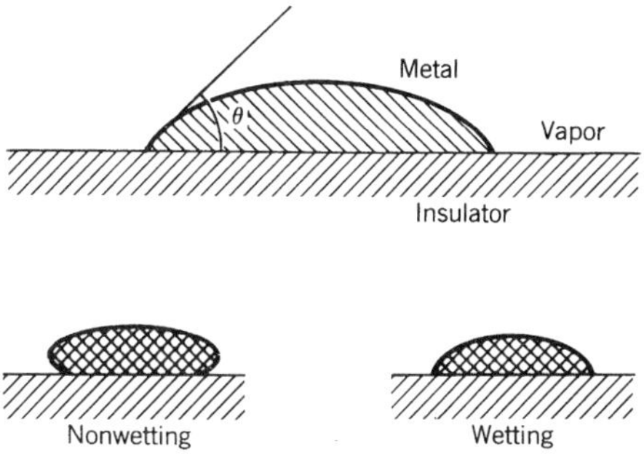

Figure 7.4 Wetting angle θ for a molten metal droplet on an insulating substrate. $\theta > \pi/2$ (e.g., most metals on MgO) means nonwetting; $\theta < \pi/2$ (e.g., most metals on TiO_2) means wetting.

In practice, surface defects and adsorbed species make interpretation difficult, for they can even reverse the sign of γ. While σ is generally positive (it costs energy to create a surface), γ can have either sign. Apparent negative surface tensions (although positive surface energies) can arise when impurities generate forces that favor a larger lattice parameter, or when adsorbed dipoles like H_2O are aligned so as to repel each other. A second consequence of the surface stress is that the pressure induced by the surface stresses can affect the formation energy of bulk defects which cause volume changes. However, it is only for the very smallest crystals that this is significant. In general, in considering the effects of stresses or fields near surfaces, one should not forget the random stresses and strains present in any bulk crystal. Indeed, the surface and bulk may have effects hard to separate. Nor should one forget that surfaces cleaned by ion-beam methods may have residual surface damage which induces stress.

7.2 SURFACE RELAXATION AND RECONSTRUCTION

The simplest model of a surface is one in which the solid merely terminates, leaving all atoms or ions in the positions they occupied in the bulk. This so-called ideal surface does not exist. It is merely a convenient reference from which to measure displacements of ions in the outer few layers of atoms. We may distinguish several different types of displacement normal to the surface, some mentioned previously. The first is *surface relaxation*, which decreases the interplanar spacing of the outer layers. The second is *rumpling* (also known as *puckering* or *buckling*). For polar solids, rumpling is measured by the amount the anions move in or out relative to the cations. Generally, for example, in both II–VI and III–V compounds, rumpling is positive, that is, the anions move outward. The third is a change in the periodicity of the surface, so that, for example, the displacements and surface structure reappear every N ideal repeat units in one direction on the surface and every M ideal repeat units in the other. This gives an $(N \times M)$ *reconstructed surface*.

Another form of reconstruction is faceting, where a particular surface may be unstable against forming adjacent regions of two or more surfaces terminating in different crystal planes and having lower surface energy. This has been seen for the (001) face of TiO_2.

The experimental determination of atomic positions at surfaces is harder than for the bulk. There are two main reasons for this. One is the problem of surface contamination, since the impurity concentration must be much less than a few percent of that equivalent to monolayer coverage. The second is that one must measure only the outer few layers, without being misled by the underlying bulk atoms; hence very low penetration is needed. X-rays are not suitable using normal methods (for $\lambda = 1$ Å and atomic number 20 the penetration depth is ~0.2 mm) and electrons are used instead. In one technique high-energy electrons (10–15 keV) are directed at a glancing angle at the specimen, giving shallow penetration depth, and the diffracted beams provide information about the structure of the outermost layers of atoms. This technique is known as reflection high-energy electron diffraction [RHEED, (Prutton 1975)] and reveals facets on surfaces as well as atomic arrangements.

The usual technique for studying the structure of surface layers involves low-energy electron diffraction (LEED). Electrons of energy 20–500 eV impinge on the surface near normal incidence and the diffracted beams may be accelerated to ~2 keV, sufficient to excite a fluorescent screen for detection. Both the penetration depth and the escape depth of the electrons are so small that they sample only atoms near the surface. However, the theoretical analysis of LEED data is much more difficult than that of X-ray diffraction. With X-rays, the atomic scattering cross sections are so small that it suffices to describe interaction with the crystal as a coherent superposition of single scattering events from each atom. This so-called kinematic theory of diffraction cannot be applied to LEED. Here the intensity scattered from an atom is almost equal to the incident intensity, leading to multiple scattering. We now require a dynamical theory of diffraction that takes into account a number of coherent scattering events for each electron (Pendry 1974).

In almost all cases studied, the actual displacements of surface atoms from an ideal termination are small, that is, less than 10% (indeed often less than 3%) of the nearest-neighbor distance. Relaxations that change interplanar spacings are found with either sign. While metals lie outside the scope of this book, it is noteworthy that they can show relaxations that are inward [e.g., fcc (110) or bcc (100) faces], outward [e.g., (111) of Al] or negligible [e.g., (001) of fcc metals]. For cubic oxides, the relaxation is very small (Prutton et al. 1979; Welton-Cook and Prutton 1980). This is confirmed by calculations on oxides (and halides), which also make the point that distortions in the surface region are important to greater depths when the anion and cation sizes differ greatly.

Rumpling is more systematic for rather general reasons noted first by Tasker. Underlying the rule that rumpling is positive (i.e., anions out, cations in) is the polarization of the outer layer of ions by the electric field of the ions in the second layer. It is simplest to use the shell model of Section 2.2.3 and to remember that LEED monitors the positions of the positively charged cores; while the electron energies are low, they are nevertheless sufficiently high to be scattered mainly by the cores. Consider a (100) surface of NaCl. Under each Na^+ ion lies a negative Cl^-, whose electric field pulls the (positive) core inward. Under each Cl^- ion lies a positive Na^+ ion, whose electric field pushes the (positive) core outward. If the shells remain unmoved (as will be the case, since the shells respond directly to the strong constraints of short-range repulsion), the net effect is that anion cores move out and cation cores move in, that is, positive rumpling occurs. As formulated, this argument is natural for ionic crystals. In fact, the same behavior is seen for the (110) faces of the fourfold coordinated II–VI and III–V semiconductors, although it is often described differently. Thus the ionic surfaces rumple to polarize in an optimum way. The polar semiconductors like CdTe or GaP transfer charge from cation to anion to give buckling, with the cation moving in to a more planar sp^2 configuration (see Section 8.4), and the anions moving out (Figure 7.5). In a band picture, the energy is lowered as the half-filled band of singly-occupied dangling bonds splits on distortion, giving a full band of doubly-occupied anion dangling bonds and an empty band (see Section 7.4).

Semiconductor surfaces have received special attention (Haneman 1982), both because of their technological importance and because of the availability of high-quality specimens. The importance of directed covalent bonds (Chapters 1 and 8) leads to extensive restructuring and to distortion over a relatively large surface region. For the nonpolar semiconductors (notably Si and Ge) the buckling mechanism described remains valid, but other mechanisms prove more important, and a variety of reconstructions is seen. Studies have concentrated on the low-index faces, notably (100), (110), and (111).

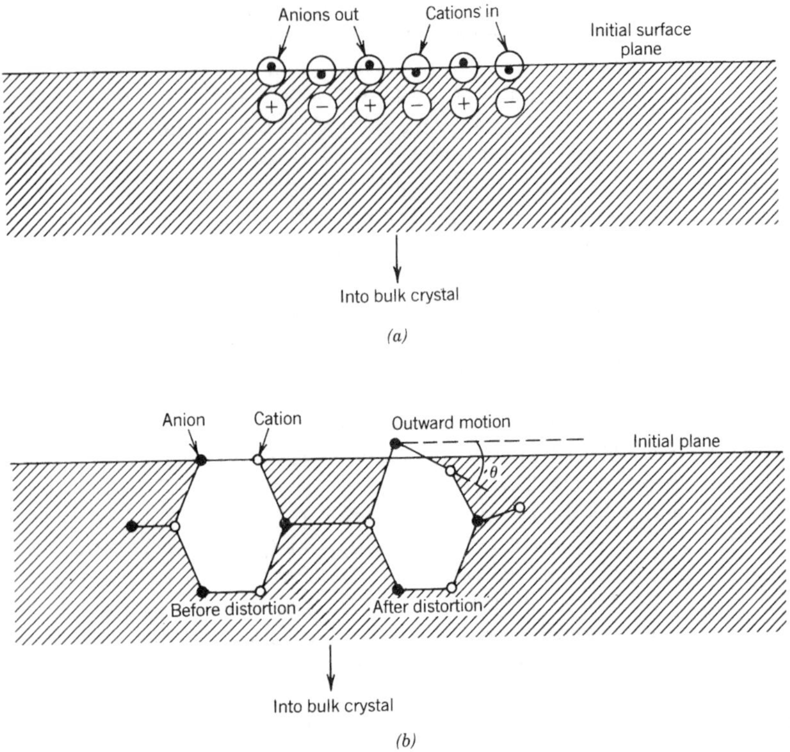

Figure 7.5 Surface distortion of polar crystals. (*a*) Rumpling of the (100) NaCl surface. Only the core motion is shown; relaxation altering mean layer spacings is omitted. (*b*) Puckering (buckling) of the (110) zincblende surface. For II–VI and III–V compounds the bond rotation (or tilt angle) θ is usually around 30°.

The (111) plane is a cleavage surface for diamond and for Si. When silicon is cleaved at 77 K in ultra-high vacuum, a (2×1) surface structure is formed. This structure appears to be metastable; when the crystal is heated in the range 200–420°C for 15 min, the surface changes irreversibly to a (7×7) structure, which is stable on cooling to room temperature. A metastable (1×1) structure is also observed after quenching from ~900°C. Germanium also shows a metastable (2×1) structure of the (111) surface after cleavage. Heating to 400–700 K leads to a (2×8) structure, which appears to be the equilibrium form.

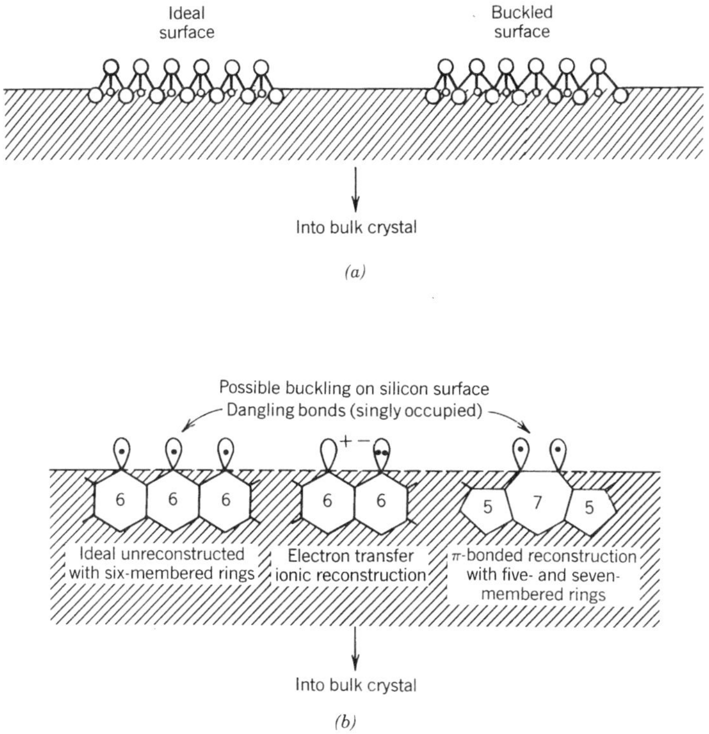

Figure 7.6 (*a*) (2 × 1) reconstruction on a semiconductor surface [after Haneman (1982)]. (*b*) Two possible mechanisms leading to the (2 × 1) reconstruction shown in (*a*), involving either charge transfer or altered coordination [after Robertson (1983)].

In the ideal unreconstructed surface of silicon, each (111) surface atom is bonded to three atoms in the layer below and has a dangling bond. Figure 7.6 shows some of the proposed mechanisms involved in reconstruction. The (2 × 1) structure is encouraged by π bonding and leads to a raising and lowering of alternate rows of silicon surface atoms following the rehybridization of the orbitals. This buckling mechanism keeps the bond lengths nearly constant. As mentioned earlier, and as we shall see later (Section 7.4), the (2 × 1) buckling splits the half-filled band associated with the dangling bonds of the ideal structure into two bands, one filled and one empty. This results in an energy gain and, since the excitation from one subband to the other is possible, the surface is semiconducting.

The structure of the silicon (7 × 7) cell is controversial, partly because a full dynamical LEED calculation is not practical (there are 49 lattice points per cell in the surface layer alone). One key factor is the energy gained by removal of the dangling bonds of the ideal structure. At the time of writing the favored models (Pandey 1983) include two levels of reconstruction, one which produces steps and another which reconstructs the steps themselves. In consequence each (7×7) unit is divided into two equal triangles, one lower than the other by a layer of puckered hexagons. Figure 7.1*e* shows an experimental image of the (7×7) surface.

LEED observations for (001) surfaces of silicon generally show a (2×1) surface unit cell. Here the ideal structure would have two broken bonds localized on the same atom. In this case it is suggested that rehybridization occurs, giving a dangling bond of the type sp_x perpendicular to the surface, and bonds of the form p_y or p_z between pairs of atoms in the plane of the surface (see Section 8.4). This pairwise bonding gives a doubled unit cell perpendicular to dimer rows. The (110) surface is not a cleavage plane, nor is it a favored surface for devices, like (100). It has therefore been less studied than (111) and (100).

7.3 SURFACE VIBRATIONS

The earliest studies of surface phonons came from elasticity theory, often from work stimulated by seismic waves created by earthquakes. Many of the same phenomena have been studied widely on a much smaller scale too, notably in ultrasonic nondestructive examination and in surface acoustic-wave devices. The surface waves and analogous modes that appear within elasticity theory are therefore quite well known and are related to the acoustic bulk modes (Chapter 2). The most important are Rayleigh waves, in which the individual atoms move in an elliptical motion in the plane of the propagation direction and the surface normal. Related are Stoneley waves, which occur when there are two media in contact, rather than a bulk–vacuum interface. Lamb waves occur in plates of finite thickness. They can be regarded as symmetric or antisymmetric combinations of Rayleigh (or Stoneley) waves from the interfaces on each side of the plate. As the thickness of the plate grows, the Lamb waves resemble more and more closely the simpler Rayleigh surface waves or the Stoneley interface waves.

Surface waves have lower velocities than the corresponding bulk waves. A simple way to understand this is to recall that the sound velocity increases if the elastic restoring forces on an atom are increased. At a surface, there are only restoring forces from one side (i.e., from the solid itself, with no contribution from the vacuum), and the lower restoring force means a lower velocity. Rayleigh wave velocities in isotropic systems are rather more than 90% of the bulk shear wave velocity, the precise fraction increasing with Poisson's ratio ν (91% for $\nu = 0.2$; 94% for $\nu = 0.33$).

The surface waves are localized, and their amplitude falls off exponentially as one moves into the bulk (clearly this is an oversimplification for plate modes, where both surfaces must be considered). However, the *total* vibrational amplitude of near-surface atoms includes both surface and modified bulk terms. The leading term in the difference between the mean square displacement at depth d from the surface and that in the bulk falls off inversely with depth d.

Surface elastic waves are important in a range of applied problems. We have already mentioned the most dramatic example of earthquakes, which, in addition to their destructive power, provide much information about the internal structure of the Earth. Seismological waves are of several types: the bulk waves comprising compression (P) waves, and the horizontally (SH) and vertically (SV) polarized shear waves, as well as the specific surface modes. Their velocities indicate the elastic properties inside the Earth. We have mentioned too the applications in nondestructive testing of engineering components. Here the important questions hinge on the ultrasonic reflectivity of internal surfaces and cracks. The ideal such defect for detection is one which diverts a large fraction of the incident energy in a directed beam into an unambiguously different signal that can be detected and analyzed.

The simplest elastic solutions omit two features important for surface-wave devices; elastic anisotropy (which leads to equations that are straightforward in principle but complex to solve in practice) and piezoelectric effects. The piezoelectric effect, which couples atomic displacements and a macroscopic electric field, usually stiffens the effective elastic constants; in $LiNbO_3$, for instance, the piezoelectric terms enhance the surface sound velocity by about 15%. The piezo terms are especially important because they can be influenced by applied electric fields. This has led to a new area of device technology. The advantages of *acousto-electronics* (see e.g., Maines and Paige 1983) are that the surface modes can be manipulated easily, and they have a velocity much less than that of light. This slow velocity means that processing in the time domain rather than frequency domain is possible; in turn, this leads to advantages of low attenuation and good signal-to-noise ratios.

Surface acoustic waves may be manipulated in many ways. They can be generated electromagnetically, using microwaves coupled to spaced metal electrodes on the surface. The waves can be guided and velocities modified by adding extra surface films (Au on Si slows surface waves, whereas Al on glass speeds them), and a whole range of techniques exists to filter, amplify, store, and compare signals. The slow velocities are of immense importance in producing lightweight, cheap delay lines. In addition to surface waves on insulators, acoustic waves in thin insulating films on metals have been of interest recently.

Our discussion so far has concentrated on elastic waves, or their equivalents, so that it has not mattered whether the solid is ionic or not. In ionic crystals, the acoustic modes correspond to the elastic waves, and, consequently, one expects the same Rayleigh and Lamb waves in ionic crystals as in nonpolar solids. The optic modes introduce new features, principally resulting from the electric field associated with longitudinal optic vibrations. The long-wavelength modes are the most important,

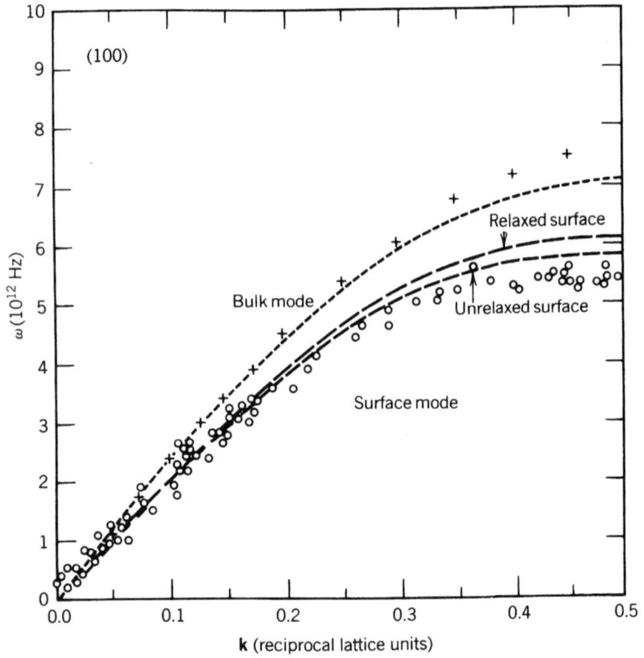

Figure 7.7 Comparison of experiment and theory for bulk transverse acoustic modes (plus signs, experiment; short dashed line, theory) and surface Rayleigh waves (circles, experiment; long-dashed lines, theory, the upper curve including lattice relaxation) in LiF [after Hoare et al. (1983), who give full references].

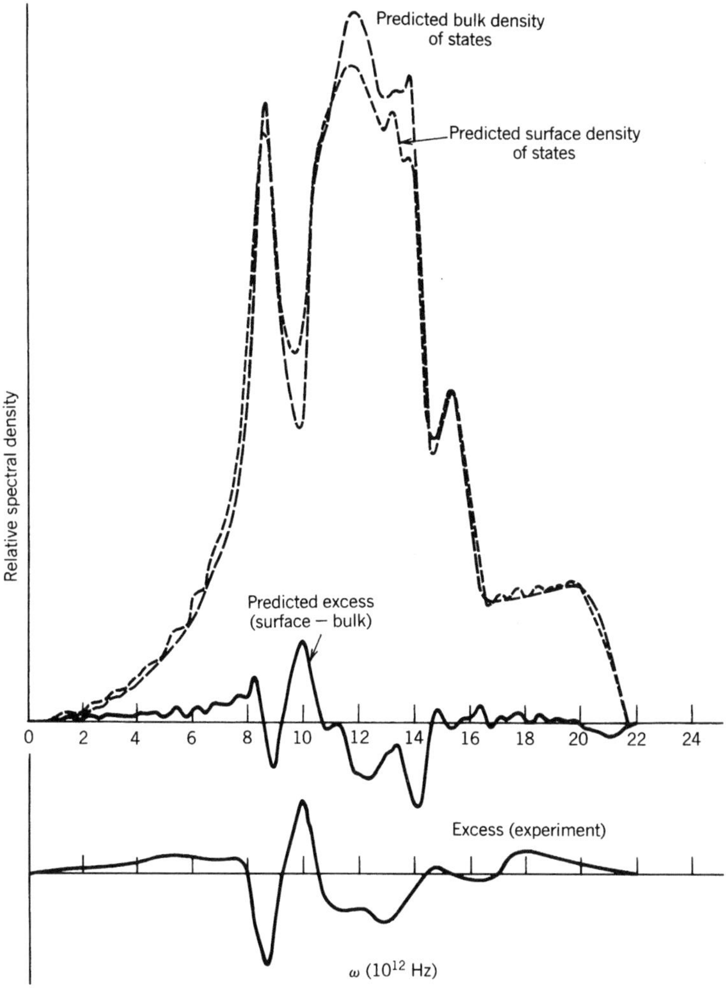

Figure 7.8 Comparison of the bulk and the (100) surface vibrational densities of states for MgO. Experiment gives the lower difference curve (Rieder 1971) and theory the upper value (Hoare et al. 1983).

since these couple to electromagnetic radiation and are directly observed. The optic-mode solutions for a thin slab of ionic crystal (either freely supported in vacuum, or mounted on a metal substrate) have been discussed by Berreman (1963), and by Fuchs and Kliewer (1965) and Kliewer and Fuchs (1966), using both conventional lattice dynamics and continuum electrodynamics. They find two solutions in addition to the usual bulk modes. For wavevector $\mathbf{k} \equiv (k_\perp, k_\parallel)$ these solutions have frequencies given by

$$\omega^2 = \omega_{LO}^2 - \tfrac{1}{2}(\omega_{LO}^2 - \omega_{TO}^2) k_\parallel L \tag{7.5}$$

$$\omega^2 = \omega_{TO}^2 + \tfrac{1}{2}(\omega_{LO}^2 - \omega_{TO}^2) k_\parallel L \tag{7.6}$$

where L is the thickness of the slab, assumed small. One important feature is that the bulk values ω_{LO} and ω_{TO} can be deduced from measurements of infrared reflection or thermal emission. Long-wavelength surface optical phonons have been observed in several oxides by high-resolution electron-energy-loss spectroscopy (EELS) (Cox et al. 1983).

We now turn from the long-wavelength surface waves to atomistic studies of surface phonons. These are especially important for thermodynamic properties of surfaces, since there is a vibrational entropy that contributes to the surface free energy and influences the relative stability of different crystal faces. In detailed studies, which follow the general approaches of Chapter 2, the complications of surface relaxation and reconstruction must be included. Experimentally, the inelastic scattering of low-energy He atoms can be exploited to give the low-frequency Rayleigh wave for all wavelengths. Figure 7.7 shows experiment and theory for LiF. If one wants the total density of states for all surface modes, infrared spectroscopy is effective, although subtraction of bulk contributions is needed. This can be done by comparing single-crystal and powder measurements. Figure 7.8 compares theory and experiment for MgO.

7.4 SURFACE ELECTRONIC STATES

Even for an ideally terminated crystal, there will be a substantial change in the potential experienced by an electron at the boundary. It is clear that the distribution of states in energy will be modified and that the changes will be concentrated near to the surface. For both ideal and realistic surfaces, one expects both bound surface states, which decay exponentially to zero in the bulk, and resonant states, which can be regarded as bulk states enhanced at the surface. The effect on the density of states is shown in Figure 7.9. Surface defects lead to another type of localized state, although these defect states are often very similar to their bulk analogs.

One of the most useful and general ways of investigating the electronic band structure of solids is by photoelectron spectroscopy (PES). Here a photon is absorbed near the surface and electrons are excited from both filled bulk and surface states to empty vacuum levels. The strength of the transition depends on the density of both the initial occupied states and the final empty states and also on the symmetry properties of the wavefunctions involved (Williams et al. 1980). The energies of the emitted electrons are analyzed to give peaks characteristic of the solid. Additional information may be obtained using the technique of angularly resolved PES (Williams et al. 1980). In these experiments the energy analysis is carried out on a fine pencil of electrons emitted at an angle θ to the normal to the surface. The momentum of the electrons emitted perpendicular to the surface changes as they pass through the surface discontinuity. However, the electron momentum parallel to the surface, $k_{\|}$, is a good quantum number, and measurement of the energy distribution of emitted electrons

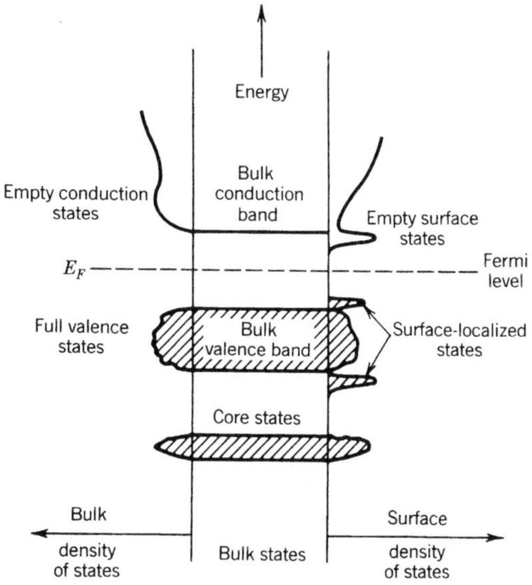

Figure 7.9 Schematic diagram of bulk and surface electronic states. Note there can be local surface states below the valence band. All states below E_F are occupied and all those above are empty.

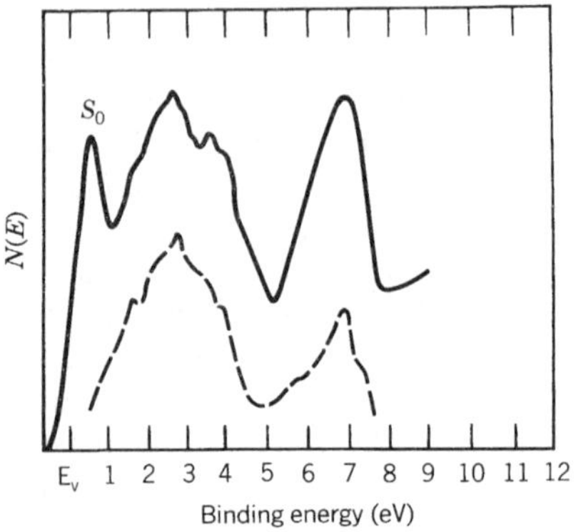

Figure 7.10 Comparison of the photoelectron spectrum (full line) for a vacuum-cleaved (111) surface of Si with an electron density of states for the bulk (dashed line) calculated by Schlüter et al. (1975). The peak labeled S_0 is assigned to surface states [after Williams (1978)].

as a function of θ gives the energy dispersion relation $E(k_{\parallel})$. Figure 7.10 shows a photoelectron spectrum of a vacuum-cleaved (111) surface of Si using an incident photon energy of 21.2 eV. Most of the emitted intensity is characteristic of the bulk crystal. However, the large peak S_0 has been assigned to surface electronic states and is probably due to transitions involving filled surface states associated with reconstructed dangling bonds (see Section 7.2 and below).

Surface electronic states may also be studied by means of electron-energy-loss spectroscopy (EELS). The electrons are energy-analyzed after scattering, showing elastic and inelastic peaks. Figure 7.11 displays the EELS spectrum obtained by scattering 50-eV electrons from a cleaved (111) surface of silicon at an angle of 56°. The most pronounced inelastic peak S_0 at 0.52 eV, corresponds to S_0 in Figure 7.10. The interpretation of the higher-energy peaks is less clearcut. It should be kept in mind that, at least for simple reconstructions, each Si on a (111) surface is bonded to three silicons in the layer below (Section 7.2). The bonds between these outer two planes, the so-called back bonds, will also be different from those in the bulk. They too, in principle, can give a contribution to the scattering recognizably different from that from the bulk. Collectively, the PES, EELS, and other data indicate localized surface states for both the (2×1) and (7×7) reconstructions [see Haneman (1982) for review].

If there is a sufficient density of surface electronic states in the bandgap of the bulk crystal, we may find that the Fermi level is pinned in the surface region (see Section 1.2.1). In bulk Si, we know that the Fermi level for the bulk crystal shifts by the bandgap energy ($\sim$1.1 eV) on going from highly p-type (where E_F is close to the valence band) to highly n-type (where E_F is close to the conduction band) crystals. However, for the same extremes

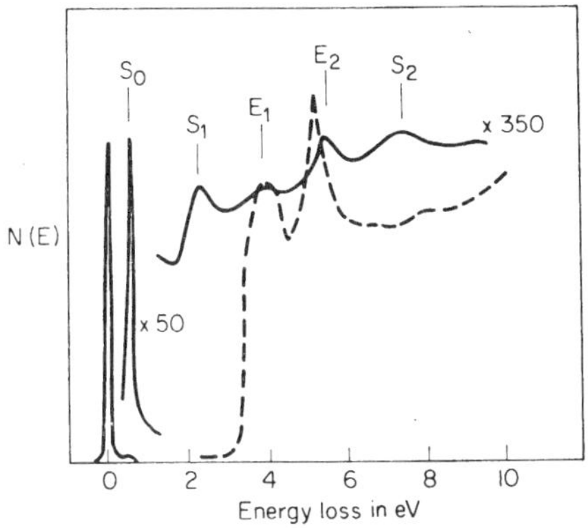

Figure 7.11 Comparison of electron energy loss spectrum for a cleaved (111) surface of Si (solid line) [after Rowe et al. (1975)] compared with excitation for the bulk (dashed line). Peaks assigned to surface states are labeled S_n with S_0 corresponding to S_0 in Figure 7.10.

of doping, the Fermi level shifts by only ~0.2 eV at a cleaved (111) (2×1) surface of Si. This implies that the full and empty surface states of Si are both close in energy and situated in the bandgap of the bulk crystal. This conclusion is consistent with the assignment of the peak S_0 (Figure 7.11) given above. By contrast, the cleaved (110) surface of GaAs produces no electronic states in the bandgap, and no pinning of E_F (Pollman 1980).

Electronic defects near a semiconductor surface play an important role in the formation of a Schottky barrier between a metal and a semiconductor surface. Metal–semiconductor junctions show many interesting features, notably rectification, since the current for a given voltage depends strongly on the direction of the voltage. Current flows most readily when carriers move from the semiconductor (electrons for *n*-type and holes for *p*-type semiconductors) to the metal. This rectification is characterized by a barrier, Φ_B, whose dependence on metal and semiconductor is still a serious challenge to the understanding of interfaces. This barrier is associated with a narrow region, perhaps 10 Å thick, at the junction itself. In addition, space charge from ionized donors and acceptors causes band bending in the semiconductor, so that the electrical effects of the interface are apparent up to about 10^3–10^4 Å from the junction.

Schottky's (1940) original model assumed equilibrium established, with no charge flow across the barrier. Thus, Fermi levels of metal and semiconductor match at large distances; the vacuum levels match at the

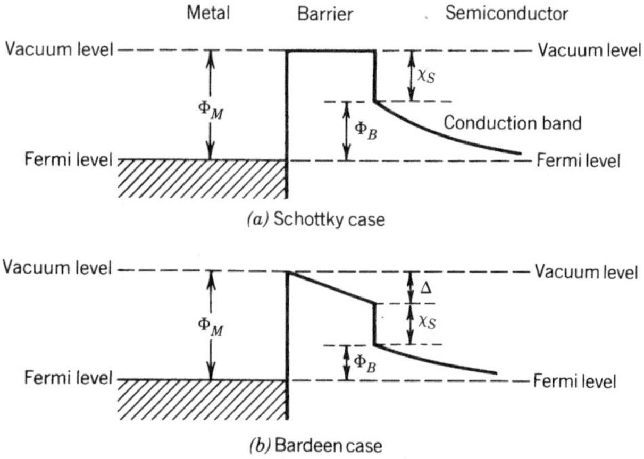

Figure 7.12 Metal/semiconductor junctions in two limiting models: (*a*) Schottky model, where there is no dipole associated with the barrier; the Fermi levels of metal and semiconductor match far from the junction. (*b*) The Bardeen model, where there is a dipole contribution Δ from surface states. Typical barrier thicknesses are 10 Å; band bending in the semiconductor occurs over 10^2–10^3 Å. The barrier height is Φ_B; Φ_M is the metal work function and χ_S is the semiconductor work function.

contact, giving the level scheme of Figure 7.12*a*. No metal–semiconductor interaction is assumed, and there is no dipole layer contributing to the barrier. It follows that

$$\Phi_B = \Phi_M - \chi_S \tag{7.7}$$

so that rectification depends only on the metal work function Φ_M and the semiconductor work function χ_S. While rare systems do seem to behave this way, most show quite different trends from metal to metal, with Φ_B frequently independent of Φ_M.

Bardeen's (1947) model showed how even a modest density of surface states (perhaps one per 1000 surface atoms) could lead to a significant dipole contribution Δ, giving a rectifying barrier essentially independent of the metal (Figure 7.12*b*). This picture, generalized in many ways, is still important, although the nature of the surface states is controversial. Heine (1965) noted they should be "decaying states" in the situation originally envisaged, corresponding to metal states falling off exponentially on the semiconductor side. Photoemission studies (Spicer et al. 1980) showed absence of the states suggested by Bardeen. Since the semiconductor Fermi level was pinned at very low metal coverages and the behavior was essentially independent of which metal was used, Spicer et al. (1980)

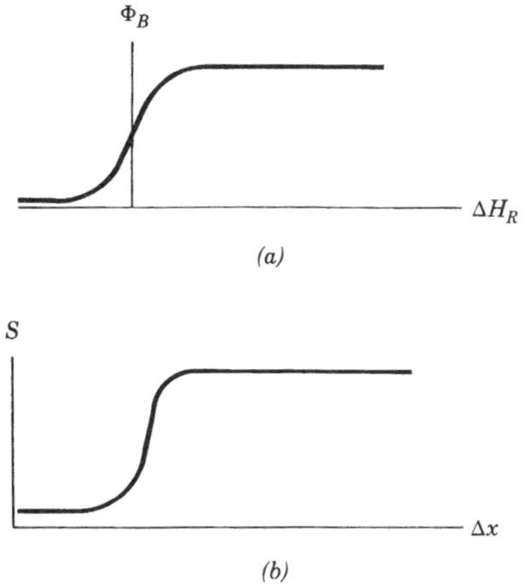

Figure 7.13 (*a*) Dependence of observed Schottky barrier heights, Φ_B, on the heat of reaction, ΔH_R, of the metal and the semiconductor (specifically with the anion for polar semiconductors). (*b*) Dependence of S $(=\delta\Phi_B/\delta x_M)$ on Δx $(=x_{SA}-x_{SC})$, the difference in electronegativity of the anion and cation in the semiconductor.

argued that secondary states were responsible, associated with subsurface atoms of the semiconductor.

A third class of models invoked reactions between the metal and the semiconductor, notably with semiconductor anions in polar systems. Phillips (1970) and Andrews and Phillips (1975) suggested that chemical bonding, either weak (as in silicides) or strong, would lead to the observed characteristic dependence of barrier height on enthalpy of formation of the presumed reaction product (Figure 7.13*a*). This picture has been developed and extended in studies by Brillson (1978).

Clearly, one significant factor in rectification is the dependence of barrier height on some property of the metal, like its electronegativity x_M. One important empirical result is that $\delta\Phi_B/\delta x_M$ is roughly constant for a range of metals on a given semiconductor, and shows a pronounced trend (Figure 7.13*b*) with the ionicity (defined here as the difference of anion and cation electronegativities) of the semiconductor. These empirical results, together with proposed explanations, are reviewed by Schlüter (1982).

7.5 NEAR-SURFACE EFFECTS

We turn now to the effects that depend on the modifications of defect or carrier concentrations near to surfaces. Enhanced concentrations (surface segregation), band bending, and space charge are among the phenomena of interest. For these *near-surface* effects, it is often the first few hundreds of atomic layers that are important, and not just a monolayer or two.

The main differences of the near-surface region from the bulk have two origins. One comes solely from intrinsic properties, that is, impurities need not play a role. In the bulk, any intrinsic reaction preserves electrical neutrality. Thus, for Schottky disorder, negative vacancies and positive vacancies will be created in equal numbers. At the surface, however, different numbers can occur, depending on their individual formation energies. Obviously, the solid is going to conserve charge overall, but a local imbalance is possible which cannot be maintained in the bulk (Section 3.1.1). The second difference concerns extrinsic effects, that is, those associated with impurities. If segregation occurs, the impurity concentration will be altered in the near-surface region. If the impurity needs charge compensation, then we expect these charge-compensating defects to dominate over the intrinsic defects at low temperatures (the so-called extrinsic region) and the thermally induced defects to dominate in the higher-temperature, intrinsic region. In particular, the net charge of the surface may change between the two regions. In AgCl and AgBr, divalent cation impurities lead to a surface that is negative relative to the bulk at low temperatures by about 0.2 V. At higher temperatures Ag vacancy generation becomes more important and there is the possibility that in this intrinsic region the surface will eventually become positive relative to the bulk. There is also the possibility of observing an *isoelectric temperature* at which there is no surface charge. Although this type of temperature behavior is seen for dislocations in silver halides no isoelectric temperature is found for AgCl or AgBr surfaces up to 200°C. The surface is found to be negative in both the intrinsic and extrinsic regions apparently because the free energy of cation interstitial formation is less than that of cation vacancy formation.

Typical space-charge layers extend for about 100–1000 Å from the surface, with a difference of potential of around 0.2 eV and an electric field of 10^4–10^5 V/cm. Such numbers correspond to charge densities of about one electron per 10^4 surface atoms. We can understand these numbers as follows. If there are charged defects present, the electrostatic potential ϕ will be determined by the Poisson equation

$$\nabla^2\phi = \frac{4\pi}{\epsilon}\rho \tag{7.8}$$

where ϵ is the dielectric constant and $\rho \equiv \sum_i c_i(r) Z_i |e|$ is the total charge density from the various species with concentrations c_i. In turn, the defect concentrations are determined by ϕ, the potential obtained from Equation (7.8), through the extra Boltzmann factor

$$c_i(\mathbf{r}) = c_{i0}(T)\exp[Z_i|e|\phi(\mathbf{r})/k_BT] \tag{7.9}$$

Here $c_{i0}(T)$ is the equilibrium concentration in the absence of ϕ, that is, the bulk concentration in thermal equilibrium. Clearly there are some constraints; in the bulk of the crystal, far from the surface responsible for inhomogeneity, one expects electrical neutrality.

Equation (7.8) and (7.9) can be solved usefully and easily in the limit that ϕ is small, when only the leading terms of the exponential are needed. The equations then reduce to a standard form $\nabla^2\phi = \phi/L^2$, where L is a screening length [see Flynn (1972, Section 9.3.2)]. If the system contains only positive and negative defects whose charges are $\pm Z|e|$ and whose bulk concentrations (with $\phi = 0$) are c_0, the screening length is given by

$$L = \{8\pi Z^2 e^2 c_0(T)/\epsilon k_B T\}^{1/2} \tag{7.10}$$

which is clearly temperature dependent. Typical values of the parameters ($\epsilon = 10$, $T = 300$ K, $c_0 = 10^{16}$ cm^{-1}) give $L \sim 25$ Å.

Two useful solutions of the equation $\nabla^2\phi = \phi/L^2$ cover most of the cases needed in practice. The first considers a test charge $Z|e|$ placed at $r = 0$ in an infinite medium. The potential, which is the sum of that of the test charge and that of the screening cloud of other charges, has the form $(Z|e|/r)\exp(-r/L)$. Positive test charges tend to attract negative charges around them so as to screen out the long-range field. This gives the so-called Debye–Hückel screening, well known in its applications to electrolytes, plasmas, and many other situations [see Flynn (1972, Section 9.3.2)].

The second solution applies to a sphere of radius R, in which defects with charge $\pm|e|$ have free energies of formation g_+, g_-, respectively. In this case the solution has the form

$$|e|\phi(r) = \tfrac{1}{2}(g^- - g^+)\left(1 - \frac{R\sinh(r/L)}{r\sinh(R/L)}\right) \tag{7.11}$$

L contains $c_0(T)$ [Equation (7.10)] which, in this case, is proportional to $\exp[-(g^+ + g^-)/2k_BT]$. This expression for $\phi(r)$ is useful in studies of fine powders. Its most common application, however, is in the form corresponding to a planar surface, where $R \to \infty$. If $x \equiv R - r$, with x remaining small (i.e., we concentrate on the near-surface region), then we obtain

$$|e|\phi(r) = \tfrac{1}{2}(g^- - g^+)[1 - \exp(-r/L)] \tag{7.12}$$

This shows two important features. First, there is an equilibrium potential difference between surface and bulk of $\tfrac{1}{2}(g^- - g^+)/|e|$. Second, this potential difference corresponds to a dipole layer which leads to a surface excess of the defect with the larger formation energy. These results have been verified in several ways, notably in the silver halides. In particular, the measured concentration profiles of low concentrations of divalent impurities near surfaces have been used to verify equation (7.12) (Farlow et al. 1983).

In AgBr, the (111) surface is negative corresponding to $\sim 2 \times 10^{11}$ charges/cm^2; for the (100) surface, the charge densities are a factor of 20 lower. These electric fields are important in photography (Section 6.3.5) for several reasons. First, they produce more cation interstitials near the surface; without the space charge there would be on average less than one interstitial per grain. Second, holes are attracted to the surface and electrons are encouraged to seek out *sensitivity centers*, like Ag_2S, which act as electron traps and thus inhibit both the formation of many subcritical metal particles and electron-hole recombination. Following trapping by the sensitivity centers processes such as

$$e^- + Ag_i^+ \rightarrow Ag_i^0 \tag{7.13a}$$

$$Ag^0 + e^- + Ag^+ \rightarrow Ag_2^0 \tag{7.13b}$$

may occur, as discussed in Section 6.3.5.

In the results outlined above, the Coulomb interactions are central, since they give long-range effects characterized by the screening length. Varying concentrations of defects near the surface can occur, even when no charged defects are present, so that the Poisson equation is not needed. For example, Ca in MgO segregates to the surface. The relevant equations to describe this effect are those of Section 3.1.1, since it is the configurational entropy that is the important factor. If f_B is the bulk ratio of concentrations $[Ca^{2+}]/[Mg^{2+}]$, then the corresponding ratio f_S for the surface layer is given by the Arrhenius form

$$f_S = f_B \exp(-h_s / k_B T) \tag{7.14}$$

The enthalpy of segregation, h_s, is estimated as ~ 0.5 eV for $MgO{:}Ca^{2+}$, both experimentally and theoretically. Analogous effects occur when gas molecules are adsorbed on surfaces, where similar expressions can be written down relating the fractional coverage to the external gas pressure.

7.6 SURFACE DIFFUSION AND SINTERING

The most common applications of surface diffusion occur in the processing of solids with very high melting points. Whereas most metals can be cast or machined, and most semiconductors can be grown from the melt or (because only thin layers are needed) by molecular-beam epitaxy, this is not the case for important oxides. These, the so-called ceramics (Davidge, 1979), are most easily obtained as powders or similar aggregates. The aggregated forms (often of mixed powders) are consolidated by heat treatments, such as sintering and hot pressing, which may exploit surface diffusion.

For a powder, or for a partly consolidated aggregate of powder grains, we can characterize the material by several different parameters. The first is the density. Suppose a given single-crystal oxide has density ρ_0, and the powder has density ρ. Then it is said that the powder has $100\rho/\rho_0$ % of the ideal density. The second is the surface area. If each unit volume is imagined as consisting of N spheres of radius R, then there is a total area $4\pi R^2 N$ and a total volume of material $\frac{4}{3}\pi R^3 N$. Thus the area, measured by gas adsorption experiments, would be $3(\rho/\rho_0)/R$, that is, several square meters per cubic centimeter for micron-sized powders. The third is porosity, which is concerned with the connectivity of the spaces between grains, and not just with the total volume of these spaces. The porosity affects apparent bulk diffusion. In many cases there is a *percolation threshold*, that is, essentially no diffusion occurs until a critical porosity is reached. Even in some single-phase systems, amorphous SiO_2 apparently being one, there are voids within the structure that contribute to gas solubility, but not to diffusion.

The processing of high-melting-point oxides or other ceramics is aimed at controlling specific physical features. One is external shape. Since these materials cannot be cast, and since they are brittle and hard to machine, it is useful to be able to form the powder in approximately the right form, and then to encourage the particles to join and consolidate under heat and pressure. A second factor is the porosity. Sometimes one wants a high surface area, as for a ceramic substrate for metal catalyst particles. At other times it is best to have zero porosity, as when a surface layer is formed to protect the underlying material from its environment. Such layers may also be decorative, as with enamels. The third feature concerns grain sizes and compositions. In magnetic ceramics, as used in recording tapes, the efficiency is very sensitive to grain size. In other systems the aim is to leave a layer of expensive, active, material on the outside of some relatively cheap material, which forms most of the grain.

Some of the processing can be achieved in the formation of the initial material. A good example is the sol–gel process, in which an oxide is precipitated from solution to form a gel, which can then be dehydrated. Doping with specific impurities can be carried out in several ways, depending on the stage at which the impurity is introduced; likewise, there is a choice between a porous or nonporous final form.

7.6.1 Surface Diffusion

Surface diffusion is studied by several methods. In the case of good electrical conductors, field-ion microscopy (Prutton 1975) allows one to monitor individual atomic motions over the surface. For semiconductors, at least in special cases, one can study how surface reconstruction reemerges after a perturbation. But most experiments use relatively coarse techniques. There are, of course, tracer methods using radioactive isotopes, although the lower numbers of atoms involved in surface (rather than bulk) diffusion create difficulties. Sintering (Section 7.6.2) in which one measures the rate of porosity or density change, gives a useful measure of surface tension, but it does not usually indicate over which surface diffusion occurs.

Three main mechanisms of surface diffusion are commonly considered. The first is the vacancy mechanism, exactly as in bulk diffusion, except that the vacancy remains at (or, in some cases merely near) the surface. One expects that both formation and motion energies will be smaller for surface vacancies, since there are weaker interactions because there are fewer neighbors. This rationalization is borne out by detailed calculations and, to a lesser extent, by experiment. Obviously, one should expect both intrinsic vacancy production as well as its counterpart, the extrinsic vacancy production arising from the presence of impurities. The second mechanism of surface diffusion involves the motion of an adatom, that is, an atom or ion moving along the surface, in contact with the outer layer of surface atoms. Here one might expect the activation energy to be related to the heat of vaporization, since the interactions to be overcome in removing the adatom are similar to those experienced in going to a saddle point. Third, we must not ignore the role of vapor transport at very high temperatures; in this process an atom leaves the surface and is readsorbed some distance away. This can lead to apparently anomalous diffusion, with very high preexponential factors arising from the long effective jump distances. Vapor-phase diffusion is especially important for internal pores in a temperature gradient, where an evaporation/condensation mechanism takes matter from one side of the pore to the other.

Grain-boundary diffusion also shows some of the above features, notably in relation to vacancy diffusion. The lower activation energy means that grain-boundary diffusion will always dominate over bulk diffusion at low temperatures (Figure 7.14*a*). Thus, in low-temperature oxidation, it is the grain boundaries that limit the rate of oxidation; the high-angle boundaries (see Section 3.7) are especially important (Figure 7.14*b*). Obviously, other factors enter too, including the density of grain boundaries and their effective thickness; in practice, all good measurements suggest thicknesses of a few angstroms.

In many oxides, like MgO and Al_2O_3, diffusion appears to be extrinsic, that is, presumably caused by vacancies present in concentrations determined by unidentified impurities. Even the dislocation-enhanced diffusion in MgO appears to be impurity controlled. In other oxides, like UO_2, which readily go nonstoichiometric (Section 3.3.1), the surface stoichiometry need not be the same as the bulk value, nor need it remain independent of temperature. In bulk UO_2, oxygen moves much faster than uranium. At the surface both move at comparable rates, with the vapor-phase mechanism taking over at 1700°C, or possibly lower temperatures.

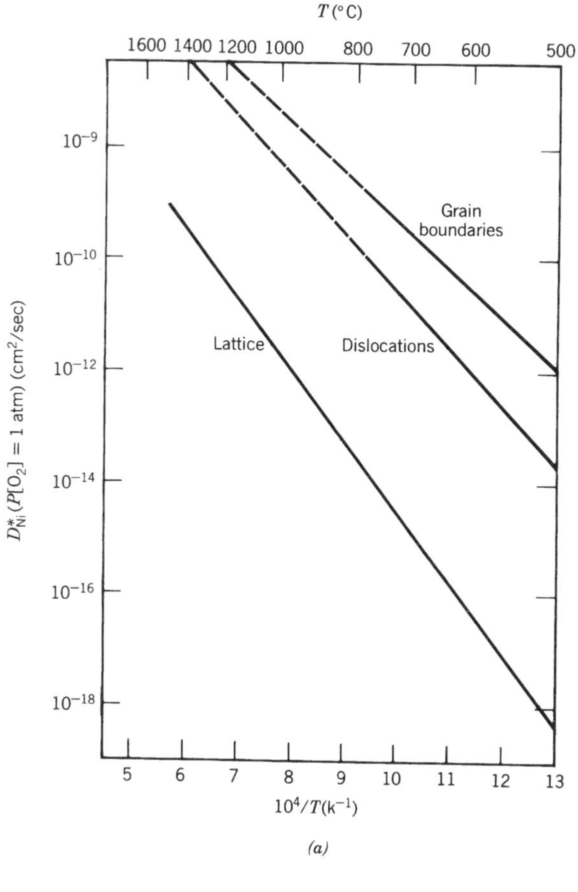

Figure 7.14(a).

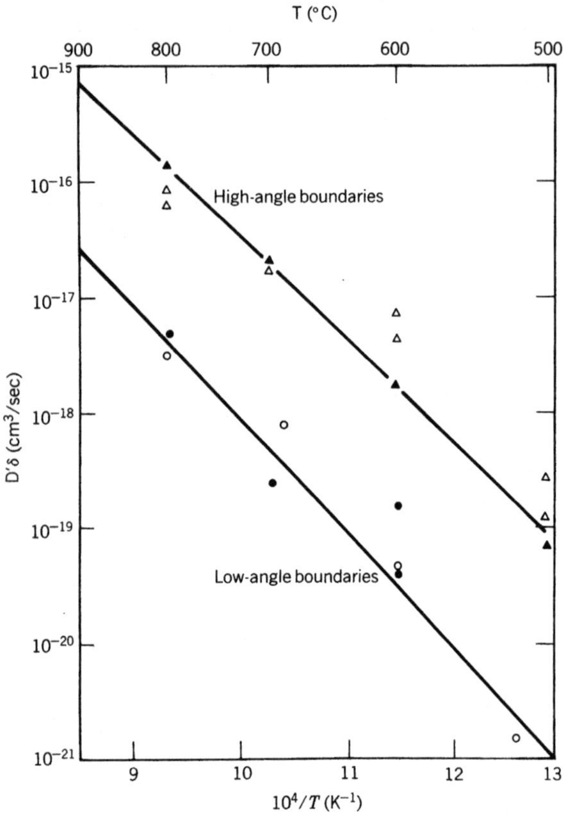

(b)

Figure 7.14 Diffusion in $Ni_{1-x}O$ [after Atkinson and Taylor (1981)]. (*a*) Relative contributions from grain boundaries, dislocations, and bulk diffusion for tracer diffusion of Ni in NiO at 1 atm oxygen pressure, derived from experiment. (*b*) Product of grain-boundary diffusion constant D_{gb} and effective boundary thickness δ for Ni tracer diffusion in $Ni_{1-x}O$ at 1 atm oxygen pressure, derived from experiment. Open and closed circles refer to subgrain (low-angle) boundaries, closed triangles refer to high-angle boundaries and open triangles to values derived from observed oxidation rates.

Diffusion at semiconductor surfaces is expected to show special features, including dependence on Fermi level, on degree of excitation, and on precise surface reconstruction. The various activation energies measured for silicon (1.1–2.0 eV) are intermediate between the high bulk activation energies for self-diffusion ($\gtrsim$4 eV; Section 3.6) and the low bulk motion energies reported for vacancies ($\lesssim\frac{1}{3}$ eV, depending on charge state; Section 5.4).

7.6.2 Sintering

As was already mentioned, the process of transforming a powder into a strong, coherent solid is one of great practical importance. The principal driving force is the elimination of the (internal) surface area of the powder, with its surface energy, replacing it by continuous solid. For 1 cm^3 of powder, with 95% nominal density and spherical grains of 1 μm diameter, and with a surface energy of 1000 erg/cm^2, full compaction releases 2.85 J of energy.

The main way in which consolidation occurs is the infilling of the narrow neck regions near points of contact (Figure 7.15). We may classify the several different mechanisms by the source (e.g., other surface regions, grain boundaries, or dislocations) of the material used in this infilling and by the type of diffusion involved (e.g., whether it is surface diffusion, bulk diffusion, or grain-boundary diffusion). We now use geometric arguments to describe how a powder compact evolves during sintering.

Consider the system of two spheres in contact, illustrated in Figure 7.15. If there is a rate dm/dt of mass transfer to the neck, then a volume

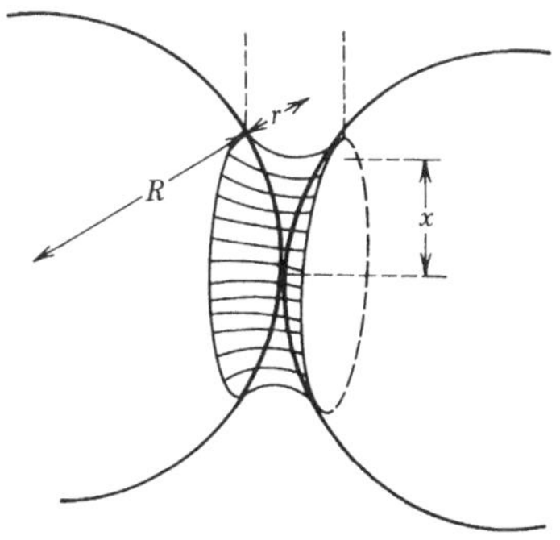

Figure 7.15 Geometry of sintering spheres. The spheres, of radius R, are joined by a neck region where the surface has radii r (concave) and x (convex).

$\rho^{-1}\,dm/dt$ of material is transferred in unit time. In terms of x, the radius of the region of contact, and R, the radius of the spheres, we have

$$\frac{dV}{dt}=\frac{d}{dt}(C_1x^4/R)=\rho^{-1}\frac{dm}{dt} \tag{7.15}$$

where C_1 is of order unity (actually $\pi/2$ in our case). Since x varies much more rapidly than R, we may write

$$\frac{dx}{dt}=\frac{R}{4C_1x^3\rho}\frac{dm}{dt} \tag{7.16}$$

to a good approximation. The contact area is πx^2, so that a corresponding expression for the rate of change of contact area can be obtained. A more interesting question is whether the centers of the spheres move closer or not. If they move closer, consolidation will lead to an increase in density; if the separations remain constant, the internal pores will merely change shape. In essence, if the matter filling in the neck regions comes from the bulk, densification will occur. If the matter merely comes from another surface, only the pore shape changes.

We have dealt with the first geometric problem, namely, how to relate changes in the neck region to dm/dt. The second component relates dm/dt to other known quantities. We use vapor transport as our example. The driving force here is the difference in vapor pressure at the neck and away from it. Continuing, as with C_1 above, we introduce constants C_2, C_3, etc. Since the radius of curvature of the neck (r in Figure 7.15) is much smaller than either the sphere radius R or the orthogonal neck radius x, we may expect the fractional vapor pressure difference relative to that for a flat surface to be

$$\frac{\Delta p}{p_0}\simeq\frac{C_2\sigma M}{k_BTr\rho} \tag{7.17}$$

where M is the molecular weight and ρ is the density of the solid. The net condensation rate at the neck is dm/dt, given by

$$\frac{dm}{dt}=\begin{pmatrix}\text{Accommodation}\\ \text{coefficient}\end{pmatrix}\begin{pmatrix}\text{Collisions per}\\ \text{unit area per}\\ \text{unit time}\end{pmatrix}(\text{Capture area})\Delta p \tag{7.18}$$

Given that the area of neck surface is approximately $(2\pi x)(\pi r)=2\pi^2xr$, and that $r\simeq x^2/2R$, we find

$$\frac{dm}{dt}=C_3\cdot\frac{x^3}{R}\cdot\frac{R}{x^2}=Vx \tag{7.19}$$

where the x^3/R comes from the capture area and R/x^2 from Δp. Our earlier equation (7.16) shows that, if $dm/dt \propto x$, one has

$$\frac{dx}{dt} \propto x^{-2} \tag{7.20}$$

so that we expect x, the radius of the filled-in region, to increase as $t^{1/3}$.

Sintering behavior is usually displayed on a *sintering diagram* (Ashby 1974). This identifies regimes in which sintering is dominated by one particular mechanism (e.g., volume diffusion from dislocations) of the many possible mechanisms. The regimes are classified by two dimensionless variables, the homologous temperature (i.e., T divided by the melting temperature) and the logarithm of the ratio (neck radius x/particle radius R). Sintering rate diagrams (Figure 7.16) identify either the behavior after a given time, or the behavior at a given rate. Such plots can be used to select heat treatments to optimize the final product. Vink (1975) surveys some of the applications of sintering to the production of optical and magnetic materials with closely controlled and optimized properties.

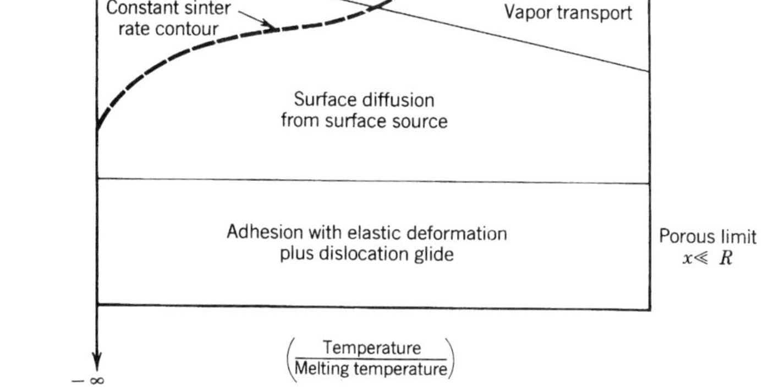

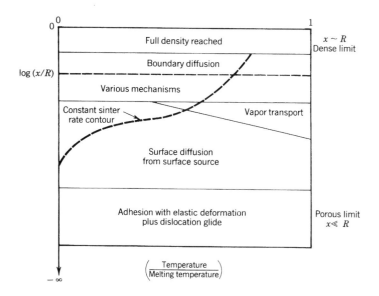

Figure 7.16 A typical sintering map [see Ashby (1974)].

7.7 ADSORPTION

In most of our discussion so far we have been concerned with either free surfaces, in which the solid is exposed to the vacuum, or with grain boundaries, where the solid is in contact with more of the same solid, but not matched as in a perfect crystal. In practical applications, the solid is frequently exposed to gases or liquids. In so-called heterogeneous catalysis, the solid–gas interface encourages reactions between different gas species. Gas sensors produce recognizable electrical signals (whether through changes in conductivity, capacitance, or temperature) when the chosen species is adsorbed. Electrodes in liquid electrolytes also show adsorption, and this may control critical electrode processes. We shall not discuss the electrolytic processes here [Bockris and Dražić (1972) and Hansen et al. (1983) do so in detail], though we note that the electrolyte can lead to modifications of the near-surface defect concentrations in electrodes. Thus, in photoelectrolytic cells, both the efficiency and the long-term stability of the electrodes (which are often oxides) can be controlled by adsorption of specific species from the electrolyte. One final point about liquid–solid interfaces concerns the rates of crystal growth or of crystal dissolution. We have already mentioned that growth (and in some cases crysstallographic form) depends on surface topography, and that screw dislocations have a dramatic effect on growth rates. Dissolution has proved more puzzling. For oxides, apart from those that have a strong vigorous chemical reaction with water, it appears that dissolution is relatively slow even when thermodynamically favorable, usually between one atomic layer per second and one atomic layer per day.

Surface species can be identified most easily at a gas–solid interface. One of the most convenient ways of identifying surface impurity atoms on solids is by use of Auger electron spectroscopy (AES) [see, for example, Riviere (1972)]. An electron of about 2 keV is incident on the solid, exciting an electron from a core level (Figure 7.17). An electron from a higher occupied level, such as the valence band, may fall into the core hole, releasing energy. This energy may excite yet another electron, in the valence band, for example, and if the imparted energy is sufficient, the electron may emerge from the solid and be energy analyzed. The energies of emitted (Auger) electrons are characteristic of the atoms involved and are generally the same, irrespective of the energy of the exciting particle. Alternatively, the released energy may appear as a soft X-ray photon, which can be measured using the techniques of soft X-ray spectroscopy (SXS). Auger spectroscopy samples two or three atomic layers at the surface and can detect less than a monolayer of adsorbed species.

In X-ray photoelectron spectroscopy (XPS), a monochromatic beam of

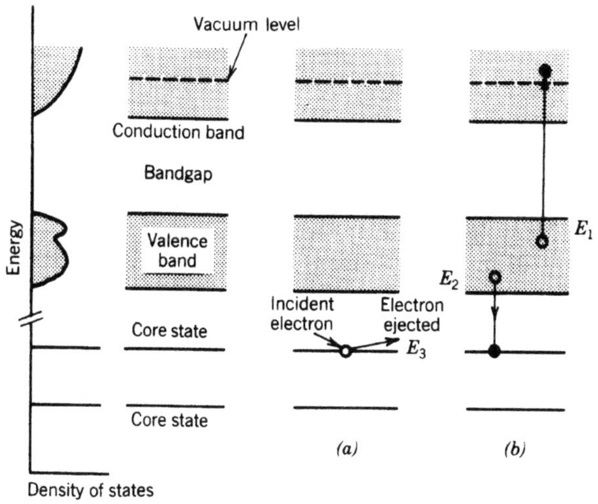

Figure 7.17 Schematic representation of stages involved in Auger electron spectroscopy. (*a*) An incident electron of energy ~2 keV ejects a core electron leaving a core hole. (*b*) An electron from the valence band of energy E_2 falls into the core hole, of energy E_3; the energy $E_3 - E_2$ may be emitted as an X-ray photon [which can be analyzed using soft X-ray spectroscopy (SXS)] or it may cause the emission of an Auger electron from the valence band with an energy $(E_3 - E_2 - E_1)$. The relaxation processes involved in Auger emission are complex.

X-rays is incident on a surface. The X-ray photon ejects an electron with a well-defined energy, usually from a core level. This characteristic energy makes identification of the emitting atom possible [see, for example, Prutton (1975)]. The XPS method is sufficiently sensitive to detect changes in the binding energy of core electrons arising from changes in the chemical state of the atom. These chemical shifts may be of order a few electron volts, so that it is in principle possible (although not always easy) to identify both the surface species and its valence state. Hence XPS is also known as ESCA (electron spectroscopy for chemical analysis). Although X-rays penetrate to a depth of order 1 mm, the escape depth of electrons from a solid is small, so that XPS provides information about the outermost three or four layers.

Atoms and molecules can be adsorbed on surfaces, held there by forces ranging from weak van der Waals forces to strong chemical bonds. Sometimes these adsorbed species are intrinsic. In particular, the paramagnetic species O^-, O_2^-, and even O_3^- are often encountered on oxide surfaces. Clearly, they can only be studied by special methods, and most observations have been of epr of powder samples after radiation damage. Adsorbed impurities are broadly classified as *physisorbed* if they are bound with an energy less than about 0.5 eV, and as *chemisorbed* if the binding is stronger. An example of *physisorption* is LiF:He, where the binding energy is less than 0.01 eV. In this case, the He–LiF potential produces four weakly-bound states (0.0058 eV, 0.0022 eV, 0.0005 eV, and 0.0001 eV). Examples of chemisorption include oxygen on graphite, Si, and Ge, where the effects have great technological importance. The heats of adsorption are around 9.5 eV (Si) and 5.3 eV (Ge). For graphite, adsorbed oxygen is partly responsible for the lubricating properties, a result that was very important in the design of airplane engines for high-altitude flying. The control of the oxidation of graphite is important in other areas, for example, in the successful practical design of graphite-moderated nuclear reactors.

Even for the weakly bound, physisorbed species, a variety of behavior is observed, for the adsorbed atoms and molecules can exist in a variety of thermodynamic forms (Nash, 1975). At submonolayer coverages two-dimensional gas, liquid, and solid phases have been identified. The structure of the solid phase is generally ordered and can be either commensurate or incommensurate with the structure of the host surface (Figure 7.18). The transition between the two-dimensional solid and liquid phases is of interest because it can in principle be second order at high enough coverages, in contrast to the first-order melting process characteristic of all three-dimensional systems.

The most detailed studies of physisorbed systems have been carried out on monolayer and submonolayer coverages of the basal plane of graphite by rare-gas atoms. For near-monolayer coverages of krypton, for example,

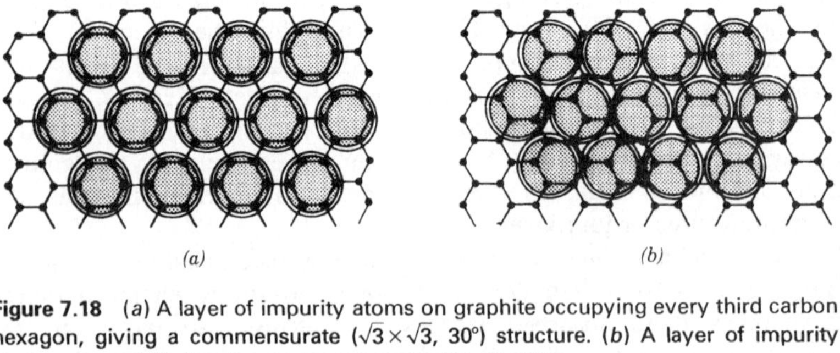

Figure 7.18 (*a*) A layer of impurity atoms on graphite occupying every third carbon hexagon, giving a commensurate ($\sqrt{3}\times\sqrt{3}$, 30°) structure. (*b*) A layer of impurity atoms on graphite forming an incommensurate structure.

one finds the $(\sqrt{3}\times\sqrt{3})$ 30° commensurate structure shown in Figure 7.18(a). These solid films melt at $\gtrsim$100 K in a second-order fashion (Horn et al. 1978). Structural studies have also been carried out on physisorbed N_2 molecules adsorbed on graphite (Eckert et al. 1979). Below ~90 K, N_2 monolayers form a liquidlike film on graphite basal planes and below ~84 K this solidifies into a $(\sqrt{3}\times\sqrt{3})$ 30° commensurate structure. This structure is orientationally disordered but Horn et al. (1978) find evidence to suggest that orientational ordering sets in on cooling to 30 K.

In chemisorption, several distinct types of adsorption are found. Some are effectively intrinsic, for example, when oxygen is adsorbed on an oxygen-deficient (or cation-excess) nonstoichiometric oxide. In microcrystals of MgO containing paramagnetic F^+ centers, the annihilation of epr can be used to monitor oxygen adsorption. Other types of adsorption give well-defined new phases. Reactive transition metals on simple oxides often form more complex oxides, like spinels. Of course, oxygen itself can be chemisorbed by a wide range of materials. The coverage of oxygen on (111) faces of Si tends to stabilize at 1.5 monolayers. This is usually interpreted as a two-step process. First, O_2 chemisorbs to give two oxygen atoms bonded to adjacent dangling bonds. (These need not exist as dangling bonds before adsorption, of course). Second, interchange of one oxygen with its nearest neighbor occurs; after rebonding, an additional oxygen atom can be added. We shall discuss the later stages of Si oxidation in Section 7.8.

Desorption of surface atoms can be stimulated by electron beams or laser excitation. Since the direction of the desorbed ion's trajectory gives information about the surface site it occupied, and since choice of excitation wavelength selects which species is stimulated, photon-induced desorption is a powerful technique for surface examination. Several desorption mechanisms have been proposed. In alkali halides, for example, it is possible that a photochemical process like that described in Section 6.2.2, will lead to a collision sequence ending with the ejection of surface atoms. Other mechanisms have been suggested too, like evaporation from a hot surface or ejection by shock waves produced by the photon beam. One of the most important suggestions is that of Knoteck and Feibelman (1978), who noted that the energy threshold for O^+ emission from TiO_2 corresponded to the energy needed to excite electrons from the Ti^{4+} $3p$ levels. These authors suggested that the $3p$ hole is filled by a $2p$ electron from a surface O^{2-} ion, and that this occurs by an Auger transition in which other electrons are ejected from the surface oxygen. The overall result of these transitions is the presence of an O^+ ion at an O^{2-} site, and the Coulomb repulsion is sufficient to lead to ejection. Similar processes can occur when the core levels of an adsorbed species are excited directly, for example, F^+ and OH^+ can be desorbed from TiO_2 surfaces.

7.8 OXIDATION AND CORROSION

7.8.1 Factors that Control Corrosion

The corrosion of metals in large engineered structures involves an enormous diversity of phenomena (Hoar 1976). Some elements are simply protective, e.g., minor alloy constituents such as Al, Cr, and Si in steels can oxidize to form layers that protect the underlying material from a hostile environment. Some effects are electrochemical, as in pitting corrosion, where crevices or pits develop within which the chemical environment becomes increasingly hostile. Microstructure matters too; grain boundaries are especially susceptible to attack, the intergrannular precipitates can also affect attack. When a highly caustic or a highly acid solution causes corrosion, one has to consider both the chemical reactions at the surface (the rate at which the products are removed and replaced by fresh reactants) and how this is influenced by any applied electric potential. It is no surprise that the roughness or smoothness of the surface is also important.

One major initial factor is simply thermodynamic stability. Many practical corrosion problems involve aqueous environments and an applied electric field, whether intended or merely the result of some unintended chemical reaction. A vast body of information has been compiled to define the thermodynamic stability of an element in these conditions, and this is displayed in Pourbaix diagrams (see Guy 1971) (Figure 7.19). These classify behavior as *corrosion* (with soluble or gaseous products), as *passive* (with a stable oxide, hydroxide, or hydride limiting loss) and as *immune* (when all reactions are endothermic). The regimes are plotted on a standard reference frame, which also shows the stability of water. The overall result is that the equilibrium (but not kinetic) electrochemical behavior of many systems in a solution of known acidity or alkalinity is defined and systematically documented. Examples appear in Figure 7.19.

We are mainly concerned here with those cases in which the behavior is controlled by an insulating film that grows during corrosion. The simplest case is oxidation in a dry oxygen atmosphere (Atkinson 1984). The four main factors that influence oxidation rates under these conditions are:

1. Diffusion through the oxide film. This may involve either oxygen or metal motion, and it may involve more than one diffusion mechanism for each mobile species.
2. Chemical equilibria at the metal–oxide interface, at the oxide–gas interface, or at any internal boundary.
3. Electronic transport, for example, in the maintenance of space charge or of impurity species in their correct charge states.
4. Internal stress fields and electric fields which may develop during oxidation (in principle, magnetic fields and thermal fluxes can be important, but only in rare cases; examples include effects of a magnetic field on oxidation near magnetic phase transitions and the effects of thermal gradients from laser-beam heating during oxidation).

We may handle these factors at several levels of detail. First, we shall show that simple use of the diffusion equation allows a very important

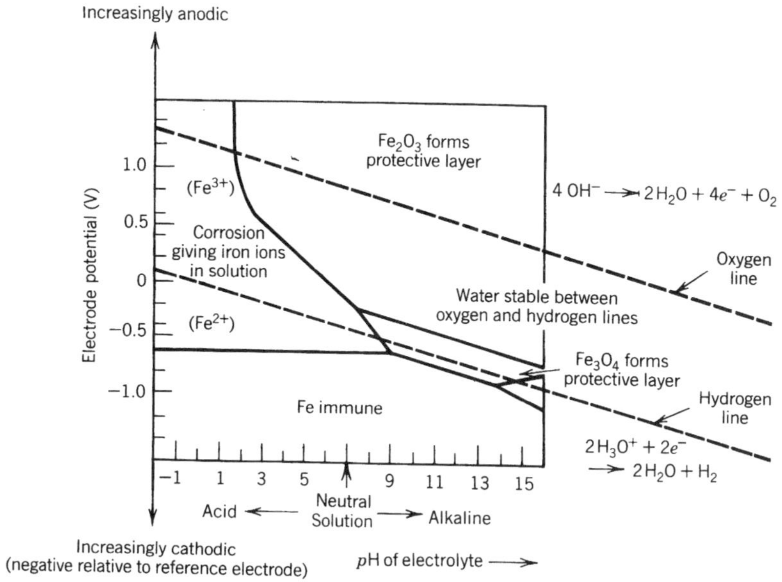

Figure 7.19 Electrochemical stability of iron: the Pourbaix diagram. This summarizes the regions of immunity (where the applied potential inhibits any reaction), of passivity (where a protective oxide grows), and of corrosion in its several forms. Water is itself unstable outside the oxygen and hydrogen lines. The electrode reactions involving OH^- (alkaline solution), H_3O^+ (acid solution) and water are given on the right.

phenomenological and quantitative description of the rate at which an oxide film develops. It has immediate application in classifying behavior, for example, whether diffusion or surface reactions control rates of oxidation. A second application is that one may wish to use one's knowledge of the controlling mechanisms to try to influence the oxidation. Here one might hope to reduce diffusion by use of suitable dopants or to control stresses so that the oxide continues to protect the metal, without *spalling* (flaking from the surface). At the time of writing, the problems of internal stresses and spalling are among the most important in oxidation; they are understood satisfactorily in only two cases namely, internal oxidation, where the oxide growing subsurface pushes against the surrounding matrix, and oxidation of curved surfaces.

7.8.2 Oxidation Kinetics

We start this section by treating the two extreme cases of pure diffusion control and of purely surface-reaction control of oxidation kinetics. The oxidation rate is determined by Fick's law [Equation (3.24)], relating diffusion flux to concentration gradient, and by a geometric factor relating the diffusion flux to the volume of oxide produced.

(a) Diffusion Control

Consider a metal covered with an oxide layer and suppose that chemical equilibrium is established at the metal–oxide plane, with concentration C_{MO} of the mobile species. Likewise, suppose that an equilibrium is established at the parallel gas–oxide interface, with concentration C_{GO}. The concentration difference across the oxide is thus $\Delta C = [C_{MO} - C_{GO}]$. (If metal atoms move outward, $C_{MO} > C_{GO}$; if oxygens move inward, $C_{MO} < C_{GO}$. In either case, the flux is such as to cause oxide growth.) For oxide thickness x, there is a mean concentration gradient of magnitude $\Delta C/x$, and we shall assume for the present that this gradient is uniform. If a volume Ω of oxide is produced per atom transferred, then the thickness increases at a rate $dx/dt \equiv \Omega x$ (diffusion flux), that is,

$$\frac{dx}{dt} = \Omega D \frac{\Delta C}{x} \tag{7.21}$$

so that, starting from zero thickness at $t = 0$, we obtain the well known *parabolic law*

$$x(t) = \sqrt{2\Omega D \Delta C}\sqrt{t} \tag{7.22}$$

This result ($x \propto \sqrt{t}$) is used widely in phenomenological analyses of oxidation. Clearly high temperatures raise the diffusion rate and will enhance oxidation rates; rapid grain-boundary diffusion also speeds oxidation. Likewise, anything that alters the concentration gradient will affect the rate of film growth.

(b) Surface Reaction Control

Suppose that chemical reaction at one of the interfaces limits oxide growth. This is more likely in very thin oxide layers, where diffusion can be rapid. The rate of growth is now constant,

$$\frac{dx}{dt} = B, \tag{7.23}$$

giving the *linear growth law*,

$$x(t) = Bt + b \tag{7.24}$$

Clearly, a basic measurement of the qualitative form of the time dependence of the thickness $x(t)$ already gives important information about the rate-determining step. Precise information, especially in more complicated cases, requires both independent measures of diffusion coefficients and knowledge of what the main defect species are (cf. Chapter 3 and below).

Even with a uniform oxide film growing on a flat metal substrate (a highly idealized situation), there are considerable complications when there is also an electronic flux, or when equilibria are not instantaneously established. The theory of these phenomena is reviewed by Fromhold (1976) and Wagner (1973). The main components of their calculations are (a) a group of differential equations (notably diffusion for ions, a similar equation for electrons but including the possibility of tunneling, and Poisson's equation relating electric field to charge distribution), (b) a group of conservation laws (for charge and for each atomic species), and (c) boundary conditions (which may involve the metal, the gas phase, and features like surface charge). Often it is necessary to resort to numerical solutions of the coupled differential equations, and even here the real system may be poorly represented.

Two additional situations are encountered where analytic forms are obtained. One is the logarithmic form

$$\frac{dx}{dt} \propto \exp(-ax) \Rightarrow x = A\ln(1 + bt) \tag{7.25}$$

This might be found if, for example, the activation energy becomes larger as the oxide thickness grows. The inverse logarithmic law [Mott (1947) as corrected by Ghez (1973)] is appropriate when electron tunneling is rate determining:

$$\frac{dx}{dt} \propto \exp(-a/x) \Rightarrow x^{-1} \propto \ln(t/x^2) + \text{const} \tag{7.26}$$

Electron tunneling is also a component of the Mott–Cabrera (1948) model of oxidation. The physical ideas behind this model are themselves sufficiently interesting for more detailed comment. Suppose a metal M is being oxidized to $MO_{n/2}$, where n is the metal valence. The metal has work function ϕ, that is, at electron at infinity, e_∞, is released from a state at the Fermi level at energy cost ϕ:

$$(NM) \to (NM)^+ + e_\infty - \phi \tag{7.27}$$

for a metal of N atoms. We assume that if $n/2$ oxygen molecules are adsorbed on the $MO_{n/2}$ oxide, they absorb n electrons e_∞, producing nO^- ions at the surface and releasing energy Δ_{el}:

$$ne_\infty - \frac{n}{2}O_2 \to nO^- + \Delta_{el} \tag{7.28}$$

This gives a chemical potential, $\Delta_{el} - \phi$, favoring electron transfer through the film from metal to oxide until a compensating space charge builds up. This space charge opposes electron motion from metal to oxide, and also drives metal ions M^{n+} through the oxide to react with the adsorbed O^- ions. This driving field has been observed directly in experiments exploiting tunneling in metal–insulator–metal junctions.

Two comments are in order. First, even a low density of surface O^- ions gives large electric fields; a value of 1 V for $\Delta_{el} - \phi$ generates a field of 10^6 V/cm for a 100-Å-thick layer, and can be achieved by about 10^{-4} of a monolayer of surface O^-. Second, there are natural comparisons with an electrochemical cell. In the cell, however, it is the ions that are mobile in the electrolyte, and the passage of electric current corresponds to the oxide layer generated in our case.

Suppose diffusion-controlled (parabolic-rule) oxidation is occurring through a coherent, single-phase, stress-free oxide. How can one alter the oxidation rate? Obviously, control of the gas phase may help; one could reduce the oxygen pressure, or include additives to the gas to make the surface reactions slow enough to limit the rate. One could also, perhaps, apply an electric field to suppress ionic motion. But if these are not practical, one is forced to modify the defect concentration (and hence ΔC; see Equation 7.21) or the diffusion rate D. If lowering the temperature to reduce D is not practical, the defect populations, which may influence ΔC through trapping, must be controlled. In practice, control of defect populations has been relatively unsuccessful, both because defects have been misidentified and because of unintended side effects of doping.

Wagner, Schottky, and Hauffe have suggested rules on the following lines for the control of defects by doping [see Kofstad (1966)]. Suppose cation vacancies dominate diffusion. Higher-valent cations (e.g., Al in NiO) should increase the number of vacancies, and hence enhance the oxidation rate; on the other hand, lower-valent cations (e.g., Li in NiO) should decrease the vacancy concentration, for charge compensation, and so reduce the growth rate. Table 7.3 summarizes this picture.

able 7.3 Wagner–Schottky–Hauffe Rules[a]

ype of Oxide	Type of Mobile Defect	Addition which Decreases Oxidation Rate	Direct Consequence of Addition	Example
xide hich conducts ainly ionically, at is, electronic rrier motion rate termining	Cation vacancy or anion interstitial	Higher-valent cation	Lower hole concentration $[h]$	Cd slows bromination* of Ag
	Anion vacancy or cation interstitial	Lower-valent cation	Lower electron concentration $[e]$	*
xide which nducts mainly ctronically, at is, ion otion is rate termining	Electrons (n-type oxide) e.g., ZnO	Higher-valent cation	Fewer ionic defects; $[e]$ grows	Al slows (and Li enhances) Zn oxidation
	Holes (p-type oxide) e.g., NiO	Lower-valent cation	Fewer ionic defects; $[h]$ grows	Li slows (and Cr enhances) Ni oxidation (cf. Section 7.6.2)

As expressed in the table, these rules (Kofstad 1966), define the alloy additions that are pected to decrease oxidation rates. The examples are cases where the rules work ccessfully.

No verified examples for oxides are known to us.

7.8.3 Microstructure and its Influence

It has long been known that fast-diffusion paths exist in oxide films. These are traditionally ascribed to dislocations, to grain boundaries, or even to channels or cracks through which molecular oxygen can move. Only in rare cases, like Ni/NiO, has this been verified in detail (Atkinson and Taylor 1981). In this instance, much was already known of the diffusing defects and their equilibrium concentrations. Direct measurements were made of grain-boundary diffusion (D_{gb}) and bulk diffusion (D_{bulk}) rates. An effective diffusion constant can also be obtained from the oxidation rate constant, $k_p \equiv x^2(t)/t$, using the parabolic rule [Equation (7.22)]. At low temperatures, grain-boundary diffusion dominates. Bulk diffusion, with its higher activation energy (2.56 eV, rather than 1.78 eV on high-angle boundaries), takes over at high temperatures. Agreement between the different experiments (Figure 7.14) is excellent.

In grain-boundary transport there remains a possibility that the boundary (or other fast-diffusion path) will change as oxidation continues. The two important possibilities are *sealing* and *cracking due to stress* (which may, of course, occur at separate sites in the oxide film). Sealing can take several forms, including self-blockage, when oxide grows to obstruct a channel, and mutual blockage, where the compressive stress from one channel affects another. Phenomenologically, [see, e.g., Smeltzer et al. (1961)] the natural assumption is that the effective diffusion constant for parabolic growth is a weighted mean $D_{\mathrm{eff}} = D_{\mathrm{bulk}}(1-f) + D_{gb}f$ of bulk and grain-boundary contributions in which the factor f depends on temperature and thickness. Thus, if f increases with time such that $df/dt \sim f$, the parabolic law gains a correction such that grain boundaries dominate for very thick oxides.

Stress is of enormous importance, since it can cause cracking and spalling of oxide films, nullifying their effectiveness as protective layers (Stringer 1970). The largest and commonest effect is purely geometric. Suppose the metal surface is convex with radius of curvature R. If the initial oxide were strain free, and if defect-free oxide crystals were to grow subsequently, to a thickness T, the outer layer would be stretched to $(R+T)/R$ of its strain-free interatomic spacing, that is, there would be a strain T/R. It is therefore part of good antioxidation design to make sure that purely geometric stresses are minimized. Obviously, some roughness will remain, and here the plasticity of the oxide is critical. Likewise, a good epitaxial match of oxide and metal is only possible if the *mismatch dislocations* described by Frank and van der Merwe (1949) are produced to compensate for a lattice parameter difference.

Rupture of surface oxide by stress is the commonly held explanation of so-called duplex layer oxides. Here an oxide is formed with two layers of distinct microstructure. The origin usually suggested is that the compact *outer* layer is formed by the conventional mechanisms we have already discussed. The inner layer grows from oxygen penetration through cracks in the outer layer, forming a coarser oxide inside the original. This fits many of the results, including some marker experiments (where an inert marker is placed on the outer surface before oxidation, and monitored thereafter) and tracer measurements. However, the cause of cracks in the outer layer is far from clear. There are many qualitative explanations, almost all of which give insignificant strain magnitudes (Stringer 1970).

7.8.4 Oxidation of Silicon

Silicon oxidation is a major step in device manufacture, giving a stable insulator that can withstand significant voltages. Its oxidation involves quite a few distinct complexities. First, we note that the permeability of the oxide depends on both a solubility factor and a diffusion constant. For the oxide, the solubility contains two quite separate components, one consisting of unconnected voids and the other of connected but unidentified sites of higher energy, which contribute to diffusion (Mott 1981). Second, oxidation of silicon in dry oxygen and the diffusion of oxygen in vitreous silica both appear to involve motion of interstitial oxygen molecules, without oxygen exchange with the oxygens in the silica matrix, as measured by radioactive tracers. However, when wet oxygen is used, there appears to be rapid isotope exchange involving OH species. Third, during oxidation, stacking faults (Section 3.7) develop in the silicon substrate from silicon interstitials produced by oxidation (Murarka 1977, 1980). These defects are reduced as the oxidation rate rises or if chlorine is present, when impurity complexes form instead.

One recent development is the formation of oxide layers in devices by implantation of oxygen ions (see Section 6.4.4).

7.8.5 Oxidation of Alloys

Suppose an alloy AB consists of a metal A, like iron, and another species B (perhaps Si) that forms a more stable oxide. Suppose further that the oxidizing conditions are such that only B oxidizes. Depending on the concentration of B, on diffusion-controlled properties, and to a lesser extent on molar volumes, three main regimes occur (see, e.g., Hughes 1982, for a straightforward account). Below a level N'_B atoms of B per unit volume, the only oxidation of B that occurs is internal. At this low level the diffusion flux of B into the surface region is too small to produce a coherent film. Below a (higher) level N''_B, a continuous oxide of B may grow, but it cannot be sustained. The net flux of B atoms into the oxide at the metal–oxide boundary cannot be kept positive. And above a (still higher) level N'''_B, only an external scale grows. The equilibrium moves progressively in favor of B in its alloyed metal form (rather than oxide) as one moves into the alloy from the metal–oxide interface. General arguments like these illustrate both the complexity of oxidation and the problems of designing alloys to grow desirable protective films.

7.9 CATALYSIS

Catalysts have two main functions. One is *enhancement* of the rate of a chemical reaction, usually with only minor indirect effects on the ultimate equilibrium. The other is *selectivity*, that is, the ability to enhance only a chosen reaction. Practical catalysts must be stable against poisoning in operation, and against degradation when subjected to the high thermal and mechanical stresses possible in an industrial chemical reactor. Science on an atomic scale and chemical engineering are thus the two extremes of catalysis.

Insulators and semiconductors are used in catalysis in two main ways, as operational catalysts, actively involved in the chemical process, and as substrates, usually for supported-metal catalysts, like platinum or its alloys. In all cases the precise details of physical and chemical constitution of catalysts are optimized with great care.

Catalysts operate in four main ways:

1. They may merely absorb reactants inertly, thus increasing the probability of them coming together (suprafacial processes). Zeolites, which are often used in the cracking of large organic molecules to smaller hydrocarbons, are an example (see Section 8.3).
2. The catalyst may provide special *active sites* that encourage one of the critical steps in the reaction. This case is harder to identify with certainty, but suggestions of surface-ledge involvement are often made. Thus, it has been argued that, in the polymerization of ethylene (C_2H_4) by $TiCl_3$, the effects of growth spirals (i.e., the characteristic spiral features when growth occurs at screw dislocations) are important.
3. The catalyst may supply one of the reactants. Oxide catalysts such as bismuth molybdate often appear to supply oxygen from their surface

for oxidation reactions; the surface oxygen is then replenished, either from bulk diffusion or from the surface diffusion of adsorbed oxygen. The reverse process is also possible, for example, oxygen vacancies are involved in the reduction of NO on perovskites by processes labeled intrafacial.

4. The catalyst may act as a source of electrons or holes. This is commonly the case for so-called cooperative catalysis by semiconductors, where electron and hole concentrations are readily controlled. An example might be the photoelectrochemical decomposition of water. Other clear examples include reactions involving charge transfer as well as the stimulation of reactions by optically generated carriers, for example, the reaction $H_2 + D_2 \rightleftharpoons 2HD$ on MgO:Fe (Harkins et al. 1969).

Catalysis by metal particles on oxide (or other nonmetal) substrates is very common, notably with the Pt-based catalysts for reducing automobile exhaust pollution. In many cases the oxide substrate is simply an inert support. In others, there is strong involvement of both metal and support (known as strong metal-support interaction or SMSI). Although the mechanisms involved are uncertain, there are obvious possibilities. For example, the metal and substrate could react to form some complex oxide, or the oxide substrate could itself catalyze some reaction among the several occurring simultaneously, or physical changes associated with the spreading of the metal over the substrate could be a feature. The possibility of physical changes correlates quite well with what is known of metal–oxide interactions. One measure of these is the shape of liquid-metal droplets (see Figure 7.4). Experimentally, the contact angle θ is most easily measured. Metals which wet their substrates ($\theta < \pi/2$, with flat disclike form) tend to show either strong metal-support interaction in catalysis, or else show chemical reaction. Metals which do not wet their substrate ($\theta > \pi/2$, with roughly spherical droplets) show neither reaction nor substrate effects in catalysis.

Catalyst stability is of major importance, since one wishes to conserve high surface area. Suppose a given volume V of catalyst is divided into N cubes. The total area will be $6N^{1/3}V^{2/3}$, that is, the area increases for finer particles. For MgO cubes of side 1000 Å, for instance, the area is 6×10^5 cm^2 in a volume of 1 cm^3 or about 16 m^2/g. Many systems (notably based on alumina or silica) achieve several hundreds of square meters per gram. One major factor is thus balancing the advantages of high surface area against the added stability of larger crystals (see Section 7.6.2).

Degradation can take a number of forms, all needing attention in working systems. First, some substrates (notably alumina) having a distinct crystal structure in their high-surface-area microcrystalline form. The preparation of γ-alumina and the prevention of a transition to the usual α-alumina are an essential part of catalyst technology. Second, one wants to prevent sintering to larger crystallites, either of the insulator or of any supported-metal catalyst. This is done by suitable stabilizers, possibly dopants that segregate to the surface in oxides. With an appropriate dopant it may be possible to reduce the surface energy of a preexisting surface so that the energy becomes negative, thus stabilizing the high surface energy required (Tasker et al. 1984). Other oxides can stabilize the metal; some 3% of Al_2O_3 effectively stabilizes Fe metal catalysts for ammonia synthesis. In much the same way MgO appears to reduce Ru losses by volatilization. Third, one may wish to avoid deposition of poisons, for example, of carbon (inevitably present in many reactions) or of lead or sulfur (which are traditional poisons). It is in cases like this that the understanding of precise reaction mechanisms is critical. One possible poisoning mechanism is formation of surface compounds that block active sites; thermodynamic stability of such surface compounds is thus important. Finally, one may wish to stabilize certain key impurities in specific charge states; this may be achieved by codoping with other impurities, known as *promoters*.

The solid-state science of catalysis has made enormous gains from the advanced methods of experimental analysis. It is clear, however, there is much more to practical catalysts than a simple extension of surface science.

CHAPTER EIGHT

Special Systems

In this chapter we briefly review results of research in rather complex systems that are important from the point of view of applications and that have attracted considerable attention in recent years. There have been great advances in our understanding of the physics of amorphous solids and glasses of late, and Section 8.1 is devoted to these materials. The very difficult area of metal–insulator transitions is dealt with in Section 8.2, with some emphasis on Mott–Hubbard transitions and Anderson localization. The Peierls transition in one-dimensional conductors is also discussed, using organic conductors for illustration.

The rather exotic compounds made by intercalating graphite and transition-metal dichalcogenides with other elements and compounds are reviewed in Section 8.3. Finally, in Section 8.4, we comment on some recent work on polymers, with emphasis on the electronic properties of polyacetylene.

We emphasize that each of the topics covered in this chapter represents a major body of knowledge and is a major area of research in its own right; our concern will be only with defect-related aspects. We shall see that there can be analogies between some defects seen in amorphous systems and those in crystalline solids, although major differences can occur. Likewise, defect processes such as photo-induced degradation are found in many systems, both crystalline and noncrystalline. Other phenomena, notably in metal–insulator or order–disorder transitions show defect processes in a rather broader context.

8.1 AMORPHOUS SOLIDS

We have noted in Chapter 1 that many solid-state properties do not depend on whether there is long-range order or not. Continuous crystalline solids have many features in common with disordered ones. Window glasses and transparent crystals have very similar optical behavior and room-temperature mechanical properties. Obviously, discontinuities matter; window glasses and bricks are both largely silicates, yet the granular structure (see Section 3.7) and the impurities in bricks give a profound change in transparency and other properties.

Almost all the discussion in Sections 1.1 and 1.2 is valid for both amorphous systems as well as for crystals.* However, there are some important differences:

*We shall sometimes use the prefix *c*- to denote crystalline, the prefix *a*- to denote amorphous, and the prefix *v*- to denote vitreous, as in earlier parts of the book. The prefixes *a*- and *v*- are often used idiosyncratically. Strictly, the term amorphous implies a well defined phase, with a latent heat associated with the crystalline-to-amorphous transition, whereas the term vitreous implies a continuous change in behavior.

1. The environment of any particular ion in a crystalline solid is determined by its position in the unit cell only and crystalline solids have long-range order. Since even in distant cells, corresponding ions in crystals have the same neighbors at the same distances, electronic transition energies have well-defined values, and symmetry-related features in the electronic spectra of crystals can sometimes be identified. By contrast, there is no unique environment in an amorphous solid and such features are generally smeared out.

2. One-electron states in a continuous band in a crystal are delocalized, that is, they can be represented as modulated plane waves $u(\mathbf{k}, \mathbf{r}) \exp(i\mathbf{k} \cdot \mathbf{r})$ extending through the whole crystal. In an amorphous solid (or, in suitable cases, a disordered solid) the states below a so-called mobility edge are localized, falling off exponentially with distance away from their main region of concentration.

3. Both amorphous and crystalline solids may have real bandgaps [see, for example, Weaire and Thorpe (1971)]. However, an amorphous solid with no real gap (i.e., no finite range of energy for which there are no electronic states; see Section 1.1) may still have a mobility gap, in which there is a finite range of energy for which there are only localized electronic states.

4. In a crystalline solid, there are clearly defined vacancy and interstitial defect centers. In an amorphous system this is not so, although one should not dismiss the ideas entirely, for there may be enough short-range order for sensible analogies to be drawn. In alkali-excess liquid alkali halides, for example, epr and optical studies show that centers like the crystalline F center exist, that is, like an electron trapped at an anion vacancy. Presumably the extra electron determines the short-range order, just as an extra hole tends to produce X_2^- halogen molecular ions (V_k centers, see Section 5.1.3) in both glasses and aqueous solutions.

8.1.1 Structural Aspects

The regularity of crystal lattices is destroyed in two main ways. In disordered systems, like mixed ionic crystals ($Na_xK_{1-x}Cl$) or semiconductors ($Al_xGa_{1-x}As_yP_{1-y}$, or Si_xGe_{1-x}), there is still a regular crystal lattice (Section 2.5). There are, of course, slight distortions, but each atom has the same number of neighbors as in the unmixed component crystals. Indeed, in $Na_xK_{1-x}Cl$ the Na and K ions have their usual first neighbors. In amorphous systems there are two main structural changes, namely, those in the number of nearest neighbors (coordination changes) and those in the connectivity of the disordered lattice. Changes in coordination are common on melting, especially where there is a large change in density, e.g., Si becomes metallic on melting with a change in mean coordination from four to about six. But the nature of the bonds between atoms often favors one particular coordination strongly. If so, the short-range order of the atoms will be closely the same in the ordered and amorphous systems. The disorder can then be topological, involving distant neighbors in some recognizable way. Thus amorphous silicon might have five- or seven-membered rings instead of the six-membered rings of the crystal form, with each atom nevertheless retaining its fourfold coordination (Figure 8.1). In such a system a dangling bond would constitute a defect. Another example is found in the polyacetylenes (see Section 8.4) where the bonds (i.e., single C—C bonds or double C=C bonds) are ordered and where an altered bond sequence along a chain can be regarded as a defect. Other systems are different. The chalcogenide glasses, for example, normally have the S, Se, or Te bonded to two other similar atoms; these show variations in structural features, for example, bond angles, as in *a*-Si and *a*-Ge. In addition, the disorder in these systems can take the form of so-called valence-alternation pairs (Section 8.1.4), with a threefold and a singly coordinated atom. Such pairs are the analogues of dangling bonds in *a*-Si.

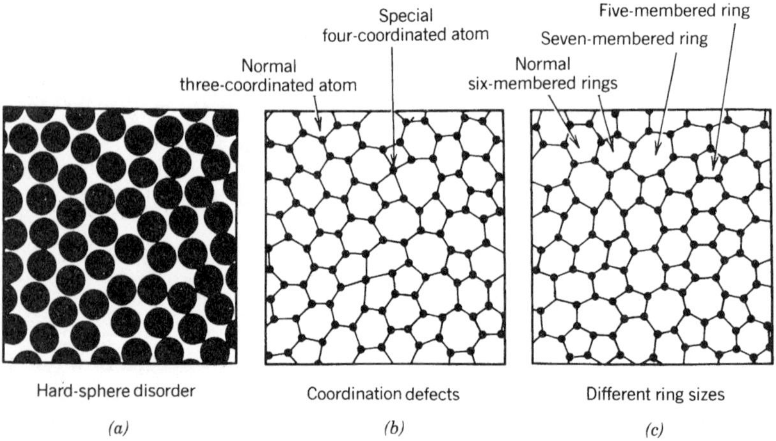

Figure 8.1 Disordered solids. Three different forms of disorder are shown: (*a*) Hard-sphere disorder, with only short-range repulsion between atoms; (*b*) coordination disorder, in which, here, a few of the atoms are bonded to one more atom than usual; and (*c*) topological disorder in which, here, rings of five and seven atoms are found as well as the usual six-atom rings.

Most models of amorphous systems adopt a simple conceptual approach which accounts for broad features of behaviour, and which may be used as a reference state in defining defects. The main models are these:

1. *Random Close Packing*. This resembles the order achieved by a box of ball bearings continuously shaken (corresponding to thermal motion). Glassy metals can exhibit this behavior and also nonmetallic glasses, for example, *a*-$ZnCl_2$ with random-close-packed chlorine and with zincs in interstices.

2. *Microcrystalline*. Here one has strong crystal-type short-range order over 10–20 Å, separated by relatively disordered regions. Highly ionic systems favor this form, although one should not assume that the same crystal structure is favored by both bulk and microcrystalline forms. Thus bulk α-Al_2O_3 is trigonal. However, the high-surface-area form, γ-Al_2O_3, used in catalyst supports (Section 7.9) is a cubic defect-spinel structure, and any amorphous form of alumina is more likely to resemble γ-Al_2O_3 then the better-known α form.

3. *Random Network*. This simplified picture supposes each species demands a given coordination. Examples include *a*-Ge and *a*-Si. Full coordination is particularly useful as a reference state for defining types of defects, against which we can compare those that are actually observed. First, there may be atoms with too few neighbors. A silicon atom with only three neighbors, for example, would be expected to have a dangling bond, occupied by a single electron that can be detected by epr. Second, there may be atoms with the right number of nearest neighbors and possibly the right number of second neighbors, but with the wrong number of more distant neighbors. This leads to classification as five-membered rings or seven-membered rings (Figure 8.1), or some similar topological description. Third, we can imagine atoms at wrong sites; in SiO_2 one could envisage Si–O–O–Si sequences instead of –Si–O–Si–. Chemical impurities allow further variety, and two broad classes are very important: (1) the network formers, which take part in the continuous network as far as possible, and (2) the network modifiers, which enter interstitially and are unbonded (see Section 8.1.5). Commercial glasses are designed by adjustments of both network formers and network modifiers to have desired properties.

Electronic structure and geometric structure are closely linked, and it is useful to indicate two examples of this. First, suppose we attempt to alloy two systems, for example, two crystalline alkali halides. Common sense suggests and experience confirms that large mismatches of ionic radii can inhibit alloy formation over wide variations in relative concentrations. But even if a good single alloy is formed, one finds two classes of behavior. In alloys that match well, all properties move smoothly with composition from one extreme to the other. When the match is poorer, the lattice vibration frequencies corresponding to one component may die away as its proportion falls, replaced by separate bands appropriate to the other component (Section 2.5). Second, systematic effects may result from longer-range geometric features even when the nearest-neighbor coordination is correct. In random-network structures, which can be considered usefully in terms of rings (e.g., six-membered for *a*-Si, with occasional five- and seven-membered forms), one effect occurs which is referred to as frustration. This effect is associated with the antibonding electronic states, in which the wavefunction attempts to alternate in sign from atom to atom. In a solid with only even-membered rings, no problems arise since each circuit around a ring matches. If the rings are odd numbered, however, there is an incompatibility, and this increases the energy of certain states.

Some of the ideas we shall describe for amorphous systems, or for special systems like polymers, also have their parallels in inorganic crystals. Crystalline ice, for example, has two important structural rules (Hobbs 1974). The first rule is that there are two hydrogens close to every oxygen, and the second rule is that, on the line joining any two adjacent oxygens, there is only a single hydrogen atom. Each of the two main types of defect breaks one of these rules. The so-called ionization defects are ones with either three or one hydrogen adjacent to a particular oxygen. Since these are usually H_3O^+ or OH^-, they correspond to C_3^+ and C_1^- of the chalcogenide glasses (Section 8.1.4). The so-called Bjerrum defects, however, place either two hydrogens (*D* defect) or none (*L* defect) between two oxygens. The rotation of an H_2O molecule in ice can produce an *L*-*D* pair. The topological disorder with zero, one, or two hydrogens on specific oxygen–oxygen links has parallels with some of the polymer systems where single, double, and treble bonds may occur out of sequence (Section 8.4).

8.1.2 General Aspects of Electronic Conduction

In crystalline nonmetals, entirely free from defects, electronic conduction occurs by motion of electrons in the conduction band or holes in the valence band. Two main mechanisms operate, the well-known propagating (large-polaron) mechanism of carriers of Si, Ge, and the III–V compounds, and the much slower diffusive or hopping conduction by small polarons in crystals with strong coupling between carriers and lattice vibrations (Section 1.2). Small-polaron systems include holes in halides such as KCl (Section 5.1.3) and AgCl (Section 6.3.1).

Defects in crystalline systems have three main effects on electronic conduction. First, they affect electron and hole concentrations (see, for example, Section 5.3). Second, they scatter carriers. At low concentrations one can superpose the contributions of the individual defects; at higher concentrations one must be careful to include space-charge effects consistently (see, for example, Section 7.5). Lattice vibrations are an extra source of scatter, with piezoelectric, Fröhlich, and deformation potential contributions (Section 1.2.2). Third, carriers can be trapped by defects. The effective drift mobility, as measured in a time-of-flight experiment, then depends on both the capture and release rates, adding an extra temperature dependence to that for the mobility of carriers in the defect-free case [see Equation (6.8)].

In amorphous and disordered systems this description of conduction needs modification. There is first a question of whether states of a given energy are localized or not. In crystalline materials states within the bands are delocalized (in a sense to be defined) and those in bandgaps, entirely associated with defects, are localized. Second, there is the question of how the carrier concentrations depend on dopant levels and on temperature. Some amorphous systems show clearly the negative U behavior (Section 8.1.4; see also Section 5.4), which tends to pin the Fermi level and so limit carrier concentrations on doping. Third, there are questions that concern the nature of disorder and the scale over which inhomogeneity occurs. These include classical ideas like percolation, in which a (possibly complex) path of low resistance can emerge through a highly disordered phase in which most carrier trajectories are blocked. Despite the immense clarification of the behavior of amorphous systems that has already occurred, there are some basic issues which still cause controversy. One example is the abruptness with which mobility changes with energy.

Near a band edge in a perfect crystalline solid one expects the density of states to take the form $\rho_c(E) \sim (E - E_c)^{1/2}$ for energies above the band edge E_c, and to be zero for $E < E_c$ (Section 1.3.1). This (Figure 8.2*a*) is a reference case to which we may relate disordered and amorphous systems. We assume that electrons are the carriers, although parallel descriptions could be given for holes. When a low concentration of defects is added, a narrow band (with width determined by defect–defect interactions) appears (Figure 8.2*b*) which, as more defects are added, broadens until it overlaps with the original band edge (Figure 8.2*c*). Note that there is still a cutoff at the lower-energy extremum of the defect band. We shall consider shortly the nature of the defect states and their

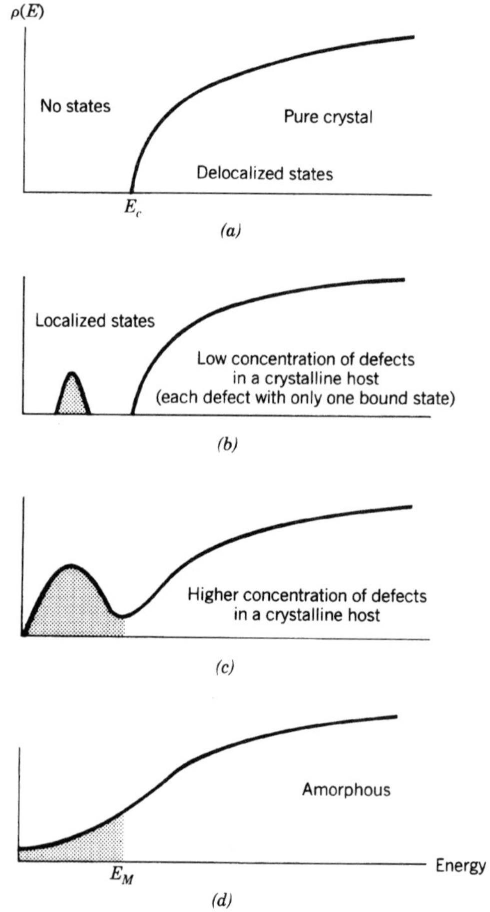

Figure 8.2 Schematic representation of the electronic density of states for concentrations of donors increasing from (*a*) to (*d*). While we have shown the bottom of the conduction band [E_c in case (*a*)] and the mobility edge [E_M in case (*d*)] at the same energy, this is a simplification.

localization. In Figure 8.2*d* we show an amorphous system with a mobility edge E_M; localized states occur below E_M and extended states above E_M. Note that the density of states for an amorphous system no longer has the $(E - E_c)^{1/2}$ dependence. This will be important later when we consider metal–insulator transitions (Section 8.2.1).

In all but the perfect crystalline case of Figure 8.2 there are regimes of energy in which states are localized (and hence make, at most, a restricted contribution to conduction) and there are higher-energy delocalized states that (at least in the absence of self-trapping) have high mobility and dominate the conduction. We may postulate the following types of behavior:

1. *Fermi Level E_F below E_M.* The conductivity should fall as T falls to zero. For crystalline systems and for amorphous solids at high temperatures, conduction is by excitation to the delocalized states; this gives a factor $\exp[-(E_M - E_F)/k_B T]$ in the conductivity. For amorphous systems at low temperatures variable range hopping occurs instead (Section 8.1.3).

2. *Fermi Level E_F above E_M.* The conductivity resembles that of metals, and is roughly independent of temperature. For $E_F \sim E_M$ there have been suggestions of a *minimum metallic conductivity*, $\sigma_{\min}$, of order $\frac{1}{20}(e^2/\hbar a)$, where a is a bond length (but see Mott 1984). This borderline situation is associated with the Yoffe–Regel rule, which implies that if the electronic mean free path is less than one lattice spacing ($\mu \lesssim 1\ \mathrm{cm^2/V\ sec}$) the notion of band conductivity breaks down. This type of behavior could in principle occur in highly disordered solids and in principle it is possible to pass from metallic to insulating behavior with increasing disorder (the so-called Anderson transition; see Section 8.2.2).

8.1.3 Electronic Conduction in Amorphous Silicon

Amorphous silicon is not a uniquely defined substance like crystalline silicon. In particular, a-Si, prepared by the vacuum evaporation of Si films and a-Si:H, prepared by the decomposition of SiH_4 in a glow discharge, have quite different properties. Further variations depend on the precise conditions of formation, for example, substrate temperature, T_s, during deposition. Infrared absorption (see Figure 2.18) shows that bonded hydrogen may be present in a-Si:H to some 40 at% for $T_s = 300$ K, falling to 20 at% for $T_s = 520$ K.* Effects of T_s also show in photoluminescence. As the substrate temperature rises a luminescence band near 1 eV is replaced by a higher-energy band nearer 1.2 eV (Figure 8.3*a*). A contrary trend is seen with temperature of observation (Figure 8.3*b*), where three peaks can be identified, the lowest energy band being most prominent at high T.

Both a-Si and a-Si:H show broad photoluminescence spectra and both exhibit strong epr signals with $g \simeq 2$. This $S = \frac{1}{2}$ spin resonance is normally assumed to be due to single dangling bonds (in crystalline

*Some hydrogen is present as a gas in voids (Chabal and Patel 1984).

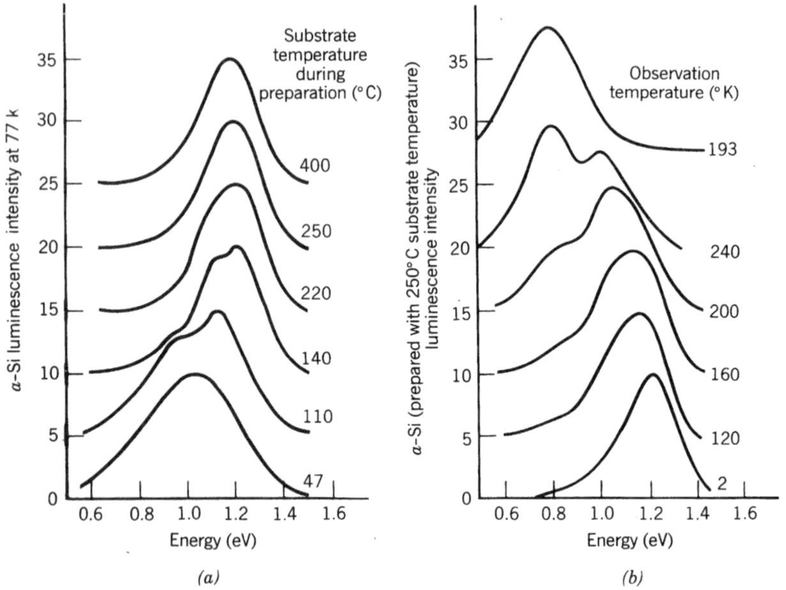

Figure 8.3 Photoluminescence intensity of glow-discharge-produced *a*-Si:H. (*a*) Dependence on substrate temperature during preparation [after Engemann and Fischer (1974)]. (*b*) Dependence on observation temperature [after Engemann and Fischer (1977)].

silicon a vacancy would give four dangling bonds with different characteristics; see Section 5.4). However, *a*-Si contains up to 10^{20} paramagnetic centers per cubic centimeter, several orders of magnitude more than *a*-Si:H, and the efficiency of photoluminescence of *a*-Si is less than that of *a*-Si:H by about two orders of magnitude, due to nonradiative recombination via the dangling bonds.

In passing it seems worth emphasizing that the silicon–hydrogen system is a good example of the difference in behavior between crystalline and noncrystalline forms in defect studies. In contrast to the behavior of *a*-Si, hydrogen is difficult to observe in *c*-Si, being present apparently as interstitial molecules (see Section 5.5.2).

The different behaviors of *a*-Si and *a*-Si:H can be rationalized by assuming the atomic hydrogen produced in the glow discharge reacts with dangling bonds. The simplest reaction

$$\gtrdot\mathrm{Si}- + \mathrm{H} \rightarrow \gtrdot\mathrm{Si}-\mathrm{H} \tag{8.1}$$

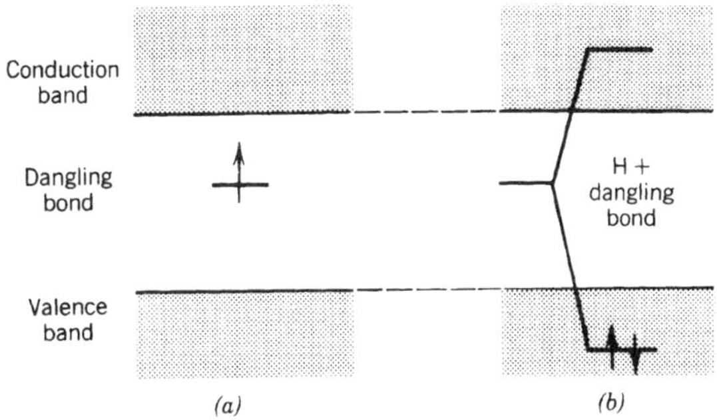

Figure 8.4 Spin pairing of hydrogen with an Si dangling bond in *a*-Si. (*a*) Ground state of the unpaired dangling bond, close to the center of the bandgap. (*b*) Ground state of the paired bond in the valence band.

appears to have a stabilization energy of about 1 eV. The resulting defect has spin zero, and so will not show epr; in essence, the bonding orbital formed from hydrogen and the dangling bond are in the valence band (Figure 8.4). The bonded hydrogen has a characteristic infrared absorption and, as expected for an atom as light as hydrogen, has a relatively high (isotope-dependent) frequency (Figure 2.18). This high frequency is one factor that makes the hydrogenated dangling bond an effective center for nonradiative recombination, since very few local-mode excitations dissipate a lot of energy (see Section 4.3). As a result, nonradiative processes in *a*-Si:H compete with luminescence, to reduce the observed efficiency.

Several techniques give a measure of the density of states in the gap. These include field-effect conductance and capacitance, deep-level transient spectroscopy (DLTS) (Section 4.3.4), and a range of optical and conduction methods. For glow-discharge *a*-Si:H the defect concentration is still quite high (Figure 8.5), although considerably less than in *a*-Si. If P donors are introduced into *a*-Si, there will be little effect on the number of conduction electrons because they fill empty gap states rather than contribute to mobile carrier concentrations. For doping to be highly effective the change in Fermi level per dopant should be as large as possible. The shift per donor is of order $1/N(E_F)$, that is, doping is most effective when there is a low density of states at the Fermi level. Clearly *a*-Si:H is potentially more suitable for doping than *a*-Si. A hydrogen concentration of around 1 at% should suffice to remove most of the dangling-bond states in the gap.

Doping may also be made ineffective by the pentavalent donors acquiring three neighbors and, less probably, by trivalent acceptors

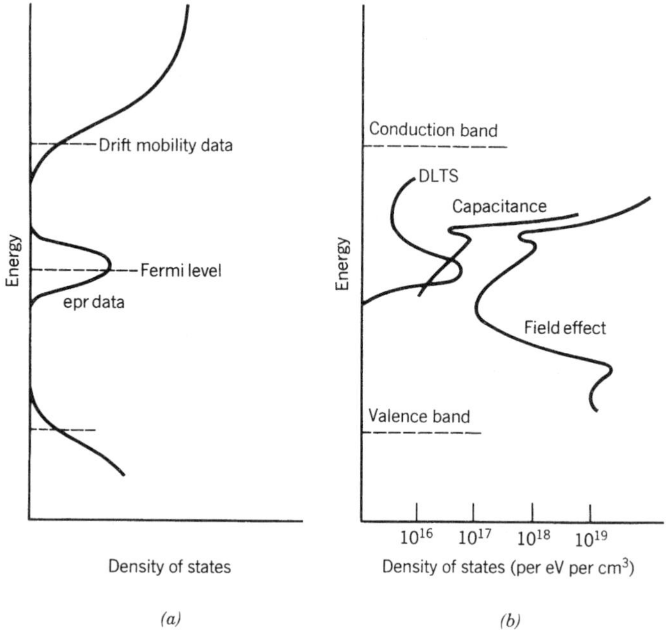

Figure 8.5 (*a*) Schematic representation of the electronic density of states in an amorphous semiconductor, showing some of the experimental methods available. (*b*) Measured density of states for *a*-Si:H from other types of experiment [after Madan et al. (1976)].

acquiring five neighbors during deposition, forming a fully saturated network with no carriers free to conduct. Arguments like this lead to the so-called 8-*N* rule; bond saturation occurs when the sum of the valency of the impurity and its coordination is 8. Similar ideas have also been used to explain resistance of amorphous solids to irradiation, since they have more possibilities of reconstruction to a fully saturated network than crystals.

The electrical conductivity of undoped *a*-Si:H at room temperature is in the range 10^{-10}–$10^{-8}\ \Omega^{-1}\ \mathrm{cm}^{-1}$, the Fermi level position being determined largely by structural defects with energies in the gap. The conductivity shows a remarkable increase when a little PH_3 gas is added to the SiH_4 in the production process. When the ratio of the concentrations $[PH_3]/[SiH_4]$ reaches about 10^{-3}, the conductivity reaches a maximum of about $10^{-2}\ \Omega^{-1}\ \mathrm{cm}^{-1}$ (Figure 8.6). Additions of diborane, B_2H_6, have a pronounced but distinct effect. Small additions decrease the conductivity

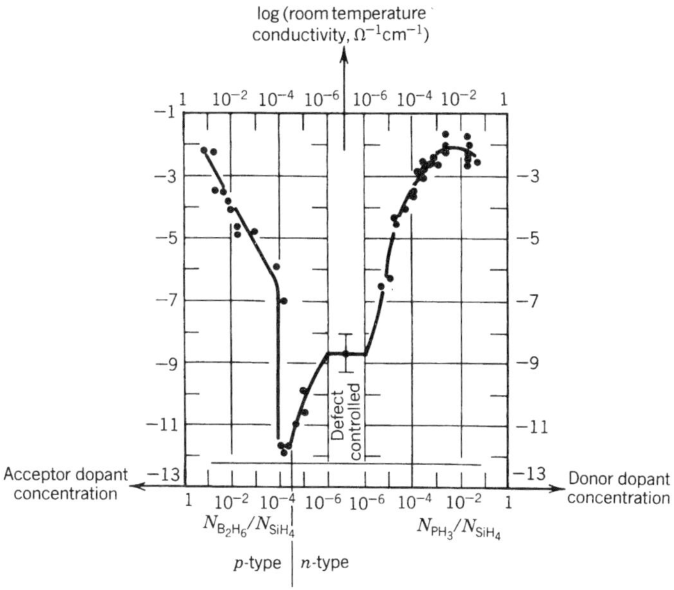

Figure 8.6 Room-temperature conductivity of doped *a*-Si:H. The conductivity is shown as a function of the fraction of donor (PH_3) or acceptor (B_2H_6) molecules compared with SiH_4 molecules in the gas mixture used for specimen preparation. The central region refers to undoped material [after Spear and LeComber (1976)].

to $\sim 10^{-12}\,\Omega^{-1}\,cm^{-1}$ as the boron compensates, reducing the number of electrons in the conduction band from the undoped level. However, as the concentration ratio $[B_2H_6]/[SiH_4]$ increases beyond $\sim 10^{-4}$, hole conduction takes over and conductivity rises by some six orders of magnitude.

While *a*-Si is relatively free of light-induced structural changes, less heavily doped *a*-Si:H shows a novel change in high-intensity light; the electrical conductivity decreases slightly and the dark conductivity falls by a factor 10^4 (Staebler and Wronski 1977). Adler (1981) has suggested a photostructural effect in which two close, unhydrogenated, dangling bonds exchange electrons, creating a donor (with the added electron and net negative charge) and acceptor (with net positive charge) that have a barrier against recombination because of local network reconstruction following the exchange. The conductivity changes arise because of asymmetry in charge trapping at the donor and acceptor levels, leading to a change in the Fermi level position. Raising the temperature restores the original conductivity. It should be said that the precise role of hydrogen is not clear. This photostructural effect has parallels in ionic crystals (see Section 6.2 and 6.3) where defects can also be formed by band-gap energy, with a barrier against recombination.

In amorphous silicon at low temperatures electrical conduction involves the hopping of electrons among random sites. We can see the general features to be expected from this type of conductivity from the expression,

$$\sigma = \frac{1}{6}\frac{ne^2}{k_BT}\langle\Gamma a^2\rangle \tag{8.2}$$

where the average reflects the statistical differences in jump rates Γ and jump distances a throughout the solid [see Equations (3.26) and (3.27)]. The factors determining the jump rate are of particular interest. There is a probability factor $p_i(1-p_f)$ that the initial site i is occupied and the final site f is empty; since the carrier energies must be nearly the same for the initial and final states, it follows that only traps with energy levels within the range $E_F \pm k_BT$ are important. If the density of states is roughly constant near E_F, there will be about $\rho(E_F)k_BT = n$ carriers per unit volume involved. There is a factor $\exp(-W/k_BT)$ which includes any energy difference E_{if}, as well as polaron terms, and there is also a preexponential frequency ω_0. Finally, there is an overlap factor $\exp(-2\alpha R)$, because the rate involves the overlap of initial- and final-state wavefunctions, both falling off exponentially with distance. We therefore expect

$$\Gamma = p_i(1-p_f)\omega_0\exp(-W/k_BT)\exp(-2\alpha R) \tag{8.3}$$

At high temperatures the temperature factor is relatively unimportant and short-range jumps, which maximize the overlap factor, dominate. However, at low temperatures the transitions that dominate have to optimize both the overlap factor and the temperature-dependent term. There will be a distribution of values of W. Within a radius R the average spacing of the energy intervals $\Delta W(R)$ will be that energy interval which contains one trap level on average, so that $\Delta W(R)$ is related to $\rho(E_F)$ by $\Delta W(R) = 1/[\rho(E_F)\cdot\frac{4}{3}\pi R^3]$. As one chooses larger and larger R one can assume, on average, that smaller and smaller values of W are possible, with smaller and smaller values of order $\Delta W(R)$. The transition probability of the dominant transitions minimizes $\{\Delta W(R)/k_BT\} + 2\alpha R$, so that the dominant range $R_M(T)$ depends on temperature, giving the name *variable range hopping*. One readily finds that

$$R_M(T) = [8\pi\alpha\rho(E_F)k_BT\}^{1/4} \tag{8.4}$$

and that $\exp[-(W/k_BT+2\alpha R)]$ takes the form $\exp[-(\theta/T)^{1/4}]$ with $k_B\theta \sim (2.1)^4\alpha^3/\rho(E_F)$. Combining Equations (8.2), (8.3), and (8.4) and, bearing in mind that n in Equation (8.2) is proportional to k_BT (see above), we find that

$$\sigma = Ae^{-B/T^{1/4}} \tag{8.5}$$

where A and B are constants. This $\ln\sigma \propto T^{-1/4}$ law is expected to hold for unhydrogenated amorphous silicon up to ~200 K.

8.1.4 Amorphous Chalcogenides and Chalcogenide Glasses

The chalcogens are the group VI elements, notably S, Se, and Te. They form compounds (chalcogenides) with elements like As, giving As_2Se_3, for instance. Both the elemental forms and the compounds exist in amorphous forms; there are also several crystalline forms of the elements, for example, trigonal and monoclinic polymorphs of Se.

Amorphous selenium has many commercial uses, both in rectifiers and in xerography. In photocopiers, the special properties we discuss in this section are not central; in effect, *a*-Se acts as highly insulating film in which light can create electron–hole pairs. The photocopying process consists of several steps, namely, (1) preparation, where a corona discharge lays positive charges on the surface of an *a*-Se film; (2) illumination, where electrons and holes are optically generated; (3) partial neutralization where, under the action of an electric field, holes are removed to a metal substrate and electrons neutralize localized zones of surface positive charge; and finally (4) a printing step, in which these zones are used to determine the inking of a page.

Structurally, Se appears to form chains or rings, with each Se bonded to two neighbors. The chains and rings are themselves bound by weaker forces, possibly the van der Waals interaction. It is convenient to relate the basic energy-level structure of *a*-Se to that of the Se atom in its $4p^4$, $^3P_2$ valence configuration. The $4p$ orbitals form σ orbitals (along the chain) of both bonding and antibonding character; in addition, there are π orbitals, directed normally to the chain. The bonding (σ) orbitals are fully occupied; two electrons occupy the lone-pair π orbitals, constituting a valence band, and the empty antibonding (σ^*) orbitals form part of the conduction band. The σ^* and σ orbitals are separated by some 7 eV (Figure 8.7).

There are major differences between the behaviors of *a*-Si and *a*-Se, and these are summarized in Table 8.1. In particular, only a very weak epr signal is seen in *a*-Se, corresponding to less than 10^{15} spins/cm^3. The conductivity of pure *a*-Se is likewise low, of order $10^{-17}\ \Omega^{-1}\ \text{cm}^{-1}$ at

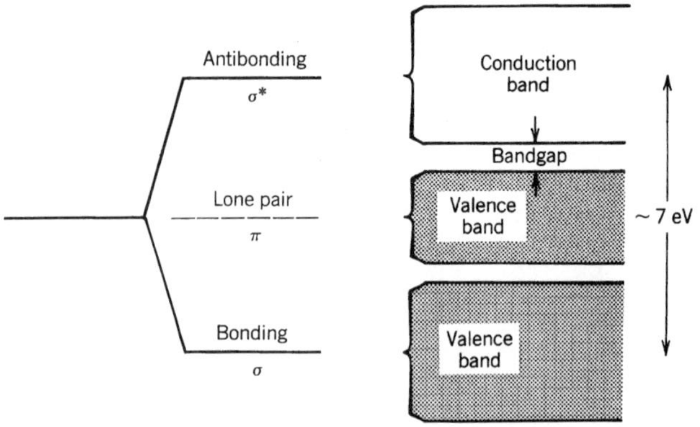

Figure 8.7 Orbitals and energy bands in *a*-Si [after Kastner (1972)].

Table 8.1 Comparison of Properties of Tetrahedrally Bonded Amorphous Systems and of Chalcogenide Glasses

Tetrahedrally Bonded Amorphous Systems		Chalcogenide Glasses
Si, Ge, GeAs, $CdGeAs_2$	Examples	Se, Te, and compounds such as As_2Se_3
Fourfold bonds	Structure	Twofold bonds (chains)
$\sigma \sim 10^{-4}\ \Omega^{-1}\ cm^{-1}$ at room temperature, variable range hopping at low temperatures with $\sigma \sim 10^{-10}$–10^{-8} $\Omega^{-1}\ cm^{-1}$	Conduction (see also Figure 1.8)	Generally bandlike with $\sigma \sim 10^{-17}$–10^{-12} $\Omega^{-1}\ cm^{-1}$ at room temperature
Shows epr ($\leq 10^{20}$ spins cm^{-3})	epr	Weak epr seen ($\leq 10^{15}$ spins cm^{-3})
Si can be doped *n* or *p* type	Doping	Apparently cannot be doped
Valency type, and coordination defects, for example, five- and seven-member rings	Defect structure	Valence-alternation pairs with *negative U* behavior

room temperature (see also Figure 1.8), principally because of the larger band gap. These two features are rationalized by the so-called negative U behavior which pins the Fermi level and eliminates unpaired spins. While the one-electron density of states near the Fermi level is finite, the electrons have an overwhelming tendency to pair, so that very stable $S=0$ defects replace $S=\frac{1}{2}$ defects (Anderson 1975). This contrasts with a-Si, where $U>0$ and pairing does not occur.

We can see the negative U effect most simply in the case of dangling bonds (Street and Mott 1975). Suppose D^0 is a dangling bond containing a single unpaired electron and hence has $S=\frac{1}{2}$. If an electron is transferred from one such center to another,

$$2D^0 \rightarrow D^+ + D^- \tag{8.6}$$

both the empty dangling bond (D^+) and the doubly occupied dangling bond (D^-) have spin zero. If the energy U needed to drive the reaction is negative, it proceeds spontaneously. There are two principal terms contributing to U, one being purely electronic. In a rigid lattice one can relate the electronic energy term U_{electr} to the electron affinity A and ionization potential I:

$$\begin{aligned} &D^0 \rightarrow D^+ + e && \text{absorbing energy } I \\ &D^0 + e \rightarrow D^- && \text{emitting energy } A \\ &2D^0 \rightarrow D^+ + D^- && \text{absorbing energy } U_{\text{electr}} = I - A \end{aligned} \tag{8.7}$$

In all known systems (e.g., all free atoms) U_{electr} is positive.

The second contribution to U comes from lattice relaxation. If a defect exerts a force F on the lattice, and if the restoring elastic force constants are K, then there will be a relaxation energy $F^2/2K$ (see Section 1.2.2). We shall suppose that the defect force F is proportional to the number of electrons in the dangling bond. In the case of the singly-occupied, neutral dangling bond, D^0, we write $F_0=f$, say; in the doubly occupied bond $F_-=2f$ and the unoccupied bond has $F_+=0$. The relaxation energy for $2D^0$ is thus $2f^2/2K$, that for D^+ is zero and that for D^- is $(2f)^2/2K$. Overall, there is a contribution to U given by

$$U_{\text{elast}} = 2\frac{f^2}{2K} - \frac{(2f)^2}{2K} - 0 = -\frac{f^2}{K} \tag{8.8}$$

The total correlation energy U is the sum of a positive (U_{electr}) and a negative term (U_{elast}). This is shown in the configuration coordinate picture in Figure 8.8. In the chalcogenides the negative term from lattice relaxation dominates. However, U is a defect property, not just a host

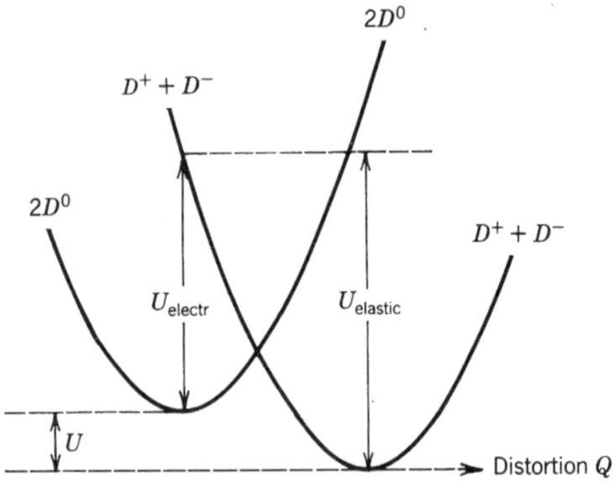

Figure 8.8 Negative U models for a-Se and similar systems. The electronic contribution U_{electr} is always positive; the polarization and distortion terms U_{elastic} lead to a net negative value of $U = U_{\text{electr}} - |U_{\text{elastic}}|$.

property. Even though the dominant defects in the tetrahedrally coordinated amorphous semiconductors have positive U, there are special cases (like V^+, the positive vacancy in crystalline silicon) that have negative U (Section 5.4).

In the chalcogenides the dangling-bond states are those shown in Figure 8.9. Here we follow the notation of Kastner and Fritsche (1978), who write C_N^{M+} for a defect at a chalcogen site with coordination N and

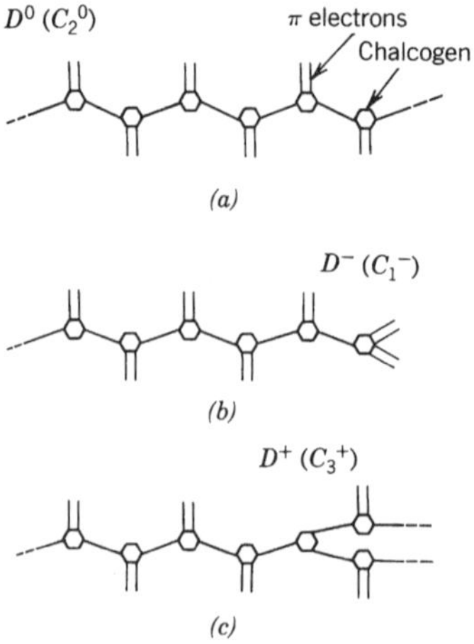

Figure 8.9 Dangling-bond states in chalcogens. The open circles represent chalcogen atoms, with lines representing bonds. In (a) the perfect chain is shown. In (b) the chain terminates, giving an empty state below the Fermi level. With an extra electron, four π electrons are associated with the end atom. In (c) an overcoordinated atom gives a state above the Fermi level; the atom has lost an electron, and has no π electrons.

charge $M+$ (a full discussion of defect notations is given in the Appendix). Thus one has three states

$$\begin{aligned} D^0 &\rightarrow C_2^0 \\ D^+ &\rightarrow C_3^+ \\ D^- &\rightarrow C_1^- \end{aligned} \tag{8.9}$$

and C_3^+ and C_1^- are sometimes known as valence-alternation pairs. These do not form in crystalline Se, although they appear to form in both amorphous and crystalline binary chalcogenides like As_2Se_3. One dramatic effect is the increase in conductivity caused by removal of D^- centers by doping. For *a*-Se, about 30 ppm of oxygen suffices to raise the conductivity by a factor of 10^6; 500 ppm of chlorine produce an increase by a factor of 10^8.

The binary chalcogenides, notably As_2X_3 with $X = S$, Se, Te, have been studied extensively, especially As_2Se_3. The basic structural model presumed is a continuous random network, with threefold-coordinated As and twofold-coordinated Se. This is an oversimplification; for example, Raman data for films of *a*-As_2Se_2 show more disorder than the bulk glass, with Se–Se bonds and molecular units [perhaps (As_4S_4)] in evidence. Annealed films largely polymerize into the continuous random network.

As-prepared *a*-As_2Se_3 has a similarly low level of epr centers to *a*-Se, less than $10^{15}\ cm^{-3}$. The bandgap is about 1.7 eV. In luminescence, the main feature (Figure 8.10) is a peak at 0.8 eV, about 0.25 eV wide at

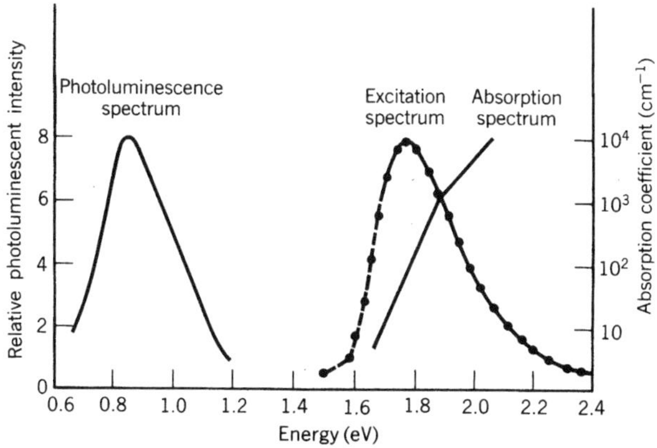

Figure 8.10 Photoluminescence, excitation, and absorption spectra for *v*-As_2Se_3 at 6 K [after Bishop and Mitchell (1973)].

half-maximum intensity. The peak is similar in both amorphous and crystalline As_2Se_3, and its shape is independent of excitation energy. Cooling from 300 to 4 K enhances the intensity of the peak by several orders of magnitude, reaching a maximum quantum efficiency of around 15%. Luminescence is most readily excited in the tail of the optical absorption at around 1.8 eV in the amorphous form; in the crystal there is no similar fall-off in efficiency at higher energies.

It seems that the luminescence in both a-As_2Se_3 and c-As_2Se_3 is caused by similar defects. Photoluminescence is generated either when these defects are excited directly, or when an electron–hole pair, optically created, recombines at the defect. The fall-off in efficiency observed at low energies in both crystalline and amorphous forms simply reflects the decreased optical absorption, with a consequent reduction in electron–hole pairs created. The difference in efficiency at higher energies between crystalline and amorphous samples, however, appears to be associated with carrier mobility. In c-As_2Se_3, carrier transport to radiative centers is more effective; indeed, the efficiency may approach unity at low temperatures. In a-As_2Se_3 there is competition, possibly due to electron–hole dissociation, which leads to nonradiative recombination. Further discussion of the origin of luminescence in both c-As_2Se_3 and a-As_2Se_3 has been given by Robins and Castner (1984).

Street (1976) suggested that the luminescence results from charged D^+ and D^- with levels near a band edge. For example, Figure 8.11*a* shows an exciton pair created by the exciting light; the exciton becomes trapped (Figure 8.11*b*) at a D^- defect, giving an excited state D^{-*} of the defect. Lattice relaxation occurs, so that the luminescence is Stokes shifted (Figure 8.11*c*) from the energy needed for direct excitation (Figure 8.11*d*) of D^-. Similar processes for D^+ can be envisaged. At

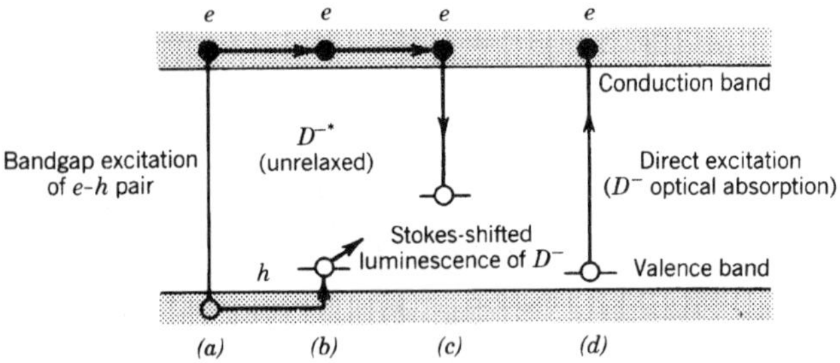

Figure 8.11 Proposed luminescence process associated with D^- in a-As_2Se_3 [after Street (1976)]. The various steps in the process are discussed in the text.

high temperatures, excitons created as a first step can dissociate, resulting in a free electron and D^0 instead of D^{-*}, followed by nonradiative recombination. The proposed radiative sequence is

$$h\nu_a \rightarrow [eh]$$
$$[eh] + D^- \rightarrow D^{-*}$$
$$D^{-*} \rightarrow D^- + h\nu_e$$

and the proposed nonradiative sequence is

$$h\nu_a \rightarrow [eh]$$
$$[eh] + D^- \rightarrow e + h + D^-$$
$$\rightarrow e + D^0$$
$$e + D^0 \rightarrow D^- + \text{heat}$$

8.1.5 Silica and Silicate Glasses

Silicate-based glasses and glassy forms of silica (SiO_2) are among the commonest and most important nonmetals. They are major components of optical fibers and of materials from bricks to window glasses. They provide both insulating layers in silicon-based semiconductor devices and protective layers on metals.

The basic, defect-free, structure of amorphous silica can be imagined in several ways. The first structural rule is that each atom has only neighbors of the other species. The second rule is that each has fixed coordination, with four oxygens surrounding each Si and two silicons bound to each oxygen. Structurally, one can imagine a continuous random network of SiO_4 tetrahedra joined by common oxygens; the structure can also be conceived (rather like a-Si) in terms of six-membered rings of (SiO) units. Oxygens that form part of the network are referred to as bridging oxygens. If an oxygen is bound to only one neighbor, it is known as nonbridging. We shall see that the structural rules are such as to give parallels in behavior both between a-SiO_2 and a-Si and between a-SiO_2 and crystalline oxides.

We shall discuss two main classes of defect in a-SiO_2, (a) those intrinsic centers in amorphous SiO_2 that are produced by electron irradiation (see also Section 6.2.3) and (b) those centers produced by additions of other oxides. It is these oxide additions that give control over glass properties such as melting temperature, resistance to chemical attack, color, and electrical conductivity.

(a) Intrinsic Defects in Irradiated a-SiO_2

Although epr has proved a useful tool in identifying point defects in a-SiO_2 caused by electron irradiation, it should be kept in mind that angular averaging of spectra in glasses may lead to uncertainty in interpretation. Among the most important models proposed are (Griscom 1975; Griscom and Friebele 1982):

1. *E′ Center.* Here an unpaired spin occupies a dangling bond on a silicon, that is, essentially in an sp^3 hybrid (see Section 8.4). It is natural to regard this center as resulting from an oxygen vacancy. Removing an oxygen atom leaves two silicons with incorrect coordination; it is assumed (Figure 8.12) that one of the silicons retains a single electron in the dangling bond. The other silicon either loses its electron to some sink, relaxing to a planar SiO_3 geometry (so that O^- has, in effect, been removed overall) or it gains an electron to give a doubly occupied dangling bond (so that O^+ has, in effect, been removed overall).

2. *Peroxy Radical.* An unpaired spin is associated with two adjacent oxygens (i.e., there is a structural defect, contrary to the Si/O ordering rules, like the Bjerrum D defect in ice (see Section 8.1.1)). Two models for formation have been suggested. One relies on a peroxy linkage already existing in thermal equilibrium (Figure 8.13a). The other suggestion involves capture of an O_2 molecule by an E' center (Figure 8.13b). Both pictures are consistent with the observed uneven spin density distribution, in which 75% of the spin density lies on the oxygen more distant from the Si.

3. *Nonbridging Oxygen Hole Center* (*NBOHC*). Again an unpaired spin is associated with oxygen, although this time (Figure 8.14) with a single nonbridging ion. The natural assumption is that the nonbridging ion is initially associated with hydrogen, unintentionally but inevitably present in most silicas; ionization leads to defect reactions that remove the hydrogen, leaving the so-called NBOHC. The unpaired spin is in a π orbital, normal to the Si–O axis.

The nonbridging oxygen center has some parallels with the V^- center (Section 5.1.2) in crystalline oxides, although there the hole is in the p_σ orbital, rather than in the p_π orbital, as here. Molecular oxygen centers, for example, O_2^-, are frequently found as surface defects in irradiated oxides. The E' center, however, does not resemble the F^+ centers one might have expected from most ionic oxides; instead, a closer parallel is

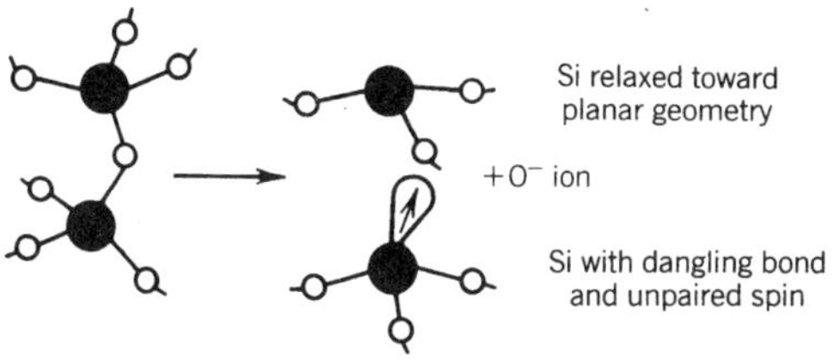

Figure 8.12 Possible formation mechanism for the E' center in irradiated a-SiO_2. Note that, while O^- is removed in the sequence shown, removal of O^+ would be equally consistent with experiment. (Solid circles, Si; open circles, O.)

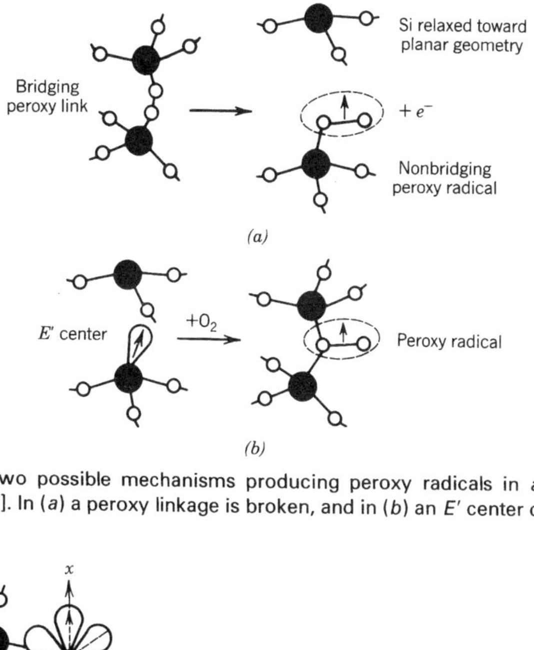

Figure 8.13 Two possible mechanisms producing peroxy radicals in a-SiO_2 [see Griscom (1975)]. In (*a*) a peroxy linkage is broken, and in (*b*) an E' center captures an oxygen.

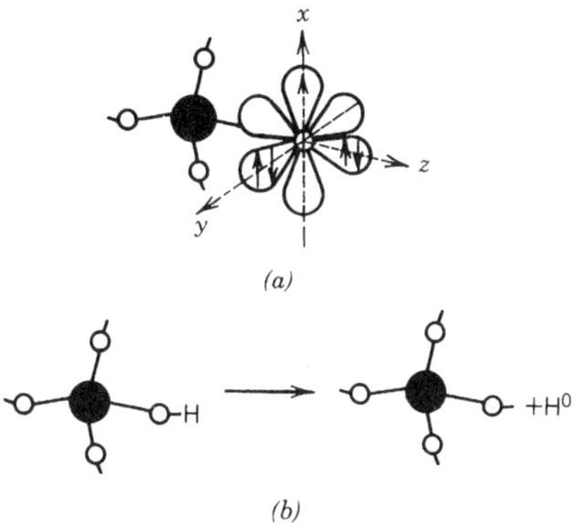

Figure 8.14 (*a*) A model for the nonbridging oxygen hole center in irradiated v-SiO_2 and (*b*) its likely formation mechanism through rupture of an OH bond. In (*a*) the hole is in a π_x orbital of the oxygen, rather than the σ orbital of the oxygen hole in the V^- center (Figure 5.6).

the positively charged vacancy, V^+, in *c*-Si (Section 5.4), where a trigonal Jahn–Teller distortion localizes an unpaired spin in an sp^3 hybrid (see Section 8.4) on a single silicon.

(b) Defects Associated with Dopants

An important division occurs with dopant ions between the network formers and the network modifiers. The former, typically group III (B, Al), group IV (Ge), or group V (P, As) elements, participate in the network, replacing silicon. Sometimes they enter with their own appropriate coordination, for example, threefold-coordinated boron; usually, however, they are compensated in some other way. Network modifiers, typically alkalis (Na, K) or, to a lesser extent, halogens, enter interstitially. They do not participate in the network; indeed, they tend to disrupt it in the sense that they cause silicate glasses to soften at lower temperatures. Clearly, the mobile alkali modifiers can provide charge compensation for nonbridging oxygens or for network formers. While the division into formers and modifiers is useful and mainly unambiguous, special care is needed in considering transition-metal ions which, in effect, form a class of their own (Doremus 1973).

In an amorphous host one cannot readily exploit anisotropy of an epr signal. There may therefore remain some uncertainty about the defect observed, especially when the g factor is close to the free-spin value. This is the case for the boron-associated oxygen hole center (BOHC), for which there are two models, both of which appear to be adequate. One model invokes an unpaired π orbital for a bridging oxygen; the other invokes an unpaired π orbital for a nonbridging oxygen (Figures 8.14*a* and 8.14*b*). The bridging version is analogous to the acceptor formed by B in *c*-Si, with the important difference that the hole is localized, rather than the highly delocalized form in *c*-Si:B.

The centers referred to in the previous paragraph are themselves members of a class of similar centers. The so-called wet oxygen center, for example (Stapelbroek et al. 1979), involves a hole on a nonbridging oxygen apparently formed by reactions in which a hydroxyl radical loses hydrogen. The phosphorus–oxygen hole center (POHC) is a more complex variant; the unpaired spin is associated with two adjacent nonbridging oxygens, each bonded to the same phosphorus. The donor analog of the POHC is the so-called P_2 defect seen in irradiated crystalline α-quartz. Here the unpaired spin is associated with a dangling bond on a fourfold-coordinated phosphorus. There is a stronger analogy of the P_2 defect with the $[PO_4]^0$ radical than with the shallow P donor in crystalline silicon.

8.2 METAL–INSULATOR TRANSITIONS

In describing the many and varied ways an insulator can transform to show metallic behavior, we shall encounter some of the ideas presented in Chapter 1. Thus the competition between the low kinetic energy for delocalized wavefunctions (band formation), effects of correlation, and the Madelung energy gains from charge transfer all appear. There are also new features, to be discussed here, like the so-called Anderson localization in heavily disordered systems and the Peierls instability mechanism for one-dimensional systems. This variety of behavior has become important in many practical areas. Metal-oxide–semiconductor junctions offer a controllable means of studying disordered systems, quite apart from any clarification of device behavior (Sham and Nakayama 1979; Laughlin, Joannopoulos, and Chadi 1980). Dielectric breakdown has long been an important and practical, if extreme, example of a metal–insulator transition. Switching at critical voltages or currents and the related aspects of memory have many possible applications.

Since the range of metal–insulator transitions itself constitutes a major field of current activity (Mott 1984), we shall merely sketch some of the contributing factors in representative systems, emphasizing those cases where defects are important. It must be stressed that there is still no general agreement on the mechanisms operating in most metal–insulator transitions. Nevertheless, the mechanisms proposed illustrate aspects of defect physics in a broader context. In Sections 8.2.1 and 8.2.2 we shall find mechanisms that relate to the number of bound states associated with isolated defects and to the additional effects of defect–defect interactions and screening. In Section 8.2.3 we find instabilities that may be compared with processes of surface reconstruction (see Section 7.2) or self-trapping (see Section 1.2.3). In Section 8.2.4, dealing with dielectric breakdown, the closest parallels are with the mechanism of nonradiative transitions (see Section 4.3), although clearly the details of the processes differ.

8.2.1 Mott–Hubbard Transitions

Consider a regular array of one-electron centers, perhaps H atoms, with a large intercenter spacing d. This will be a nonmetal since a finite energy is needed for electron transfer between the centers (Section 1.1.3). As the spacing is decreased, two changes occur: the energy gain from forming extended states increases and screening reduces the energy needed for electron transfer. At a critical spacing $d_{\text{crit}} \simeq \alpha a_{\text{H}}^*$, where a_{H}^* is the effective Bohr radius for the one-electron centers, the metallic state is favored. This type of relation between d_{crit} and a_H^* occurs for several distinct mechanisms, often with similar values of α.

The Hubbard model (Section 1.1.3) gives an especially useful analytic treatment of the behavior described above. A key energy is the intra-atomic correlation energy, U, which we encountered earlier in Section 8.1.4. This is the extra electron–electron repulsion energy required to put a second electron on one of the centers. If an extra electron is added to the array of one-electron centers, it can move through the lattice with band width $2B_1$. The gain in kinetic energy from delocalization in this upper Hubbard band is therefore B_1. If an electron is removed from one of the one-electron centers, the hole can move through the lattice. The resulting lower Hubbard band has width $2B_2$, with gain in energy B_2 from delocalization. When the total delocalization energy $(B_1 + B_2)$ exceeds U (see Figure 1.7), we can expect a transition from the insulating to a conducting state. Such transitions are referred to as Mott–Hubbard transitions.

The Hubbard model has been applied to many systems, notably oxides of transition metals, including UO_2 and NiO. NiO is worth special comment for another reason, namely, that simple band theory fails dramatically. In the NiO crystal the $3d$ orbitals split into those of t_{2g} and of e_g symmetry (Section 5.2). The eight $3d$ electrons fill the orbital triplet (six electrons), leaving two electrons which half-fill the orbital doublet. Thus, simple band theory would argue that one has a half-filled band with metallic conduction. Even without going into the complex escape routes attempted within conventional band theory, one can see how important it is to remember that crystals are built of atoms (manifest in correlation, as in the Hubbard U parameter, for instance) and not merely controlled by translational symmetry. In the case of NiO the delocalization energy associated with the relatively narrow $3d$ bands (width ~ 1 eV) is not enough to overcome the correlation energy, and NiO is an insulator.

Changes between metallic and nonmetallic behavior under hydrostatic pressure are also referred to as Mott–Hubbard transitions if they occur without abrupt changes in volume or structure. In the Mott–Hubbard picture it is the relative values of U and of the bandwidths B_1 and B_2 that control behavior. One can imagine applied pressure increasing B_1 and B_2 but leaving the intraatomic U essentially unaltered, until a transition occurs. This appears to be the case for iodine at 150 kbar, for example. Metallic ytterbium is an unusual case, since it becomes insulating at 20 kbar; the explanation is uncertain, although there have been arguments to suggest that there is a discontinuity in the number of free electrons in the transition.

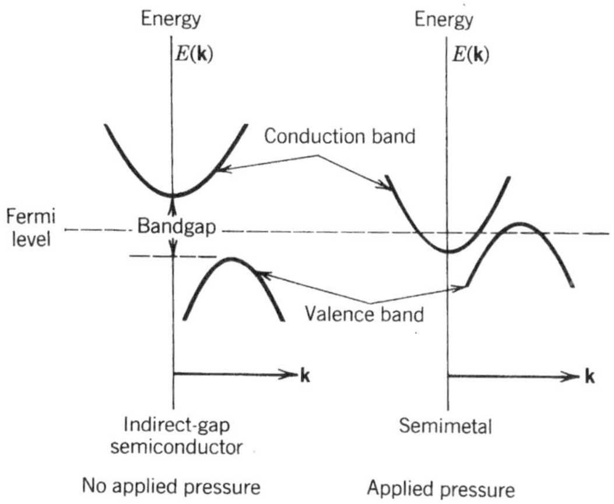

Figure 8.15 Metal–insulator transition in an indirect-gap semiconductor caused by pressure.

Other types of metal–insulator transition can occur under pressure, notably band-crossing transitions. Suppose the valence-band maximum and the conduction-band minimum are at very different points in the Brillouin zone and that compressing the solid moves them closer in energy (Figure 8.15). Ultimately the two bands cross without merging. Clearly a metal–insulator transition can occur, because, after crossing, there will be excitations of either electrons or holes needing negligible energy. The nature of the transition may be complex in detail, both experimentally and theoretically. In theory, one should consider the balance between positive kinetic-energy terms (if there are n electrons and n holes, these contribute an energy $An^{2/3}$ in a crystal, with $A \equiv 6\hbar^2/5m_{\text{eff}}$, where m_{eff} is a suitably averaged effective mass) and the attractive Coulomb interaction, $-Bn^{1/3}$, between electrons and holes. B also contains electron–hole correlation terms, and there have been suggestions of an insulating superfluid of Bose-condensed electron–hole pairs under suitable conditions. However, for our purposes, we simply need to note that as the ratio B/A increases there will be a discontinuous change of n from zero to a value that minimizes $An^{2/3} - Bn^{1/3}$. Experimentally, there are problems in detecting the transition from finite (insulator) to zero (metal) activation energy, partly because of the contribution of thermally produced carriers and partly because of the likely discontinuous change in volume and, possibly, in crystal structure. In amorphous systems, when there is an exponential tail to the density of states, the $An^{2/3}$ term is replaced by one varying more rapidly with n, and it is not necessary for there to be a discontinuity in n.

8.2.2 Heavily Doped Semiconductors and Anderson Localization

In a semiconductor like silicon, doping with phosphorus is a controllable way of introducing carriers. At sufficiently high doping concentrations, transition to metallic conductivity occurs. Details of behavior as doping levels change depends both on actual concentrations of donors (N_D) and acceptors (N_A) and on the degree of compensation ($N_D - N_A$). At low concentrations thermal ionization of the donors is the important process, with an activation energy ϵ_1. Clearly, ϵ_1 is essentially the donor ionization energy, E_D, although corrections have to be made for two factors at finite concentrations. First, there may be sufficient overlap of normally bound excited states on different donors to give conduction without the full ionization energy. Second, the other donors are polarizable and, basically because the effective dielectric constant changes, ϵ_1 decreases from E_D as the donor concentration rises.

A second conduction process can be identified from the Hubbard mechanism. There should be transport with activation energy $\epsilon_2 \sim \frac{1}{2}[U-(B_1+B_2)]$, involving an upper band of width $2B_1$, in which two electrons are associated with a given donor, and a lower band of width $2B_2$ in which the donor is unoccupied (see Figure 1.7).

Neither of these mechanisms takes any account of the randomness in space of the donors or acceptors, and this has several consequences. There will, for example, be random variations in energy and electric field from site to site and variable-range hopping (Section 8.1.3) at low temperatures. We may therefore anticipate a third activation energy, ϵ_3, associated with the spread of energies from site to site when nearest-neighbor jumps dominate.

Another possible consequence of randomness is Anderson localization (Anderson 1958) to which we turn now. This phenomenon is one of several that can be invoked in describing the metal–insulator transition in heavily doped semiconductors, and it should be stressed that there are still open questions as to which type of transition actually occurs in these systems.

Anderson considered a regular lattice in which there was an isolated band of width $2B = 2zJ$ on which were superimposed random perturbations; here z is the coordination number and J is the transfer integral (see Section 1.1). These perturbations, characterized by a mean square value $U_0^2 \equiv \langle U^2 \rangle$, shift the energy of a carrier localized at any one of the sites (this is the so-called diagonal disorder, rather than off-diagonal disorder that would affect the transfer integrals). Anderson's theorem asserts that, if a carrier is placed on an atom at $\mathbf{r} = 0$ at time $t = 0$, it will not diffuse away provided U_0/J exceeds a critical value. In this case all states (except for a statistically negligible fraction) are localized, that is, their wavefunctions contain an attenuating factor, $\exp(-\alpha|\mathbf{r} - \mathbf{r}_0|)$, localizing them near some $\mathbf{r}_0$.

Localization also appears in a distinct but related context when one considers trapping by an isolated defect. If there is an attractive perturbation A at one site only, then, provided A is large enough (strictly A/J greater than a critical value), a bound state emerges. This leads to the possible localization of a small number $[O(1)]$ of the many $[O(N)]$ states in the crystal, and is the usual phenomenon of defect trapping. This behavior is not itself Anderson localization, although we can connect the Anderson and trapping limits by imagining the perturbations switched on in the Anderson case at only a fraction of the sites. As this fraction increases, one moves from isolated trapping into the Anderson localization regime. Thus, if an exciton is formed at random, or if a particle like a positive muon is injected into the crystal, its behavior depends on how many sites are perturbed. When only a few sites have altered energies, the particle may be localized at the first trap (i.e., the first site for which there is a localized state) that it encounters. This is distinct from Anderson localization because there is not a stochastic variation of energy from site to site nor need all the states be localized (Figure 8.16).

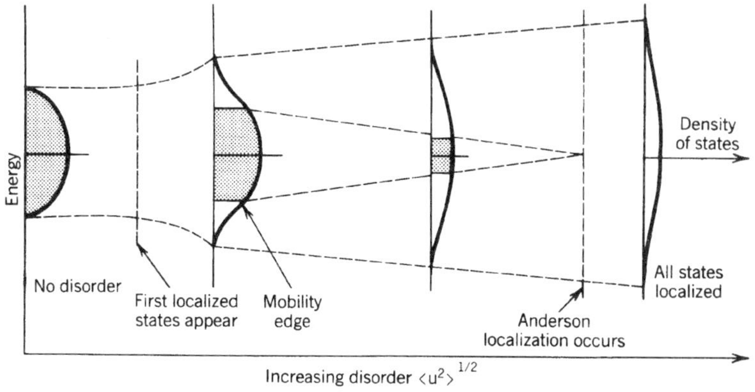

Figure 8.16 Development of Anderson localization with increasing disorder in a tight-binding band. The densities of states of localized (unshaded) and delocalized (shaded) states are shown.

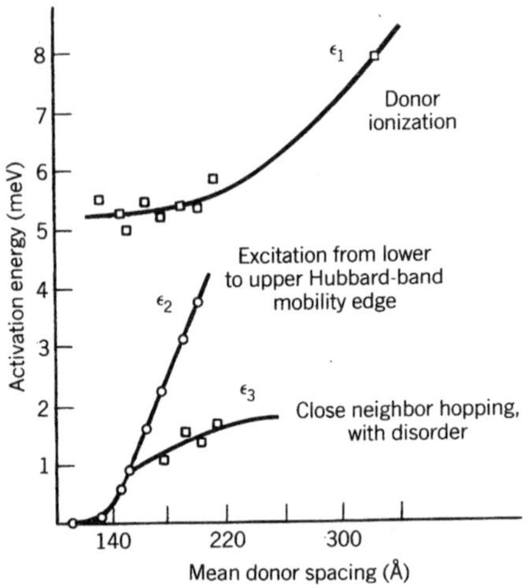

Figure 8.17 Activation energies for heavily doped *n*-type Ge as a function of donor concentration [after Davis and Compton (1965)].

Whichever mechanism causes the metal–insulator transition in heavily doped semiconductors, we can anticipate that it will happen when the orbital radius a_{H}^{*} of the shallow donors or acceptors [$a_{\mathrm{H}}^{*} \sim \epsilon_0(m^*/m_e)$ Bohr radii, where ϵ_0 is the dielectric constant and m^* is the effective mass] and their average spacing L [where $3/(4\pi L^3)$ is the number per unit volume] are comparable. For Si and Ge this suggests a transition for concentrations around $10^{18}\ \mathrm{cm}^{-3}$, with rather lower concentrations for Ge and higher for Si, and a modest dependence on which donor or acceptor is involved. Experiment (Figure 8.17) shows activation energies corresponding to ϵ_1, ϵ_2, and ϵ_3, all with a concentration dependence; their different magnitudes at a given concentration result in mechanisms changing as the temperature is altered. At the highest temperatures straightforward ionization (ϵ_1 mechanism) dominates. At the lowest temperatures one finds variable-range hopping [Equation (8.5)], or thermally activated motion with activation energy ϵ_3. It is in the intermediate temperature regime (associated with ϵ_2, where the Hubbard model is appropriate) that Anderson localization appears to apply.

While heavily doped semiconductors do not strictly fulfill the conditions under which Anderson localization can be demonstrated, the

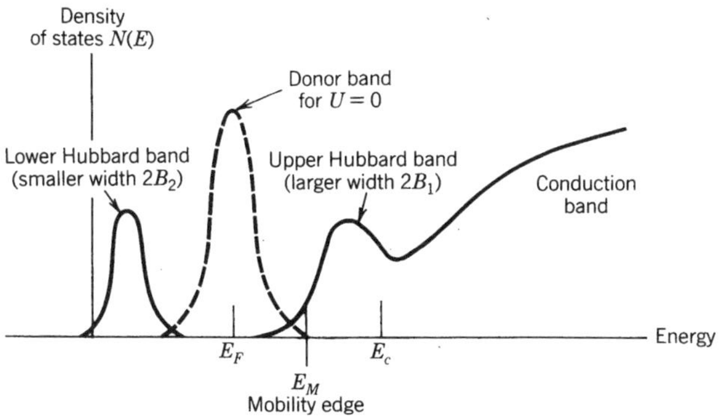

Figure 8.18 Upper and lower Hubbard bands for a high concentration of donors in a semiconductor (see also Figure 1.7).

phenomenon appears to occur in these systems in two ways. First, the narrower, lower, Hubbard band (associated with empty donors) does appear to be Anderson localized. Second, the broader, upper Hubbard band (associated with doubly occupied donors) may have Anderson localization in its tail. The mobility edge is shown as E_M in Figure 8.18. Thus, we expect an activation energy $E_M - E_F \equiv \epsilon_2$ in the relevant temperature range. As the donor concentration increases, the localized states are the first to overlap. It is only when the extended states overlap that the insulator-to-metal transition results. This leads (as do several other mechanisms) to a critical concentration of donors inversely proportional to the cube of the effective Bohr radius a_{H}^* (Figure 8.19). The situation just described is typical of uncompensated semiconductors; for compensated semiconductors Mott (1982) has argued that the Anderson transition is again involved.

Rather similar behavior can occur in the very different narrow-band oxides. In $SrTiO_3$, for instance, where the Ti $3d$ orbitals are empty, excess metal acts as a donor, contributing electrons to the narrow d band. The strong electron–lattice interaction modifies many details; the static dielectric constant is around 220, and the carrier effective mass is around $10m_e$, presumably including a strong polaron effect. The metal–insulator transition occurs at around 5×10^{18} excess metal ions per cubic centimeter, corresponding to the value expected for an effective Bohr radius of 10–15 Å.

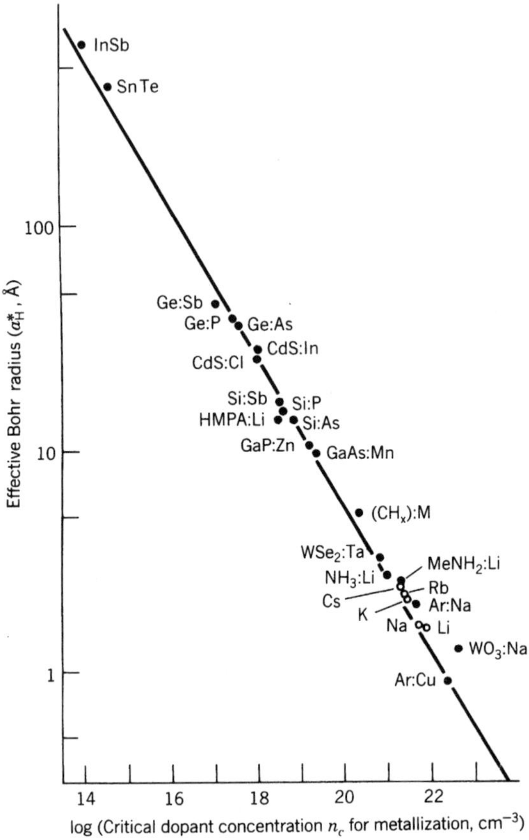

Figure 8.19 Critical concentration n_c (cm^{-3}) of donors for the insulator–metal transition as a function of effective Bohr radius a_H^* (Å). The straight line corresponds to $n_c^{1/3} a_H^* = 0.26$ [after Edwards and Sienko (1981)].

8.2.3 One-Dimensional Conductors and the Peierls Transition

In molecular solids the molecules retain their individual identity, weakly held together by van der Waals forces. The equilibrium intermolecular spacings are still quite large, resulting in bandwidths of only 0.01–0.1 eV. Such organic molecules, for example, anthracene, are insulators, with fully saturated bonds and, so, without partly filled bands. However, one can produce organic salts which, at least in single-crystal form, are good electronic conductors (Jerome and Schulz 1982). In some cases they give rise to superconductivity at low temperatures. For structural reasons we

Table 8.2 Molecular Crystals Showing One-Dimensional Conductivity

TTF (TCNQ)	Tetrathiafulvalene (7,7,8-tetracyano-*p*-quinodimethane)
NMP (TCNQ)	*N*-methyl phenazinium (7,7,8-tetracyano-*p*-quinodimethane)
KCP	$K_2Pt(CN_4)_4Br_{0.3}\cdot 3H_2O$

shall see that they usually have highly anisotropic conduction, in which carriers move essentially along a preferred axis. This one dimensionality, in turn, favors a certain type of metal–insulator transition.

One dimensionality is shown by molecular crystals known as TTF–TCNQ, as KCP, and as NMP–TCNQ. The full chemical description is given in Table 8.2, and some molecular components are illustrated in Figure 8.20. Both TTF–TCNQ and NMP–TCNQ grow in a herringbone

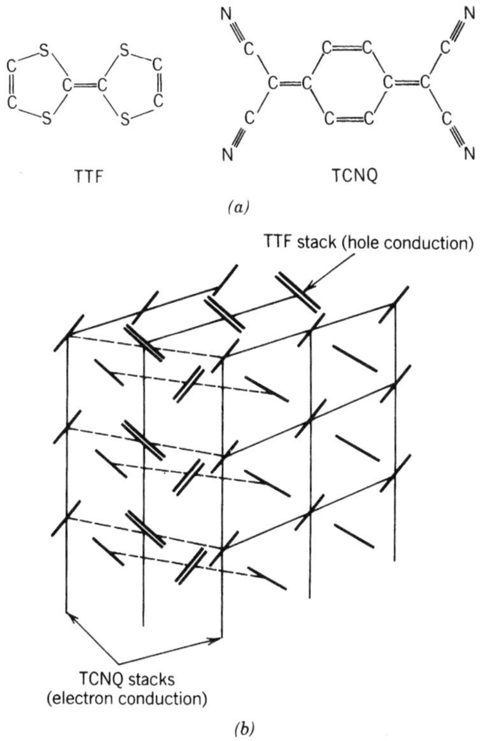

Figure 8.20 (*a*) Molecular forms of TTF and TCNQ. (*b*) Structure (shown schematically) of TTF–TCNQ. High conductivity is along the stacks, that is, vertically in the diagram.

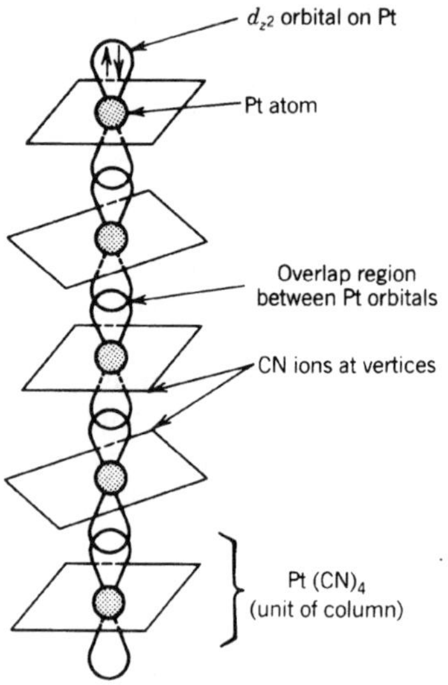

Figure 8.21 Schematic representation of one-component chain in the structure of KCP.

structure, with planes of stacked, parallel molecular units. TTF–TCNQ generally grows as black needles with the axial conductivity, $\sigma_{\|}$, much larger than the perpendicular value, $\sigma_{\perp}$. The electrical conduction arises from two factors. One is that the orbitals normal to each planar unit in the stack can overlap with similar orbitals along the axis of the stack. The second factor is electron transfer; the TTF molecule acts as a donor, contributing about 0.6 electron per molecule to the acceptor TCNQ. Electronic conductivity occurs, in both stacks, in the partly filled bands that result.

Both the structure and general properties of KCP differ from TTF–TCNQ, although again there is intermolecular overlap along an axis. In KCP (Figure 8.21) overlap of the $5d\,(3z^2 - r^2)$ orbitals of the planar $Pt(CH_4)_4$ complexes is involved. Normally these orbitals would be fully occupied, but here Br acts as an acceptor, removing about 0.3 electron from each Pt complex. Again, conduction results from the partly filled band of states overlapping along an axis.

The conductivity shows several interesting features as a function of temperature (Figure 8.22). In KCP and in NMP–TCNQ the conductivity falls as the temperature is lowered. Thermal activation over barriers between conducting axial segments may be part of the explanation, although the precise temperature dependence is not of the Arrhenius

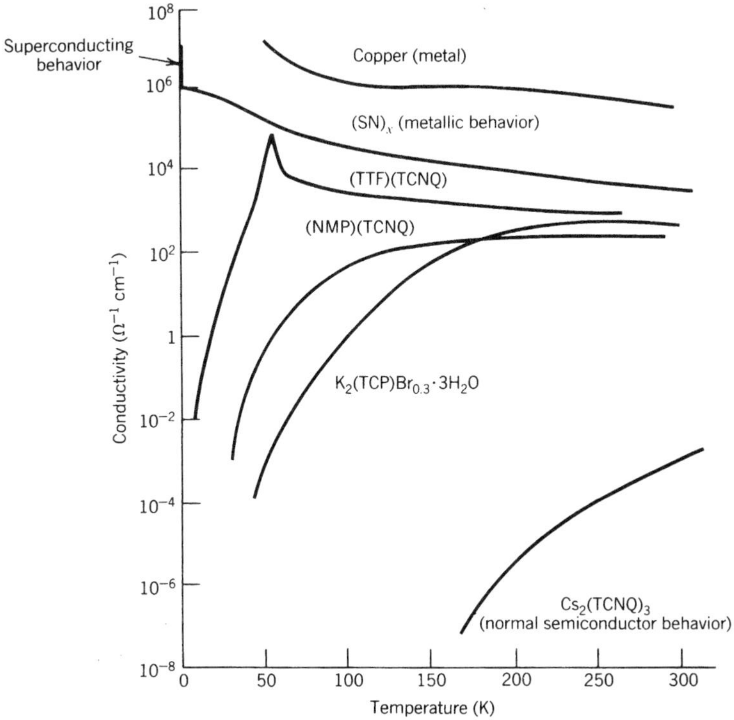

Figure 8.22 Temperature dependence of conductivity in one-dimensional conductors and in Cu. Note the differences (e.g., the sharp peak for TTF–TCNQ) from normal metallic and normal semiconductor behavior and also the superconductivity behaviour of $(SN)_x$. Figure 1.8 gives conductivities for other systems, for comparison.

type. While there is a broad maximum in σ for NMP–TCNQ, TTF–TCNQ has a very sharp peak of $\sim 10^4\,\Omega^{-1}\,\text{cm}^{-1}$ at 54 K, decreasing slowly to about $650\,\Omega^{-1}\,\text{cm}^{-1}$ at high temperatures and to much lower values at low temperatures.

It is generally agreed that the so-called Peierls (1955) distortion is part of the explanation of the transition from the conducting to the insulating state. The important element here is that a lattice distortion will always exist for one-dimensional systems so as to open a gap of energy 2Δ at a wavevector corresponding to the Fermi wavevector k_F of the undistorted state. Suppose we have an undistorted one-dimensional crystal with a band partly filled up to wavevector k_F. If there is a periodic distortion with wavelength λ, where

$$2\pi/\lambda = 2\mathbf{k}_F \tag{8.10}$$

then the superlattice distortion Q at position R along the axis will be proportional to $\cos(2k_F R + \phi)$. This will give a perturbation that opens up a gap, lowering by Δ the energy of carriers with $k \lesssim k_F$ (Figure 8.23). In fact there are two terms in the energy, an increase in elastic energy, proportional to Q^2, where Q is the distortion amplitude, and a decrease in electronic energy proportional to $Q^2 \ln Q$; the balance of these terms fixes Q. This reduction of energy from electron–lattice interaction has parallels with our earlier discussions of polarons (Section 1.2.3) and the Jahn–Teller effect (Section 4.2.3). However, it is important to note that the Peierls distortion is incommensurate as a rule, since there is no reason why $2\pi/k_F$ should be simply related to a lattice spacing.

While this mechanism seems to give a simple insulating state, this is an oversimplification. In an ideal one-dimensional system at finite temperatures, the large effects of thermodynamic fluctuations lead to a range of energies in which the density of states is low, rather than a simple gap. Long-range order is not possible in such ideal one-dimensional systems at finite temperatures. A gradual metal-to-insulator transition results as the temperature falls below $T \sim \Delta/k_B$. In real systems, which can only be quasi one-dimensional, three-dimensional Coulomb and strain interactions modify the picture further.

Another feature to stress is that the distortion (with $\lambda \sim \pi/k_F$) gives rise to a so-called charge-density wave (CDW) from the effects of the displacements on the charge density (see also Section 8.3.2). One suggestion for the behavior of TTF–TCNQ is that as the temperature is lowered to 54 K the conductivity increases as the charge-density waves become more mobile. Below 54 K interactions with defects and with charge-density waves in neighboring chains inhibits this mobility and lowers the conductivity. This effect is referred to as the pinning of charge-density waves.

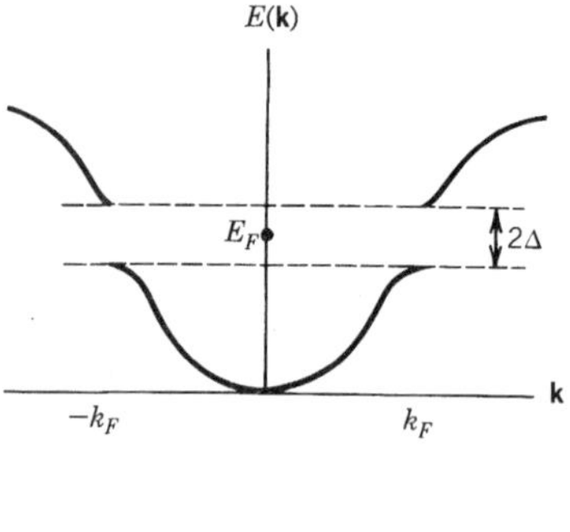

Figure 8.23 Peierls instability, showing an energy gap 2Δ opening at $\mathbf{k} = \pm\mathbf{k}_F$.

While TTF–TCNQ and KCP are especially striking examples, we emphasize that the Peierls transition is by no means confined to organic compounds. It is found, for example, in the one-dimensional blue bronze $K_{0.3}MoO_3$ which has a metal-to-semiconductor phase transition at 180 K (Travaglini et al. 1981).

8.2.4 Electrical Breakdown in Insulators

The dramatic (and often destructive) electrical breakdown of insulators is just one class of a range of high-electric-field phenomena. These include voltage-controlled and current-controlled negative resistance behavior (Figure 8.24) as well as semipermanent switching (Dearnaley et al. 1970). Insulating ceramics are used widely in capacitors and as insulators in large electrical equipment. The specifications and operating conditions are tightly confined for, while the best systems survive to 10^7 V/cm, breakdown below 10^5 V/cm can easily occur. For reference we note that $1\ \mathrm{eV/Å} \equiv 1.44 \times 10^9$ V/cm, that is, the maximum fields sustained (10^7 V/cm) corresponds to (1 phonon energy)/(1 interatomic spacing). Breakdown itself falls into three main categories: (1) thermal breakdown, (2) intrinsic (avalanche) breakdown, and (3) breakdown optically induced or induced by ionizing radiation. The relative importance of these categories depends on many factors, notably environmental (like temperature and whether oxygen is present) and the extent to which the

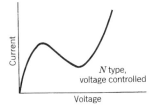

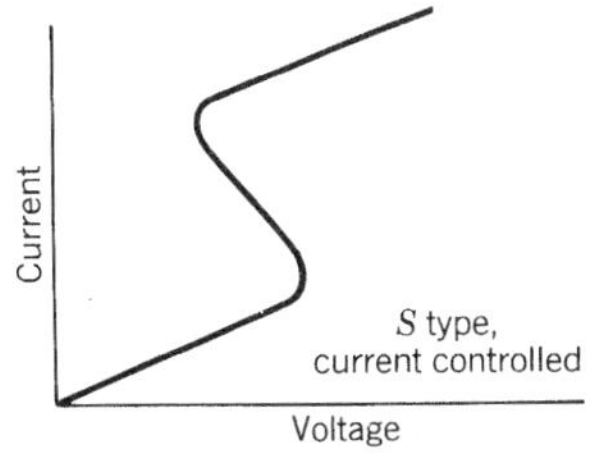

Figure 8.24 *N*-type and *S*-type instabilities giving differential negative resistance. The *N*-type instability is known as voltage controlled, since each voltage defines a unique current; the *S*-type instability is known as current controlled, because each current defines a unique voltage. For a discussion of the physical mechanisms underlying these instabilities see Dearnaley et al. (1970) and Ridley (1963).

applied field is uniform (e.g., asperities on the electrodes applying the field and focusing of light). There is inevitably a statistical aspect to breakdown behavior because, for example, of minor features of surface topography (see Figure 7.1).

Thermal breakdown results from Joule heating by the small currents that flow even in good insulators. The temperature rise generates carriers either intrinsically (thermal excitation across the bandgap) or extrinsically (thermal ionization of defects) and may cause melting. Local overheating can occur more readily because of special local conditions. Thermal breakdown occurs preferentially at high temperatures ($T \gtrsim 150°C$) and at high frequencies, when the Joule heating is greater.

Intrinsic (or avalanche) breakdown is mostly noted at lower temperatures, $T \lesssim -50°C$, and is almost independent of temperature in this range. Typically, intrinsic breakdown is rapid, with a time scale of tens of nanoseconds, and is faster than thermal breakdown.

Intrinsic breakdown occurs when a local electric field exceeds a characteristic breakdown field. The mechanism is usually assumed to be an avalanche mechanism, that is, when a carrier's velocity exceeds a critical velocity, it can generate additional carriers, either by defect ionization or by excitation over the bandgap. We can make this more precise. The velocity of a carrier driven by the field F is $v = (e/m)F\tau(E)$, where the time τ characterizes scattering for a carrier of energy $E = \frac{1}{2}mv^2$. The energy gain between scattering events is therefore $e^2F^2(E)/m$. If secondary electrons are to be generated, a critical velocity v_c (energy E_c) must be exceeded. This can only happen if $e^2F^2\tau(E_c)$ exceeds the corresponding losses in generating heat, L_{ph}. In ionic crystals both the losses and the scattering can be estimated using the Fröhlich coupling we have discussed in connection with polaron behavior (Section 1.2.2). At high carrier energies (say a few eV) the resulting mean free paths are 100–1000 Å, with $\tau^{-1} \propto E^{-1/2}$. The temperature dependence of τ depends on whether $\hbar\omega \lesseqgtr E$, where $\hbar\omega$ is the energy of an optical phonon. If $E > \hbar\omega$, scatter can occur by either absorption or emission of phonons ($\tau^{-1} \propto 2n+1$, where n is the phonon occupation number). At lower energies, $E < \hbar\omega$, phonon absorption alone is possible, so that $\tau^{-1} \propto n$. At low energies the mean free path is proportional to $[\exp(\theta/T) - 1](T/\theta)^{1/2}$ for carriers in thermal equilibrium, where $\theta \equiv \hbar\omega/k_B$; there is a corresponding mobility proportional to $[\exp(\theta/T) - 1]$. All electrons do not have the average thermal energy, of course, and sometimes the few high-energy electrons in the thermal distribution are sufficient to cause breakdown.

Breakdown induced by laser pulses shares many features with avalanche breakdown. Again one defines a critical electric field (here laser induced) which, because of statistical variations, is usually taken to correspond to the optical intensity required to give damage 50% of the time. There are, however, some special factors. One is self-focusing, where the laser beam alters the refractive index through its dependence on the electric field (electrostriction being one cause), so that focusing of the beam is induced. However, if an electron–hole plasma is generated, there is some defocusing. The resulting focal volume V is clearly an important parameter, as is the pulse duration τ_p. Experiment shows that the laser-induced breakdown field E_{LB} varies as

$$E_{LB} \propto AV^{-1}\tau_p^{-1/4} + B \tag{8.11}$$

where A and B are constants of the material.

8.3 INTERCALATES

Intercalation implies reversible insertion of guest chemical species into a host structure without altering the general structural features of the host. The range of intercalation compounds is very extensive (Whittingham and Jacobson 1982), including insulators such as zeolites, the semimetal graphite, semiconductors, semimetals such as transition-metal dichalcogenides, and intermetallic compounds such as $LaNi_5$. The electrical conductivity can change markedly on intercalation, depending on electron transfer between host and guest species, and the bonding between host and guest can vary from van der Waals to ionic and metallic. One feature of interest in intercalates is the ordering that occurs when there are large interstitial concentrations.

Zeolites are crystalline aluminosilicates whose structure contains channels and cages filled with exchangeable cations and water molecules, with the general formula

$$M_xD_y(Al_{x+2y}Si_{n-(x+2y)}O_{2n}) \cdot nH_2O,$$

where M is a monovalent cation such as H^+, Na^+, or K^+ and D is a divalent cation such as Ba^{2+}, Ca^{2+}, or Mg^{2+}. When these materials are dehydrated, their channels and cages become accessible to molecules whose dimensions and shape permit passages through the host structure. Hence they are called molecular sieves and are selective adsorbents. Molecules moving through the zeolite structure can meet other molecules in catalytically active sites leading to chemical reaction; the cracking of paraffins in the presence of hydrogen is an example (see Section 7.9).

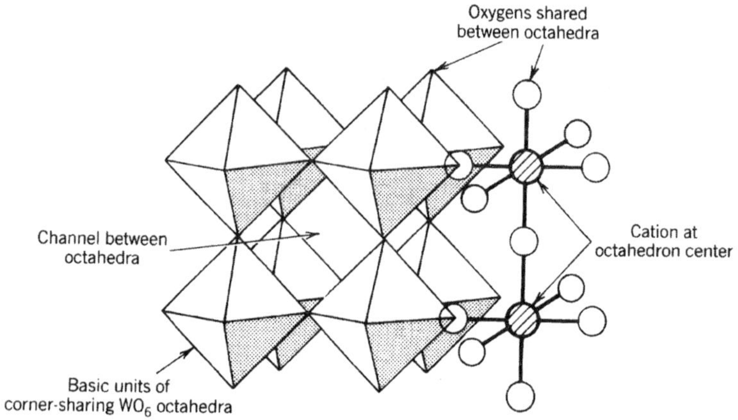

Figure 8.25 The ReO_3 structure, adopted by WO_3, for example. Alkali intercalation in the channels of WO_3 gives some of the tungsten bronzes.

Tungstic oxide (WO_3) is an insulating compound. At room temperature it adopts a distorted monoclinic form of the rhenium oxide (ReO_3) structure (Figure 8.25), which consists of vertex-sharing chains of Re–O octahedra. This structure gives a three-dimensional array of square tunnels and the cavities may be filled with ions of suitable size. Occupation of all tunnel sites gives the well-known perovskite structure AMO_3 adopted by a large number of mixed-metal compounds, for example $KMgF_3$. Only H and Li have sufficient mobility in the WO_3 lattice to be intercalated at room temperature. Hydrogen may be introduced by electrochemical reduction. The ease and reversibility of reduction, together with a dramatic color change to dark blue, makes WO_3 potentially useful in electrochromic devices. For H_xWO_3 with $0.5 < x < 0.6$ a cubic phase occurs with hydrogen randomly attached to the oxygen atoms as hydroxyl groups. With introduction of alkali ions, giving, for example, Li_xWO_3, the crystals gradually become metallic and are referred to as tungsten bronzes.

The study of intercalated compounds is important because of potential ability to design properties into a material at the synthetic level for use, for example, in electrical batteries, or for catalysis. However, intercalated compounds are complex, and their physical properties are not well understood. We shall outline in more detail some of the physical properties of graphite intercalated compounds in Section 8.3.1 and of layer dichalcogenides in Section 8.3.2 [see Levy (1979); Pietronero and Tosatti (1981)].

8.3.1 Graphite Intercalated Compounds

Graphite is a black semimetal and is a highly anisotropic material. The carbon atoms are bonded covalently in a hexagonal arrangement within layers (Figure 8.26). The in-plane bonding is associated with sp^2 hybridization of the carbon orbitals (see Section 8.4) and is very strong. In purely two-dimensional graphite the remaining valence electron ($2pz$) would give rise to a completely filled π-bonded valence band, the antibonding π^* states would be empty and the material would be an insulator. However, weak interlayer interaction between the $2pz$ electrons give rise to an overlap of ~0.04 eV between the valence and conduction bands leading to semimetallic behavior, with about 10^{-4} carriers for each C atom. The weak van der Waals-type bonding between the layers makes intercalation possible because the strain energy associated with dilation, necessary to accommodate the intercalant, is small. The in-plane C–C separation of 1.42 Å changes by less than 1% on intercalation, whereas the interlayer separation increases from 3.35 Å to as much as 8 Å, to accommodate some intercalant layers.

The carbon layers in graphite are arranged in an alternating stacking sequence *ABABAB*... (Figure 8.26). Intercalation generally introduces stacking faults (see Figure 3.24), changing the stacking sequence such that a layer of intercalate *I* is flanked by equivalent carbon layers giving the arrangement *ABAIABAB*.... This rearrangement requires a shear displacement of neighboring carbon layers. Graphite has the unusual

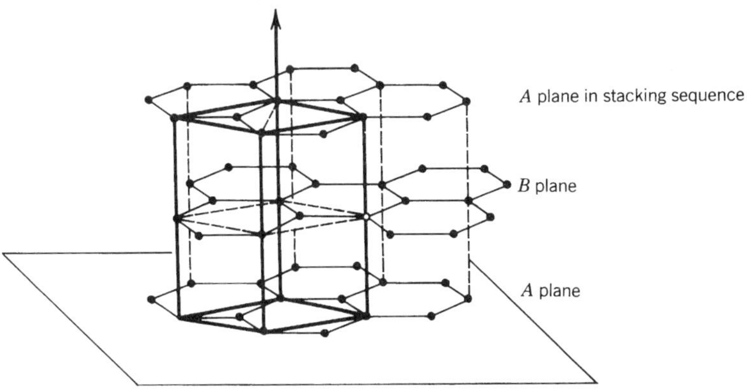

Figure 8.26 Graphite structure, showing the stacking sequence and the basal plane. The unit cell of graphite is indicated by bold lines.

property that a series of stoichiometric compositions, referred to as stages, can be obtained with a given intercalate (Figure 8.27). The stage number, defined as the number of carbon layers between nearest intercalate layers, decreases as the concentration of intercalate increases.

A wide variety of chemical species can be intercalated into graphite. The alkali metals, alkaline earths, and lanthanides act as electron donors, whereas halogens such as Cl_2 and Br_2 and acids such as HNO_3 and H_2SO_4 act as acceptors. Intercalation with alkali metals has received the most attention. The heavier metals—K, Rb, and Cs—react readily with graphite at temperatures of $\lesssim 200°C$; the ions Li and Na react less readily. This difference of behavior is partly a question of electronegativity (Section 1.1), since the ionization potentials of K (4.34 eV), Rb (4.18 eV), and Cs (3.89 eV) are less than the electron affinity of graphite (~4.6 eV), whereas those of Li (5.39 eV) and Na (5.1 eV) are greater. The low-stage alkali intercalates are yellow in color.

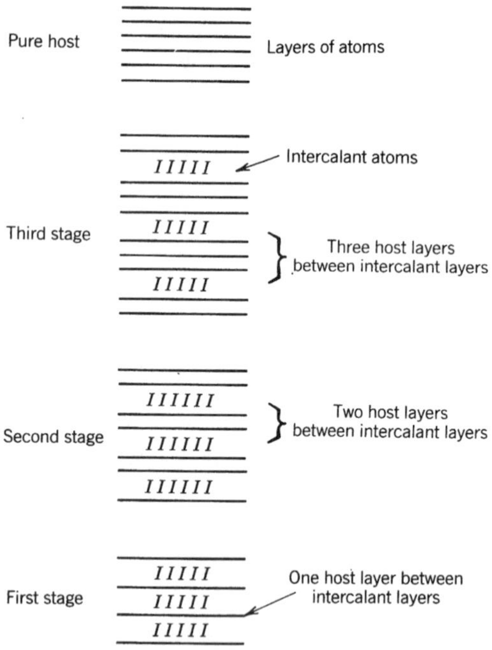

Figure 8.27 Stages of intercalation of graphite. At the *n*th stage, there are *n* planes of carbon atoms between each intercalant plane.

Most graphite intercalation compounds exhibit long-range order in the basal plane at sufficiently low temperature. The intercalate layers form a centered, close-packed hexagonal arrangement, the area of the hexagons being a simple multiple of the area of the carbon hexagons; the intercalate arrangement is said to be commensurate with that of the graphite crystal. A $\sqrt{3} \times \sqrt{3}$ basal plane superlattice is favored by the small Li ion (Figure 8.28*a*); the unit cell of the superlattice is three times larger than the graphite hexagon and for a stage 1 compound the overall stoichiometry is LiC_6. A 2×2 basal plane superlattice gives an area ratio of 4:1 between intercalate and C hexagons (Figure 8.28*b*); this type of basal plane is formed by potassium and for stage 1 the stoichiometry is KC_8. There are four equivalent sites α, β, γ, δ in the unit cell, but only one is occupied in each layer. The stacking sequence of intercalate planes along the c axis for KC_8 and RbC_8 is $\alpha\beta\gamma\delta\alpha\ldots$, but other stacking sequences can occur for other intercalates.

Ordered sequences of intercalate layers can occur along the c axis for separation of intercalate layers as large as 30 Å (stage 9). The origin of the long-range order of intercalate layers is not fully understood [see Millman and Kirczenow (1982)] but is presumably similar to the origin of ordering of crystallographic shear planes (Section 3.3.3). The long-range order does not persist either in the intercalate layer or along the c axis at high temperatures. We have already pointed out that in KC_8 there are four inequivalent K layers, and these persist up to 700 K. Above 700 K the c axis collapses, indicating two inequivalent K layers. Above 750 K all Bragg reflections from K layers disappear, signaling onset of long-range disorder in the K layers.

Zone-folding effects (see Section 2.2.2) are relevant in the comparison of the electronic and vibrational properties of low-stage, intercalated

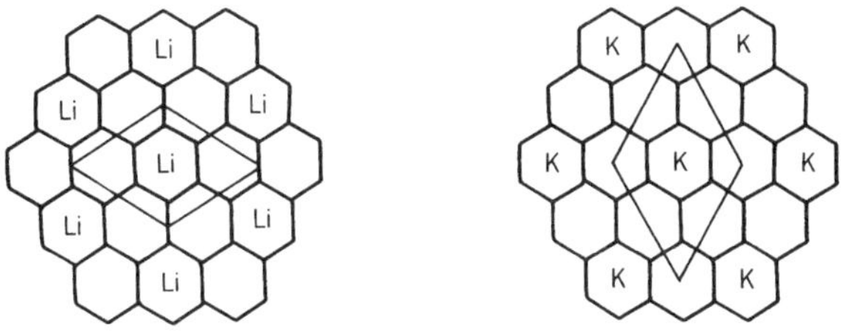

(*a*) (*b*)

Figure 8.28 Schematic diagrams of the arrangements of alkali atoms intercalated into graphite. In each case the rhombus outlines the (two-dimensional) repeat unit. (*a*) C_6X structure, as for X = Li; (*b*) C_8X structure, as for X = K, Rb, and Cs. The stacking sequence of planes is the same for K and Rb but different for Cs.

graphite compared with the properties of pristine graphite. Taking KC_8 as an example, we find that since the real-space layer unit cell is four times larger than that of graphite (Figure 8.28*b*) the reciprocal-space unit cell must be four times smaller. The electronic structure of the intercalated graphite must be constructed from intercalate orbitals and from graphite orbitals placed in the reduced zone scheme of the intercalated compound.

In-plane zone folding is relevant for the interpretation of infrared and Raman studies of phonons. Only zone-center phonons are excited in single-quantum excitations. Since zone-folding brings new modes to the zone center (see also Figure 2.1), additional Raman and infrared structures can be observed. The observation of zone-folding effects in optical studies is complicated by breakdown of selection rules associated with disorder, inevitably present in intercalated materials (Pietronero and Tosatti 1981, p. 187).

Because the c-axis interactions are weak compared to in-plane interactions, the electronic and vibrational properties of KC_8, for example, are determined primarily by a truncated unit cell consisting of a single graphite layer and the intercalate layer. However, if full account were taken of the stacking sequence $\alpha\beta\gamma\delta\alpha\ldots$, we would have a unit cell in the reciprocal lattice 16 times smaller than that of graphite.

The room-temperature electrical conductivity σ of graphite is only about a factor of 25 less than that of copper. This indicates a very high carrier mobility within the layers, since the number of carriers is $\sim 10^4$ less than that of copper. The value of σ parallel to the c axis (σ_c) is much smaller than the value of σ in the planes (σ_a), and the ratio σ_a/σ_c can either be increased or decreased by intercalation. At 300 K σ_a/σ_c varies from ~ 30 in the donor compound KC_8, to $\sim 10^3$ in pure graphite, to $\sim 10^6$ with acceptor intercalates, and the changes in these ratios are due primarily to changes in σ_c. Although very extensive studies have been made of electrical conductivity of intercalated graphite, the results are not well understood.

Although neither graphite nor alkali metals are superconductors, the compounds KC_8, RbC_8, and CsC_8 become superconducting at low temperatures (T_c for KC_8 is in the range 0.08–0.55 K). The explanation of superconductivity in KC_8, for example, involves an assumption of two kinds of carriers, one with two-dimensional and the other with nearly isotropic motion. If $f \sim 0.5$ is on average the charge transfer from each alkali atom to the graphite layers, then each K becomes K^{f+} and each C becomes $C^{(f/8)-}$. The electric fields of longitudinal optic modes associated with the relative motion of the charged planes couple strongly to the nearly isotropic carriers to give the net attractive electron–electron interaction required for superconductivity.

8.3.2 Intercalated Transition-Metal Dichalcogenides

Intercalation compounds of group IVb (Ti, Zr, Hf), Vb (V, Nb, Ta), and VIb (Cr, Mo, W) transition-metal dichalcogenides have been prepared with the majority of metals in the periodic table as intercalants. The dichalcogenides (with the formula MX_2, where M is a transition metal and X is S or Se) are sandwichlike structures consisting of a layer of transition-metal atoms between two layers of chalcogen atoms (Figure 8.29*a*). The sandwiches are loosely bound to each other by chalcogen–chalcogen van der Waals forces and are readily cleaved. These materials are generally rich in polytypes. The polytypes differ from each other in the stacking arrangement of the layers and by the metal coordination within a given layer. The two observed coordinations in which the chalcogens are arranged above each other in each layer are octahedral (Figure 8.29*b*) and trigonal prismatic (Figure 8.29*c*). The simplest polytypes are the 1*T* polytype which has one sandwich per unit cell, octahedral coordination, and overall trigonal symmetry, and the 2*H* polytype, which has two sandwiches per unit cell, trigonal prismatic coordination, and overall hexagonal symmetry. The designations 1*T* and 2*H* come from the number of sandwiches per unit cell and the initial letters of the overall symmetry classes. More complex stacking arrangements along the *c* axis, for example, 4*H*, also occur. Examples of materials with

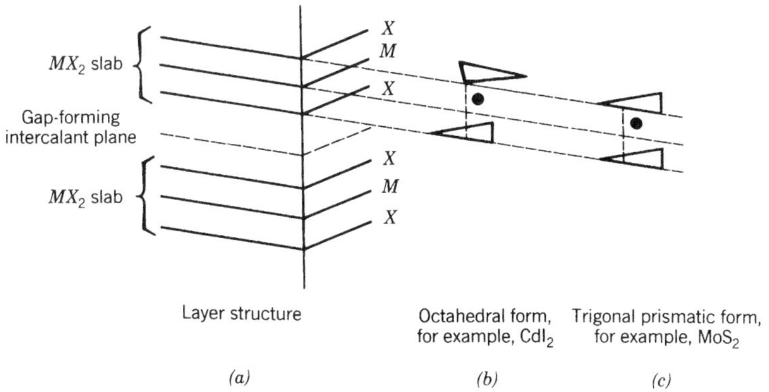

Figure 8.29 (*a*) Metal dichalcogenide structure, showing [(*b*), (*c*)] the different variants of metal coordination geometry.

octahedral coordination are TiS_2 and TaS_2 ($1T$) and of trigonal prismatic coordination are NbS_2, TaS_2 ($2H$), MoS_2, and WS_2.

The narrow *d* orbitals of the transition-metal ions lie between wide bands constructed from filled bonding and empty antibonding *s* and *p* orbitals. Depending on whether the *d* orbitals are partly or completely filled, these materials are metals or semiconductors (e.g., MoS_2 and WS_2 are semiconductors). Structural phase transitions are observed in many polytypes of group Vb disulphides and diselenides due to charge-density-wave (CDW) instability (see Section 8.2.3). In most cases an incommensurate lattice distortion is stable in some temperature interval, below which a commensurate distortion occurs. The CDW instability in $2H$-$NbSe_2$ is always incommensurate, the wavelength being about 3% greater than three times the Nb–Nb intralayer separation of 3.3 Å. At 4.2 K the amplitude of the Nb lattice distortion is only 0.05 Å, so that the effect of the distortion on physical properties is small. A CDW instability will be likely to occur in those materials where the shape of the Fermi surface is such that many states on the surface can be connected by the same wavevector **Q**. Such surfaces are referred to as *nested* surfaces. A periodic distortion having wavevector **Q** will then produce gaps at the portions of the Fermi surface connected by **Q**. As in the case of one-dimensional compounds (Section 8.2.3), a distortion will occur if the energy gained by creating gaps at the Fermi surface is greater than the energy lost in creating the distortion. The lattice instability of transition-metal dichalcogenides is associated with a marginally stable metal–metal and metal–chalcogenide bond geometry.

The group Vb compounds $2H$-$NbSe_2$ and $2H$-TaS_2 have been most widely studied as hosts for intercalation. The electron affinity of group VIb hosts is lower than that of group Vb hosts by 1.5–2.0 eV and hence they intercalate only with the most electropositive elements. The behavior of the chalcogens is different from that of graphite in various respects. Thus the dichalcogenides can form stage 1 compounds with broad homogeneity ranges, for example, Li_xTiS_2 with $0 < x < 1$. In addition, the dichalcogenides, in contrast to graphite, do not intercalate inorganic acids; transition metal dichalcogenides only accept intercalate species that donate electrons to the host transition-metal *d* band. The electrostatic forces between positively charged layers of donor intercalates and the negatively charged MX_2 sandwiches is a stabilizing influence; although CrS_2 is not stable, stoichiometric alkali intercalation compounds of CrS_2 exist.

CDWs are suppressed for high concentrations of intercalates. In $2H$-H_xTaS_2 the CDW instability disappears for $x > 0.11$. When the larger alkali atoms such as Na and K are intercalated, structural changes develop in the stacking sequences with increasing x. For example, in Na_xTiS_2 the sodium ions occupy octahedral sites at low x, move to a trigonal prismatic site at intermediate x and at the highest x move into another site not yet identified. These changes are associated with large expansion of the c axis.

8.4 POLYMERS

The types of material referred to as polymers or plastics have a wide range of mechanical properties, from the elasticity of synthetic rubbers to the rigidity of perspex. Some polymers may also be prepared with a wide range of electrical properties, varying from insulating to metallic, and it is these latter properties that will mostly concern us here. At the outset we shall say a little about the synthesis of polymers [see Bloor (1982)].

A polymer molecule is composed of subgroups called monomers and may be produced by two different kinds of chemical reaction:

1. Addition reactions, requiring the presence of a catalyst or free radical to break a carbon–carbon bond in a monomer, so that new bonds can link, forming a polymer chain. This process is illustrated in Figure 8.30*a* for formation of polyethylene from ethylene.
2. Condensation reactions in which two molecules are linked to form a larger one, with a by-product such as water. The process is illustrated in Figure 8.30*b* for formation of nylon-66 from hexamethylene diamine and adipic acid.

The product of these reactions will be long polymer molecules. The individual chain lengths will not be the same in any one sample, but the average molecular weight can be controlled over wide ranges. This is one important factor affecting physical properties. For a molecular weight of around 200, polyethylene is a viscous fluid at room temperature. With a molecular weight of around 400, the melting temperature is 62°C; the melting point reaches 137°C when the molecular weight is 10,000.

The C–C bonds forming the backbone of a polymer chain are strong; the forces holding chains together are much weaker. In polyethylene, and some other cases, the interchain forces are sufficient to bring about crystallization from a melt or solution. The chains minimize energy by folding back on themselves to form microscopic lamellar crystals, with the chain axes normal to the plane of the lamella. Crystallization is only one of the strengthening mechanisms of polymers. Chain stiffening is a

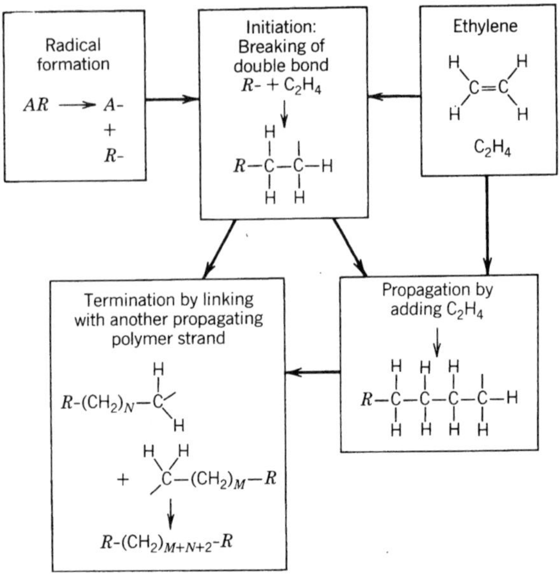

(a)

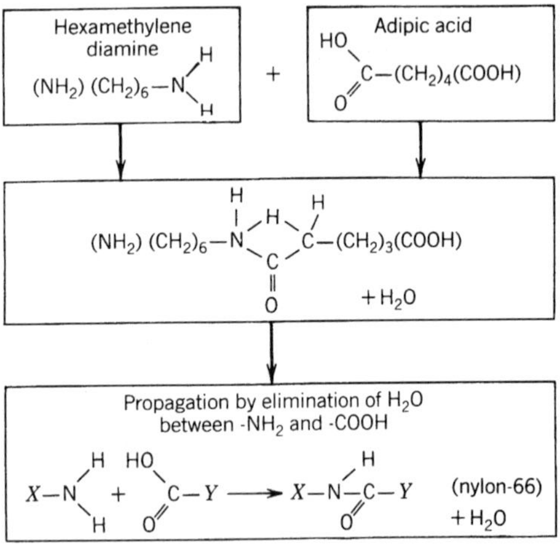

(b)

Figure 8.30 Polymerization mechanisms. (*a*) Addition polymerization of ethylene to give polyethylene. (*b*) Condensation polymerization giving nylon-66.

second mechanism, where the chains resist the folding which is important in crystallization. Examples are polyimides or ladder molecules. Cross-linking is a third mechanism where, as in phenolformaldehyde, chains may bifurcate, linking themselves to each other by bridges of polymer strands.

Defects in polymers are of two main types. There are defects that affect the way the chains are organized relative to one another, for example, by altering the degree of crystallinity or the extent of the cross-linking. There are also defects that affect individual polymer strands. These defects are often the result of a chemical or photochemical reaction, and are responsible for the main degradation mechanisms in practice. Some examples are given in Figure 8.31.

Thermal degradation takes two main forms. One is *scission* (Figure 8.31*a*), in which a break occurs randomly in the chain. This leads to a

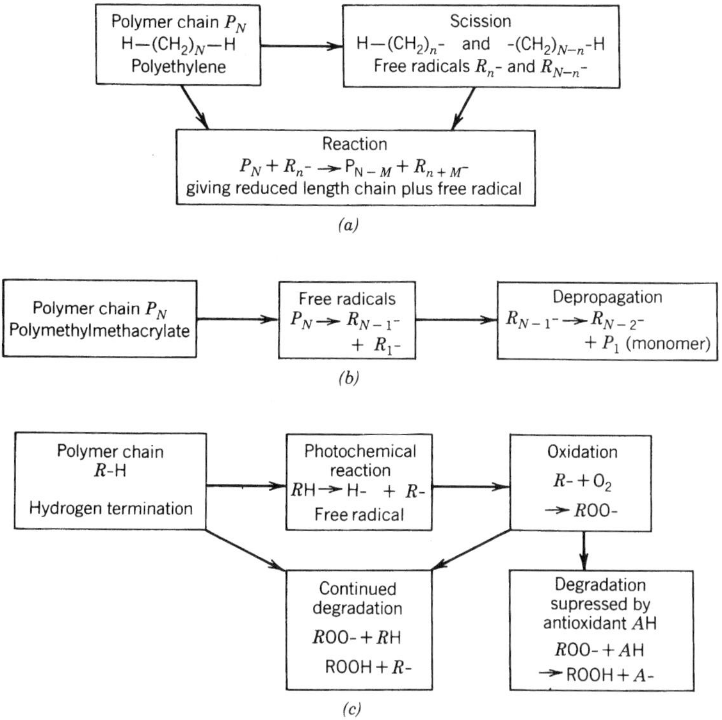

Figure 8.31 Polymer degradation mechanisms: (*a*) scission, (*b*) depropagation, (*c*) photochemical reactions in the presence of oxygen.

rapid reduction in average chain length, and is seen for polyethylene. The second form of thermal degradation, the *depropagation reaction* (Figure 8.31*b*), is found in polymethylmethacrylate, producing the monomer. The combination of ultraviolet light and oxygen is effective in degrading polymers even at room temperature (Figure 8.31*c*). Here the first step is photochemical bond breaking, perhaps removing a hydrogen that terminates a polymer chain. The second step involves oxygen in a reaction that allows the degradation to continue. Such reactions are inhibited in practice by including antioxidants like amines or phenols.

Before proceeding to discuss the electronic conductivity of polymers, we shall make some comments on the chemical bonding of carbon. Although the ground state of the C atom is $1s^2 2s^2 2p^3$ (^3P, with two unpaired electrons), the characteristic valence of carbon is four, as in methane (CH_4), or in the diamond crystal, for example. To reconcile carbon chemistry with the valence structure of the C atom, we consider a configuration $1s^2 2s 2p_x 2p_y 2p_z$, that is, an sp^3 configuration rather than s^2p^2. We now create four equivalent, mutually orthogonal, localized orbitals by taking linear combinations (*hybrids*) of $2s$ and $2p$ orbitals:

$$\begin{aligned}
\psi_1 &= \tfrac{1}{2}\psi(2s) + \frac{1}{\sqrt{2}}\psi(2p_x) + \tfrac{1}{2}\psi(2p_z) \\
\psi_2 &= \tfrac{1}{2}\psi(2s) - \frac{1}{\sqrt{2}}\psi(2p_x) + \tfrac{1}{2}\psi(2p_z) \\
\psi_3 &= \tfrac{1}{2}\psi(2s) + \frac{1}{\sqrt{2}}\psi(2p_y) - \tfrac{1}{2}\psi(2p_z) \\
\psi_4 &= \tfrac{1}{2}\psi(2s) - \frac{1}{\sqrt{2}}\psi(2p_y) - \tfrac{1}{2}\psi(2p_z)
\end{aligned} \tag{8.12}$$

If we refer our wavefunction axis to the axis of a regular tetrahedron (Figure 8.32*a*), the wavefunctions (8.12) can be rewritten

$$\begin{aligned}
\psi_1 &= \tfrac{1}{2}\psi(2s) + \tfrac{1}{2}\sqrt{3}\,\psi(2p_A) \\
\psi_2 &= \tfrac{1}{2}\psi(2s) + \tfrac{1}{2}\sqrt{3}\,\psi(2p_B) \\
\psi_3 &= \tfrac{1}{2}\psi(2s) + \tfrac{1}{2}\sqrt{3}\,\psi(2p_C) \\
\psi_4 &= \tfrac{1}{2}\psi(2s) + \tfrac{1}{2}\sqrt{3}\,\psi(2p_D)
\end{aligned} \tag{8.13}$$

where $\psi(2p_A)$ has the form of a $2p$ function directed from the carbon

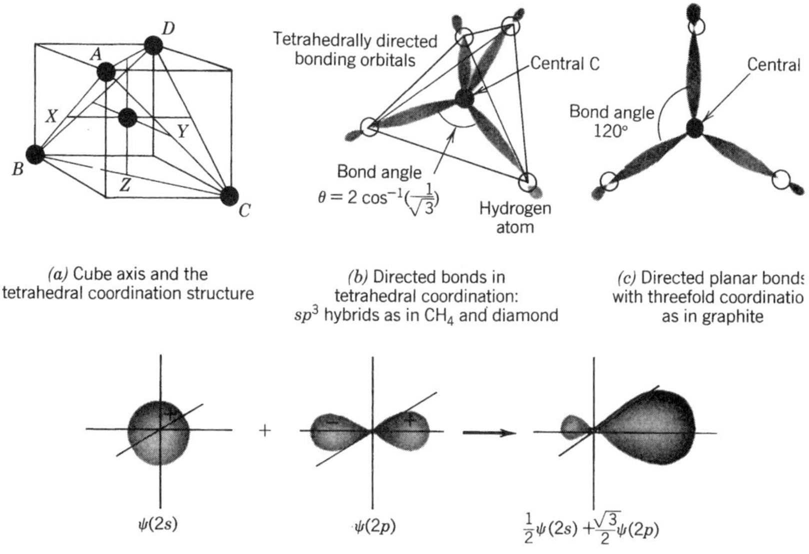

(a) Cube axis and the tetrahedral coordination structure

(b) Directed bonds in tetrahedral coordination: sp^3 hybrids as in CH_4 and diamond

(c) Directed planar bonds with threefold coordinatio as in graphite

(d) Formation of a hybrid

Figure 8.32 Hybridization in tetrahedral geometry (*a*, *b*). The trigonal planar form of hybridization is shown in (*c*) and linear hybridization in (*d*).

atom to the vertex A (Figure 8.32*a*). The four orbitals (8.12) are inclined at equal angles of 109°28′ to each other, and all are equivalent.

If a hydrogen atom is placed on the vertex, new orbitals of the type

$$\begin{aligned} \bar{\psi}_1 &= C_{11}\psi_1(C; 2s, 2p_A) + C_{12}\psi(H_A; 1s) \\ \bar{\psi}_2 &= C_{21}\psi_1(C; 2s, 2p_A) + C_{22}\psi(H_A; 1s) \end{aligned} \tag{8.14}$$

form, one bonding and one antibonding. Similar pairs of orbitals for each of the other three vertices and linear combinations of the four bonding and also of the four antibonding orbitals may be chosen that are consistent with the tetrahedral symmetry of CH_4 (Figure 8.32*b*).

The mixing of s and p orbitals in (8.13) is referred to as sp^3 hybridization. The energy required to promote the $2s$ electron in C is more than recovered by the gain in binding energy associated with the directionality of the hybrid orbitals. The tetrahedral bonding that occurs in diamond and in silicon is also an example of sp^3 hybridization.

If we hybridize $2s$ with $2p_x$ and $2p_y$ only, leaving $2p_z$ unchanged, we can construct three equivalent orbitals in the xy plane, pointing to the corners of an equilateral triangle. These are

$$\psi_1 = \frac{1}{\sqrt{3}}\psi(2s) + \sqrt{\tfrac{2}{3}}\psi(2p_A)$$
$$\psi_2 = \frac{1}{\sqrt{3}}\psi(2s) + \sqrt{\tfrac{2}{3}}\psi(2p_B) \tag{8.15}$$
$$\psi_3 = \frac{1}{\sqrt{3}}\psi(2s) + \sqrt{\tfrac{2}{3}}\psi(2p_C)$$

The orbitals (8.15) are called trigonal hybrids (Figure 8.32*c*) and are important for the in-plane bonding of graphite (see Section 8.3.1).

Similarly, if we mix $2s$ with $2p_x$ only and leave $2p_y$ and $2p_z$ unchanged, we can obtain two equivalent orbitals, referred to as digonal hybrids, strongly localized in the positive and negative x directions (Figure 8.32*d*). It is not difficult to show that to obtain hybrid orbitals directed toward the corners of a regular octahedron a combination of the form d^2sp^3 must be chosen.

The bonding in ethylene (C_2H_4) involves only partial hybridization. Taking the carbon and hydrogen atoms of this planar molecule to be in the xy plane (Figure 8.33) we find that carbon forms an sp^2 hybrid in this plane, with one of the hybrid bonds directed at the other carbon atom and the remaining two directed at hydrogen atoms. The $2p_z$ orbitals on the carbon atoms form a π-bonding molecular orbital, constituting a second, weaker C–C bond.

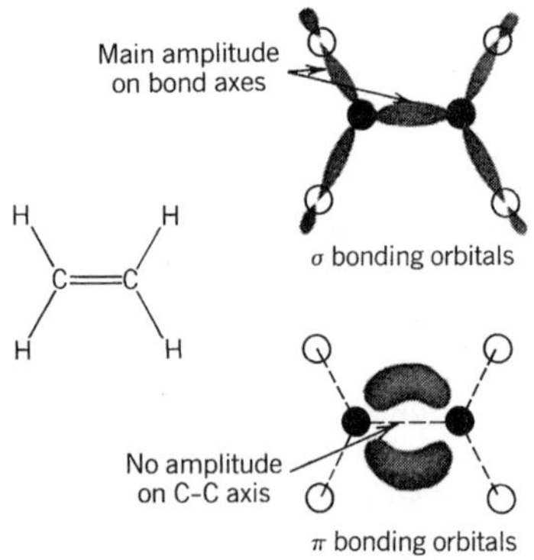

Figure 8.33 Charge densities associated with σ and π bonding orbitals in a simple ethylene (monomer) molecule.

Polymers are generally chemically inert and electrically insulating because all the valence electrons are paired in localized bonds. In any but the most perfect crystalline polymers the energy of an electron injected into the polymer will vary from site to site. The rms deviation from the mean site energy $\bar{\epsilon}$ of the injected electron is

$$\Delta = (\langle(\epsilon_1 - \bar{\epsilon})^2\rangle_{av})^{1/2} \tag{8.16}$$

and measures what is referred to as diagonal disorder. There are also variations from the mean value of the hopping integral, $\bar{t}$, determining motion from site to site, and this is referred to as off-diagonal disorder (see also Section 8.2.2). In polymers both the diagonal and off-diagonal disorder are largely static in origin. This is in contrast to the situation for molecular crystals such as anthracene where the disorder is largely dynamic in origin. Electrons injected into a material are localized if

$$\Delta > cNJ = cB \tag{8.17}$$

where c is a dimensionless number of order unity that depends on the dimensionality of the conductor and on the extent of off-diagonal disorder, N is the coordination number of the sites, and $2B$ is an average bandwidth [see Equation (1.5)]. In most polymers the inequality (8.17) is satisfied so that electrons and holes injected into a polymer form localized states. This is again in contrast to molecular crystals such as anthracene where $\Delta \sim 0$ and band photoconductivity, for example, is readily observed (Silinsh 1980).

Only one polymer, the inorganic material $(SN)_x$, where x is a large number, is metallic at room temperature; it is also superconducting below 0.3 K (Figure 8.22). The presence of different atoms, S and N, in the polymer backbone causes overlap of valence and conduction bands. However, some organic polymers, such as polyacetylene, $(CH)_x$, have an unsaturated (or conjugated) electronic structure with alternating single and double bonds along the chain (Figure 8.34).

In $(CH)_x$ three of the four valence electrons of C are in sp^2 hybridized orbitals. Two of these σ-type bonds form the backbone of the polymer chain, and the third bonds with hydrogen. The 120° bond angle between the three coplanar orbitals of the sp^2 hybrid can be satisfied by two possible arrangements of carbons, forming *cis*-$(CH)_x$ and *trans*-$(CH)_x$ (Figure 8.34*a* and 8.34*b*) with four and two CH monomers per unit cell, respectively. In both *trans*- and *cis*-isomers, the fourth valence electron of each carbon has $2p_z$-type symmetry (forming π bonds), the z direction being perpendicular to the plane defined by the sp^2 hybrid (see earlier comments on ethylene). If all bond lengths remained equal, $(CH)_x$ would be a one-dimensional metal with a half-filled band. However, the system is

(*a*) *Cis*-polyacetylene

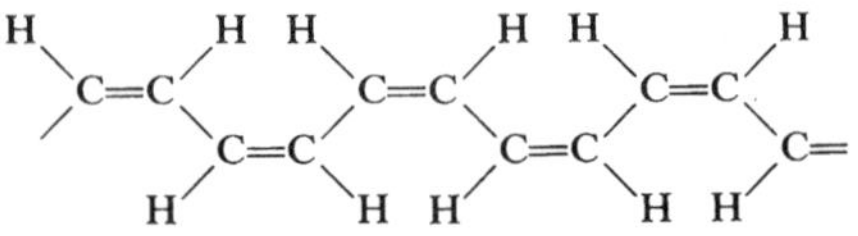

(*b*) *Trans*-polyacetylene

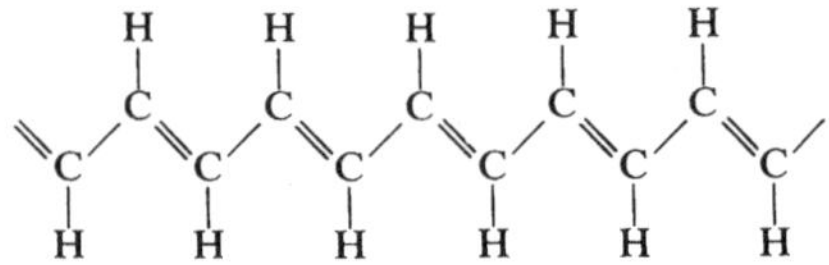

(*c*) Kink (neutral soliton) in *trans*-$(CH_2)_x$

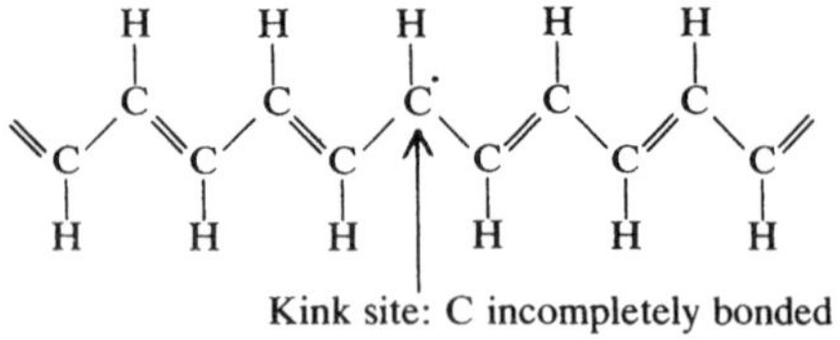

(*d*) Charged kink with nearby acceptor

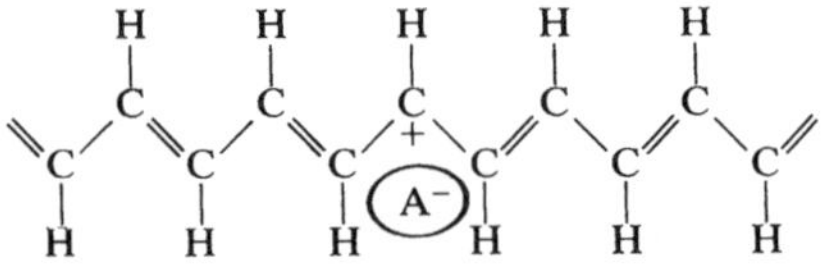

Figure 8.34 Polyacetylene and its defect structure: (*a*) *cis*-polyacetylene, (*b*) *trans*-polyacetylene, (*c*) a neutral kink (neutral soliton) in *trans*-polyacetylene; this bond-alternation defect is associated with an unpaired spin, (*d*) a charged kink with a nearby acceptor A^- in doped *trans*-polyacetylene.

unstable against a Peierls instability (see Section 8.2.3) in which adjacent CH monomers move toward each other forming alternating short (double) and long (single) bonds. This bond-alternation distortion creates a gap in the conduction band, and $(CH)_x$ is an insulator (Longuet-Higgins and Salem, 1959).

The *trans* isomer is the thermally stable form of $(CH)_x$. Any *cis–trans* ratio can be maintained at low temperatures. Almost complete isomerization from *cis* to *trans* can be accomplished after synthesis by heating the film to temperatures above 150°C for a few minutes. However, the study of $(CH)_x$ is made difficult by rapid degradation on exposure to oxygen in the atmosphere.

Interest in $(CH)_x$ grew in recent years with the discovery that it can be doped chemically or electrochemically at room temperature to form *n*- or *p*-type semiconductors; alkali metals act as donors and molecules such as Br_2 and AsF_5 as acceptors. The electrical conductivity can be varied in a controlled way over about 12 orders of magnitude, up to values of $\gtrsim 3 \times 10^3\ \Omega^{-1}\ \mathrm{cm}^{-1}$. The transition from semiconducting to metallic behavior occurs for dopant concentrations of $\gtrsim 1\%$. The basic difference between $(CH)_x$ and other organic materials such as anthracene and

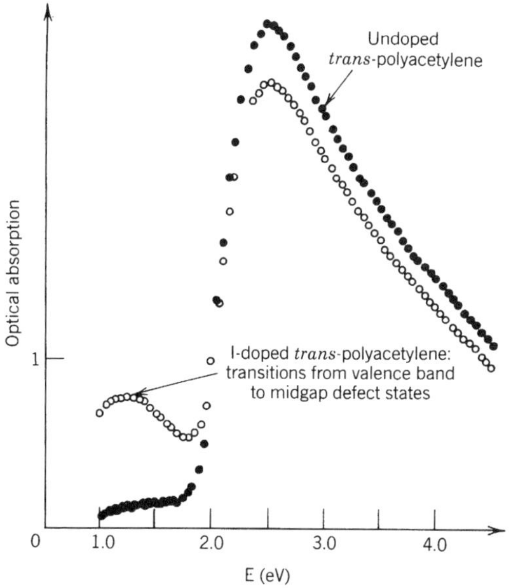

Figure 8.35 Optical absorption in undoped and iodine-doped *trans*-polyacetylene. Both curves show the band-to-band absorption, but only the doped specimen shows the lower-energy transition to midgap defect states [after Suzuki et al. (1980)].

polyethylene arises from the fact that the bandwidth associated with unsaturated π electrons in $(CH)_x$ is large, $\gtrsim 8$ eV, comparable to inorganic semiconductors. The effect of interaction with nearest-neighbor chains is ~0.1 eV, so that although $(CH)_x$ may be regarded in a first approximation as a one-dimensional conductor (see Section 8.2.3), interchain transport cannot be ignored.

The band edge for undoped *trans*-$(CH)_x$ is close to 1.5 eV and for undoped *cis*-$(CH)_x$ it is close to 1.9 eV. The divergence in the joint density of states expected for a one-dimensional system is smeared out by effects of disorder and interaction between chains, giving a broad absorption band (Figure 8.35).

It seems that when pure *cis*-$(CH)_x$ isomerizes to *trans*-$(CH)_x$, the isomerization process may occur at different parts of a single chain giving rise to two neighboring single bonds (Figure 8.34*c*) associated with an unpaired spin (Pople and Walmsley 1962). About 1 in 10^3 of the CH units in *trans*-$(CH)_x$ has an unpaired spin and can be observed by epr. There is evidence to suggest that some of these spins may be associated with the bond-alternation defect referred to above, sometimes described as a soliton. The possibility of the existence of solitons in *trans*-$(CH)_x$ is a consequence of the existence of a two-fold structural degeneracy of the chain, resulting in a symmetry of the chain about the soliton site (Figure 8.34*c*); this degeneracy is not present in *cis*-$(CH)_x$ chains. When the unpaired electron is removed by acceptor doping (Figure 8.34*d*), the soliton state is positively charged, with electron spin $S = 0$. Similarly, double occupancy induced by donor doping gives $S = 0$. It should be realized that unpaired spins in *trans*-$(CH)_x$ could also arise from cross-linking between chains.

Although the soliton is shown localized on one carbon site in Figure 8.34, it is expected that minimization of energy will result in a spread over ~15 (CH) monomers. Motion of the soliton requires only minor shifts (~0.01 Å) of the CH units, and Su et al. (1980) estimate that the neutral soliton has an effective mass comparable to that of the free-electron mass and hence should be highly mobile.

At the time of writing research into the properties of polyacetylene is still being actively pursued with much of the effort directed at establishing consequences of the soliton concept. Effects of correlation on electronic properties, (cf. Section 1.1.3) and the nature of the 'mid-gap' states (Figure 8.35) are also receiving attention. The study of the electronic properties of other polymers such as polydiacetylene (Hattori et al. 1984a), polypyrrole (Kaufman et al. 1984) and polythiophene (Hattori et al. 1984b) is also being pursued. These materials do not have the structural degeneracy of *trans*-$(CH)_x$ and hence cannot support solitons. However, in long-chain polymers generally, localized distortions associated with trapped charges can exist known as polarons (singly charged with $S = \frac{1}{2}$) and bipolarons (doubly charged with $S = 0$) (Brédas et al. 1982; Fesser et al. 1983). These polarons are analogous to the polarons described earlier for ionic solids (Section 1.2.3) and have been observed in chemically-doped polypyrrole (Kaufman et al. 1984) and polythiophene (Hattori et al. 1984b).

APPENDIX

Notation for Defects

Many different schemes have been developed for defect labels. Some are systematic, some make it clear that untidiness is inevitable in active research areas, and others are simply inadequate. We hope this Appendix will make current usage clear.

We start with those schemes first used to label an observed spectrum before the precise nature of the defect responsible was known.

1. *Historical Labels.* Here one labels a particular optical band or epr signal. When the defect responsible is identified, it too receives the same label. Ionic crystals are often characterized this way. Examples are the *F*, *M*, *R*, and *N* centers in alkali halides (Section 5.1.1). These defects are based on anion vacancies containing trapped electrons. Traditionally, hole centers in ionic crystals are called *V* centers (Section 5.1.2). Note that the *H* center (neutral halogen interstitial), the *F* center (neutral halogen vacancy), and the *V* center have nothing to do with hydrogen (H), fluorine (F), or vanadium (V). In covalent crystals, examples include the *GR*1 center (neutral vacancy) in diamond and the *E* center (vacancy associated with a phosphorus donor) in silicon.

2. *Laboratory Labels.* Here each laboratory gives itself a prefix (e.g., we might use OX for Oxford, HAR for Harwell) and simply number spectra as they are found. Subsequently, one may discover, for instance, that OX-7 is a bound exciton, etc. This method has been used mainly for epr spectra in semiconductors and diamond. Examples in silicon are the *G*2 center (negatively charged vacancy), the *B*1, *G*3, and *G*4 centers (vacancy-oxygen complexes), and the *G*9 center (aluminum–vacancy pair).

3. *Energy Labels.* Obviously many optical lines simply become labeled "the 2.37 eV emission" or an epr spectrum may be assigned to "the $g = 2.07$, $S = \frac{1}{2}$ center". However, two other types of label are found. One is the DLTS type (Section 4.3.4) where the electron traps are labeled $E1, E2, \ldots$, in order of increasing electron binding; correspondingly, hole centers are $H1, H2, \ldots$, with increasing hole binding. The other case refers to defect annealing stages after radiation damage. Stages I, II, III

(and perhaps substages IA, . . .) occur as the temperature is raised. The mobile defect at each anneal stage is called the "stage I defect," etc.

We now turn from labels for experimental signals to labels for known or proposed defect structures.

4. *Quantum Chemical Labels.* Here only the carriers are important, and the net charge of the defect is secondary. The core is simply written ⊕, ⊖ (or similar), defining its net charge, and the carriers are indicated explicitly. Thus a neutral donor is [⊕e], a neutral acceptor [⊖h], and an exciton bound to an isoelectronic defect is [⊙eh] (see Section 5.3.3).

5. *Kröger–Vink Notation.* Here we encounter the recurring problem of whether one wishes to give the net charge of a defect explicitly, or whether one wishes to define the electronic structure clearly. For example, in alkali halides the anion vacancy with one trapped electron (F center) has no net charge and is referred to as a neutral center. The anion vacancy with one trapped electron in an oxide has a net charge $+|e|$. Do we give both centers the same notation, since they may have similar epr or optical spectra? Or do we concentrate on the net charge, since this is important in transport and defect equilibria?

The Kröger–Vink notation comes down firmly on a notation which gives net charge relative to a perfect-host solid. It uses three superscripts, namely, $^{\cdot}$ for net positive charge, x for neutral, and $'$ for net negative charge. These can be cumulative, for example, $X^{\cdot\cdot}$ has net charge $2|e|$ and Y''' has net charge $-3|e|$. Some examples are listed below:

a. Vacancy centers in alkali halides and alkaline-earth oxides:

Center	Notation for Alkali Halides (e.g., KCl)	Notation for Alkaline Earth Oxides (e.g., MgO)
Anion vacancy; no trapped carriers	α center, $V_{Cl}^{\cdot}$	$V_0^{\cdot\cdot}$
Anion vacancy; one trapped electron	F center, V_{Cl}^{x}	F^+ center, $V_0^{\cdot}$
Anion vacancy; two trapped electrons	F' center, V_{Cl}'	F^0 center, V_0^{x}
Cation vacancy; one trapped hole	V_F center, V_K^{x}	V^- center, V_{Mg}'
Cation vacancy; two trapped holes (not found)	$V_K^{\cdot}$	V^0 center, V_{Mg}^{x}

Note that many early papers on oxides call $V_0^{\cdot}$ the F center and many recent papers call V_O^x the F center. This confusing notation should be discouraged, i.e., one should write F^0 or F^+ (or even F') but never just F center.

b. The one-electron oxygen center in Gap is O_P^x and the two-electron center is O'_P (see Section 5.5).

c. Transition-metal ions. Here the commonest system of notation does not stress the net defect charge, but instead emphasizes the number of $3d$ electrons. In ionic crystals, there is no real problem. Consider Cr in MgO, for instance. The free atom has (ignoring core electrons) the ground configuration $3d^5 4s^1$. The ionized substitutional impurity with configuration $3d^3$ is naturally called Cr^{3+}. But what is done in covalent semiconductors, where the charge state is less well defined? The convention is to continue to say that $3d^3$ corresponds to Cr^{3+}. Any wary scientist will define charge states operationally; for example "remove Ga^0 from GaP and add Cr^0" would give a neutral defect (assumed stable), that is, Cr_{Ga}^x in Kröger–Vink notation. The defect is, nevertheless, known as Cr^{3+}. The argument is that free Ga^0 has configuration $3d^{10}4s^2 4p^1$, and three electrons $[4s^2 4p^1]$ participate in bonding. When Cr^0 is substituted, three of its $[3d^5 4s^1]$ electrons participate in bonding, leaving a configuration $3d^3$.

6. *Amorphous Semiconductors*. Here we have to accommodate different coordinations as well as different charges. As a result, a convention is used with implies a specific type of bonding. This means that there can be problems in comparing, say, a relatively ionic II–VI oxide (like BeO) and a relatively covalent oxide like SiO_2. Whereas, for the II–VI oxide one can start from either an ionic picture (Be^{2+}, O^{2-}) or a covalent picture (sp^3 bonds, which give Be^{2-}, O^{2+} when electrons are evenly assigned), this choice is impractical with the notation chosen for amorphous semiconductors. For such systems, three species are considered, namely, (1) tetrahedral-bonding group IV elements like Si and Ge, labeled T; (2) pnictides like N, As, and Sb, labeled P; and (3) chalcogenides, like O, S, Se, Te, labeled C. Each of these atoms is given a subscript, N, describing its coordination, and a superscript describing the charge Q relative to a reference state. Thus P_N^Q is a N-fold coordinated pnictide with charge Q. The charge is relative to a standard state, namely T_4^0 (sp^3 state), P_3^0 (s^2p^3 state), or C_2^0 (s^2p^4 state), and indicates the number of valence electrons n other than the N in bonding orbitals. Thus, for chalcogenides $n = 4 + Q - N$, the 4 corresponding to p^4 [for further discussion see Robertson (1983)]. There is one further complication. Suppose we have antisite defects, that is, instead of a given atom having all neighbors of another species, it has several of the same species next to it. The convention is to use primes to show this. Thus, for a chalcogen with twofold coordination, C_2, C'_2, and C''_2 indicate, respectively, all unlike neighbors, one like neighbor, and two like neighbors.

Physical Constants: Units and Conversion Factors

While precise values of physical constants are not of special importance in this book, there are many cases in practice where these constants and their conversions from one set of units to another are required. We collect here some useful values and some of the less common but, in our experience, helpful conversions.

PHYSICAL CONSTANTS

Velocity of light in vacuum	c	2.9979×10^{10} cm/sec
Planck's constant	h	6.623×10^{-27} erg sec
	$\hbar = h/2\pi$	1.0545×10^{-27} erg sec
		0.657×10^{-15} eV sec
		0.5308×10^{-11} cm^{-1} sec
Boltzmann's constant	k_B	0.8617×10^{-4} eV K^{-1}
		1.3807×10^{-16} ergs K^{-1}
Avogadro's number	N	6.024×10^{23} molecules/mole
Gas constant	$R = Nk_B$	8.314 ergs K^{-1}/mole
Proton charge	e	4.8030×10^{-10} esu
		1.602×10^{-19} C
Electronic mass	m_e	0.9109×10^{-27} g
Atomic mass unit	amu	1.6597×10^{-24} g
Bohr magneton	β (or μ_B)	4.6704×10^{-5} cm^{-1} G^{-1}
		1.3997 MHz G^{-1}
		5.7884×10^{-9} eV G^{-1}
Fine structure constant	$\alpha = \hbar c/e^2$	1/137

UNITS AND CONVERSION FACTORS

Energies

1 cm^{-1}	$k_B \times 1.439$ K
	1.9855×10^{-16} ergs
	1.2395×10^{-4} eV
	258.4 cm/sec (Mössbauer velocity shift)
1 eV	8068.3 cm^{-1}
	1.602×10^{-12} ergs
	23.06 kcal/mole
	$k_B \times 11{,}610$ K
1 au	27.2116 eV
$e^2/1$ Å	14.4 eV
$(h^2/2m_e)(1/1\ \text{Å})^2$	3.806 eV
$\beta^2/(1\ \text{Å})^3$	1.298 10^4 MHz
1 calorie	4.184 J

The wavelengths of excitations in the infrared are measured in microns (μm) (1 μm $= 10^{-3}$ mm) and in the visible and ultraviolet in angstrom units (Å) (1 Å $= 10^{-7}$ mm $= 10^{-8}$ cm) or in nanometers (nm) or millimicrons (mμm) (1 nm = 10 Å).

The frequency ν' and wavenumber ν of excitations are connected by Planck's relation $E = h\nu' = h\nu c$, and ν is taken as a measure of energy in spectroscopy. The electron volt (eV) is also used as a measure of energy in spectroscopy and is equivalent to 8068.3 cm^{-1} or 1.602×10^{-12} erg or 23.06 kcals/mole. A quantum of light of wavelength $L \times 10^2$ nm has energy $(12.395/L)$ eV.

Lengths

1 au	0.5291 Å
1 Å	1.889 au

Pressures and Stresses

1 N/m^2 (Pa)	10^{-5} bar
	0.98692×10^{-5} atm
	10 $dyne/cm^2$
	7.5006×10^{-3} torr

Supplementary Reading

CHAPTER 1

Coles, B. R., and Caplin, A. D. (1976), *The Electronic Structure of Solids*, Edward Arnold, London.

Coulson, C. A. (1961), *Valence*, Clarendon Press, Oxford.

Haken, H. and Nikitine, S., Eds. (1975), *Excitons at High Density*, *Springer Tracts in Modern Physics*, Vol. 73, Springer-Verlag, New York.

Jones, W. and March, N. H. (1973), *Theoretical Solid State Physics*, Wiley, New York, Vols. I and II.

Kittel, C. (1964), *Quantum Theory of Solids*, Wiley, New York.

Knox, R. S. (1963), *Theory of Excitons*, Academic Press, New York.

Rashba, E. I. and Sturge, M. D., Eds. (1982), *Excitons*, North Holland, Amsterdam.

Stoneham, A. M. (1975), *Theory of Defects in Solids*, Clarendon Press, Oxford, England.

Wells, A. F. (1975), *Structural Inorganic Chemistry*, Clarendon Press, Oxford, England.

Ziman, J. M. (1964), *Principles of the Theory of Solids*, Cambridge University Press, England.

CHAPTER 2

Born, M. and Huang, K. (1954), *Dynamical Theory of Crystal Lattices*, Clarendon Press, Oxford, England.

Brüesch, P., Ed., *Phonons: Theory and Experiment I*, *Springer Series in Solid State Science*, Vol. 34, Springer-Verlag, Heidelberg.

Cochran, W. (1973), *The Dynamics of Atoms in Crystals*, Edward Arnold, London.

Hardy, J. R. and Karo, A. M. (1979), *The Lattice Dynamics and Statics of Alkali Halide Crystals*, Plenum Press, New York.

Hayes, W. and Loudon, R. (1978), *Scattering of Light by Crystals*, Wiley-Interscience, New York.

Reissland, J. A. (1973), *The Physics of Phonons*, Wiley, New York.

CHAPTER 3

Catlow, C. R. A. and Mackrodt, W. C. (1982), Editors, *Computer Simulation in Solids*, Springer-Verlag, New York.

Flynn, C. P. (1972), *Point Defects and Diffusion*, Clarendon Press, Oxford, England.

Girifalco, L. A. (1973), *Statistical Physics of Materials*, Wiley, New York.

Henderson, B. and Hughes, A. E., Eds., (1975), *Defects and Their Structure in Non-Metallic Solids*, Plenum Press, New York.

Kittel, C. (1958), *Elementary Statistical Physics*, Wiley, New York.

Kofstad, P. (1972), *Nonstoichiometry, Diffusion and Electrical Conductivity in Binary Metal Oxides*, Wiley, New York.

Mott, N. F. and Gurney, R. (1964), *Electronic Processes in Ionic Crystals*, Dover, New York.

Schmalzried, H. (1974), *Solid State Reactions*, Academic Press, New York.

Shewmon, P. G. (1963), *Diffusion in Solids*, McGraw-Hill, New York.

Sørensen, O. T., Ed., (1981), *Nonstoichiometric Oxides*, Academic Press, New York.

CHAPTER 4

Abragam, A. and Bleaney, B. (1970), *Electron Paramagnetic Resonance*, Clarendon Press, Oxford, England.

Agranovich, V. M. and Galanin, M. D. (1982), *Electronic Excitation Energy Transfer in Condensed Matter*, North Holland, Amsterdam.

Engeman, R. (1979), *Non-Radiative Decay of Ions and Molecules in Solids*, North Holland, Amsterdam.

Rebane, K. K. (1970), *Impurity Spectra of Solids*, Plenum Press, New York.

Standley, K. J. and Vaughan, R. A. (1969), *Electron Spin Relaxation Phenomena in Solids*, Adam Hilger, London.

Stoneham, A. M. (1975), *Theory of Defects in Solids*, Clarendon Press, Oxford, England.

Wooten, F. (1972), *Optical Properties of Solids*, Academic Press, New York.

Yen, W. M. and Selzer, P. M., Eds. (1981), *Laser Spectroscopy of Solids, Topics in Applied Physics*, Springer-Verlag, New York, Vol. 49.

CHAPTER 5

Crawford, J. H. and Slifkin, L. M., Eds., *Point Defects in Solids*, Plenum Press, New York, Vol. 1 (1972) and Vol. 2 (1975).

Fowler, W. B., Ed. (1968), *Physics of Color Centres*, Academic Press, New York.

Jaros, M. (1982), *Deep Levels in Semiconductors*, Adam Hilger, London.

Orgel, L. E. (1960), *An Introduction to Transition Metal Chemistry*, Methuen, London.

Pantelides, S. T. (1978), *Rev. Mod. Phys.*, **50**, 797.

Seeger, K. (1973), *Semiconductor Physics*, Springer-Verlag, New York.

Smith, R. A. (1978), *Semiconductors*, Cambridge University Press, England.

Stoneham, A. M. (1975), *Theory of Defects in Solids*, Clarendon Press, Oxford, England.

CHAPTER 6

Baldereschi, A., Czaja, W., Tosatti, E. and Tosi, M., Eds., (1983), *The Physics of Latent Image Formation in Silver Halides*, World Scientific, Singapore.

Hasiguti, R. R. (1980), *Defects and Radiation Effects in Semiconductors*, Institute of Physics Conference, London, Proc. No. 59.

Hughes, A. E. and Pooley, D. (1975), *Real Solids and Radiation*, Wykeham Publication, London.

Lannoo, M. and Bourgoin, J. C. (1981), *Point Defects in Semiconductors*, Springer-Verlag, New York, Vols. 1 and 2.

CHAPTER 7

Blakeley, J. M. (1973), *Introduction to the Properties of Crystal Surfaces*, Pergamon Press, Oxford, England.

Kingery, W. D., Bowen, H. K., and Uhlmann, D. R. (1976), *Introduction to Ceramics*, Wiley, New York.

Prutton, M. (1983), *Surface Physics* (2nd Ed.), Oxford University Press, England.

Somorjai, G. A. (1972), *Principles of Surface Chemistry*, Prentice-Hall, London.

CHAPTER 8

Brodsky, M. H., Ed. (1979), *Amorphous Semiconductors*, Topics in Applied Physics, Springer-Verlag, New York, Vol. 36.

Devreese, J. T., Evrard, R. P., and van Doren, V. E., Eds. (1979), *Highly Conducting One-Dimensional Solids*, Plenum Press, New York.

Guy, A. G. (1971), *Introduction to Materials Science*, MgGraw-Hill, New York.

Mort, J. and Pfister, G. P. (1982), *Electronic Properties of Polymers*, Wiley-Interscience, New York.

Mott, N. F. (1974), *Metal-Insulator Transitions*, Taylor and Francis, London.

Mott, N. F. and Davis, E. A. (1979), *Electronic Processes in Non-Crystalline Materials*, 2nd ed., Clarendon Press, Oxford, England.

Phillips, W. H., Ed., (1981), *Amorphous Solids*, Springer-Verlag, New York.

Pietronero, L. and Tosatti, E., Eds. (1981), *The Physics of Intercalation Compounds*, Springer-Verlag, New York, Vol. 38.

Pope, M. and Swenberg, C. E. (1982), *Electronic Processes in Organic Crystals*, Clarendon Press, Oxford, England.

Seanor, D. A. (1982), *Electrical Properties of Polymers*, Academic Press, New York.

Whittingham, M. S. and Jacobson, A. J. (1982), *Intercalation Chemistry*, Academic Press, New York.

Young, R. J. (1981), *Introduction to Polymers*, Chapman and Hall, London.

References

Abragam, A. and Bleaney, B. (1970), *Electron Paramagnetic Resonance of Transition Ions*, Clarendon Press, Oxford.

Adler, D. (1981), *J. Phys.*, **42** (Supp. 10), C-4-3.

Alben, R., Weaire, D., Smith, Jr., J. E., and Brodsky, M. H. (1975), *Phys. Rev. B* **11**, 2271.

Alig, R. C. and Bloom, S. (1975), *Phys. Rev. Lett.* **35**, 1522.

Allen, J. P., Colvin, J. T., Stinson, D. G., Flynn, C. P., and Stapleton, H. O. (1982), *Biophys. J.* **38**, 299.

Anderson, A. C. (1981), in *Amorphous Solids*, W. A. Phillips Ed., Springer-Verlag, New York, p. 65.

Anderson, J. S. (1970a), in *Problems of Non-Stoichiometry*, A. Rabenau, Ed., North Holland, Amsterdam, p. 1.

Anderson, J. S. (1970b), in *The Chemistry of Extended Defects*, L. Eyring and M. O'Keefe, Eds., North Holland, Amsterdam, p. 1.

Anderson, P. W. (1958), *Phys. Rev.* **109**, 1492.

Anderson, P. W. (1959), *Phys. Rev.* **114**, 1002.

Anderson, P. W. (1975), *Phys. Rev. Lett.* **34**, 953.

Andrews, J. M. and Phillips, J. C. (1975), *Phys. Rev. Lett.* **35**, 56.

Ashby, M. F. (1974), *Acta Met.* **22**, 275.

Atkinson, A. (1984), *Rep. Prog. Phys*, in press and AERE Report R11293 (1984).

Atkinson, A. and Taylor, R. I. (1981), *Phil. Mag. A* **43**, 979.

Baetzold, R. (1973), *Photog. Sci. Eng.* **17**, 78.

Baetzold, R. (1975), *Photog. Sci. Eng.* **19**, 11.

Baraff, G. A. and Schlüter, M. (1978) *Phys. Rev. Lett.* **41**, 892.

Baraff, G. A., Kane, E. O., and Schlüter, M. (1980), *Phys. Rev. B* **21**, 3563.

Baranowski, J. M., Allen, J. W., and Pearson, G. L. (1968), *Phys. Rev.* **167**, 758.

Bardeen, J. (1947), *Phys. Rev.* **71**, 717.

Barker, A. S. and Sievers, A. J. (1975), *Rev. Mod. Phys.* **47**, Suppl. 2.

Barr, L. W. and Lidiard, A. B. (1970), *Phys. Chem.* **10**, 152.

Bartram, R. H. and Stoneham, A. M. (1975), *Sol. St. Comm.* **17**, 1593.

Bartram, R. H. and Stoneham, A. M. (1983), *Semicond. Insulators* **5**, 297.

Bartram, R. H., Stoneham, A. M., and Gash, P. W. (1968), *Phys. Rev.* **176**, 1014.

Bar-Yom, Y. and Joannopoulos, J. D., (1984), *Phys. Rev. Lett.*, **52**, 1129.

Beaumont, J. H., Harmer, A. L., and Hayes, W. (1972), *J. Phys. C, Sol. St. Phys* **5**, 257.

Beniere, F. and Catlow, C. R. A., Ed. (1983), *Mass Transport in Solids*, Plenum Press, New York.

Berreman, D. W. (1963), *Phys. Rev.* **130**, 2193.

Bessent, R. G., Hayes, W., and Hodby, J. W. (1967), *Proc. Roy. Soc. A* **297**, 376.

Binnig, G. and Rohrer, H. (1983), *Helv. Phys. Acta* **56**, 482.

Bishop, S. G. and Mitchell, D. L. (1973), *Phys. Rev. B* **8**, 5696.

Black, J. L. and Halperin, B. I. (1978), *Phys. Rev. B* **16**, 2879.

Bloembergen, N., Purcell, E. M., and Pound, R. V. (1948), *Phys. Rev.* **73**, 679.

Bloor, D. (4 March 1982), *New Scientist*, p. 577.

Bockris, J. O'M. and Dražić, D. M. (1972), *Electrochemical Science*, Taylor and Francis, London.

Bonch-Bruevich, V. L. and Landsberg, E. G. (1968), *Phys. Stat. Sol.* **29**, 9.

Born, M. and Huang, K. (1954), *Dynamical Theory of Crystal Lattices*, Clarendon Press, Oxford.

Bosomworth, D. R., Hayes, W., Spray, A. R. L., and Watkins, G. D. (1970), *Proc. Roy. Soc. A* **317**, 133.

Bourgoin, J. C. (1984), *J. Phys. C*, in press.

Bowen, S. P., Gomez, M., Krumhansl, J. A., and Matthews, J. A. D. (1966), *Phys. Rev. Lett.* **16**, 1105.

Boyce, J. B. and Hayes, T. M. (1949), in *Physics of Superionic Conductors*, M. B. Salamon, Ed., Springer-Verlag, New York, Vol. 15, p. 5.

Brédas, J. L., Chance, R. R. and Silbey, R. (1982), *Phys. Rev. B* **26**, 5843.

Brillouin, L. N. (1953), *Wave Propagation in Periodic Structures*, Dover, New York.

Brillson, J. B. (1978), *J. Vac. Sci. Tech.* **15**, 1378.

Brinkman, W. F., Rice, T. H. and Bell, B. (1973), *Phys. Rev. B* **8**, 1570.

Brown, F. C., Cavenett, B. C., and Hayes, W. (1967), *Proc. Roy. Soc. A* **300**, 78.

Brya, W. J. and Wagner, P. E. (1967), *Phys. Rev.* **157**, 400.

Buisson, J. P., Sadoc, A., Taurel, L., and Billardon, M. (1975), in *Light Scattering in Solids*, M. Balkanski, R. C. C. Leite, and S. P. S. Porto, Eds., Flammarion, Paris, p. 587.

Bursill, L. A. and Hyde, B. G. (1971), *Phil. Mag.* **23**, 3.

Cardona, M. (1963), *Phys. Rev.* **129**, 69.

Carr, R., Kelly, P. J., Oshiyama, A. and Pantelides, S. T., (1984), *Phys. Rev. Lett.*, **52**, 1814.

Castaing, J. (1984), Proc. 'Conf. Prop. Binary Oxides', University of Seville Press, Seville.

Catlow, C. R. A. (1977), *Proc. Roy. Soc.* **A353**, 533.

Catlow, C. R. A. (1982), *Phys. & Chem. of Refract. Oxides* (Ed. P. Thévenard), Plenum Press, New York.

Catlow, C. R. A., Corish, J., Jacobs, P. W. M., and Lidiard, A. B. (1981), *J. Phys. C, Sol. St. Phys.* **14**, L 121.

Catlow, C. R. A., Diller, K., and Hobbs, L. W. (1980), *Phil. Mag. A* **42**, 123.

Catlow, C. R. A. and Fender, B. F. (1975), *J. Phys. Sol. St. Phys.* **8**, 3267.

Catlow, C. R. A. and Macrodt, W. C., Eds. (1982), *Computer Simulation of Solids*, Springer-Verlag, New York.

Catlow, C. R. A., Macrodt, W. C., and Norgett, M. J. (1977), *Phil. Mag.* **35**, 177.

Catlow, C. R. A. and Norgett, M. J. (1973), *J. Phys. C, Sol. St. Phys.* **6**, 1325.

Catlow, C. R. A. and Stoneham, A. M. (1981), *J. Am. Ceram. Soc.* **64**, 234.

Catlow, C. R. A. and Stoneham, A. M. (1983), *J. Phys. C, Sol. St. Phys.*, **16**, 4321.

Chabal, Y. J. and Patel, C. K. N. (1984), *Phys. Rev. Lett.*, **53**, 1771.

Challis, L. J., Rampton, W. W. and Wyatt, A. F. G. (1976), Eds. *Phonon Scattering in Solids*, Plenum Press, New York.

Chang, I. F. and Mitra, S. S. (1968), *Phys. Rev.* **172**, 924.

Cheetham, A. K. and Taylor, J. C. (1977), *J. Sol. St. Chem.* **21**, 253.

Clerjaud, B., Gelineau, A., Gendron, F., Porte, C., Baranowski, J. M., and Liro, Z. (1984), *J. Phys. C, Sol. St. Phys.*, **17**, 3837.

Cochran, W. (1971), *Crit. Rev. Sol. St. Sci.* **2**, 1.

Cochran, W. (1973), *The Dynamics of Atoms in Crystals*, Edward Arnold, London.

Cohen, M. L. and Heine, V. (1961), *Phys. Rev.* **122**. 1821.

Colbourn, E. A. and Macrodt, W. C. (1981), *Sol. St. Comm.* **40**, 265.

Collins, A. T. (1981), *J. Phys. C., Sol. St. Phys.* **14**, 289.

Cooper, M. and Leake, J. A. (1967), *Phil. Mag.* **15**, 1201.

Corbett, J. W., Bourgoin, J. C., Cheng, L. J., Corelli, J. C., See, Y. H., Mooney, P. M. and Weigel, C. (1976), *Inst. of Phys. Conf. Ser.* **31**, 1.

Coulson, C. A. (1956), *Electricity*, Oliver and Boyd, London.

Coulson, C. A. (1961), *Valence*, Clarendon Press, Oxford, England.

Coulson C. A. and Kearsley M. J. (1957) *Proc. Roy. Soc. A* **241**, 433.

Cousins, C. S. G. (1982), *J. Phys. C., Sol. St. Phys.* **15**, 1857.

Cowley, R. A. and Cochran, W. (1962), *J. Phys. Chem. Sol.* **23**, 447.

Cox, P. A., Eydell, R. G., and Naylor, P. D. (1983), *J. Elec. Spec. and Rel. Phen.* **29**, 247.

Crawford, J. H. and Slifkin, L. M., Eds. (1972), *Point Defects in Solids*, Plenum Press, New York, Vol. 1.

Davidge, R. W. (1979), *Mechanical Behaviour of Ceramics*, Cambridge University Press, England.

Davies, J. J. (1976), *Cont. Phys.* **17**, 275.

Davis, E. A. and Compton, W. D. (1965), *Phys. Rev. A* **140**, 2183.

Dawson, R. K. and Pooley, D. (1969), *Phys. Stat. Sol.* **35**, 95.

Dean P. J. and Herbert, D. C. (1979), in *Excitons*, K. Cho, Ed., *Topics in Current Physics*, No. 14, Springer-Verlag, New York, p. 55.

Dean, P. J., Schönherr, E. G., and Zetterstrom, R. B. (1970), *J. Appl. Phys.* **41**, 3434.

Dearnaley, G., Freeman, J. H., Nelson, R. S., and Stephen, J. (1973), *Ion Implantation*, North Holland, Amsterdam.

Dearnaley, G., Stoneham, A. M., and Morgan, D. V. (1970), *Rep. Prog. Phys.* **33**. 1129.

Delbecq, C. J., Hayes, W., O'Brien, M. C. M., and Yuster, P. H. (1963), *Proc. Roy. Soc. A* **271**, 243.

Delbecq, C. J., Hayes, W., and Yuster P. H. (1961), *Phys. Rev.* **121**, 1043.

Dexter, D. L., Klick, C. C., and Russell, G. A. (1956), *Phys. Rev.* **100**, 603.

Dick, B. G. and Overhauser, W. A. (1958), *Phys. Rev.* **112**, 90.

Dieterich, W., Fulde, P. and Peschel, I. (1980), *Adv. Phys.* **29**, 527.

Dingle, R., Wiegmann, W., and Henry, C. H. (1974), *Phys. Rev. Lett.* **33**, 827.

Dixon, M. and Gillan, M. J. (1980), *J. Phys. C, Sol. St. Phys.* **13**, 1901.

Doremus, R. H. (1973), *Glass Science*, Wiley, New York.

Dow, J. D. and Redfield, D. (1972), *Phys. Rev. B* **5**, 594.

Duke, C. B. and Mahan, G. D. (1965), *Phys. Rev.* **139**, A1965.

Eby, J. E., Teegarden, K. J., and Dutton, D. B. (1959), *Phys. Rev.* **116**, 1091.

Eckert, J., Ellinson, W. D., Hastings, J. B., and Passell, L. (1979), *Phys. Rev. Lett.* **43**, 1329.

Edwards, P. P. and Sienko, M. J. (1981), *J. Am. Chem. Soc.* **103**, 2967.

Elliott, R. J. (1962), in *Polarons and Excitons*, C. G. Kuper and G. D. Whitfield, Eds., Oliver and Boyd, London, p. 269.

Elliott, R. J., Hayes, W., Jones, G. D., Macdonald, H. F., and Sennett, C. T. (1965), *Proc. Roy. Soc. A* **289**, 1.

Elliott, R. J. and Pfeuty, P. (1967), *J. Phys. Chem. Sol.* **28**, 1789.

Emin, D. and Holstein, T. (1976), *Phys. Rev. Lett.* **36**, 323.

Enderby, J. E. and Neilson, G. W. (1981), *Rep. Prog. Phys.* **44**, 593.

Engemann, D. and Fischer, R. (1974), in *Proceedings of the 12th International Conference on the Physics of Semiconductors*, M. H. Pilkuhn Ed., Teubner, Stuttgart, p. 1042.

Engemann, D. and Fischer, R. (1977), *Phys. Stat. Sol.* (*b*) **79**, 195.

Eshelby, J. D. (1955), *Acta Met.* **3**, 487.

Etemad, S., Heeger, A. J., and MacDiarmid, A. G. (1982), *Ann. Rev. Phys. Chem.* **33**, 443.

Euwema, R. N., Wilhite, D. L., and Surratt, G. T. (1973), *Phys. Rev. B* **7**, 818.

Farlow, G., Blose, A., Sister, J., Feldott, B., Sounsberry, B. and Slifkin, L. (1983), *Rad. Effects* **75**, 1.

Feldman, D. W., Parker, J. H., Choyke, W. J., and Patrick, L. (1968), *Phys. Rev.* **173**, 787.

Fender, B. E. F. (1973), in *Chemical Applications of Thermal Neutron Scattering*, B. T. M. Willis, Ed., Oxford University Press, England, p. 250.

Fesser, K., Bishop, R. and Campbell, D. K. (1983), *Phys. Rev. B*, **27**, 4804.

Fitchen, D. B., Silsbee, R. H., Fulton, T. A., and Wolf, E. L. (1963), *Phys. Rev. Lett.*, **11**, 275.

Flynn, C. P. (1972), *Point Defects and Diffusion*, Clarendon Press, Oxford.

Förster, T. (1948), *Ann. Phys.* (*Paris*) **2**, 55.

Fowler, W. B. (1968), in *Physics of Color Centers*, W. B. Fowler, Ed., Academic Press, New York, p. 54.

Frank, F. C. and van der Merwe, J. (1949), *Proc. Roy. Soc.* **198**, 205, 216.

Fröhlich, H. (1958), *Theory of Dielectrics*, Clarendon Press, Oxford.

Fromhold, A. T. (1976), *Theory of Metal Oxidation*: *I Fundamentals*, North Holland, Amsterdam.

Fuchs, R. and Kliewer, K. L. (1965), *Phys. Rev.* **140**, A2076.

Funke, K. and Jost, A. (1971), *Ber. Bunsenges Phys. Chem.* **75**, 436.

Fussgänger, K., Martienssen, W. and Bilz, H. (1965), *Phys. Stat. Sol.* **12**, 383.

Gebhardt, W. and Kühnert, H. (1964), *Phys. Lett.* **11**, 15.

Gebhardt, K. F., Soper, P. D., Merski, J., Balle, T. J., and Flygare, W. H. (1980), *J. Chem. Phys.* **72**, 272.

Gellermann, W., Lüty, F., and Pollock, C. R. (1981), *Opt. Comm.* **39**, 391.

Gellermann, W. and Pollock, C. R. (1981), *Opt. Comm.* **39**, 391.

Ghez, R. (1973), *J. Chem. Phys.* **58**, 1838.

Gillan, M. J. (1981), *Phil. Mag. A* **43**, 301.

Gordon, R. E. and Strange, J. H. (1978), *J. Phys. C, Sol. St. Phys.* **11**, 3213.

Gourary, B. S. and Adrian, F. J. (1957), *Phys. Rev.* **105**, 1180.

Gourary, B. S. and Adrian, F. J. (1960), *Sol. St. Phys.* **10**, 127.

Greenwood, N. N. (1968), *Ionic Crystals, Lattice Defects and Nonstoichiometry*, Butterworths, London.

Griscom. D. L. (1975), in *Radiation Damage Processes in Materials*, C. H. D. Dupuy, Ed., Noordhoff, Groningen, p. 209.

Griscom, D. L. and Friebele, E. J. (1982), *Rad. Effects* **65**, 63.

Guinier, A. (1963), *X-ray Diffraction*, Freeman, San Francisco.

Guy, A. G. (1971), *Introduction to Materials Science*, McGraw-Hill, New York.

Haken, H. and Nikitine, S., Eds. (1975), *Excitons at High Density*, Springer Tracts in Modern Physics, Springer-Verlag, New York, Vol. 73.

Ham, F. S. (1965), *Phys. Rev.* **138**, A1727.

Hemstreet, L. (1980), *Phys. Rev. B* **22**, 4590.

Hanamura, E. (1973), *Sol. St. Comm.* **12**, 951.

Haneman, D. (1982), *Adv. Phys.* **31**, 165.

Hansen, W. N., Kolb, D. M., and Lynch, D. W., Eds. (1983), *Electronic and Molecular Structure of Electrode–Electrolyte Interfaces*, Elsevier, Amsterdam.

Harkins, C. G., Chang, W. W., and Leland, T. W. (1969), *J. Phys. Chem.* **73**, 130.

Harper, P. G., Hodby, J. W., and Stradling, R. A. (1973), *Rep. Prog. Phys.* **36**, 1.

Hattori, T., Hayes, W. and Bloor, D. (1984a), *J. Phys. C, Sol. St. Phys.* **17**, L881.

Hattori, T., Hayes, W., Wong, K., Kaneto, K. and Yoshino, Y. (1984b), *J. Phys. C, Sol. St. Phys.* **17**, L803.

Haydock, R., Heine, V. and Kelly, M. J. (1972), *J. Phys. C., Sol. St. Phys.* **5**, 2845.

Hayes, W. (1982), in *Light Scattering in Solids III*, M. Cardona and G. Güntherodt, Eds., Springer-Verlag, New York, Vol. 51, p. 43.

Hayes, W. (1983), *Semicond. Insulators* **5**, 333.

Hayes, W., Kane, M., Salminen, O., Wood, R. L. and Doherty, S. P. (1984), *J. Phys. C., Sol. St. Phys.* **17**, 2943.

Hayes, W. and Loudon, R. (1978), *Scattering of Light by Crystals*, Wiley-Interscience, New York.

Hayes, W., Owen, I. B., and Walker, P. J. (1977), *J. Phys. C., Sol. St. Phys.* **10**, 1751.

Hayes, W. and Stott, J. P. (1967), *Proc. Roy. Soc. A* **301**, 313.

Haynes, J. R. (1960), *Phys. Rev. Lett.* **4**, 361.

Heine, V. (1965), *Phys. Rev.* **138**, A1689.

Henderson, B. (1972), *Defects in Crystalline Solids*, Edward Arnold, London.

Henderson, B. (1981), *Opt. Lett.*, **6**, 437.

Henderson, B., Chen, Y., and Sibley, W. A. (1972), *Phys. rev. B* **6**, 4060.

Henry, C. H. (1980). *J. Appl. Phys.* **51**, 3051.

Henry, C. H. and Logan, R. A. (1977), *J. Appl. Phys.* **48**, 3902.

Henry, C. H. and Nassau K. (1970), *Phys. Rev. B* **1**, 1628; *B* **2**, 997.

Henry, C. H., Schnatterly, S. E., and Slichter, C. P. (1964), *Phys. Rev. Lett.* **13**, 130.

Hensel, J. C., Phillips, T. G., and Thomas, G. A. (1977), *Sol. St. Phys.* **32**, 88.

Hirai, M. and Wakita, S. (1983), *Semicond. Insulators* **5**, 231.

Hoar, T. P. (1976), *Proc. Roy. Soc. A* **348**, 1.

Hoare, J. P., Masri, P., and Tasker, P. W. (1983), AERE Report R10752.

Hobbs, L. W., Hughes, A. E., and Pooley, D. (1973), *Proc. Roy. Soc. A* **332**, 167.

Hobbs, P. (1974), *Ice Physics*, Oxford University Press, Oxford, England.

Hoffman, R. (1963), *J. Chem. Phys.* **39**, 1397.

Hooper, A. (1978), *Cont. Phys.* **19**, 147.

Hopfield, W. J. (1964), *Proc. Paris Semicond. Conf.* M. Hulin, Ed., Dunod, Paris, p. 725.

Horn, P. M., Birgeneau, R. J., Heiney, P., and Hammonds, E. M. (1978), *Phys. Rev. Lett.* **41**, 961.

Howard, R. E. and Lidiard, A. B. (1964), *Rep. Prog. Phys.* **27**, 161.

Hubbard, J. (1963), *Proc. Roy. Soc. A* **276**, 238.

Hudson, J. A. (1980), *The Excitation and Propagation of Elastic Waves*, Cambridge University Press, England.

Hughes, A. E. (1978), *Comm. Sol. St. Phys.* **8**,83.

Hughes, A. E. (1982), *Corros. Sci.* **22**, 103.

Hughes, A. E. and Jain, S. C. (1979), *Adv. Phys.* **28**, 717.

Hughes, R. C. (1977), *Phys. Rev. B* **15**, 2012.

Hunklinger, S. and Schickfus, M. (1981), in *Amorphous Solids*, W. A. Phillips, Ed., Springer-Verlag, New York, p. 81.

Hutchings, M. T., Clausen, K., Dickens, M. H., Hayes, W., Kjems, J. K., Schnabel, P. G., and Smith, C. (1984), *J. Phys. C, Sol. St. Phys.* **17**, 3903.

Itoh, N. (1976), *Nucl. Instr. Methods* **132**, 201.

Itoh, N. (1982), *Adv. Phys.* **31**, 491.

Itoh, N., Stoneham, A. M., and Harker, A. H. (1977), *J. Phys. C., Sol. St. Phys.* **10**, 4197.

Jäckle, J. (1981), in *Amorphous Solids*, W. A. Phillips, Ed., Springer-Verlag, New York p. 135.

Jaros, M. and Dean, P. J. (1983), *Phys. Rev. B* **28**, 6104.

Jagannath, C. and Ramdas, A. K. (1981), *Phys. Rev. B* **23**, 4426.

Jaenicke, W. (1977), *Adv. Electrochem and Electrochem. Eng.* **10**, 91.

Jerome, D. and Schulz, H. J. (1982), *Adv. Phys.* **31**, 299.

Jonscher, A. K. (1977), *Nature* **267**, 673.

Jonscher, A. K. (1980), *Phys. Thin Films* **11**, 206.

Jorgensen, C. (1958), *Disc. Farad. Soc.* **26**, 110.

Kabler, M. N. (1964), *Phys. Rev.* **136**, A1296.

Kastner, M. (1972), *Phys. Rev. Lett.* **28**, 335.

Kastner, M. and Fritsche, H. (1978), *Phil. Mag.* **37**, 199.

Karim, D. P. and Aldred, A. T. (1970), *Phys. Rev. B* **20**, 2255.

Kaufman, J. H., Colaneri, N., Scott, J. C. and Street, G. B. (1984), *Phys. Rev. Lett.*, **53**, 1005.

Keldysh, L. V. (1968), *Proceedings of the 9th International Conference on Semiconductors*, Moscow, Nauka, Leningrad, p. 1303.

Kelly, M. J. (1983), *G.E.C. J. Res.* **1**, 15.
Kirk, D. L. and Pratt, P. L. (1967), *Proc. Brit. Ceram. Soc.* **9**, 215.
Kiss, Z. J. (1970, Jan.), *Physics Today*, p. 42.
Kittel, C. H. (1976), *Introduction to Solid State Physics*, 1st ed. 1953; 3rd ed., 1966; 5th ed., 1976; Wiley, New York.
Kliewer, K. L. and Fuchs, R. (1966), *Phys. Rev.* **150**, 573.
Klug, D. D. and Whalley, E. (1984), *Phys. Rev.* **B25**, 5543.
Knoteck, M. L. and Feibelman, P. J. (1978), *Phys. Rev. Lett.* **40**, 964.
Koch, F. and Cohen, J. B. (1969), *Acta Cryst.* **B25**, 275.
Kofstad, P. (1966), *High-Temperature Oxidation of Metals*, Wiley, New York.
Kohn, W. (1957), *Sol. St. Phys.* **5**, 285.
Kunz, A. B. (1978), in *Excited States in Quantum Chemistry*, C. A. N. Nicolaides and D. R. Leck, Eds., Reidel, Dordrecht, p. 471.
Lampert, M. A. (1958), *Phys. Rev. Lett.* **1**, 450.
Landauer, R. (1952), *J. Appl. Phys.* **23**, 779.
Lang, D. V. (1974), *J. Appl. Phys.* **45**, 3023.
Lang, D. V. (1976), *Inst. Phys. Conf. Ser.* **31**, 70.
Laughlin, R. B., Joannopoulos, J. D., and Chadi, D. J. (1980), *Phys. Rev. B* **21**, 5733.
Layne, C. B., Lowdermilk, W. H., and Weber, M. J. (1977), *Phys. Rev. B* **16**, 10.
Leigh R. S., Szigeti, B., and Tewary, V. K. (1971), *Proc. Roy. Soc. A* **320**, 505.
Levine, R. and Jortner, J. (1976), *Molecular Energy Transfer*, Wiley, New York.
Levy, F., Ed. (1979), *Intercalated Layered Materials*, Vol. 6, Reidel, Dordrecht.
Lewis, M. F. and Stoneham, A. M. (1967), *Phys. Rev.* **164**, 271.
Lidiard, A. B. (1974), in *Crystals with the Fluorite Structure*, W. Hayes, Ed., Clarendon Press, Oxford, p. 101.
Lidiard, A. B. and Norgett, M. J. (1972), in *Computational Solid State Science*, F. Herman, N. W. Dalton, and T. R. Koehler, Eds., Plenum Press, New York, p. 385.
Lindhard, J. (1965), *KGL Danske Videnskat Selskat Matt-Fys. Med.* **34**, No. 14.
Lipari, N. O. and Baldereschi, A. (1970), *Phys. Rev. Lett.* **25**, 1660.
Longuet-Higgins, H. C., Öpik, U., Pryce, M. H. L., and Sack, R. A. (1958), *Proc. Roy. Soc. A* **244**, 1.
Longuet-Higgins, H. C. and Salem, L. (1959), *Proc. Roy. Soc. A* **251**, 172.
Luty, F. (1960), *Z. Phys.* **160**, 1.
Luty, F. (1983), *Semicond. Insulators* **5**, 245.
Luty, F. and Mort, J. (1964), *Phys. Rev. Lett.* **12**, 45.
Macfarlane, R. M. and Shelby, R. M. (1979), *Phys. Rev. Lett.* **42**, 788.
Madan, A., Le Comber, P. G. and Spear, W. E. (1976), *J. Noncryst. Sol.* **20**, 239.
Maines, J. D. and Paige, E. G. S., (1973), *Proc. IEEE*, **120**, 1078.
Mainwood, A. M., Larkins, F. P., and Stoneham, A. M. (1978), *Sol. St. Electronics* **21**, 1431.
Mainwood, A. M. and Stoneham, A. M. (1983), *Physica B* **116**, 101.
Maradudin, A. A., Montroll, E. W., Weiss, C. H., and Ipatova, I. P. (1971), *Sol. St. Phys.* Suppl. 3.
Margerie, J. and Romestain, R. (1964), *Comp. Rend.* **258**, 4490.

Marquardt, C. L., Williams, R. T., and Kabler, M. N. (1971), *Sol. St. Comm.* **9**, 2285.

Masri, P. M., Harker, A. H., and Stoneham, A. M. (1983), *J. Phys. C., Sol. St. Phys.* **16**, L613.

McClure, D. S. (1959), *Sol. St. Phys.* **9**, 399.

McFeely, F. R., Kowalczyk, S. P., Ley, L., Cavell, R. G., Pollak, R. A., and Shirley, D. (1974), *Phys. Rev. B* **9**, 5628.

McGeehin, P. and Hooper, A. (1977), *J. Mat. Sci.* **12**, 1.

McKeighan, R. E. and Koehler, J. S. (1971), *Phys. Rev. B* **4**, 462.

Merz, J. L., Faulker, R. A., and Dean, P. J. (1969), *Phys. Rev.* **188**, 1228.

Messmer, R. and Watkins, G. D. (1972), in *Radiation Damage and Defects in Semiconductors*, J. E. Whitehouse, Ed., Institute of Physics, London, p. 255.

Millman, S. E. and Kirczenow, G. (1982), *Phys. Rev. B* **26**, 2310.

Mitchell, A. C. G. and Zemansky, M. W. (1934), *Resonance Radiation and Excited Atoms*, Cambridge University Press, Cambridge, England.

Mizakawa, T. and Dexter, D. L. (1970), *Phys. Rev. B* **1**, 2961.

Mollenauer, L. F., Viera, N. D., and Szeto, L. (1982), *Opt. Lett.* **7**, 9.

Mott, N. F. (1947), *Trans. Farad. Soc.* **43**, 429.

Mott, N. F. (1977), *Adv. Phys.* **26**, 363.

Mott, N. F. (1981), *Proc. Roy. Soc. A* **376**, 207.

Mott, N. F. (1982), *Proc. Roy. Soc. A* **382**, 1

Mott, N. F. (1984), *Rep. Prog. Phys.* **47**, 909.

Mott, N. F. and Cabrera, N. (1948), *Rep. Prog. Phys.* **12**, 163.

Mott, N. F. and Gurney, R. W. (1948), *Electronic Processes in Ionic Crystals*, 2nd ed., Clarendon Press, Oxford.

Mott, N. F. and Littleton, M. J. (1938), *Trans. Far. Soc.* **34**, 485.

Murarka, S. P. (1977), *Phys. Rev. B* **16**, 2849.

Murarka, S. P. (1980), *Phys. Rev. B* **21**, 692.

Naidich, J. V. (1981), *Prog. Surf. Membr. Sci.* **14**, 354.

Nash, J. G. (1975), *Films on Solid Surfaces*, Academic Press, New York.

Nassau, K. (1980, October), *Sci. American*, p. 106.

Newman, R. C. (1974), *Infrared Studies of Crystal Defects*, Taylor and Francis, London.

Newman, R. C. (1984), *J. Elect. Mat.* in press.

Norgett, M. J., Pathak, A. P., and Stoneham, A. M. (1977), *J. Phys. C., Sol. St. Phys.* **10**, 555.

Norgett, M. J. and Stoneham, A. M. (1973), *J. Phys. C., Sol. St. Phys.* **229**, 238.

Nowick, A. S. and Heller, W. R. (1963), *Adv. Phys.* **12**, 25. (1965), *Adv. Phys.* **14**, 101.

O'Brien, M. C. M. (1964), *Proc. Roy. Soc. A* **281**, 323.

O'Brien, M. C. M. (1976), *J. Phys. C., Sol. St. Phys.* **9**, 3153.

Ong, C. K., Song, K. S., Monnier, R., and Stoneham, A. M. (1979), *J. Phys. C., Sol. St. Phys.* **12**, 4641.

Painter, G. S., Ellis, D. E., and Lubinsky, A. R. (1971), *Phys. Rev. B* **4**, 3610.

Pandey, K. C. (1983), *Physica* **117/118** (B and C), Part 1, p. 761.

Patel, J. L., Davies, J. J., Nichols, J. E., and Lunn, B. (1981), *J. Phys. C., Sol. St. Phys.* **14**, 4717.

Peierls, R. E. (1955), *Quantum Theory of Solids*, Oxford University Press, Oxford, England, p. 108.

Peisl, H. (1975), in *Defects and their Structure in Nonmetallic Solids*, B. Henderson and A. E. Hughes, Eds., Plenum Press, p. 381.

Pendry, J. B. (1974), *Low Energy Electron Diffraction*, Academic Press, New York.

Perram, J. W., Ed. (1983), *The Physics of Superionic Conductors and Electrode Materials*, Plenum Press, New York.

Petroff, P. M. and Kimerling L. C. (1976), *App. Phys. Lett.* **29**, 461.

Peyghambarian, N., Chase, L. L., and Mysyrowicz, A. (1983), *Phys. Rev. B* **27**, 2325.

Phillips J. C. (1970), *Rev. Mod. Phys.* **42**, 317.

Pietronero, L. and Tosatti, E., Eds. (1981), *Physics of Intercalation Compounds*, Springer-Verlag, New York.

Ploog, K. and Döhler, G. H. (1983), *Adv. Phys.* **32**, 285.

Pointon, A. J. (1982), *IEE Proc. A* **129**, 285.

Pollman, J. (1980), *Festkörperprobleme*, **20**, 117.

Pople, J. A. and Beveridge, D. L. (1970), *Approximate Molecular Orbital Theory*, McGraw-Hill, New York.

Pople, J. A. and Walmsley, S. H. (1962), *Mol. Phys.* **5**, 15.

Prutton, M. (1975), *Surface Physics*, Oxford University Press, Oxford, England.

Prutton, M., Ramsey, J. A., Walker, J. A., and Welton-Cook, M. R. (1979), *J. Phys. C., Sol. St. Phys.* **12**, 5271.

Ramdas, A. K. and Rodriguez, S. (1981), *Rep. Prog. Phys.* **44**, 1297.

Rasagni, G., Vernier, F., Palmari, J. P., Mayani, N., Rasigni, N., and Llebaria, A. (1983), *Opt. Comm.* **46**, 294.

Rashba, E. (1982), "Self Trapping of Excitons," in *Excitons*, E. I. Rashba and M. D. Sturge, Eds., North Holland, Amsterdam.

Reynolds, D. C. and Collins, T. C. (1981), *Excitons, Their Properties and Uses*, Academic Press, New York.

Rice, T. M. (1977), *Sol. St. Phys.* **32**, 1.

Ridley, B. K. (1963), *Proc. Phys. Soc.* **82**, 945.

Rieder, K. (1971), *Surf. Sci.* **26**, 637.

Riviere, J. C. (1972), *Cont. Phys.* **14**, 513.

Roberts, R. A. and Walker, W. C. (1967), *Phys. Rev.* **161**, 730.

Robertson, J. (1983), *Adv. Phys.* **32**, 361.

Robins, L. H. and Kastner, M. A. (1984), *Phil. Mag.* **B50**, 29.

Rowe, J. E., Ibach, H. and Froitzheim, H. (1975), *Surf. Sci.* **48**, 44.

Salamon, M. B., Ed. (1979), *Physics of Superionic Conductors*, Springer-Verlag, Heidelberg, Vol. 15.

Sangster, M. J. L. (1974), *Sol. St. Comm.* **15**, 471.

Sawicka, B. D. (1980), Nuclear Physics Methods in Materials Research, Eds. K. Bethge, H. Baumann, H. Jex and F. Rauch, Vieweg Braunschweig, Wiesbaden, p. 216.

Schäfer, G. (1960), *J. Phys. Chem. Sol.* **12**, 233.

Schenck, A. (1976), in *Nuclear and Particle Physics at Intermediate Energies*, J. B. Warren, Ed., Plenum Press, New York, p. 159.

Schiff, L. (1955), *Quantum Mechanics*, 2nd ed., McGraw Hill, New York.

Schlüter, M. (1982), *Thin Sol. Films* **93**, 3.

Schlüter, M., Chelikowsky, J. R., Lowie, S. G., and Cohen, M. L. (1975), *Phys. Rev. B* **12**, 4200.

Schnatterly, S. E. (1965), *Phys. Rev.* **140**, A1364.

Schottky, W. (1940), *Phys. Zeit.* **41**, 570.

Schröter, W. and Nölting, J. (1979), *J. Phys. C* **6**, 20.

Seitz, F. (1938), *J. Chem. Phys.* **6**, 150.

Seitz, F. (1940), *Modern Theory of Solids*, McGraw-Hill, New York.

Sell, D. D. (1972), *Phys. Rev. B* **6**, 3750.

Sham, L. J. and Nakayama, M. (1979), *Phys. Rev. B* **20**, 734.

Shaw, D. (1973), *Atomic Diffusion in Semiconductors*, Plenum Press, New York.

Shelby, R. M., Tropper, A. C., Harley, R. T., and Macfarlane, R. M. (1983), *Opt. Lett.* **8**, 304.

Shewmon, P. G. (1963), *Diffusion in Solids*, McGraw-Hill, New York.

Sievers, A. J. (1964), *Phys. Rev. Lett.* **13**, 310.

Sievers, A. J. and Takeno, S. (1965), *Phys. Rev.* **140**, A1030.

Silinsh, E. A. (1980), *Organic Molecular Crystals*, Springer-Verlag, New York.

Silsbee, R. H. (1957), *J. App. Phys.* **28**, 1246.

Slater, J. C. (1958), *Rev. Mod. Phys.* **30**, 197.

Slater, J. C. (1963), *Quantum Theory of Atoms and Molecules*, McGraw-Hill, New York, Vol. 1.

Smeltzer, W. W., Haering, R. R., and Kircaldy, J. S. (1961), *Acta Met.* **9**, 880.

Smith, D. Y. and Spinolo, G. (1965), *Phys. Rev.* **140**, A2121.

Sørensen, O. T., Ed. (1981), *Nonstoichiometric Oxides*, Academic Press, New York.

Spear, W. E. and Le Comber, P. T. (1976), *Phil. Mag.* **33**, 935.

Spicer, W. F., Lindau, I., Skeath, P. R., Su, C. Y., and Chye, P. W. (1980), *Phys. Rev. Lett.* **44**, 420.

Staebler, D. L. and Wronski, C. R. (1977), *Appl. Phys. Lett.* **31**, 292.

Stanek, J., Kowalski, J., Sawicka, B. D. and Sawicki, J. A., (1981), *Nucl. Inst. and Methods*, **182/183**, 801.

Stapelbroek, M., Griscom, D. L., Friebele, E. J., and Sigel, G. H. (1979), *J. Non-Cryst. Sol.* **32**, 313.

Stevenson, R. W. H., Ed. (1966), *Phonons in Perfect Lattices and in Lattices with Point Imperfections*, Oliver and Boyd, London.

Stoneham, A. M. (1965), *Proc. Phys. Soc.* **86**, 1163.

Stoneham, A. M. (1969), *Rev. Mod. Phys.* **41**, 82.

Stoneham, A. M. (1975), *Theory of Defects in Solids*, Clarendon Press, Oxford.

Stoneham, A. M. (1979), in *The Physics of Diamond*, J. E. Field, Ed., Academic Press, New York.

Stoneham, A. M. (1981), *Rep. Prog. Phys.* **44**, 1251.

Stoneham, A. M. and Durham, P. J. (1973), *J. Phys. Chem. Sol.* **34**, 2127.

Stoneham, A. M. and Harding, J. H. (1982), in *Computer Simulation of Solids*, C. R. A. Catlow and W. C. Mackrodt, Eds., Springer-Verlag, New York, p. 162.

Stoneham, A. M. and Harker, A. M. (1975), *J. Phys. C., Sol. St. Phys.* **8**, 1109.

Stoneham, A. M. and Lannoo, M. (1969), *J. Phys. Chem. Soc.* **30**, 1769.
Stoneham, A. M., Wade, E., and Kilner, J. A. (1979), *Mat. Res. Bull.* **14**, 661.
Stott, J. P. and Crawford, J. H. (1971), *Phys. Rev. Lett.* **26**, 384.
Street, R. A. (1976), *Adv. Phys.* **25**, 397.
Street, R. A. and Mott, N. F. (1975), *Phys. Rev. Lett.* **35**, 1293.
Stringer, J. (1970), *Corros. Sci.* **10**, 513.
Stryer, L. and Haugland, R. P. (1967), *Proc. Nat. Acad. Sci., USA* **59**, 719.
Sturge, M. D. (1967), *Sol. St. Phys.* **20**, 92.
Su, W. P., Schrieffer, J. R., and Heeger, A. J. (1980), *Phys. Rev. B* **22**, 2099.
Subbarao, E. C. (1980), *Solid Electrolytes*, Plenum Press, New York.
Sumi, H. and Toyozawa, J. (1971), *J. Phys. Soc. Jpn.* **31**, 342.
Süptitz, P. and Teltow, P. (1967), *Phys. Stat. Sol.* **23**, 9.
Suzuki, N., Ozaki, M., Etemad, S., Heeger, A. J., and MacDiarmid, A. G. (1980), *Phys. Rev. Lett.* **45**, 1209.
Swanson, M. L. (1982), *Rep. Prog. Phys.* **45**, 47.
Sze, S. M. (1981), *Physics of Semiconductors*, 2nd ed., Wiley, New York.
Tanimura, K., Tanaka, T., and Itoh, N. (1983), *Phys. Rev. Lett.* **51**, 423.
Tasker, P. W. (1980), *J. Phys. (Paris) C6* **41**, 488.
Tasker, P. W. (1983a), private communication.
Tasker, P. W. (1983b), *Sol. St. Ionics* **8**, 233.
Tasker P. W. (1983c), *Mass Transport in Solids*, F. Beniere and C. R. A. Catlow, Eds., Plenum Press, New York, p. 457.
Tasker, P. W. and Duffy, D. (1983), *Phil. Mag. A* **47**, 817.
Tasker, P. W., Mackrodt, W. C. and Colbourn, E. A. (1984), *J. Am. Ceram. Soc.*, in press.
Thomas, K. (1975) *Phys. State. Sol.* **68**, K9.
Thomas, T. R. (1982), *Phys. Bull.* **33**, 327.
Thompson, M. W. (1969), *Defects and Radiation Damage in Metals*, Cambridge University Press, Cambridge, England.
Thornber K. K. and Feynman R. P. (1970), *Phys. Rev. B* **1**, 4099.
Thurmond, C. D. and Struthers, J. D. (1953), *J. Phys. Chem.* **57**, 831.
Toyozawa, Y. (1981), *J. Phys. Soc. Jpn.* **50**, 1861.
Travaglini, G., Wachter, P., Marcus, J., and Schlenker, C. (1981), *Sol. St. Comm.* **37**, 599.
Tubandt, C. and Lorenz, E. (1914), *Z. Phys. Chem.* **87**,513.
Ueta, M. and Nishina, Y., Eds. (1976), *Physics of Highly Excited States in Solids*, Lecture Notes in Physics, Springer-Verlag, New York, Vol. 57.
Urbach, F. (1953), *Phys. Rev.* **92**, 1325.
Van Vechten, J. A. (1975), *J. Electrochem. Soc.* **122**, 423.
Vink A. T., van der Heyden, R. L. A., and van der Does de Bye (1973), *J. Lum.* **8**, 105.
Vink, H. J. (1975), *Berich. der bunsenges* **79**, 957.
Vogl, P. (1981), *Festkörperprobleme* **21**, 191.
Vook, F. L. (1972), *Inst. Phys. Conf. Ser.* **16**, 60.
Wagner, C. (1975), *Prog. Sol. St. Chem.* **10**, 3.
Watkins, G. D. (1964), in *Effects des Rayonnements sur les Semiconducteurs*, Dunod, Paris, p. 97.

Watkins, G. D. (1967), in *Radiation Effects on Semiconductor Components*, *J. d'Electronique*, Toulouse, Vol. 1, Paper A1.

Watkins, G. D. (1976), *Defects and Their Structure in Nonmetallic Solids*, B. Henderson and A. E. Hughes, Eds., Plenum Press, New York, p. 203.

Watkins, G. D. (1977), *Inst. Phys. Conf. Ser.* **31**, 95.

Weaire, D. and Thorpe, M. F. (1971), *Phys. Rev. B* **4**, 2508, 3518.

Weber, E. R. (1983), *App. Phys. A* **30**, 1.

Welton-Cook, M. R. and Prutton, M., (1980), *J. Phys. C., Sol. St. Phys.* **13**, 3993.

Wertheim, G. K. (1971), *The Electronic Structure of Point Defects*, Eds. G. K. Wertheim, A. Hausmann and W. Sander, North Holland, Amsterdam.

West, C. (1980), Thesis, Oxford, England.

Whittingham, M. S. and Jacobson, A. J., Eds., (1982), *Intercalation Chemistry*, Academic Press, New York.

Wiech, E. and Zöpf, E. (1969), *J. Nat. Bur. Standards* **323**, 335.

Williams, R. H. (1978), *Cont. Phys.* **19**, 389.

Williams, R. H., Srivastava, G. P. and McGovern, I. T. (1980), *Rep. Prog. Phys.* **43**, 1357.

Willis, B. T. M. (1963), *Nature* **197**, 755.

Winterling, G. (1975), in *Light Scattering in Solids*, M. Balkanski, R. C. C. Leite, and S. P. S. Porto, Eds., Flammarion, Paris, p. 663.

Winterling, G., Senn, W., Grimsditch, M. and Katiyar, R. (1977), in *Lattice Dynamics*, M. Balkanski, Ed., Flammarion, Paris, p. 553.

Wolfe, J. P. (Dec., 1980), *Physics Today*, p. 44.

Yin, M. and Cohen, M. L., cited in G. B. Bachelet, H. S. Greenside, G. Baraff, and M. Schluter (1981), *Phys. Rev. B* **24**, 4745.

Young, R. T. and Wood, R. F. (1982), *Ann. Rev. Met. Sci.* **12**, 323.

Zeller, R. C. and Pohl, R. O. (1971), *Phys. Rev. B* **4**, 2029.

Ziman, J. M. (1960), *Electrons and Phonons*, Clarendon Press, Oxford.

Ziman, J. M. (1979), *Models of Disorder*, Cambridge University Press, Cambridge, England.

Index

A CATALOG OF SELECTED

DOVER BOOKS

IN ALL FIELDS OF INTEREST

STICKLEY CRAFTSMAN FURNITURE CATALOGS, Gustav Stickley and L. & J. G. Stickley. Beautiful, functional furniture in two authentic catalogs from 1910. 594 illustrations, including 277 photos, show settles, rockers, armchairs, reclining chairs, bookcases, desks, tables. 183pp. 6½ x 9¼. 23838-5

AMERICAN LOCOMOTIVES IN HISTORIC PHOTOGRAPHS: 1858 to 1949, Ron Ziel (ed.). A rare collection of 126 meticulously detailed official photographs, called "builder portraits," of American locomotives that majestically chronicle the rise of steam locomotive power in America. Introduction. Detailed captions. xi+ 129pp. 9 x 12. 27393-8

AMERICA'S LIGHTHOUSES: An Illustrated History, Francis Ross Holland, Jr. Delightfully written, profusely illustrated fact-filled survey of over 200 American lighthouses since 1716. History, anecdotes, technological advances, more. 240pp. 8 x 10¾. 25576-X

TOWARDS A NEW ARCHITECTURE, Le Corbusier. Pioneering manifesto by founder of "International School." Technical and aesthetic theories, views of industry, economics, relation of form to function, "mass-production split" and much more. Profusely illustrated. 320pp. 6⅛ x 9¼. (Available in U.S. only.) 25023-7

HOW THE OTHER HALF LIVES, Jacob Riis. Famous journalistic record, exposing poverty and degradation of New York slums around 1900, by major social reformer. 100 striking and influential photographs. 233pp. 10 x 7⅞. 22012-5

FRUIT KEY AND TWIG KEY TO TREES AND SHRUBS, William M. Harlow. One of the handiest and most widely used identification aids. Fruit key covers 120 deciduous and evergreen species; twig key 160 deciduous species. Easily used. Over 300 photographs. 126pp. 5⅜ x 8½. 20511-8

COMMON BIRD SONGS, Dr. Donald J. Borror. Songs of 60 most common U.S. birds: robins, sparrows, cardinals, bluejays, finches, more–arranged in order of increasing complexity. Up to 9 variations of songs of each species.
Cassette and manual 99911-4

ORCHIDS AS HOUSE PLANTS, Rebecca Tyson Northen. Grow cattleyas and many other kinds of orchids–in a window, in a case, or under artificial light. 63 illustrations. 148pp. 5⅜ x 8½. 23261-1

MONSTER MAZES, Dave Phillips. Masterful mazes at four levels of difficulty. Avoid deadly perils and evil creatures to find magical treasures. Solutions for all 32 exciting illustrated puzzles. 48pp. 8¼ x 11. 26005-4

MOZART'S DON GIOVANNI (DOVER OPERA LIBRETTO SERIES), Wolfgang Amadeus Mozart. Introduced and translated by Ellen H. Bleiler. Standard Italian libretto, with complete English translation. Convenient and thoroughly portable–an ideal companion for reading along with a recording or the performance itself. Introduction. List of characters. Plot summary. 121pp. 5¼ x 8½. 24944-1

TECHNICAL MANUAL AND DICTIONARY OF CLASSICAL BALLET, Gail Grant. Defines, explains, comments on steps, movements, poses and concepts. 15-page pictorial section. Basic book for student, viewer. 127pp. 5⅜ x 8½. 21843-0

THE CLARINET AND CLARINET PLAYING, David Pino. Lively, comprehensive work features suggestions about technique, musicianship, and musical interpretation, as well as guidelines for teaching, making your own reeds, and preparing for public performance. Includes an intriguing look at clarinet history. "A godsend," *The Clarinet,* Journal of the International Clarinet Society. Appendixes. 7 illus. 320pp. 5⅜ x 8½. 40270-3

HOLLYWOOD GLAMOR PORTRAITS, John Kobal (ed.). 145 photos from 1926-49. Harlow, Gable, Bogart, Bacall; 94 stars in all. Full background on photographers, technical aspects. 160pp. 8⅜ x 11¼. 23352-9

THE ANNOTATED CASEY AT THE BAT: A Collection of Ballads about the Mighty Casey/Third, Revised Edition, Martin Gardner (ed.). Amusing sequels and parodies of one of America's best-loved poems: Casey's Revenge, Why Casey Whiffed, Casey's Sister at the Bat, others. 256pp. 5⅜ x 8½. 28598-7

THE RAVEN AND OTHER FAVORITE POEMS, Edgar Allan Poe. Over 40 of the author's most memorable poems: "The Bells," "Ulalume," "Israfel," "To Helen," "The Conqueror Worm," "Eldorado," "Annabel Lee," many more. Alphabetic lists of titles and first lines. 64pp. 5$\frac{3}{16}$ x 8¼. 26685-0

PERSONAL MEMOIRS OF U. S. GRANT, Ulysses Simpson Grant. Intelligent, deeply moving firsthand account of Civil War campaigns, considered by many the finest military memoirs ever written. Includes letters, historic photographs, maps and more. 528pp. 6⅛ x 9¼. 28587-1

ANCIENT EGYPTIAN MATERIALS AND INDUSTRIES, A. Lucas and J. Harris. Fascinating, comprehensive, thoroughly documented text describes this ancient civilization's vast resources and the processes that incorporated them in daily life, including the use of animal products, building materials, cosmetics, perfumes and incense, fibers, glazed ware, glass and its manufacture, materials used in the mummification process, and much more. 544pp. 6⅛ x 9¼. (Available in U.S. only.) 40446-3

RUSSIAN STORIES/RUSSKIE RASSKAZY: A Dual-Language Book, edited by Gleb Struve. Twelve tales by such masters as Chekhov, Tolstoy, Dostoevsky, Pushkin, others. Excellent word-for-word English translations on facing pages, plus teaching and study aids, Russian/English vocabulary, biographical/critical introductions, more. 416pp. 5⅜ x 8½. 26244-8

PHILADELPHIA THEN AND NOW: 60 Sites Photographed in the Past and Present, Kenneth Finkel and Susan Oyama. Rare photographs of City Hall, Logan Square, Independence Hall, Betsy Ross House, other landmarks juxtaposed with contemporary views. Captures changing face of historic city. Introduction. Captions. 128pp. 8¼ x 11. 25790-8

AIA ARCHITECTURAL GUIDE TO NASSAU AND SUFFOLK COUNTIES, LONG ISLAND, The American Institute of Architects, Long Island Chapter, and the Society for the Preservation of Long Island Antiquities. Comprehensive, well-researched and generously illustrated volume brings to life over three centuries of Long Island's great architectural heritage. More than 240 photographs with authoritative, extensively detailed captions. 176pp. 8¼ x 11. 26946-9

NORTH AMERICAN INDIAN LIFE: Customs and Traditions of 23 Tribes, Elsie Clews Parsons (ed.). 27 fictionalized essays by noted anthropologists examine religion, customs, government, additional facets of life among the Winnebago, Crow, Zuni, Eskimo, other tribes. 480pp. 6⅛ x 9¼. 27377-6

THE BEST TALES OF HOFFMANN, E. T. A. Hoffmann. 10 of Hoffmann's most important stories: "Nutcracker and the King of Mice," "The Golden Flowerpot," etc. 458pp. 5⅜ x 8½. 21793-0

FROM FETISH TO GOD IN ANCIENT EGYPT, E. A. Wallis Budge. Rich detailed survey of Egyptian conception of "God" and gods, magic, cult of animals, Osiris, more. Also, superb English translations of hymns and legends. 240 illustrations. 545pp. 5⅜ x 8½. 25803-3

FRENCH STORIES/CONTES FRANÇAIS: A Dual-Language Book, Wallace Fowlie. Ten stories by French masters, Voltaire to Camus: "Micromegas" by Voltaire; "The Atheist's Mass" by Balzac; "Minuet" by de Maupassant; "The Guest" by Camus, six more. Excellent English translations on facing pages. Also French-English vocabulary list, exercises, more. 352pp. 5⅜ x 8½. 26443-2

CHICAGO AT THE TURN OF THE CENTURY IN PHOTOGRAPHS: 122 Historic Views from the Collections of the Chicago Historical Society, Larry A. Viskochil. Rare large-format prints offer detailed views of City Hall, State Street, the Loop, Hull House, Union Station, many other landmarks, circa 1904-1913. Introduction. Captions. Maps. 144pp. 9⅜ x 12¼. 24656-6

OLD BROOKLYN IN EARLY PHOTOGRAPHS, 1865-1929, William Lee Younger. Luna Park, Gravesend race track, construction of Grand Army Plaza, moving of Hotel Brighton, etc. 157 previously unpublished photographs. 165pp. 8⅞ x 11¾. 23587-4

THE MYTHS OF THE NORTH AMERICAN INDIANS, Lewis Spence. Rich anthology of the myths and legends of the Algonquins, Iroquois, Pawnees and Sioux, prefaced by an extensive historical and ethnological commentary. 36 illustrations. 480pp. 5⅜ x 8½. 25967-6

AN ENCYCLOPEDIA OF BATTLES: Accounts of Over 1,560 Battles from 1479 B.C. to the Present, David Eggenberger. Essential details of every major battle in recorded history from the first battle of Megiddo in 1479 B.C. to Grenada in 1984. List of Battle Maps. New Appendix covering the years 1967-1984. Index. 99 illustrations. 544pp. 6½ x 9¼. 24913-1

SAILING ALONE AROUND THE WORLD, Captain Joshua Slocum. First man to sail around the world, alone, in small boat. One of great feats of seamanship told in delightful manner. 67 illustrations. 294pp. 5⅜ x 8½. 20326-3

ANARCHISM AND OTHER ESSAYS, Emma Goldman. Powerful, penetrating, prophetic essays on direct action, role of minorities, prison reform, puritan hypocrisy, violence, etc. 271pp. 5⅜ x 8½. 22484-8

MYTHS OF THE HINDUS AND BUDDHISTS, Ananda K. Coomaraswamy and Sister Nivedita. Great stories of the epics; deeds of Krishna, Shiva, taken from puranas, Vedas, folk tales; etc. 32 illustrations. 400pp. 5⅜ x 8½. 21759-0

THE TRAUMA OF BIRTH, Otto Rank. Rank's controversial thesis that anxiety neurosis is caused by profound psychological trauma which occurs at birth. 256pp. 5⅜ x 8½. 27974-X

A THEOLOGICO-POLITICAL TREATISE, Benedict Spinoza. Also contains unfinished Political Treatise. Great classic on religious liberty, theory of government on common consent. R. Elwes translation. Total of 421pp. 5⅜ x 8½. 20249-6

MY BONDAGE AND MY FREEDOM, Frederick Douglass. Born a slave, Douglass became outspoken force in antislavery movement. The best of Douglass' autobiographies. Graphic description of slave life. 464pp. 5⅜ x 8½. 22457-0

FOLLOWING THE EQUATOR: A Journey Around the World, Mark Twain. Fascinating humorous account of 1897 voyage to Hawaii, Australia, India, New Zealand, etc. Ironic, bemused reports on peoples, customs, climate, flora and fauna, politics, much more. 197 illustrations. 720pp. 5⅜ x 8½. 26113-1

THE PEOPLE CALLED SHAKERS, Edward D. Andrews. Definitive study of Shakers: origins, beliefs, practices, dances, social organization, furniture and crafts, etc. 33 illustrations. 351pp. 5⅜ x 8½. 21081-2

THE MYTHS OF GREECE AND ROME, H. A. Guerber. A classic of mythology, generously illustrated, long prized for its simple, graphic, accurate retelling of the principal myths of Greece and Rome, and for its commentary on their origins and significance. With 64 illustrations by Michelangelo, Raphael, Titian, Rubens, Canova, Bernini and others. 480pp. 5⅜ x 8½. 27584-1

PSYCHOLOGY OF MUSIC, Carl E. Seashore. Classic work discusses music as a medium from psychological viewpoint. Clear treatment of physical acoustics, auditory apparatus, sound perception, development of musical skills, nature of musical feeling, host of other topics. 88 figures. 408pp. 5⅜ x 8½. 21851-1

THE PHILOSOPHY OF HISTORY, Georg W. Hegel. Great classic of Western thought develops concept that history is not chance but rational process, the evolution of freedom. 457pp. 5⅜ x 8½. 20112-0

THE BOOK OF TEA, Kakuzo Okakura. Minor classic of the Orient: entertaining, charming explanation, interpretation of traditional Japanese culture in terms of tea ceremony. 94pp. 5⅜ x 8½. 20070-1

LIFE IN ANCIENT EGYPT, Adolf Erman. Fullest, most thorough, detailed older account with much not in more recent books, domestic life, religion, magic, medicine, commerce, much more. Many illustrations reproduce tomb paintings, carvings, hieroglyphs, etc. 597pp. 5⅜ x 8½. 22632-8

SUNDIALS, Their Theory and Construction, Albert Waugh. Far and away the best, most thorough coverage of ideas, mathematics concerned, types, construction, adjusting anywhere. Simple, nontechnical treatment allows even children to build several of these dials. Over 100 illustrations. 230pp. 5⅜ x 8½. 22947-5

THEORETICAL HYDRODYNAMICS, L. M. Milne-Thomson. Classic exposition of the mathematical theory of fluid motion, applicable to both hydrodynamics and aerodynamics. Over 600 exercises. 768pp. 6⅛ x 9¼. 68970-0

SONGS OF EXPERIENCE: Facsimile Reproduction with 26 Plates in Full Color, William Blake. 26 full-color plates from a rare 1826 edition. Includes "The Tyger," "London," "Holy Thursday," and other poems. Printed text of poems. 48pp. 5¼ x 7. 24636-1

OLD-TIME VIGNETTES IN FULL COLOR, Carol Belanger Grafton (ed.). Over 390 charming, often sentimental illustrations, selected from archives of Victorian graphics–pretty women posing, children playing, food, flowers, kittens and puppies, smiling cherubs, birds and butterflies, much more. All copyright-free. 48pp. 9¼ x 12¼. 27269-9

PERSPECTIVE FOR ARTISTS, Rex Vicat Cole. Depth, perspective of sky and sea, shadows, much more, not usually covered. 391 diagrams, 81 reproductions of drawings and paintings. 279pp. 5⅜ x 8½. 22487-2

DRAWING THE LIVING FIGURE, Joseph Sheppard. Innovative approach to artistic anatomy focuses on specifics of surface anatomy, rather than muscles and bones. Over 170 drawings of live models in front, back and side views, and in widely varying poses. Accompanying diagrams. 177 illustrations. Introduction. Index. 144pp. 8⅜ x11¼. 26723-7

GOTHIC AND OLD ENGLISH ALPHABETS: 100 Complete Fonts, Dan X. Solo. Add power, elegance to posters, signs, other graphics with 100 stunning copyright-free alphabets: Blackstone, Dolbey, Germania, 97 more–including many lower-case, numerals, punctuation marks. 104pp. 8⅛ x 11. 24695-7

HOW TO DO BEADWORK, Mary White. Fundamental book on craft from simple projects to five-bead chains and woven works. 106 illustrations. 142pp. 5⅜ x 8. 20697-1

THE BOOK OF WOOD CARVING, Charles Marshall Sayers. Finest book for beginners discusses fundamentals and offers 34 designs. "Absolutely first rate . . . well thought out and well executed."–E. J. Tangerman. 118pp. 7¾ x 10⅝. 23654-4

ILLUSTRATED CATALOG OF CIVIL WAR MILITARY GOODS: Union Army Weapons, Insignia, Uniform Accessories, and Other Equipment, Schuyler, Hartley, and Graham. Rare, profusely illustrated 1846 catalog includes Union Army uniform and dress regulations, arms and ammunition, coats, insignia, flags, swords, rifles, etc. 226 illustrations. 160pp. 9 x 12. 24939-5

WOMEN'S FASHIONS OF THE EARLY 1900s: An Unabridged Republication of "New York Fashions, 1909," National Cloak & Suit Co. Rare catalog of mail-order fashions documents women's and children's clothing styles shortly after the turn of the century. Captions offer full descriptions, prices. Invaluable resource for fashion, costume historians. Approximately 725 illustrations. 128pp. 8⅜ x 11¼. 27276-1

THE 1912 AND 1915 GUSTAV STICKLEY FURNITURE CATALOGS, Gustav Stickley. With over 200 detailed illustrations and descriptions, these two catalogs are essential reading and reference materials and identification guides for Stickley furniture. Captions cite materials, dimensions and prices. 112pp. 6½ x 9¼. 26676-1

EARLY AMERICAN LOCOMOTIVES, John H. White, Jr. Finest locomotive engravings from early 19th century: historical (1804–74), main-line (after 1870), special, foreign, etc. 147 plates. 142pp. 11⅜ x 8¼. 22772-3

THE TALL SHIPS OF TODAY IN PHOTOGRAPHS, Frank O. Braynard. Lavishly illustrated tribute to nearly 100 majestic contemporary sailing vessels: Amerigo Vespucci, Clearwater, Constitution, Eagle, Mayflower, Sea Cloud, Victory, many more. Authoritative captions provide statistics, background on each ship. 190 black-and-white photographs and illustrations. Introduction. 128pp. 8⅞ x 11¾. 27163-3

LITTLE BOOK OF EARLY AMERICAN CRAFTS AND TRADES, Peter Stockham (ed.). 1807 children's book explains crafts and trades: baker, hatter, cooper, potter, and many others. 23 copperplate illustrations. 140pp. 4⅝ x 6. 23336-7

VICTORIAN FASHIONS AND COSTUMES FROM HARPER'S BAZAR, 1867–1898, Stella Blum (ed.). Day costumes, evening wear, sports clothes, shoes, hats, other accessories in over 1,000 detailed engravings. 320pp. 9⅜ x 12¼. 22990-4

GUSTAV STICKLEY, THE CRAFTSMAN, Mary Ann Smith. Superb study surveys broad scope of Stickley's achievement, especially in architecture. Design philosophy, rise and fall of the Craftsman empire, descriptions and floor plans for many Craftsman houses, more. 86 black-and-white halftones. 31 line illustrations. Introduction 208pp. 6½ x 9¼. 27210-9

THE LONG ISLAND RAIL ROAD IN EARLY PHOTOGRAPHS, Ron Ziel. Over 220 rare photos, informative text document origin (1844) and development of rail service on Long Island. Vintage views of early trains, locomotives, stations, passengers, crews, much more. Captions. 8⅞ x 11¾. 26301-0

VOYAGE OF THE LIBERDADE, Joshua Slocum. Great 19th-century mariner's thrilling, first-hand account of the wreck of his ship off South America, the 35-foot boat he built from the wreckage, and its remarkable voyage home. 128pp. 5⅜ x 8½. 40022-0

TEN BOOKS ON ARCHITECTURE, Vitruvius. The most important book ever written on architecture. Early Roman aesthetics, technology, classical orders, site selection, all other aspects. Morgan translation. 331pp. 5⅜ x 8½. 20645-9

THE HUMAN FIGURE IN MOTION, Eadweard Muybridge. More than 4,500 stopped-action photos, in action series, showing undraped men, women, children jumping, lying down, throwing, sitting, wrestling, carrying, etc. 390pp. 7⅞ x 10⅝. 20204-6 Clothbd.

TREES OF THE EASTERN AND CENTRAL UNITED STATES AND CANADA, William M. Harlow. Best one-volume guide to 140 trees. Full descriptions, woodlore, range, etc. Over 600 illustrations. Handy size. 288pp. 4½ x 6⅜. 20395-6

SONGS OF WESTERN BIRDS, Dr. Donald J. Borror. Complete song and call repertoire of 60 western species, including flycatchers, juncoes, cactus wrens, many more–includes fully illustrated booklet. Cassette and manual 99913-0

GROWING AND USING HERBS AND SPICES, Milo Miloradovich. Versatile handbook provides all the information needed for cultivation and use of all the herbs and spices available in North America. 4 illustrations. Index. Glossary. 236pp. 5⅜ x 8½. 25058-X

BIG BOOK OF MAZES AND LABYRINTHS, Walter Shepherd. 50 mazes and labyrinths in all–classical, solid, ripple, and more–in one great volume. Perfect inexpensive puzzler for clever youngsters. Full solutions. 112pp. 8⅛ x 11. 22951-3

PIANO TUNING, J. Cree Fischer. Clearest, best book for beginner, amateur. Simple repairs, raising dropped notes, tuning by easy method of flattened fifths. No previous skills needed. 4 illustrations. 201pp. 5⅜ x 8½. 23267-0

HINTS TO SINGERS, Lillian Nordica. Selecting the right teacher, developing confidence, overcoming stage fright, and many other important skills receive thoughtful discussion in this indispensible guide, written by a world-famous diva of four decades' experience. 96pp. 5⅜ x 8½. 40094-8

THE COMPLETE NONSENSE OF EDWARD LEAR, Edward Lear. All nonsense limericks, zany alphabets, Owl and Pussycat, songs, nonsense botany, etc., illustrated by Lear. Total of 320pp. 5⅜ x 8½. (Available in U.S. only.) 20167-8

VICTORIAN PARLOUR POETRY: An Annotated Anthology, Michael R. Turner. 117 gems by Longfellow, Tennyson, Browning, many lesser-known poets. "The Village Blacksmith," "Curfew Must Not Ring Tonight," "Only a Baby Small," dozens more, often difficult to find elsewhere. Index of poets, titles, first lines. xxiii + 325pp. 5⅜ x 8¼. 27044-0

DUBLINERS, James Joyce. Fifteen stories offer vivid, tightly focused observations of the lives of Dublin's poorer classes. At least one, "The Dead," is considered a masterpiece. Reprinted complete and unabridged from standard edition. 160pp. 5$\frac{3}{16}$ x 8¼. 26870-5

GREAT WEIRD TALES: 14 Stories by Lovecraft, Blackwood, Machen and Others, S. T. Joshi (ed.). 14 spellbinding tales, including "The Sin Eater," by Fiona McLeod, "The Eye Above the Mantel," by Frank Belknap Long, as well as renowned works by R. H. Barlow, Lord Dunsany, Arthur Machen, W. C. Morrow and eight other masters of the genre. 256pp. 5⅜ x 8½. (Available in U.S. only.) 40436-6

THE BOOK OF THE SACRED MAGIC OF ABRAMELIN THE MAGE, translated by S. MacGregor Mathers. Medieval manuscript of ceremonial magic. Basic document in Aleister Crowley, Golden Dawn groups. 268pp. 5⅜ x 8½. 23211-5

NEW RUSSIAN-ENGLISH AND ENGLISH-RUSSIAN DICTIONARY, M. A. O'Brien. This is a remarkably handy Russian dictionary, containing a surprising amount of information, including over 70,000 entries. 366pp. 4½ x 6⅛. 20208-9

HISTORIC HOMES OF THE AMERICAN PRESIDENTS, Second, Revised Edition, Irvin Haas. A traveler's guide to American Presidential homes, most open to the public, depicting and describing homes occupied by every American President from George Washington to George Bush. With visiting hours, admission charges, travel routes. 175 photographs. Index. 160pp. 8¼ x 11. 26751-2

NEW YORK IN THE FORTIES, Andreas Feininger. 162 brilliant photographs by the well-known photographer, formerly with *Life* magazine. Commuters, shoppers, Times Square at night, much else from city at its peak. Captions by John von Hartz. 181pp. 9¼ x 10¾. 23585-8

INDIAN SIGN LANGUAGE, William Tomkins. Over 525 signs developed by Sioux and other tribes. Written instructions and diagrams. Also 290 pictographs. 111pp. 6⅛ x 9¼. 22029-X

ANATOMY: A Complete Guide for Artists, Joseph Sheppard. A master of figure drawing shows artists how to render human anatomy convincingly. Over 460 illustrations. 224pp. 8⅜ x 11¼. 27279-6

MEDIEVAL CALLIGRAPHY: Its History and Technique, Marc Drogin. Spirited history, comprehensive instruction manual covers 13 styles (ca. 4th century through 15th). Excellent photographs; directions for duplicating medieval techniques with modern tools. 224pp. 8⅜ x 11¼. 26142-5

DRIED FLOWERS: How to Prepare Them, Sarah Whitlock and Martha Rankin. Complete instructions on how to use silica gel, meal and borax, perlite aggregate, sand and borax, glycerine and water to create attractive permanent flower arrangements. 12 illustrations. 32pp. 5⅜ x 8½. 21802-3

EASY-TO-MAKE BIRD FEEDERS FOR WOODWORKERS, Scott D. Campbell. Detailed, simple-to-use guide for designing, constructing, caring for and using feeders. Text, illustrations for 12 classic and contemporary designs. 96pp. 5⅜ x 8½. 25847-5

SCOTTISH WONDER TALES FROM MYTH AND LEGEND, Donald A. Mackenzie. 16 lively tales tell of giants rumbling down mountainsides, of a magic wand that turns stone pillars into warriors, of gods and goddesses, evil hags, powerful forces and more. 240pp. 5⅜ x 8½. 29677-6

THE HISTORY OF UNDERCLOTHES, C. Willett Cunnington and Phyllis Cunnington. Fascinating, well-documented survey covering six centuries of English undergarments, enhanced with over 100 illustrations: 12th-century laced-up bodice, footed long drawers (1795), 19th-century bustles, 19th-century corsets for men, Victorian "bust improvers," much more. 272pp. 5⅝ x 8¼. 27124-2

ARTS AND CRAFTS FURNITURE: The Complete Brooks Catalog of 1912, Brooks Manufacturing Co. Photos and detailed descriptions of more than 150 now very collectible furniture designs from the Arts and Crafts movement depict davenports, settees, buffets, desks, tables, chairs, bedsteads, dressers and more, all built of solid, quarter-sawed oak. Invaluable for students and enthusiasts of antiques, Americana and the decorative arts. 80pp. 6½ x 9¼. 27471-3

WILBUR AND ORVILLE: A Biography of the Wright Brothers, Fred Howard. Definitive, crisply written study tells the full story of the brothers' lives and work. A vividly written biography, unparalleled in scope and color, that also captures the spirit of an extraordinary era. 560pp. 6⅛ x 9¼. 40297-5

THE ARTS OF THE SAILOR: Knotting, Splicing and Ropework, Hervey Garrett Smith. Indispensable shipboard reference covers tools, basic knots and useful hitches; handsewing and canvas work, more. Over 100 illustrations. Delightful reading for sea lovers. 256pp. 5⅜ x 8½. 26440-8

FRANK LLOYD WRIGHT'S FALLINGWATER: The House and Its History, Second, Revised Edition, Donald Hoffmann. A total revision–both in text and illustrations–of the standard document on Fallingwater, the boldest, most personal architectural statement of Wright's mature years, updated with valuable new material from the recently opened Frank Lloyd Wright Archives. "Fascinating"–*The New York Times*. 116 illustrations. 128pp. 9¼ x 10¾. 27430-6

PHOTOGRAPHIC SKETCHBOOK OF THE CIVIL WAR, Alexander Gardner. 100 photos taken on field during the Civil War. Famous shots of Manassas Harper's Ferry, Lincoln, Richmond, slave pens, etc. 244pp. 10⅝ x 8¼. 22731-6

FIVE ACRES AND INDEPENDENCE, Maurice G. Kains. Great back-to-the-land classic explains basics of self-sufficient farming. The one book to get. 95 illustrations. 397pp. 5⅜ x 8½. 20974-1

SONGS OF EASTERN BIRDS, Dr. Donald J. Borror. Songs and calls of 60 species most common to eastern U.S.: warblers, woodpeckers, flycatchers, thrushes, larks, many more in high-quality recording. Cassette and manual 99912-2

A MODERN HERBAL, Margaret Grieve. Much the fullest, most exact, most useful compilation of herbal material. Gigantic alphabetical encyclopedia, from aconite to zedoary, gives botanical information, medical properties, folklore, economic uses, much else. Indispensable to serious reader. 161 illustrations. 888pp. 6½ x 9¼. 2-vol. set. (Available in U.S. only.) Vol. I: 22798-7
Vol. II: 22799-5

HIDDEN TREASURE MAZE BOOK, Dave Phillips. Solve 34 challenging mazes accompanied by heroic tales of adventure. Evil dragons, people-eating plants, blood-thirsty giants, many more dangerous adversaries lurk at every twist and turn. 34 mazes, stories, solutions. 48pp. 8¼ x 11. 24566-7

LETTERS OF W. A. MOZART, Wolfgang A. Mozart. Remarkable letters show bawdy wit, humor, imagination, musical insights, contemporary musical world; includes some letters from Leopold Mozart. 276pp. 5⅜ x 8½. 22859-2

BASIC PRINCIPLES OF CLASSICAL BALLET, Agrippina Vaganova. Great Russian theoretician, teacher explains methods for teaching classical ballet. 118 illustrations. 175pp. 5⅜ x 8½. 22036-2

THE JUMPING FROG, Mark Twain. Revenge edition. The original story of The Celebrated Jumping Frog of Calaveras County, a hapless French translation, and Twain's hilarious "retranslation" from the French. 12 illustrations. 66pp. 5⅜ x 8½. 22686-7

BEST REMEMBERED POEMS, Martin Gardner (ed.). The 126 poems in this superb collection of 19th- and 20th-century British and American verse range from Shelley's "To a Skylark" to the impassioned "Renascence" of Edna St. Vincent Millay and to Edward Lear's whimsical "The Owl and the Pussycat." 224pp. 5⅜ x 8½. 27165-X

COMPLETE SONNETS, William Shakespeare. Over 150 exquisite poems deal with love, friendship, the tyranny of time, beauty's evanescence, death and other themes in language of remarkable power, precision and beauty. Glossary of archaic terms. 80pp. 5³⁄₁₆ x 8¼. 26686-9

THE BATTLES THAT CHANGED HISTORY, Fletcher Pratt. Eminent historian profiles 16 crucial conflicts, ancient to modern, that changed the course of civilization. 352pp. 5⅜ x 8½. 41129-X

THE WIT AND HUMOR OF OSCAR WILDE, Alvin Redman (ed.). More than 1,000 ripostes, paradoxes, wisecracks: Work is the curse of the drinking classes; I can resist everything except temptation; etc. 258pp. 5⅜ x 8½. 20602-5

SHAKESPEARE LEXICON AND QUOTATION DICTIONARY, Alexander Schmidt. Full definitions, locations, shades of meaning in every word in plays and poems. More than 50,000 exact quotations. 1,485pp. 6½ x 9¼. 2-vol. set.
Vol. 1: 22726-X
Vol. 2: 22727-8

SELECTED POEMS, Emily Dickinson. Over 100 best-known, best-loved poems by one of America's foremost poets, reprinted from authoritative early editions. No comparable edition at this price. Index of first lines. 64pp. 5³⁄₁₆ x 8¼. 26466-1

THE INSIDIOUS DR. FU-MANCHU, Sax Rohmer. The first of the popular mystery series introduces a pair of English detectives to their archnemesis, the diabolical Dr. Fu-Manchu. Flavorful atmosphere, fast-paced action, and colorful characters enliven this classic of the genre. 208pp. 5³⁄₁₆ x 8¼. 29898-1

THE MALLEUS MALEFICARUM OF KRAMER AND SPRENGER, translated by Montague Summers. Full text of most important witchhunter's "bible," used by both Catholics and Protestants. 278pp. 6⅝ x 10. 22802-9

SPANISH STORIES/CUENTOS ESPAÑOLES: A Dual-Language Book, Angel Flores (ed.). Unique format offers 13 great stories in Spanish by Cervantes, Borges, others. Faithful English translations on facing pages. 352pp. 5⅜ x 8½. 25399-6

GARDEN CITY, LONG ISLAND, IN EARLY PHOTOGRAPHS, 1869–1919, Mildred H. Smith. Handsome treasury of 118 vintage pictures, accompanied by carefully researched captions, document the Garden City Hotel fire (1899), the Vanderbilt Cup Race (1908), the first airmail flight departing from the Nassau Boulevard Aerodrome (1911), and much more. 96pp. 8⅞ x 11¾. 40669-5

OLD QUEENS, N.Y., IN EARLY PHOTOGRAPHS, Vincent F. Seyfried and William Asadorian. Over 160 rare photographs of Maspeth, Jamaica, Jackson Heights, and other areas. Vintage views of DeWitt Clinton mansion, 1939 World's Fair and more. Captions. 192pp. 8⅞ x 11. 26358-4

CAPTURED BY THE INDIANS: 15 Firsthand Accounts, 1750-1870, Frederick Drimmer. Astounding true historical accounts of grisly torture, bloody conflicts, relentless pursuits, miraculous escapes and more, by people who lived to tell the tale. 384pp. 5⅜ x 8½. 24901-8

THE WORLD'S GREAT SPEECHES (Fourth Enlarged Edition), Lewis Copeland, Lawrence W. Lamm, and Stephen J. McKenna. Nearly 300 speeches provide public speakers with a wealth of updated quotes and inspiration–from Pericles' funeral oration and William Jennings Bryan's "Cross of Gold Speech" to Malcolm X's powerful words on the Black Revolution and Earl of Spenser's tribute to his sister, Diana, Princess of Wales. 944pp. 5⅜ x 8⅜. 40903-1

THE BOOK OF THE SWORD, Sir Richard F. Burton. Great Victorian scholar/adventurer's eloquent, erudite history of the "queen of weapons"–from prehistory to early Roman Empire. Evolution and development of early swords, variations (sabre, broadsword, cutlass, scimitar, etc.), much more. 336pp. 6⅛ x 9¼.
25434-8

AUTOBIOGRAPHY: The Story of My Experiments with Truth, Mohandas K. Gandhi. Boyhood, legal studies, purification, the growth of the Satyagraha (nonviolent protest) movement. Critical, inspiring work of the man responsible for the freedom of India. 480pp. 5⅜ x 8½. (Available in U.S. only.) 24593-4

CELTIC MYTHS AND LEGENDS, T. W. Rolleston. Masterful retelling of Irish and Welsh stories and tales. Cuchulain, King Arthur, Deirdre, the Grail, many more. First paperback edition. 58 full-page illustrations. 512pp. 5⅜ x 8½. 26507-2

THE PRINCIPLES OF PSYCHOLOGY, William James. Famous long course complete, unabridged. Stream of thought, time perception, memory, experimental methods; great work decades ahead of its time. 94 figures. 1,391pp. 5⅜ x 8½. 2-vol. set. Vol. I: 20381-6 Vol. II: 20382-4

THE WORLD AS WILL AND REPRESENTATION, Arthur Schopenhauer. Definitive English translation of Schopenhauer's life work, correcting more than 1,000 errors, omissions in earlier translations. Translated by E. F. J. Payne. Total of 1,269pp. 5⅜ x 8½. 2-vol. set. Vol. 1: 21761-2 Vol. 2: 21762-0

MAGIC AND MYSTERY IN TIBET, Madame Alexandra David-Neel. Experiences among lamas, magicians, sages, sorcerers, Bonpa wizards. A true psychic discovery. 32 illustrations. 321pp. 5⅜ x 8½. (Available in U.S. only.) 22682-4

THE EGYPTIAN BOOK OF THE DEAD, E. A. Wallis Budge. Complete reproduction of Ani's papyrus, finest ever found. Full hieroglyphic text, interlinear transliteration, word-for-word translation, smooth translation. 533pp. 6½ x 9¼. 21866-X

MATHEMATICS FOR THE NONMATHEMATICIAN, Morris Kline. Detailed, college-level treatment of mathematics in cultural and historical context, with numerous exercises. Recommended Reading Lists. Tables. Numerous figures. 641pp. 5⅜ x 8½. 24823-2

PROBABILISTIC METHODS IN THE THEORY OF STRUCTURES, Isaac Elishakoff. Well-written introduction covers the elements of the theory of probability from two or more random variables, the reliability of such multivariable structures, the theory of random function, Monte Carlo methods of treating problems incapable of exact solution, and more. Examples. 502pp. 5⅜ x 8½. 40691-1

THE RIME OF THE ANCIENT MARINER, Gustave Doré, S. T. Coleridge. Doré's finest work; 34 plates capture moods, subtleties of poem. Flawless full-size reproductions printed on facing pages with authoritative text of poem. "Beautiful. Simply beautiful."–*Publisher's Weekly*. 77pp. 9¼ x 12. 22305-1

NORTH AMERICAN INDIAN DESIGNS FOR ARTISTS AND CRAFTSPEOPLE, Eva Wilson. Over 360 authentic copyright-free designs adapted from Navajo blankets, Hopi pottery, Sioux buffalo hides, more. Geometrics, symbolic figures, plant and animal motifs, etc. 128pp. 8⅜ x 11. (Not for sale in the United Kingdom.) 25341-4

SCULPTURE: Principles and Practice, Louis Slobodkin. Step-by-step approach to clay, plaster, metals, stone; classical and modern. 253 drawings, photos. 255pp. 8⅛ x 11. 22960-2

THE INFLUENCE OF SEA POWER UPON HISTORY, 1660–1783, A. T. Mahan. Influential classic of naval history and tactics still used as text in war colleges. First paperback edition. 4 maps. 24 battle plans. 640pp. 5⅜ x 8½. 25509-3